Combustion in High-Speed Flows

ICASE/LaRC Interdisciplinary Series in Science and Engineering

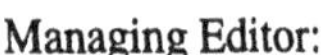

Volume 1

Combustion in High-Speed Flows

edited by

John Buckmaster

Department of Aeronautical and Astronautical Engineering,
University of Illinois,
Urbana-Champaign, U.S.A.

Thomas L. Jackson

Institute of Computer Applications in Science and Engineering (ICASE),
Hampton, Virginia, U.S.A.

and

Ajay Kumar

Theoretical Flow Physics Branch,
NASA Langley Research Center,
Hampton, Virginia, U.S.A.

KLUWER ACADEMIC PUBLISHERS

DORDRECHT / BOSTON / LONDON

A C.I.P. Catalogue record for this book is available from the Library of Congress

DOI 10.1007/978-94-011-1050-1

Published by Kluwer Academic Publishers,
P.O. Box 17, 3300 AA Dordrecht, The Netherlands.

Kluwer Academic Publishers incorporates
the publishing programmes of
D. Reidel, Martinus Nijhoff, Dr W. Junk and MTP Press.

Printed on acid-free paper

CONTENTS

UNSTEADY BEHAVIOR

PANEL DISCUSSION

PREFACE

This volume contains the proceedings of the Workshop on Combustion, sponsored by the Institute for Computer Applications in Science and Engineering (ICASE) and the NASA Langley Research Center (LaRC). It was held on October 12-14, 1992, and was the second workshop in the series on the subject. The first was held in 1989, and its proceedings were published by Springer-Verlag under the title "Major Research Topics in Combustion," edited by M. Y. Hussaini, A. Kumar, and R. G. Voigt. The focus of the second workshop was directed towards the development, analysis, and application of basic models in high speed propulsion of particular interest to NASA. The exploration of a dual approach combining asymptotic and numerical methods for the analysis of the models was particularly encouraged. The objectives of this workshop were

i) the genesis of models that would capture or reflect the basic physical phenomena in SCRAMJETs and/or oblique detonation-wave engines (ODWE), and

ii) the stimulation of a greater interaction between NASA experimental research community and the academic community.

The lead paper by D. Bushnell on the status and issues of high speed propulsion relevant to both the SCRAMJET and the ODWE parallels his keynote address which set the stage of the workshop. Following the lead paper were five technical sessions with titles and chairs: Experiments (C. Rogers), Reacting Free Shear Layers (C. E. Grosch), Detonations (A. K. Kapila), Ignition and Structure (J. Buckmaster), and Unsteady Behaviour (T. L. Jackson). The session chairs were responsible for the planning of their session. In each of the sessions formal discussion papers were given by experts followed by a general discussion involving the participants. This volume contains the manuscripts presented in each session. In addition, a panel discussion was held on the last day, chaired by J. Buckmaster. The panel members were D. Bushnell, P. Givi, A. Hertzberg, A. Kapila, J. H. S. Lee, P. Libby, and C. Rogers. Since the panel discussion was viewed by many as an essential part of the workshop, it was transcribed in its entirety and is included at the end of this proceedings without much editing.

It is a pleasure to acknowledge M. Y. Hussaini who initiated and contributed to the conception and planning of the Workshop. We wish to thank him for his support and encouragement. In addition, we gratefully acknowledge Professor J. Clarke for his advice and help in the initial planning stages. Finally, we are deeply grateful to Douglas Dwoyer whose support and encouragement was essential during all stages of the workshop.

The preface would not be complete without gratefully mentioning Emily Todd who took care of the arrangements, correspondence, collection of the manuscripts for the Workshop, and did an overall excellent job as Conference Manager; Barbara Stewart who formatted all the manuscripts; Shelly Millen who took care of the figures; Etta Blair who handled reimbursements for the participants; and finally, but not at all the least, to Lori Rowland who painstakingly and with great patience transcribed the entire panel discussion from video tape. We are pleased to express our sincere appreciation to all of them. Thanks are also due to the staff of Kluwer Academic Publishers for their assistance in bringing out this volume.

J. Buckmaster
T. L. Jackson
A. Kumar

CONTRIBUTORS

Griffin Y. Anderson
Mail Stop 168
NASA Langley Research Center
Hampton, VA 23681-0001

John Buckmaster
Department of Aeronautical
and Astronautical Engineering
University of Illinois
104 South Wright Street
Urbana, IL 61809

Dennis M. Bushnell
Mail Stop 197
NASA Langley Research Center
Hampton, VA 23681-0001

J. William Dold
School of Mathematics
University of Bristol
Bristol BS8 1TW
ENGLAND

J. Philip Drummond
Mail Stop 156
NASA Langley Research Center
Hampton, VA 23681-0001

John I. Erdos
General Applied Science
Laboratories, Inc. (GASL)
77 Raynor Avenue
Ronkonkoma, NY 11779

Bruce C. R. Ewan
Department of Mechanical
and Process Engineering
The University of Sheffield
P.O. Box 600, Mappin Street
Sheffield S1 4DU
ENGLAND

Francis E. Fendell
Center for Propulsion Technology
and Fluid Mechanics
TRW Space & Electronics Group
One Space Park
Redondo Beach, CA 90278

Sharath S. Girimaji
Analytical Services and
Materials Inc.
Mail Stop 156
NASA Langley Research Center
Hampton, VA 23681-0001

Peyman Givi
Department of Mechanical
and Aerospace Engineering
State University of New York
Buffalo, NY 14260

Chester E. Grosch
Department of Computer Science
Old Dominion University
Norfolk, VA 23529

Abraham Hertzberg
120 Aerospace and Engineering
Research Building
Mail Stop FL-10
University of Washington
Seattle, WA 98195

Rupert Klein
Institut für Technische
Mechanik, RWTH
Templergraben 64, 5100 Aachen
GERMANY

John H. S. Lee
Department of Mechanical
Engineering
McGill University
817 Sherbrooke Street West
Montreal, Quebec
CANADA H3A 2K6

Amable Liñán
Universidad Politecnica de Madrid
E.T.S.I. Aeronáuticos, UPM
Plaza del Cardenal Cisneros 3
Madrid 28040
SPAIN

Richard G. Morgan
Department of Mechanical
Engineering
The University of Queensland
Brisbane QLD
AUSTRALIA, 4072

Robert W. Pitz
Department of Mechanical
Engineering
Vanderbilt University
Box 54, Station B
Nashville, TN 37235

Joseph M. Powers
Department of Aerospace
and Mechanical Engineering
University of Notre Dame
Notre Dame, IN 46556-5637

James J. Quirk
Institute for Computer
Applications in Science
and Engineering
Mail Stop 132C
NASA Langley Research Center
Hampton, VA 23681-0001

Joseph E. Shepherd
Department of Mechanical
Engineering
Rensselaer Polytechnic Institute
Troy, NY 12180-3590

Paul A. Thibault
Combustion Dynamics Ltd.
203, 132 4th Avenue, S.E.
Medicine Hat, Alberta T1A 8B5
CANADA

Cesar Treviño
Universidad Politecnica de Madrid
E.T.S.I. Aeronáuticos, UPM
Plaza del Cardenal Cisneros 3
Madrid 28040
SPAIN

KEYNOTE ADDRESS

MIXING AND COMBUSTION ISSUES IN HYPERSONIC AIR-BREATHING PROPULSION

Dennis M. Bushnell

NASA Langley Research Center
Hampton, Virginia 23681-0001

1. Introduction

Hypersonics is defined, herein and conventionally, as flight at Mach numbers of 6 or greater. Historically, hypersonic vehicles have been rocket boosted/propelled with well-known performance limits in terms of specific impulse. Contemporary attempts to increase Rocket Isp include the Air Force Phillips Laboratory High Energy Density Materials Program (primarily creation/storage of metastable states) and anti-matter and fusion approaches (e.g., Froning, 1989). These efforts have possibly exciting payoffs, but as currently envisaged, such payoffs are problematical and will only be available in the long term. In the meantime, exploration and utilization of space, both near earth and within the solar system, as well as, eventually, interstellar, is a continuing national interest. A major key to the effective utilization and exploration of space is an efficient launch system into near earth orbit. While major strides are being made in payload miniaturization, launcher effectiveness and efficiency will remain, for the foreseeable future, a critical issue. There are two obvious approaches to reducing the cost of access to space, an inexpensive (conventional) reusable rocket system specifically designed to reduce the "standing army" and overhead associated with current launch systems and various advanced technology options. These advanced technology possibilities cover the spectrum from probably do-able to problematical, even in terms of the requisite physics. Such advanced options include metastable/excited state high energy density fuels, nuclear (fission), rockets, various fusion schemes, ground based laser propulsion, anti-matter fuel and air-breathing. Of these options Nuclear Rockets and air breathing are probably the nearest term. Conventional diffusive burning air-breathing scramjets may provide significantly increased propulsive efficiency in the Mach 6 to 15 range (e.g., Swithenbank et al., 1989) compared to Hydrogen-Oxygen rockets, due primarily to the use of atmospheric oxygen as the oxidiser. The importance of such performance improvements

J. Buckmaster et al. (eds.), Combustion in High-Speed Flows, 3–16.

are indicated by NASA Langley performance estimates (J. L. Hunt) which indicate that up to 40 percent vehicle gross weight reductions may be possible via the order of a 10 percent average Isp increase (for single stage to orbit air breathers.)

For airbreathers above Mach 15 another propulsion approach is required, e.g., greatly enhanced (reduced loss) conventional scramjets, detonation wave engines and/or rocket augmentation. Both single stage and multiple stage airbreathing options are possible, where for the latter, the first stage is airbreathing and the second stage is either rocket (e.g., Freeman et al., 1990), air-breathing or a hybrid thereof. A particularly interesting approach is a two-stage airbreathing system where both stages airbreathe and the second stage is an oblique detonation wave engine. Research in Menees et al. (1989) indicates that the performance of the oblique detonation wave approach is greater than the conventional scramjet in precisely the range where the latters performance becomes less interesting, Mach numbers greater than 0(14). The detonation wave configuration is so different from the diffusive burning scramjet that a two-stage approach is almost mandated if airbreathing via the detonation wave approach is envisioned into the hypervelocity regime ($M > 0(14)$, where much of the orbital energy is imparted to the vehicle). Significant enhancements in diffusive-burning airbreathing hypervelocity performance may also be obtainable, which would allow a reduction in the required rocket augmentation for an SSTO airbreather. Airbreathing propulsion performance is, in fact, required for a reusable SSTO which would incorporate airplane-like features such as self-ferry, rapid turnaround, all azimuth and inclination launch, powered descent and landing and autonomous operation with excellent operability, maintainability and refurbishability.

Airbreathing hypersonic engines may also have applications, particularly via use of endothermic hydrocarbon fuels, to various military, and possibly even civilian, hypersonic *cruise* missions. The civilian (Mach number of order 6) cruise machine constitutes a possible adjunct to a high speed civil transport in that the high altitude of such a device (above the ozone layer) may enable high speed (hypersonic) operation over the Eurassian land mass at "acceptable" sonic boom levels.

The design of airbreathing hypersonic cruise and accelerator machines is highly integrated with the airframe constituting portions (initial inlet, external nozzle) of the propulsive system/flow path. A particularly critical component of hypersonic airbreathing vehicles is the combustor which is, historically, heavy due to size (length) and aero/thermal loading and is responsible for major friction, heat transfer and mixing losses. The purpose of the present paper is to briefly summarize the mixing and combustion research issues associated with hypersonic scramjet combustors, along with other problem areas, such as nozzle efficiency and ground facility fidelity, where chemical kinetics is a major concern.

2. Overview of Hypersonic Air-breathing Propulsion

Excellent summaries of the overall physics and operation of supersonic combustion ramjet engines are available, for example, in Waltrup (1987), Ferri (1964), and Swithenbank (1966). Supersonic combustion (combustion with supersonic mean airstream speed) is required for airbreathers at high Mach number to reduce the unacceptable losses associated with inlet deceleration to subsonic conditions. Depending upon detailed design, the performance "cross over region" between subsonic and supersonic combustion is in the neighborhood of Mach 6. To maximize scramjet performance, the vehicle nose compression surface is utilized as an "inlet spike," and the vehicle afterbody is incorporated as part of the engine nozzle expansion process i.e. scramjets are not "podded" engines but rather fully integrated into the airframe with multiple modules wrapped around the mid-region of the vehicle.

Design studies yield the following nominal scramjet combustor conditions: mean velocity the order of (slightly less than) free stream; pressure level the order of one-half atmosphere; static temperature in the range of 2,000 to 4,000°R; Mach number order of one-third free stream and constant to mildly diverging cross-section. A pervasive feature of scramjet combustors is the presence/influences of shock waves generated in the inlet, from the fuel injection processes, front-side cooling techniques (if utilized) and any flow separations. These shock waves interact with the mixing and combustion processes within the combustor and can propagate downstream into the nozzle. For the "lower" Mach number range (6 to 0(12)) transverse fuel injection can be employed as well as in-stream injectors. At the higher Mach numbers, streamwise injection is increasingly employed to recover the thrust from the fuel injection process. Since the combustor is (at least) regeneratively cooled utilizing the fuel flow, the fuel injection velocity can be large. In the case of hydrogen heated in this fashion, fuel exit velocities can be the order of 12,000 ft./sec. or greater. This fuel exit velocity can be greater, or less than, the local stream velocity depending upon flight Mach number, and therefore has a first order effect upon mixing rate.

There are several "figures of merit" for scramjet combustors. These include rapid ignition (a particular problem with hydrocarbon fuels) and minimal combustor length to "mix and burn." The latter minimizes combustor weight (a significant fraction of the empty weight of the vehicle) as well as heat transfer and shear losses thereby reducing the excess fuel required for front-side cooling at high Mach number. Minimal combustor length obviously requires the invention and development of techniques for enhanced and *low* (additional) *loss* penetration and micro-mixing. Additional scramjet combustor "figures of merit" include operability, especially for RAM-Scram conversion, synergistic operation with integral LOX/LAIR/Rocket systems at high Mach number and mixing/combustion distant from the near wall region to further reduce heat transfer losses and attendant heat protection system weight.

The global issues associated with scramjet operation across the Mach number range are well-known. In the Mach 6 to 12 range, airbreathing engines (scramjets) produce adequate thrust margin and therefore transverse injection may be utilized, providing appreciable mixing enhancement. However, ignition delay may be a problem due to the relatively low-static temperature (compared to the high Mach number case). As Mach number increases engine thrust and performance levels decrease due primarily to the decreasing "value added" associated with Hydrogen-air (or hydrocarbon-air) combustion compared to the (increasing) stream kinetic energy. Reasonable performance at high Mach number requires a two-fold combustor design approach, utilization of thrust from the fuel addition process and invention/development/employment of loss mitigation approaches for the various loss mechanisms, such as mixing, heat transfer, friction, shock and disassociation. In a very real sense, the high Mach number airbreathing engine is distinctly different from the $M \sim 6$ to $0(12)$ machine, due primarily to the difficulty in extracting adequate net thrust at high flight energy levels. In fact, the airbreathing engine reverts, at hypervelocities, increasingly toward rocket operation, either via enhanced fuel thrust or combined cycle (airbreather-rocket) or both.

There are then 4 distinct classes of hypersonic air-breathing propulsion systems, two at $M < 0(12)$ (endothermic hydrocarbon and hydrogen-fueled "conventional" scramjets) and two at $M > 0(12)$. (Detonation wave engines/vehicle and advanced/combined cycle hydrogen fueled scramjets.) Although significant research is required for the "lower" speed range, the major research *frontiers and needs* are in the hypervelocity arena, *which has never been seriously addressed.* Even the facility base is deficient for the high Mach number case.

3. Hydrocarbon (Storable Fuel) Scramjet Research Issues

The fundamental rationale for considering hydrocarbon-fueled scramjets, in the face of their reduced specific impulse (less than 50% that of hydrogen), is the possibility for combined endothermic vehicle cooling and conventional fuel storage at "spaceports." The endothermic aspect is under investigation by the U.S. Air Force and involves "cracking" of the fuel during combustor cooling, thereby adding a "chemical" heat sink to the usual "sensible" heat sink capability. This endothermic process can be accomplished either in a separate heat exchanger (requires a 2-fluid cooling system) or directly within the engine cooling passages. This additional cooling capacity may allow operation with hydrocarbons into the Mach 6 to 8 range. The "conventional" fuel storage (and handling) aspect of hydrocarbon fuels is non-trivial. Cryogenic hydrogen is an expensive fuel to store and handle, with cost estimates in the hundreds of millions of dollars per "spaceport" for hydrogen fuel capability. In the case of the military, hydrocarbon fuels allow use of "conventional" logistics, dispersed launch sites (for strike/Recce vehicles/missiles) and smaller, lighter (empty weight) vehicles with faster vehicle turnaround and launch.

The major research issues associated with endothermic hydrocarbon fueled air-breathing propulsion systems include piloting/ignition, flameholding, coking, catalysis, fuel injection/mixing and RAM-Scram mode change. The fuel is in the gaseous state for the endothermic case, and therefore, fuel atomization and vaporization are not first order problems within the combustor as they would be if straight hydrocarbon fuels were used. The ignition delay time for most hydrocarbons are one or more orders of magnitude greater than hydrogen and this, coupled with the comparatively low Mach number/combustor static temperatures of endothermic hydrocarbon scramjet applications makes ignition and flameholding a first order issue. Suggested required research areas for this ignition problem include pyrophorics, "doped" free radical population, favorable flow interactions and "preburners," the latter being small scale imbedded combustor sub-regions which serve a piloting function. Candidate endothermic-hydrocarbon fuels include CRJ-6, JP-7 and 10, JP-900, MCH, Decalin, Heptane, Methanol and Norpar-12. The "chemical" heat sink capacity of most of these materials is in the neighborhood of 1,000 Btu/lb. These candidate fuels should be, and are being, evaluated against such figures of merit as high volumetric heating value, low ignition temperature and pressure, minimal coolant passage coking/contamination and acceptable environmental impact. Fuels additives research areas include metal slurries with wetting agents for slurry formation and gelling agents to improve storage and designer catalysts. Of all the hypersonic airbreathing propulsion options, chemical kinetics is probably the most critical in the case of the endothermic-hydrocarbon scramjet.

4. Oblique Detonation Wave Propulsion

Rather than oblique detonation wave, shock-assisted combustion is probably a more accurate description of this prospective hypervelocity air-breathing propulsion scheme. The early work on detonation wave engines (e.g., Sargent, 1960; Gross and Chinitz, 1960; and Morrison, 1980) considered either normal detonation waves, and/or Mach number less than 10. The former entails losses which are too large for efficient propulsion at high Mach number while the latter involves a speed range where the conventional diffusive burning scramjet usually has a higher performance. More recent work on oblique detonation wave engines (e.g., Menees, 1989; and O'Brien and Kobayashi, 1986) indicates that such an approach may be more efficient than the diffusive-burning scramjet in the hypervelocity range ($M_\infty \geq 0(13)$). Also, the oblique "over-driven" "detonation" wave, i.e. shock-assisted combustion, appears to yield a stable combustion process (e.g., Singh et al., 1991).

The essential features of such an engine includes even more extreme integration with the "airframe." The fuel is injected from the vehicle forebody/nose region, inside the shock layer but outside the boundary layer (to avoid/delay preignition). This injection process, both as to general approach and details, constitutes probably the single greatest uncertainty/opportunity in this entire arena. Overall considerations/concerns include penetration across the shock layer (for the high Mach no./"shock on lip" case), forebody-region mixing prior to shock interaction/combustion, predetonation/preignition, fuel injection thrust maximization, wave stability, overall performance and the current lack of an applicable experimental data base at high Mach number. Penetration may be enhanced via injection as a liquid or even a quasi-solid (i.e. hydrogen "slush balls"), with probable attendant reduction in fuel injection thrust. The forebody mixing problem for the liquid/solid injection case may be mitigated by the suggestion of Prof. Swithenbank of the United Kingdom that shock-assisted combustion is feasible using a 2-phase mixture of fuel and air (evaporated hydrogen, liquid hydrogen fuel droplets and air). Preignition delay is fostered by utilizing a weak forebody shock and the high Mach number of the application, leading to lower static pressure levels within the forebody shock layer.

The advantages of a "shock-assisted combustion" hypervelocity propulsion system includes lower inlet losses and significant reductions in combustor weight, size, heat transfer and skin friction losses and cooling requirement. These advantages are accrued due to forebody region fuel/air mixing which obviates the need to contain the fuel-air mixing process within a long enclosed channel at high pressure. These advantages are particularly crucial at hypervelocity conditions, but the thrust component due to fuel injection must be optimized to recover reasonable performance. Obviously the "hypersonic freeze" principle aids the application of such an engine in the high Mach number regime by reducing the extent to which variable geometry is required.

5. Hypervelocity Diffusive Burning Scramjet

Anderson, Kumar, and Erdos (1990) provide an excellent discussion of the essentials of hypervelocity scramjet design. In the flight Mach number range above 0(14), the scramjet combustor Mach number itself becomes hypersonic, hence the term "hypersonic combustion." As stated previously, the "value-added" from the combustion of even such an energetic fuel as hydrogen now becomes a decreasing fraction of the vehicle energy level. For this situation, mitigation/control of losses of all types, as well as fuel thrust enhancement, become overriding engine design principles/features (e.g., Czysz, 1988; Czysz and Murthy, 1989). Shock waves must be weak and their presence minimized. The losses (shock/heat transfer) associated with in-stream injectors and transverse fuel injection can no longer be tolerated and approximations to axial fuel injection are utilized for thrust enhancement. Also, inlet boundary layer transition is delayed as much as possible and leading edge radii reduced. In the nozzle, rapid expansions are used to ensure relaminarization of the wall boundary layer. Unfortunately, such a rapid expansion exacerbates a large performance loss mechanism, nozzle disassociation. Approaches to reduction of such disassociation losses include combustor temperature reduction, perhaps via improved mixing with hydrogen front-side coolant, and use of metal oxides or other recombination catalysts.

At these high Mach numbers, the inlet shock waves propagate into the combustor/nozzle, i.e. the propulsion unit becomes a "wave engine," and therefore, the inlet and combustors should be designed as a unit to make maximum use of inlet wave trains and wall turbulence for mixing enhancement. Another possibly attractive performance enhancement approach is to configure the combustor such that the inlet waves are "cancelled" via favorable interference (ala the "Busemann Biplane" approach). Obviously, fuel heating will increase fuel thrust within obvious limitations such as material/containment temperature limits and the heat transfer to the vehicle. The former can be circumvented by utilizing a "rotating wave superheater," a "Gatling-gun" arrangement of shock tubes which heats the fuel immediately before injection. Additional fuel heating could be obtained from on-bound additive energy sources such as nuclear fission reactors. Finally, combined cycle (e.g., integral rocket-scramjet) combustion chambers could be developed using synergisms from the rocket flow such as high shear for enhanced mixing and species for catalytic recombination as well as the innate additive thrust levels (see "Rocket-Based Combined Cycle (RBCC) Propulsion Technology Workshop," 1992; and Martin et al., 1991).

6. Mixing Issues for Hypersonic Air-Breathers

Air-fuel mixing rate is especially crucial for hypervelocity engines which are of necessity sensitive to all types of losses. Enhanced mixing would allow, for the diffusive-burning scramjet, shorter combustors with consequent reductions in weight and heat transfer and shear losses. Obviously, the figure of merit is increased micro/molecular mixing, mere large scale momentum "sloshing," unless accompanied by enhanced small scale mixing is of little use and probably would result in enhanced losses. The mixing task is to enhance micro-mixing via techniques which do not increase the overall combustor losses. Mixing enhancement would also increase the performance of most combined cycle engines.

The overall scramjet combustor problem is obviously non-simple. The mean flow entering from the inlet is highly 3-D and nonuniform, replete with (unsteady) shock and expansion waves, a thick boundary layer eminating from the vehicle forebody/inlet combination and embedded vortical entities (e.g., Reddy et al., 1989). The combustor itself must contain multiple fuel injection sites whose complex 3-D flow fields interact with each other and with the inlet flow just mentioned. Added to this are additional flow complexities and physics associated with front-side cooling techniques (at high Mach numbers) and reaction/combustion. Such "real combustor" flows, particularly the combinations of effects, are obviously quite different from those employed in most fundamental mixing enhancement studies (see Dimotokis, 1989). The real combustor problem must be worked in terms of complex interacting flow fields hopefully in a synergistic manner via a variety of enhancement approaches which both satisfy the restriction concerning additional (enhancement process induced) losses appropriate to the specific application speed range and make maximum use of existing flow features, e.g., inlet shocks and thick turbulent boundary layers, etc., which propagate into the combustor from upstream.

Mixing enhancement for engine performance improvement is the totality of increased (from enhancement of small scale turbulence) micro/molecular mixing, increased contact area between air and fuel (from "stirring" and multiple injection sites), and increased path length (e.g., from Swirl). Known parameters having a first order effect upon turbulent shear layers and which therefore may provide enhancement options/approaches include pressure gradient(s), flow curvature(s), energy release/chemical reaction, proximity to transition/Reynolds number, shock/expansion interaction, 3-D mean flow field details, especially mean shear, compressibility, adjacent surfaces, and stream fluctuation fields. In scramjet combustors, fuel distribution through the flow field can be influenced by injection location(s) and approaches and action of essentially inviscid wave fields as well as by turbulent mixing, (e.g., Schetz et al., 1990), but the small scales associated with turbulence are required for efficient micro/molecular mixing and combustion (Dimotokis, 1989). From these considerations an approach to scramjet mixing enhancement can be suggested which involves synergistic utilization of a) pre-existing (incoming from inlet) flow features; b) nearly inviscid "stirring" to increase contact area; and c) low loss (at high Mach no.) turbulence enhancement approaches to increase micro mixing which are instigated away from bounding surfaces to avoid friction and heat transfer loss enhancement and loss of cooling effectiveness. All of this in the context of injection techniques consistent with achieving required fuel penetration and fuel/coolant flow thrust requirements.

Indications from scramjet combustor research thus far indicate the following: a) relatively large streamwise vortices can increase macro-mixing by a factor of at least 2 and significantly increase micro mixing; b) swirl per se does not significantly increase mixing, c) shock-vortex interaction can (locally) increase mixing rate by a factor of 3; d) the effects of shock-mixing layer interaction range from no influence to a 40% increase; and e) usually combustion per se does not seem to affect mixing rate beyond the usual mean density/temperature influence (however, turbulence can significantly enhance reaction rates). There is a plethora of potential/candidate mixing enhancement approaches which require further research. These include wave interactions (both shocks/expansions, steady and unsteady (Kumar et al., 1989)) upon 2-D and 3-D shear layers as a function of the shear layer mean vorticity structure, "low loss" longitudinal vortex influences, additional turbulization of the fuel exit flow and the external (local) air stream and, perhaps most importantly and interestingly, set-up and excitation of innate instabilities within turbulent flows to locally augment the mixing rate via "self-activation." These instabilities include, but are not limited to, Rayleigh, Gortler, Kelvin/Helmholtz, azimuthal, crossflow and the effects of Baroclinic Torque and supersonic wall modes. Obviously, *combinations* of both techniques and instabilities should also be investigated.

7. Chemical Kinetic Issues Associated with Hypersonic Propulsion Ground Facilities

The basic difficulties associated with ground simulation of hypersonic propulsion stem from the necessity for a "complete" flow path simulation. This requirement includes combustor static pressure (the order of 1/2 atmosphere), combustor static temperature and velocity (and therefore flight energy) corresponding to the flight Mach number, a run time corresponding to at least 3 flow lengths for attached flow and 40 separated flow lengths (for flow establishment) with "clean air" and reasonable size/Reynolds number to allow establishment of turbulent wall flows, measurement/hardware accuracy and inlet/combustor integration. Fuel heating is also required for simulation of the correct mixing behavior.

The air heaters/heating processes utilized to produce these test conditions in ground facilities varies with flight Mach number. For Mach 8 and below (0(4800°R) or less) "steady-state" air heating is feasible and various techniques are employed including ceramic brick/pebble "storage" heaters, electrical resistance heaters, arc heaters and several types of combustion/vitiated (O_2 replenishment) heaters which burn fuels such as kerosene, methane, butane, propane, and hydrogen. Each of these heaters produce a characteristic molecular contamination whose effects upon the scramjet combustion process kinetics is essentially, at least thus far, unknown and which probably vary with the engine fuel used (H_2, Hydrocarbon). These contaminants include sand (storage heaters), carbon (some elec. resistance heaters), copper and NO (arc heaters), H_2O and NO (H_2 vitiation), H_20, CO_2, CO and NO (methane and propane vitiation). As an example, at Mach 7 conditions for H_2 vitiation the test flow includes as much H_2O as it does O_2 (by mass fraction). The few studies which are available indicate large-to-negligible differences between clean(er) and more contaminated facilities. The issue of facility contamination effects for the Mach less than 0(8) scramjet ground facilities is an extremely fertile arena for further research regarding the chemical kinetics of both the facility and engine flows. NASA is considering the establishment of a comparative data base obtained via a "round robin" test series on the same sub-scale engine geometry in the entire spectrum of agency facility heater approaches including Ames 3.5 ft. (sand), Lewis HTF (carbon), LaRC AHSTF (copper, NO), LaRC CHSTF (H_2O, NO), and LaRC 8'HTT (H_2O, CO_2,

CO, NO). Many of these facilities may also contain high levels of free stream turbulence produced by expansion of stagnation enthalpy fluctuations (especially arc/combustion heaters).

Ground simulation at flight Mach numbers greater than 8 currently employs "pulse" facilities with test times in the millisecond range and enhanced capability arcs. The chemical kinetics issues for pulse facilities include flow establishment and free stream disassociation. Pulse facilities include reflected shock tunnels with various drivers which stagnate the flow, and an expansion tube, which does not. Once the flow is stagnated, the test gas tends to "freeze" in the facility nozzle expansion process, of particular importance at Mach numbers greater than 0(13). At Mach 18, 40% of the free stream oxygen can still be in the form of NO and O rather than O_2. Tests indicate that this steam disassociation significantly (and erroneously) increases the *apparent* performance of the engine at high Mach number.

Techniques to mitigate this stream dissociation (in order of increasing effectiveness) include increased pressure level, much more gradual nozzle expansion, scale, and utilization of an expansion tube/tunnel which does not stagnate the flow, i.e. energy is added, via expansion waves, to the test gas while it is in motion.

8. Concluding Remarks

The detailed design and optimization of hypersonic airbreathing propulsion devices is at an early stage. There are essentially four intrinsically different devices currently under consideration, two at Mach numbers less than 0(10) and two for the greater than 0(10) "hypervelocity" case. The lower Mach no. engines are both diffusive burning devices, one utilizing endothermic hydrocarbon fuels and the other cryogenic hydrogen. In the higher Mach number regime a shock-assisted combustion device is under study as well as a (loss mitigated, thrust enhanced) diffusive-burning engine, both with hydrogen fuel. Fuel-air mixing enhancement approaches, with a particular emphasis upon low (additional) loss methods at high Mach number, are of interest for all of these devices. Specific chemical kinetics issues for hypersonic airbreathing propulsion include the following: a) emissions, including the effects of the heated airframe as well as the engine efflux upon the ozone depletion and "greenhouse" problems; b) free stream disassociation (at high Mach numbers) and free stream contamination (at lower Mach numbers) for ground facilities; c) nozzle disassociation losses, and their mitigation particularly at high Mach number; d) flame holding and ignition, particularly at lower Mach number and for the endothermic hydrocarbon case; e) kinetics/combustion influences upon turbulent mixing and (perhaps more important) vice versa; f) wall catalysis influences as they affect combustor wall heat transfer/survivability/losses; and g) shock-assisted combustion for a two-phase mixture of gaseous hydrogen/air and liquid hydrogen droplets.

References

Anderson, G., Kumar, A., and Erdos, J., 1990. "Propulsion hypersonic combustion technology with computation and experiment," AIAA Paper 90-5254.

Czysz, P., 1988. "Thermodynamic spectrum of airbreathing propulsion," SAE Paper 88-1203.

Czysz, P. and Murthy, S., 1989. "Energy analysis of propulsion systems for high speed vehicles," AIAA 89-0182.

Dimotokis, P. E., 1989. "Turbulent free shear layer mixing," AIAA Paper 89-0262.

Ferri, A., 1964. "Review of problems in applications of supersonic combustion," *J. Royal Aeron. Soc.* **68**(645), pp. 575-595.

Freeman, D. C., Talay, T. A., Stanley, D. O., and Wilhiet, A. W., 1990. "Design options for advanced manned launch systems (AMLS)," AIAA Paper 90-3816.

Froning, H. D., Jr., 1989. "Investigation of very high energy rockets for future SSTO vehicles," *Acta Astronautica* **19**(4), pp. 321-330.

Gross, R. and Chinitz, W., 1960. "A study of supersonic combustion," *J. Aero/Space Sciences* **27**(7), pp. 519-525.

Kumar, A., Bushnell, D., and Hussaini, M., 1989. "Mixing augmentation technique for hypervelocity scramjets," *J. of Propulsion and Power* **5**(5), pp. 514-522.

Martin, J., Kabis, H., and Hunt, J., 1991. "The necessity of on-board oxygen for airbreathing single-stage-to-orbit vehicles," AIAA Paper 91-5016.

Menees, G. P., 1989. "Wave combustors for trans-atmospheric vehicles," ISABE paper 11-5, presented at 9th ISABE meeting, Athens, Greece, Calso NASA TM-102238.

Menees, G. P., Adelman, H. G., Cambier, J.-L., and Bowles, J. W., 1989. "Wave combustors for transition-atmospheric vehicles," NASA TM 102239.

Morrison, R. B., 1980. "Oblique detonation wave ramjet," NASA Contractor Report 159192.

O'Brien, C. J. and Kobayashi, A. C., 1986. "Advanced-to-orbit propulsion concepts," AIAA Paper 86-1386.

Reddy, D. R., Smith, G. E., Liou, M. F., and Bensen, T. J., 1989. "Three-dimensional viscous analysis of a hypersonic inlet," AIAA Paper 89-0004.

"Rocket-based combined cycle (RBCC) propulsion technology workshop." NASA CP 10090, 1992.

Sargent, W. H. and Gross, R. A., 1960. "Detonation wave hypersonic ramjet," *ARS J.*, pp. 543-549.

Schetz, J., Billig, F., Favin, S., and Gilreath, H., 1990. "Effects of pressure mismatch on slot injection in supersonic flow," AIAA Paper 90-0092.

Singh, D., Carpenter, M., and Kumar A., 1991. "Numerical simulation of shock-induced combustion/detonation in a premixed H_2-air mixture using Navier-Stokes equations," AIAA Paper 91-3359.

Swithenbank, J., 1966. "Hypersonic airbreathing propulsion," in *Progress in Aeronautical Sciences* **8**, pp. 229-294, Pergamon Press, NY.

Swithenbank, J., Eames, I., Chin, S., Ewan, B., Yang, Z., Cao, J., and Zhao, X., 1989. "Turbulent mixing in supersonic combustion systems," AIAA Paper 89-0260.

Waltrup, P. J., 1987. "Hypersonic airbreathing propulsion: Evolution and opportunities," Paper 12 in Agard CP-428.

EXPERIMENTS

HYPERSONIC COMBUSTION - STATUS AND DIRECTIONS

Griffin Y. Anderson

Mail Stop 168
NASA Langley Research Center
Hampton, Virginia 23681-0001

ABSTRACT

Man's imagination has long been sparked by flight. Today sustained manned flight at hypersonic speed remains the largest unexplored region of the possible flight envelope. True aerospace planes flying at near orbital speeds will require a new form of airbreathing engine – the supersonic combustion ramjet or scramjet – and will also be required to "break the thermal barrier" with active fuel cooling of the engine and vehicle to survive. This paper will briefly review the internal flow environment of scramjets and discuss some of the relevant issues which arise in attempts to produce meaningful ground simulations of these flows. Some recent computational studies of the potential influence of injector design variables on performance will be reviewed as an example of the application of CFD to hypersonic combustion modeling. In addition, some projections to the future as to where emphasis should be placed in developing experimental capability and in extending computer modeling will be put forward.

1. Introduction

The terminology "hypersonic combustion" was perhaps first used with a special meaning by Professor Ray Stalker (Stalker and Morgan, 1988). By his definition, the term specifically refers to combustion in the flight speed regime of a supersonic combustion ramjet (scramjet) engine where the combustor entrance airflow is at hypersonic Mach number, and the combustor bulk flow remains hypersonic throughout the fuel injection, mixing, and burning process. If the definition of hypersonic is taken to be Mach number greater than five, then hypersonic combustion is relevant to scramjets at speeds approaching flight Mach number of 20 and above. In general terms, hypersonic combustion describes the airbreathing engine internal flow at speeds approaching orbital velocity.

J. Buckmaster et al. (eds.), Combustion in High-Speed Flows, 19–51.

1.1. The physics of hypersonic combustion

At such speeds the kinetic energy of the free-stream air entering the scramjet propulsion cycle is large compared to the energy released by reaction of the oxygen content of air with fuel (say hydrogen). Thus, the effects of reaction at Mach 25 speeds, where heat release from combustion may be 10 percent of the total enthalpy of the working fluid, will be small compared to Mach 8 flight where the air kinetic energy and potential combustion heat release are roughly equal. Hypersonic combustion, perhaps, corresponds closely to the gradual diffusive mixing and burning process described by Ferri (1973). Flow deflections due to heat release are small – a few degrees at most – and flow boundaries are conceived as contoured to control the pressure rise at the location of the flame and eliminate the possibility of strong shock formation.

To the contrary, at speeds of Mach 8 and below, combustion in ducted flows can generate large local pressure rise, flow deflection and separation. The behavior of this "upstream interaction" has been studied extensively by Billig (Billig and Dugger, 1969) and is characteristic of supersonic combustion in constant area channels below flight speeds of about Mach 8. In such flows involving local separation and a bulk Mach number near one, local wall static pressure is representative of the pressure across the entire flow at a given axial station. Therefore, a one-dimensional approximation to the flow can provide a reasonable description of the flow behavior. In hypersonic combustion, however, local Mach number remains high, Mach angles are quite shallow, and significant variations in static pressure are likely to occur across the combustor flow field at a given axial station. Typically, a fully three-dimensional representation of the flow field will be required; and as pointed out by Stalker (1989), special care must be taken to properly relate the implications of one-dimensional calculations to fully three-dimensional experimental data in a meaningful way.

So hypersonic combustion flows differ from supersonic combustion flows in that the flow remains hypersonic throughout in a bulk sense, the effects of heat release are smaller, and the pressure field is fully three dimensional. Common features of hypersonic and supersonic combustion flows include: real gas effects, nonadiabatic wall boundaries, finite strength shock waves, dissimilar gas injection, turbulent mixing, finite rate chemical reaction, flow separation, etc.

Also, because aerodynamic, fluid mixing, and chemical rate processes are all expected to be important, experimental simulation requires near full-size hardware and duplication of flight conditions. Simulation requirements are discussed in more detail in the next section of this paper.

Hypersonic combustion raises some additional uncertainty and concerns. First, the effects of (extreme) compressibility on turbulence generation and mixing are not well known or understood. Second, at about Mach 12 the velocity of the injected (hydrogen) fuel stream equals the velocity of the combustor air stream, and at higher flight speeds the air velocity exceeds the fuel velocity. The behavior of fuel-air mixing under these conditions is also not well known. In fact, compared to flight at Mach 8 and below, there is very little data on which to base confidence in our understanding of or ability to model hypersonic combustion flows.

1.2. Simulation requirements

The primitive variables available which describe the flow in a hypersonic combustor are:

P pressure (or ρ density)

T temperature

u velocity

L model length

ν_i gas composition

In principle, the composition can be manipulated in any fashion that results in duplication of the flight values of certain well-known dimensionless groups, i.e., simulation parameters. The first-order simulation parameters are:

M Mach number

Re Reynolds number

St Stanton number

D_1 Damkohler's first number

D_2 Damkohler's second number

GW wall enthalpy ratio

To these may be added certain second-order parameters such as:

Pr Prandtl number

Sc Schmidt number

The physical interpretation of the first three fluid dynamics and heat-transfer parameters is also well known; i.e., the Mach number represents the ratio of kinetic to thermal energy, the Reynolds number the ratio of inertial to viscous forces and the Stanton number the ratio of heat flux to inviscid energy flux. However, the interpretation of the last three is less well known as they are more specific to hypersonic reacting flows. These three parameters are the ratios of flow transit time through the combustor to chemical reaction time (D_1), the ratio of heat added by reaction to the stagnation enthalpy of the inviscid flow (D_2) and the ratio of enthalpy at the wall temperature to stagnation enthalpy of the inviscid flow (GW). The second-order parameters represent gas properties, e.g., the ratio of viscosity to thermal conductivity and the ratio of viscosity to diffusivity. However, to the extent that these parameters reflect turbulence characteristics, they are more dependent on the first-order simulation parameters than on molecular properties of the gas. The first-order parameters can, in turn, be related to the primitive variable as follows:

$$M \sim \frac{u}{\sqrt{T}} \tag{1}$$

$$Re \sim \frac{\rho u L}{\sqrt{T}} \sim \rho L M \tag{2}$$

$$St \sim \frac{q_w}{\rho u H} \tag{3}$$

$$D_1 \sim \frac{L}{u t_c} \tag{4}$$

$$D_2 \sim \frac{\eta_c \Delta h_c}{c_p T + \frac{u^2}{2}} \tag{5}$$

$$GW \sim \frac{c_p T_w}{c_p T + \frac{u^2}{2}} \tag{6}$$

where t_c is a characteristic combustion time, η_c is a combustion efficiency, Δh_c is the heat of combustion, c_p is a characteristic specific heat and T_w is the temperature of the combustor wall. The overall reaction time for combustion processes is generally proportional to pressure (or density) to an exponent around 1.75 and exponentially dependent on temperature. In certain restricted conditions wherein only binary (two-body) reactions occur, the combustion time is linear in density, leading to a direct relationship between Reynolds number and Damkohler's first number:

$$\begin{aligned} D_1 &\sim \rho L/u \exp(-T) \\ &\sim Re\frac{\sqrt{T}}{u^2}\exp(-T). \end{aligned} \tag{7}$$

Hence, if velocity and temperature were to be duplicated, then simulation of Reynolds number would also satisfy the requirements for simulation of binary reaction time, and vice versa. However, even in the simplest of situations, it is virtually impossible to manipulate the temperature, velocity and model length in a fashion that will simultaneously preserve the values of Mach number, Reynolds number, and Damkohler's numbers.

Therefore, it is clear that, in general, in hypersonic combustion experiments it is necessary to duplicate the primitive variables, including model length and gas composition, to ensure a faithful representation of the coupled chemical and flow processes. This automatically satisfies all of the simulation parameter requirements, with the possible exception of wall temperature and wall reactivity simulation.

2. Experimental Simulation

As pointed out above, combustion heat release in air produces about the same energy increment as the kinetic energy of flight at Mach 8. Thus, the experimental simulation of supersonic combustion flow conditions for propulsion studies in ground facilities frequently utilizes so-called direct-combustion heating with oxygen replenishment or vitiation heating as a means of generating a test environment. Essentially the test is conducted in (fuel-lean) combustion products where the amount of free oxygen is adjusted to represent the free-oxygen content of air. Of course, the remaining non-oxygen content of the test gas is not principally nitrogen as in air but will consist of some carbon dioxide and/or water vapor, etc., depending on the choice of fuel and oxidizer. The National Aero-Space Plane (NASP) "Engine Test Facility" (ETF) at Aerojet is an example of a Mach 8 vitiated propulsion facility based on a storable propellant (nitrogen tetroxide/monomethylhydrazine) rocket gas generator (Hooper, 1989). The NASA Langley 8-Foot High-Temperature Tunnel (8′ HTT) is an example of a Mach 7 air/methane/oxygen direct-combustion heated propulsion facility (Puster, Reubush, and Kelly, 1987).

Contamination of the test media with combustion products is, of course, a concern because the contaminants are not as inert and do not have the same thermodynamic properties as the nitrogen they replace. Suffice it to say that interpretation of the test data from vitiated facilities (and from any facility, in fact) requires detailed consideration of the actual test media as it affects the results in comparison with free flight in air. As the speed of the ground test simulation increases, in addition to the turbulence (velocity fluctuation) intensity of the test flow, other (turbulence) factors like temperature and oxygen concentration fluctuation levels, chemical contamination, the degree of dissociation of contaminants and of molecular oxygen, etc., all may become issues which must be addressed.

Other sources of energy such as storage heaters or electric arc heaters can also provide high-temperature test conditions. Contamination (with the addition of contamination from either dust particles or NO_x formation, respectively) is still an issue which must be addressed. To their credit, arc heaters have the potential of reaching energy levels corresponding to much higher flight speeds than combustion-heated facilities and, in fact, have been used successfully to simulate the orbital reentry flow environment. However, for propulsion flow simulation, conventional arc heaters lack the ability to operate at stagnation pressures adequate for hypersonic combustion simulation. Typically, arc heaters can achieve operating pressures of 20 to 50 atmospheres where pressures of 2000 to 5000 atmospheres and higher are desired.

The enthalpy requirements for hypersonic combustion simulation are summarized in Figure 1. The sensible total enthalpy of flight is shown plotted against Mach number. The right-hand-most curve shows the free-stream Mach number along the flight path. The forebody flow field and inlet compression process reduce the local Mach number and raise the flow static pressure along a nearly constant total enthalpy path as indicated by the arrow. The shaded band represents the range of Mach number and total enthalpy representative of combustor entrance conditions. The local static temperature, stagnation temperature, and stagnation pressure representative of the combustor entrance condition are shown at the points indicated for flight Mach numbers of 10, 15, 20, and 25. The static pressure is typically 0.5 to 1 atm. Note that the static temperature remains in the range of 3000° to 4000°R while the stagnation temperature quickly reaches levels where significant dissociation will occur. Also note that the stagnation pressure rises exponentially with Mach number from achievable levels (100 atm) at Mach 10 to extreme levels exceeding a million psi at Mach 25. The scale at the far right shows a level of oxygen dissociation expected in a facility which adds energy to the test gas at rest (such as an arc heater or a reflected shock tunnel) as a function of stagnation enthalpy. Pulse facilities, which generate high pressure, hypervelocity flows for a short time, represent the only presently proven means to approach simulation of these hypersonic combustion flow conditions on the ground.

Figure 2 shows the total enthalpy simulation capability of selected pulse facilities on the same coordinates used in Figure 1 (data from Rogers (1990)). Aerodynamic simulation of Mach number and Reynolds number is achieved in the Calspan reflected shock tunnels with ambient temperature or moderately heated (600°F) helium as a driver gas, but duplication is limited to a flight Mach number of about 10 for a test duration of 1 msec. The reflected shock tunnels T4 and T5 take advantage of the higher temperature achieved by free-piston compression of helium driver gas to achieve energy approaching orbital velocity. The Ames 16" shock tunnel is intermediate in energy simulation capability with a hydrogen combustion-heated driver. The expansion tube (HYPULSE), which is essentially two shock tubes in tandem, can achieve Mach 16+ energy with an ambient temperature helium driver.

Only one type of pulse facility, namely a free-piston-driven expansion tube, has the potential capability to duplicate both total

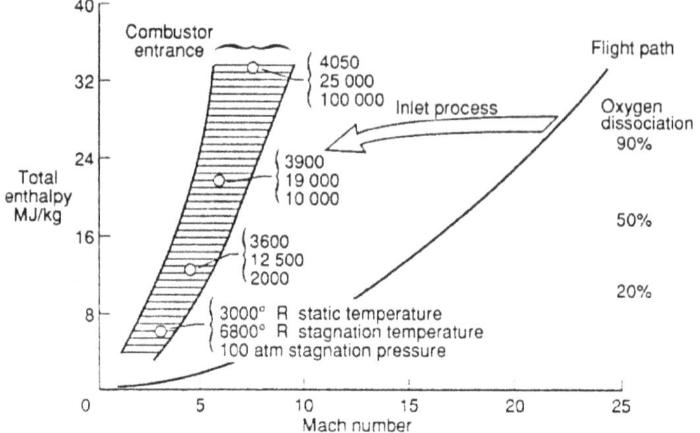

Figure 1. Total enthalpy requirements for hypersonic combustion simulation.

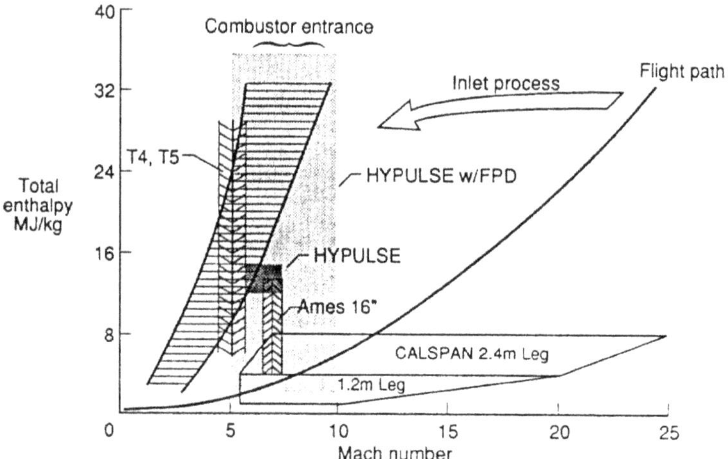

Figure 2. Total enthalpy capability of selected pulse facilities for hypersonic combustion simulation.

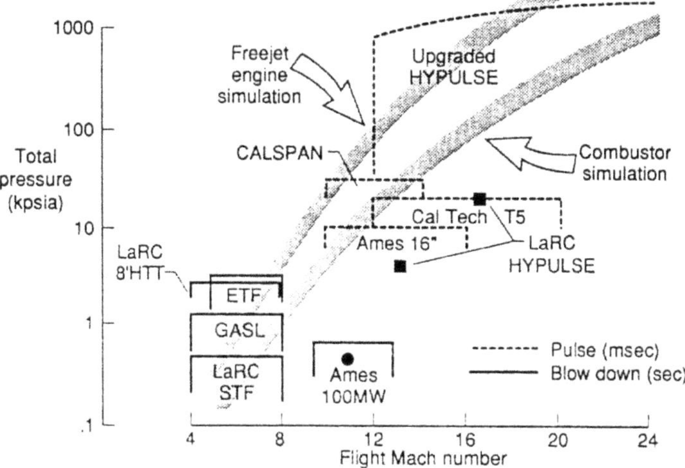

Figure 3. Total pressure capability of selected facilities for hypersonic combustion simulation.

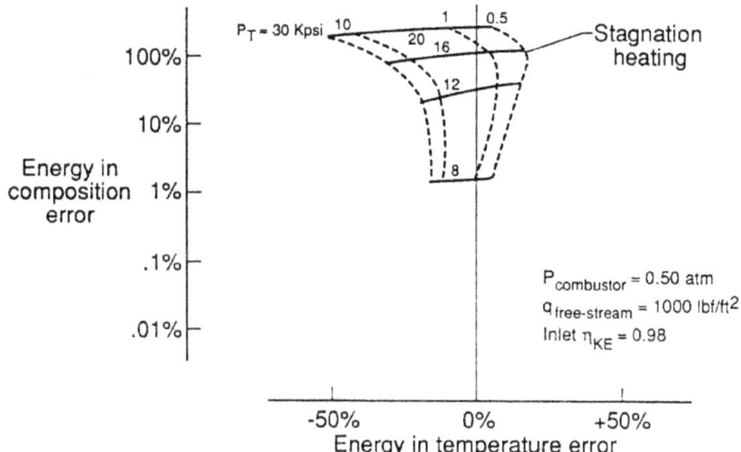

Figure 4. Ground simulation of scramjet flight combustor conditions with stagnation heating.

enthalpy and total pressure, as well as velocity and gas composition, above about Mach 16 (Tamagno et al., 1990). This capability has thus far only been demonstrated in a pilot-scale (1-1/2-inch inside diameter) facility (Paull, Stalker, and Stringer, 1988), however, addition of a free-piston driver to the 6-inch inside diameter NASA expansion tube (HYPULSE) located in Ronkonkoma, Long Island, New York, and operated by GASL is planned, and should increase the operating envelope as indicated in Figure 2.

The total pressure required for hypersonic combustion simulation is shown in Figure 3. The shaded bands show pressure requirements for either freestream engine simulation or internal combustor flow simulation along with the capabilities of the selected facilities. Blow down facilities with run times on the order of 100's of seconds like the Langley 8′ HTT and the NASP ETF's are limited to 2 to 3 kpsi. Higher enthalpy facilities like the Ames 100 MW Arc-Heated Facility have even less pressure capability because of the facility nozzle throat cooling problem. Pulse facilities like the Calspan, Ames 16", and Cal Tech T5 reflected shock tunnels (Rogers, 1990) can produce and contain significantly higher pressures (up to 30 kpsi) largely because of shorter flow times on the order of milliseconds. Note that at Mach numbers above 12, the total pressure requirement approaches a million psi, and only the expansion tube is capable of producing those pressures. The unique capability of the expansion tube is a result of the fact that the acceleration process in the expansion tube adds velocity directly to the flow without stagnating it. This means that the facility need not contain the stagnation pressure of the flow that it generates, and this feature is a considerable advantage when duplication of flight conditions is required.

The fundamental difficulty in generating hypersonic flows in a ground facility is related to putting the energy into the proper mode in the test gas that is generated. Figure 4 is an attempt to explain this in more detail (Private Communication, Auslender). The energy required for duplication of flight along a constant dynamic pressure trajectory of 1000 psf is taken as a reference with the generation of a combustor pressure of 0.5 atm and the assumption of an adiabatic inlet process with a kinetic energy efficiency of 98 percent. The facility process assumes stagnation heating with isentropic expansion from the stagnation conditions to the required combustor entrance condition and chemical freezing of the gas composition at the throat of the facility nozzle. The horizontal axis represents the error in energy in static temperature compared to flight of this particular facility stagnation pressure and total enthalpy, where the energy scale has been made dimensionless with respect to the amount of energy in stoichiometric combustion of hydrogen and air. Similarly, the vertical axis is the error in energy due to composition. That is, if the gas is dissociated at the stagnation energy level, this dissociation persists in expansion through the nozzle, and energy is tied up in NO and atomic oxygen which would not be present in clean air in flight. Various flight Mach numbers are shown by the solid lines, and various total pressure levels are indicated by the dashed lines. Since flight in clean air is taken as the reference, in Figure 4 "goodness" is towards the center and bottom of the figure with a small energy error due to a static temperature mismatch and a small energy error in dissociation. Note that as Mach number increases, the amount of energy error in composition increases substantially. Also, as Mach number increases, generally the required stagnation pressure for simulation of flight significantly increases. Figure 5 adds a similar carpet plot of energy error for a facility like the expansion tube which generates velocity directly without heating at stagnation conditions. As can be seen in Figure 5, the level of energy in composition error is approximately an order of magnitude smaller for the expansion tube or non-stagnating facility compared to a stagnation heating facility.

Another way to assess facility contamination level is in terms of the amount of molecular oxygen in the test flow compared to the oxygen content of clean air in flight. The performance of a number of facilities is shown in Figure 6 as a function of flight Mach number (Private Communication, Rogers). The hatched band shows the performance typical of reflected shock tunnels with particular facilities such as T5 and the Ames 16-inch tunnel shown by symbols on the figure. The expansion tube HYPULSE is shown by the circular symbols. As Mach number increases beyond Mach 12 the expansion tube facility shows an increasing advantage over the reflected shock tunnels in terms of available oxygen to simulate flight in clean air.

The comparatively moderate investment required to generate the desired test conditions in pulse facilities is achieved at the expense of flow duration. Producing adequate flow duration to establish a sensibly steady flow representative of the "real" steady flow becomes an issue which (like the contamination issue) must be addressed in facility design and in the interpretation of data. However, short flow duration has an advantage in high-energy flow tests in that the model

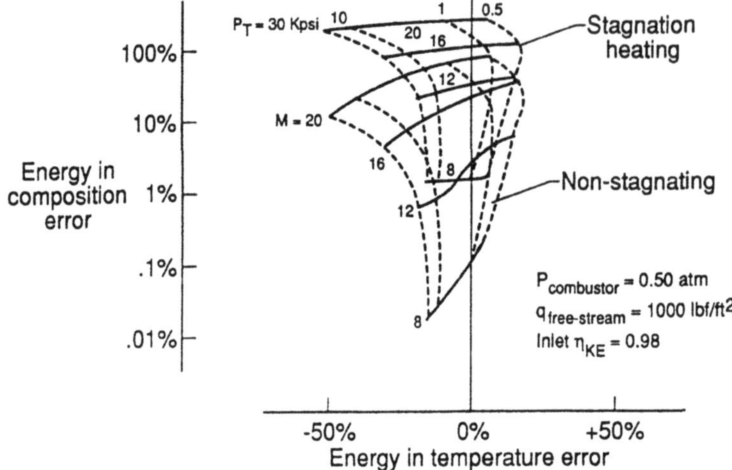

Figure 5. Ground simulation of scramjet flight combustor conditions comparing stagnation and non-stagnating heating.

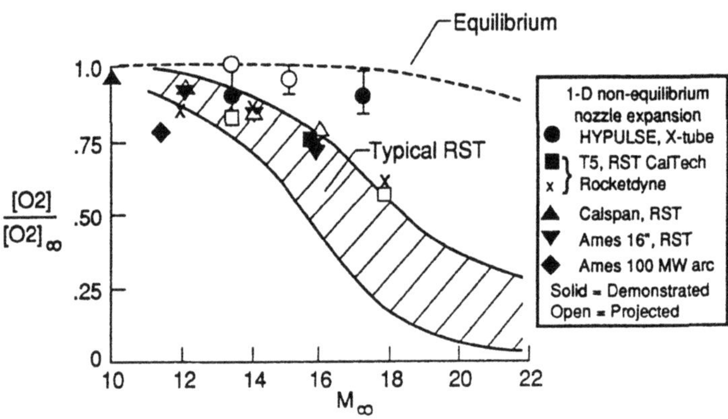

Figure 6. Test gas oxygen dissociation level at simulated combustor entrance conditions.

cooling requirement can be met by simple heat sink approaches. The short flow duration also makes some types of measurements, like local heat flux, much easier to make accurately in pulse flows than in steady flows. On the other hand, all measurements must have very high frequency response – on the order of 10^6 Hz – in order to provide meaningful data from test times of 10^{-3} sec or less. In this regard, it may be noted that hypersonic tests conducted in a pulse facility almost inevitably involve an unheated (room temperature) model. Hence the wall enthalpy ratio, GW, is typically on the order of $1/10$ or less, whereas in flight it might be as high as $2/10$. However, in point of fact, the heat-transfer and skin friction coefficients both become virtually independent of GW for values less than about $3/10$.

Of greater concern than duplication of wall temperature, per se, is the simulation of wall reactivity. This consists of both the tendency of the wall to catalyze gas phase reactions and the extent to which the gas phase reactions are thermally quenched (or promoted) at the wall. The catalytic efficiency is primarily dependent on certain gross characteristics of the wall materials, i.e., metals tend to be "fully catalytic" and glassy materials (e.g., quartz) tend to be "non-catalytic." Thus, use of a model composed of metal walls with glass windows may present a potential local disruption of wall catalyticity effects. However, the extent of the problem is easily determined by using metal blanks in place of windows. The effects of thermal quenching at a model wall temperature that is substantially less than exists in flight is more difficult to determine experimentally, as wall-mounted instrumentation may not function at the true wall temperature. However, this would appear to be an instance where computational fluid dynamics (CFD) simulations could be trusted to correct the data for wall temperature effects. If necessary, the model could be preheated (e.g., radiatively) and nonintrusive or noncontacting instrumentation (e.g., exit plane surveys) used to assess wall temperature effects.

2.1. Comparison of combustion data

This section describes an experimental assessment of the effects of the level of contamination by dissociation of test gas on a mixing and combustion experiment at Mach 17 flight energy conditions. The intention is only to show the effects of the state of the test gas produced by different facilities on the experimental results and not to imply that one facility type should be pursued to the exclusion of another. In fact, as noted earlier, increased contamination goes along with higher energy (Fig. 1). In general, all hypervelocity data need to be assessed in light of the test media involved including the initial composition and finite rate chemistry which may be important in the flow field. Figure 7 shows the experimental apparatus. Basically the configuration is a circular tube 1-1/2 inches in inside diameter with a sharp leading edge. The tube is about 36 inches long, and an annular injector is located about 7 inches from the entrance of the tube to inject hydrogen into the high velocity air captured at the entrance of the tube. Hydrogen is supplied to the injector by a fast-acting valve from a Ludwig tube, and the hydrogen flow is initiated such that hydrogen is flowing into the model when the facility flow is established and initiates the combustion simulation.

Two identical models were constructed. One was tested in the NASA HYPULSE facility at Mach 17 flight conditions. The other was tested in the reflected shock tunnel T4 at the University of Queensland in Brisbane, Australia. Typical flight conditions and the wind tunnel test conditions are shown in Table 1. Note that in flight only molecular oxygen would be expected to be present in the test stream, but at Mach 17 energy level the reflected shock tunnel T4 has significant dissociation compared to the clean flight condition. Wall pressure measured in the tube downstream of the injection point in these tests is shown in Figure 8. The combustion pressure rise parameter shown is the pressure with injection into air minus the pressure for injection into nitrogen divided by the pressure with injection into nitrogen. Presenting the data in this way shows the maximum effect of combustion on wall pressure and tends to remove some of the wave structure which otherwise produces scatter in the data. The open symbols show data from the reflected shock tunnel T4 and the solid symbols show data from the expansion tube HYPULSE. Clearly, in Figure 8, considerably more pressure rise is generated in the reflected shock tunnel T4 which has a level of dissociation approximately equal to half of the molecular oxygen compared to freestream undissociated air.

Jachimowski (1992) presents an analysis to understand the reasons for this difference in measured pressure. The flow in the tube is modeled with a simple one-dimensional flow conservation equation that includes three temperatures for the mixing/reacting fuel and air flowing down the tube. Initial temperature for the fuel and air

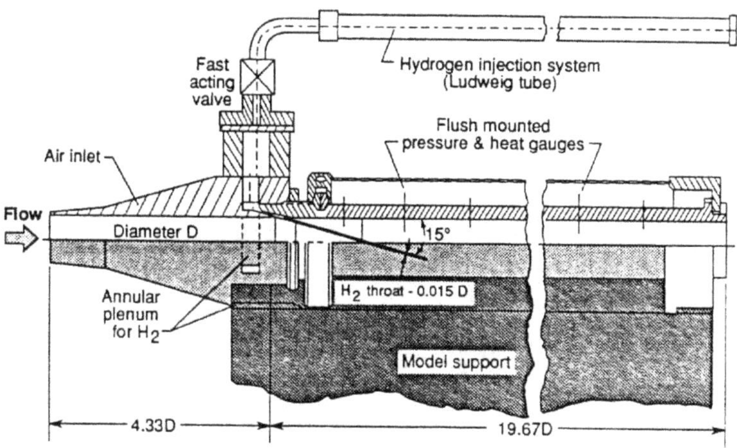

Figure 7. Mach 17 combustion experimental apparatus - axisymmetric model.

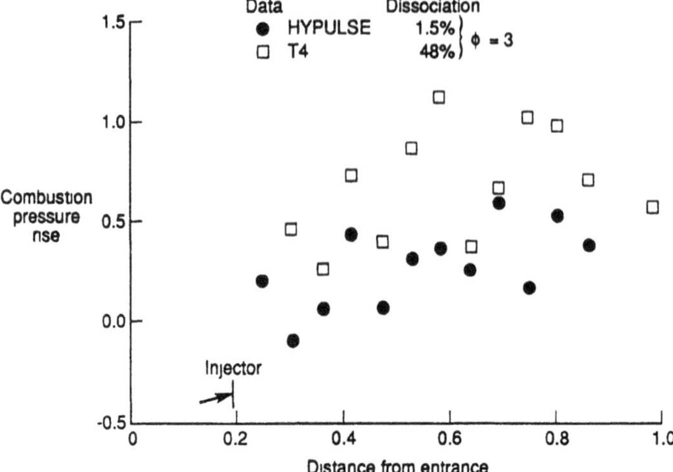

Figure 8. Comparison of combustion pressure rise.

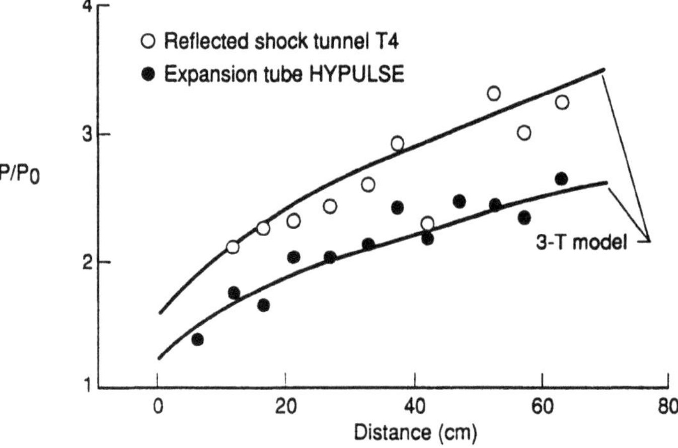

Figure 9. Assessment of experimental Mach 17 combustion pressure rise.

Table 1

HYPERSONIC COMBUSTION CONDITIONS
Mach 17 flight simulation

	P	T	V	O_2	Dissociation	M
HYPULSE	3.29	3758	16,660	.225	1.5	5.75
$T4$	2.11	3717	15,450	.108	48	5.17
Typical flight	7-20 psia	3600 $^{\circ}R$	17,000 ft/sec	.237 Mass fraction	0 %	7

- Deficiencies of simulation
  - Nonequilibrium chemical contamination (O and NO)
  - Low total pressure and Reynolds number
- Other potential differences between facilities
  - Turbulence intensity
  - Contamination from diaphragm material, etc.

are defined by the initial conditions in the experiment, and a mixed stream of fuel and air with an independent temperature generated by a finite-rate kinetic computation follows development of the flow. Wall friction, shock losses, and the rate of mixing in the tube are modeled based on the expansion tube wall pressure data and are held constant with the different initial composition of the reflected shock tunnel flow (Table 1) to derive the predicted pressure distribution which is compared with the data in Figure 9. Excellent agreement between the prediction for the reflected shock tunnel T4 is achieved with data using this three-temperature scheme with the finite-rate kinetic model. Apparently, even though a significant amount of the molecular oxygen is withheld from reaction with hydrogen fuel as NO, the initial level of atomic oxygen present in the dissociated test gas increases heat release and the amount of reaction resulting in the greater pressure rise.

Figure 10 shows results from the same computation in terms of the energy yield predicted in the experiment. Results for the expansion tube are shown by the solid lines and results for the reflected shock tunnel are shown by the dashed lines. The lowest pair of curves shown is for the actual pressure and temperature of the experiment. Due to the initial dissociation in the reflected shock tunnel, the energy yield is somewhat higher than for the expansion tube. If the pressure were raised to a level of 1 atmosphere, which is more representative of a typical flight trajectory, an even greater difference would be observed; and if the temperature was also reduced to a level of 1200 K (again, more representative of flight), a still greater energy yield would be achieved with a larger difference between the reflected shock tunnel and expansion tube resulting. The top curves show the computed result if the flow were in local equilibrium as it mixed along the length of the tube. The energy level is considerably above the result for the pressure and temperature of the experiment, indicating that finite-rate chemistry plays a dominant role for the conditions of the test. Again, the intent of this comparison is to emphasize the need to account for the composition of the test media in hypervelocity combustion simulations in order to understand the implications of ground simulations on flight performance.

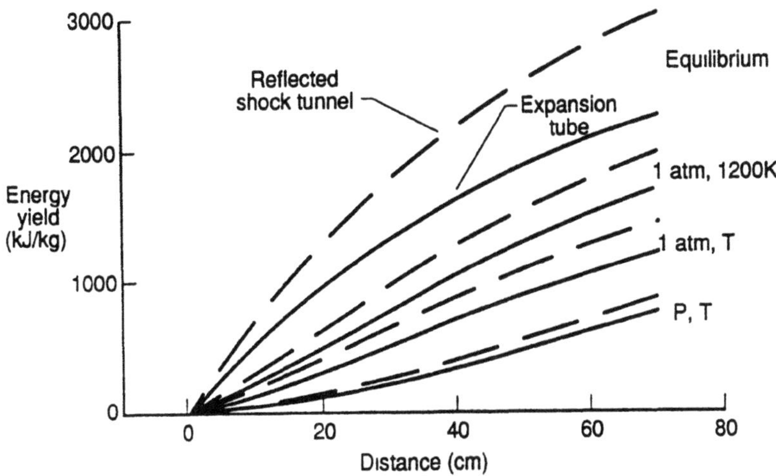

Figure 10. Computed energy yield for Mach 17 combustion experiment.

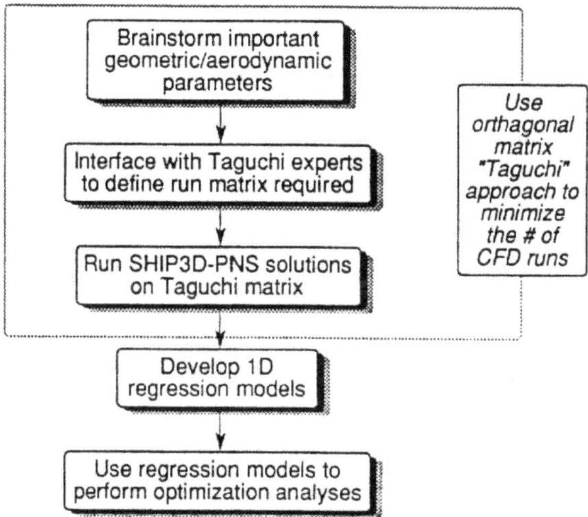

Figure 11. A Taguchi-based approach to a CFD parametric study of hypersonic fuel injector designs.

3. Computational Simulation

Much has been written about the computer age and the impact of modern computational fluid dynamics (CFD) on engineering in general and on aeronautics in particular. CFD has an important role to play in the understanding of hypersonic combustion. However, it must be recognized at the outset that some of the important physics of hypersonic combustion flows are either not included or at best only crudely modeled in even the most advanced computer codes. For instance, chemical reaction occurs at the molecular level but CFD treats the flow of a continuous fluid. Further, turbulence is generally modeled by some physical analogy rather than being computed directly and any interaction between turbulence and chemical reaction is generally ignored. Also, the physical modeling of turbulence is based entirely on empiricism derived from velocity fluctuation measurements in incompressible flows; but it is known that compressibility introduces some effects that tend to suppress the generation of turbulence, and that the assumption of simple analogous behavior of mass, momentum and energy transport is inadequate for combustion flows.

In spite of these factors, CFD predictions of complex fully three-dimensional hypersonic combustion flow fields are possible today on a routine basis. Examination of three-dimensional CFD solutions can provide valuable insight to complex hypersonic combustion flows. If these computational results can be "calibrated" with data from experimental simulations of hypersonic combustion at the appropriate conditions, then in spite of incomplete physics in the analysis, the code results can become an important design tool. Once a code is "bench marked" against data, it can be used to conduct parametric variations in design variables to help improve and refine a given configuration.

It should be recognized that since the actual physical processes going on in a hypersonic combustion flow are much more complex than the CFD models, the most complex CFD model should not necessarily be the tool of first choice for analysis. For instance, the added computational complexity of finite-rate chemistry may not be justified in a hypersonic combustion flow field where reactions may not begin to proceed until the scale of the turbulence has reached a level where adequate molecular collisions of fuel and oxidizer can occur, as suggested by Swithenbank et al. (1989). Following this, local

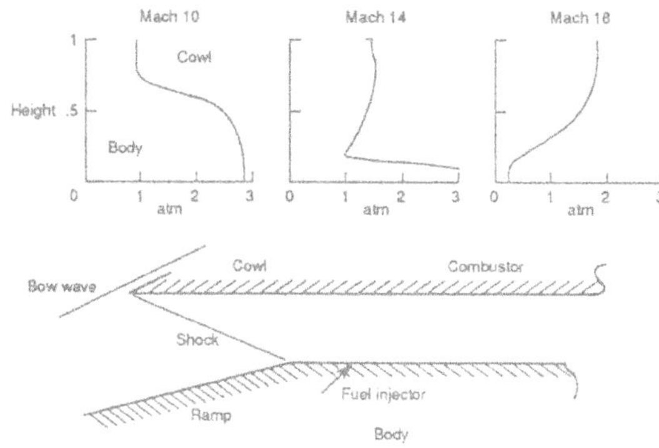

Figure 12. Combustor entrance flow profiles.

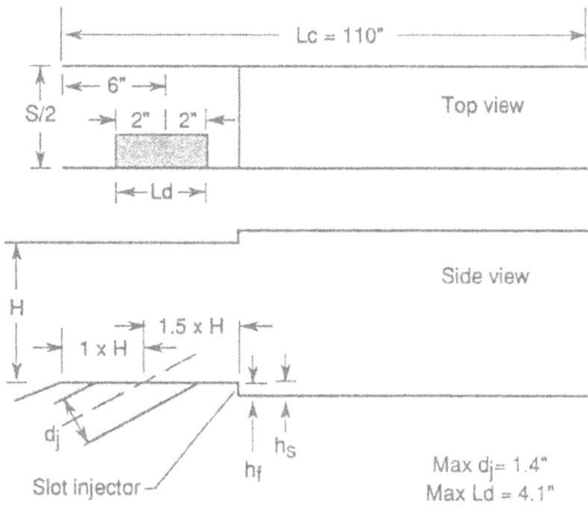

Figure 13. Schematic of flush wall/slot fuel injector geometry.

chemical equilibrium may prevail, as in the diffusive burning model for supersonic combustion envisioned by Ferri (1973). A computational approach employing a combination of chemically frozen flow and a simple complete reaction model, much like the sudden freezing model developed by Bray (1963) in the 1960's for the reverse process in nozzle flows, could be more accurate than a very complex finite-rate reaction scheme that models turbulence-chemistry interactions as augmented laminar diffusion. The point is that the complexity of the various parts of the computational model should be balanced to achieve efficient engineering results, as well as consistency with the real physical processes.

3.1. Hypersonic fuel injector design study

This section of the paper describes a CFD study of injector designs for a hypersonic combustor. The study used an orthogonal matrix or Taguchi approach in the design of the computational study which is shown schematically in Figure 11. The potential importance of planning the parametric computational study in this manner is that it may allow one to obtain more information about the design space of interest from a smaller number of computations. The computations were performed at Mach 10, 14, and 18 with theoretical combustor entrance profiles computed for a realistic inlet-type flow. Representative static pressure profiles are shown in Figure 12. Note that at every Mach number, there is a significant profile; and static pressure varies across the flow due to the reflected shock wave off the cowl which is nearly canceled by the shoulder on the body side of the engine flow field. The injection process in this example assumes flush-wall injectors on both the top and bottom walls of the combustor located immediately downstream of the inlet shoulder. The combustor height was fixed at 5 inches and the combustor length was fixed at 110 inches.

The injector geometry is shown schematically in Figure 13. Hydrogen fuel is injected at a Mach number of 2 with a total temperature of 2000 *R*. Immediately downstream of the discrete hydrogen injector there is a film injection slot. Injector design parameters are shown in Table 2 and include the split of fuel between the slot and the discrete injector, the injector spacing, the fuel equivalence ratio, flight Mach number, injection angle, and total pressure of the injectant. In the orthogonal matrix design, a nominal value of each

Table 2

FLUSH WALL INJECTOR MATRIX

- L-16 matrix (29 cases)
- Parameters

Fuel Split	Fs	(0, .35, .70)
Injector spacing ($H/2/S$)	HS	(.5, .75, 1.)
Fuel equivalence ratio	Phi	(1.0, 1.5, 2.0)
Flight Mach number	M	(10, 14, 18)
Injection angle	Th	($20°, 30°, 40°$)
Jet total pressure	Pth	(500, 900, 1300 psia)

- Interactions

$Fs * HS$	$Fs * Phi$
$Fs * Th$	$HS * Th$
$M * Th$	$HS * Phi$

- Nonlinear parameters

  M

Table 3

SUMMARY OF HIGH-SPEEC INJECTOR DESIGN STUDY RESULTS

Injector Type	N # of Design Variables	C # PNS Cases ($C/3^N * 100$)	RMS Force Deviation
Flush wall	6	29 (3.98%)	4%
Ramp	9	83 (.422%)	11%
Strut	8	33 (.503%)	8%

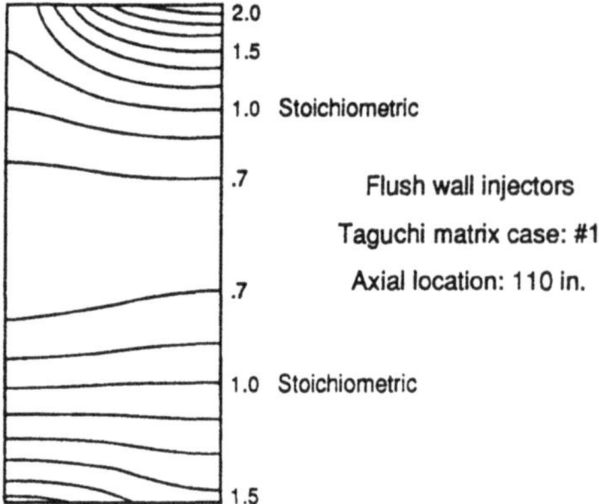

Figure 14. Example of total fuel mass fraction contours.

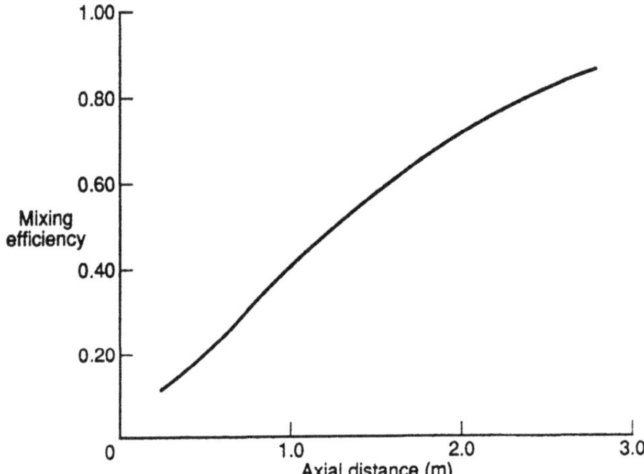

Figure 15. Mixing efficiency distribution for flush wall injectors; Taguchi matrix case 1.

parameter is chosen with a plus or minus delta to give three evenly spaced values of the design factors to define the design space of the computational study. These values are also shown in Table 2. Specific cases are chosen to examine selected interactions between the design parameters as shown in Table 2 and nonlinear effects expected. The CFD code chosen for the study is the SHIP/PNS code described in Spalding et al. (1974). The flight Mach number range chosen for the study was specifically selected to be above Mach 10 so that recirculation regions within the combustor would not be dominant, and therefore the PNS representation of the flow would be adequate to describe the mixing and combustion process.

A typical result is shown in Figure 14. Contours of total fuel mass fraction are shown for a cross plane in the computational domain. The cross plane extends from the centerline of one injector to the centerline of the opposite injector and covers the height of the duct from the cowl to the body. For the contour shown, which corresponds to the end of the combustor duct, rich regions are seen to exist at locations near the centerline of the injectors and the concentration of fuel in the middle of the duct approaches 0.7 times stoichiometric. The overall injected equivalence ratio for this case is stoichiometric so mixing has progressed quite far toward uniformity at this downstream location. Figure 15 shows an integral measure of the degree to which mixing is complete as a function of distance down the duct. The mixing efficiency shown in the figure represents the amount of the least available reactant (oxygen for an overall equivalence ratio of 1 or greater) which has been consumed by that location in the flow. As can be seen in the figure, mixing starts from near 0 at the injector location and increases monotonically to near 80 percent by the end of the combustor. Results similar to those in Figures 14 and 15 were generated for each of the 29 cases in the study. The level of mixing achieved at the end of the duct was correlated by a regression equation which follows:

$$\begin{aligned}
\eta_m &= 2.25 - 0.33 * Fs + 0.189 * HS + 0.388 * Phi - 0.252 * M \\
&\quad -0.0056 * Th + 0.00007 * Pth + 0.185 * Fs * HS \\
&\quad -0.07 * Fs * Phi + 0.0052 * Fs * Th - 0.238 * HS * Phi \\
&\quad +0.000436 * M * Th + 0.0077 * M * M.
\end{aligned}$$

Here each of the factors in the original matrix appears as a linear term in the equation along with interactions chosen in Table 2 which turned out to be significant. The effectiveness of this regression equation in representing the computed results is assessed in Figure 16. Here the mixing efficiency predicted by the regression equation is plotted as the vertical scale against the mixing efficiency from the SHIP calculation along the horizontal scale. As can be seen in the figure, the regression is an excellent representation of the PNS data with an accuracy of 5 percent or better for all of the cases in the study.

The flush-wall injector type in the preceding discussion is only one of several types of injector included in the overall design study. In addition, intrusive injectors of both ramp and strut types were also analyzed. The overall results of this study are summarized in Table 3. For each injector type, the design space described by the factors chosen in the study was well represented by regressions generated from a number of cases substantially less than a full parametric study of all the design variables would have entailed. In Table 3 the response parameter chosen to define performance for the injector type is the force produced by a typical nozzle process at the end of the combustor computation (Riggins et al., 1992). As shown in Table 3, regressions based on the limited number of computations were successful in representing the thrust produced by these combustor/nozzle combinations to within about 10 percent over the entire design space for the injectors. Since a full parametric over the entire design space would have required thousands of PNS computations, the results achieved were only possible with application of the orthogonal matrix approach to the design study.

4. On to the Future

While a computational study like that described in the previous section can produce any desired result-force, mixing, wall pressure, heat transfer, and so on-in an experimental situation, direct measurement of "real performance" is much more difficult to achieve. Particularly for conditions of hypervelocity combustion, the effect of combustion on a relatively easy to measure quantity such as wall pressure can be relatively small. Also, due to the short duration of the flow, direct measurement of integrated force or thrust is generally not possible. Some other means to determine "where the injected

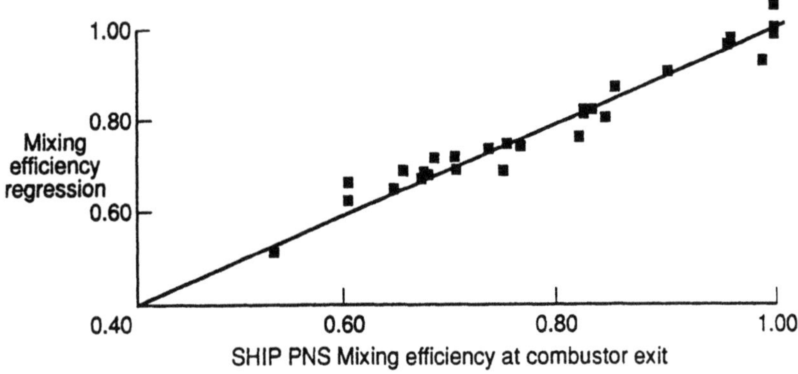

Figure 16. Assessment of flush wall injector mixing efficiency regression.

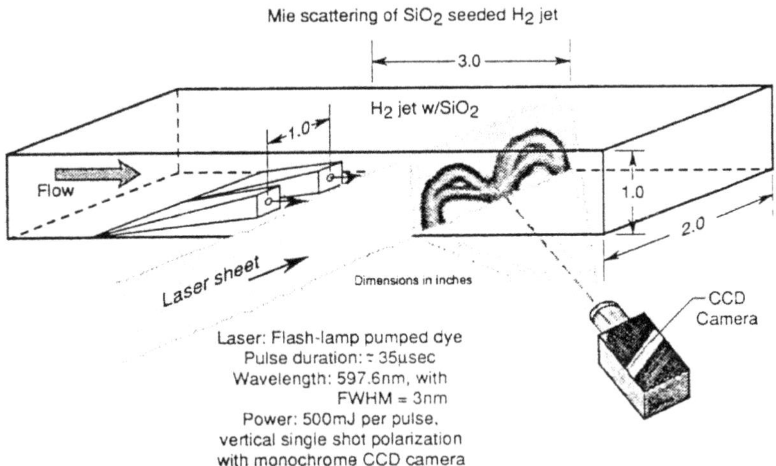

Figure 17. Schematic of fuel plume imaging in HYPULSE.

fuel goes and how much mixing and combustion is accomplished" is required. One approach, which will be described in the following paragraphs, is to use tracing of the injected fluid with particles. In this technique, the hydrogen fuel to be injected has a small fraction, say 3 percent of silane (SiH_4) added. Silane is very reactive and burns on contact with air or oxygen even at room temperature and in very dilute mixtures. Immediately prior to injection into the air flow, oxygen is added to this silane/hydrogen mixture sufficient to burn the silane present and produce silicon dioxide particles. These particles, typically on the order of 0.2 microns in diameter, are small enough to follow the fuel as it is injected, mixes and burns in the flow. Illumination of a cross plane in the flow with laser light can then produce a Mie scattering image of the particles which can be recorded with a fast electronic camera and thus visualize the fuel distribution in the experiment. This process is shown schematically in Figure 17 for a typical combustor duct with ramp-type fuel injectors.

The process of correcting the raw image and processing it to produce a true cross plane in the flow is described in Figure 18. Both an image with injection in the flow and a reference background luminosity image of the "region of interest" (ROI in Figure 18) are recorded. These images are subtracted and then stretched to represent a true cross plane. This image is further corrected to reflect the actual intensity profile of the laser sheet illumination and provide an accurate picture of particle concentration in a cross plane in the flow. If particles are neither created nor destroyed in the mixing/reacting flow as it proceeds downstream, then the total amount of scattered light in the corrected image from each plane should be approximately the same. Thus, the local amount of reflected light in each image, nondimensionalized by the average reflected light in each plane, becomes a measure of the local fuel concentration relative to the average concentration in the overall bulk flow. Thus, this technique allows a quantitative measure of the fuel distribution to be determined with successive pictures at successive planes downstream from the injection location in the duct. Data obtained in this way can be used to calibrate computations of the type described in the preceding section. The overall intent is to use this approach to provide a calibration of CFD codes to determine the level of hypersonic combustion performance achievable in flight.

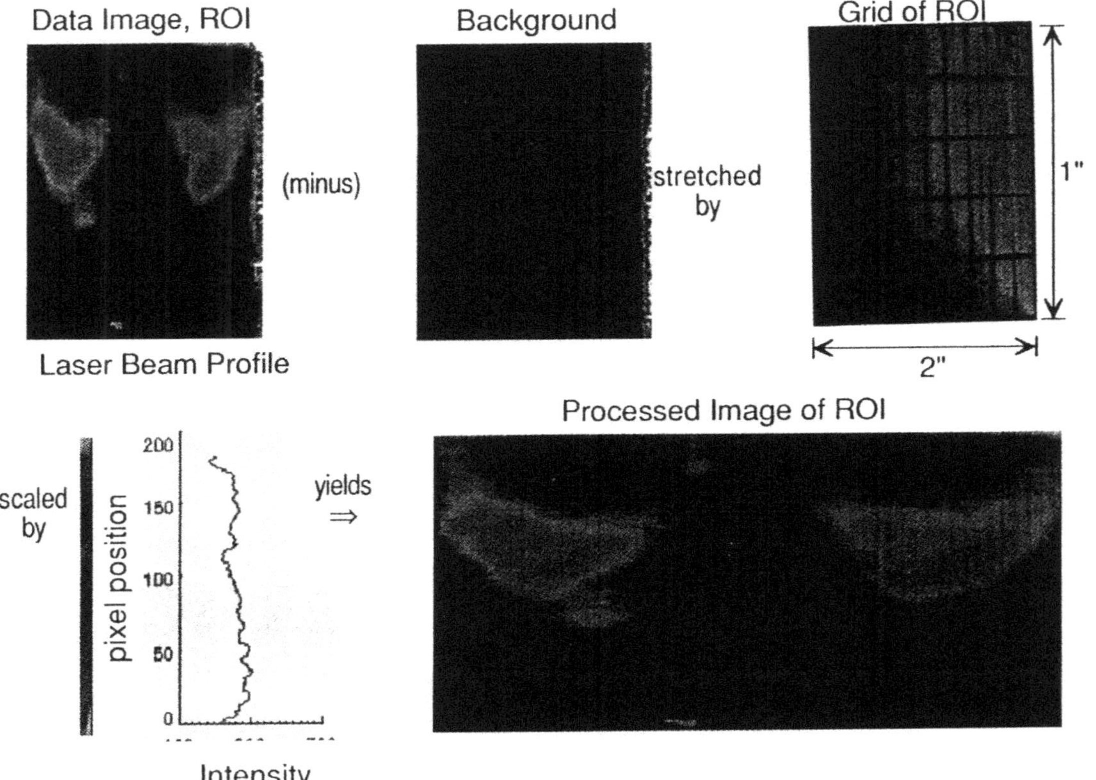

Figure 18. Fuel plume image processing procedure for quantitative Mie scattering.

5. Concluding Remarks

The purpose of this paper has been to present an assessment of the present state of the art of simulation of the hypersonic combustion environment in ground facilities. In addition, we have looked at the ability to describe these flows with simple engineering tools, such as the SHIP/PNS computational procedure. An example of a new technique intended to measure quantitative mixing behavior from ground simulation of these flows has also been described. Clearly, the tools are now in hand to make successful computational and experimental evaluations of injection/combustion situations for hypersonic combustion flows. The challenge now is to apply these tools and press on with comparative simulations in both reflected and nonreflected shock/expansion facilities. Improved calibration of these facilities, including direct measurements of the test gas that they produce, are required and should be a high priority. In addition, considerable effort is needed to continue to develop and improve measurement techniques for these types of flows. In particular, parameters that should be addressed include wall skin friction, optical techniques to assess the static temperature in the flow, and the refinement of techniques, such as the Mie scattering technique described above, to trace fuel mixing and combustion as the flow develops in the combustor. The emphasis in this process should be on measurement of first-order combustor performance variables in order to generate detailed performance data from these high velocity mixing/reacting flows which can calibrate CFD codes suitable for engineering trade studies for actual flight designs.

Acknowledgment

The author is pleased to acknowledge the numerous contributions of his colleagues at Langley and in industry to the ideas and accomplishments described in this paper. While all are recognized at least in part by direct reference where possible, specific additional mention is due Kim Erland for orchestrating the Taguchi matrix planning and Pradeep Kamath for the PNS computations in the CFD parametric study described in Section 3.1. Also, credit for the idea to use smoke and mirrors (Mie scattering) in Section 4 is due Chuck McClinton and Clay Rogers at Langley, and for successful implementation of the technique, credit is due to the HYPULSE team at GASL including at least John Erdos, Jose Tamagno, Rich Trucco, and the dedicated staff who support them.

Nomenclature

c_p specific heat at constant pressure, J/kg

D_1 Damkohler's first number (see text)

D_2 Damkohler's second number (see text)

f/a fuel air ratio

GW wall-to-stream stagnation enthalpy ratio (see text)

H stagnation enthalpy, J/kg

h static enthalpy, J/kg

Δh_c heat of combustion, J/kg

L length, m

M Mach number (see text)

P static pressure, kPa

Pr Prandtl number (see text)

q wall heat flux, J/m^2sec

Re Reynolds number (see text)

St Stanton number (see text)

T static temperature, K

t time, sec

u velocity, m/sec

η efficiency

θ angle, degrees

ν_i species volume fractions

ρ density, kg/m^3

ϕ fuel equivalence ratio $\equiv (f/a)/(f/a)$ stoichiometric

Subscripts

c combustor or combustion

m mixing

w wall value

References

Billig, F. S. and Dugger, G. L., 1969. "The interaction of shock waves and heat addition in the design of supersonic combustors," Twelfth Symposium (International) on Combustion, The Combustion Institute, Pittsburgh, Pennsylvania, pp. 1125-1134.

Bray, K. N. C., 1963. "Chemical reactions in supersonic nozzle flows," Ninth Symposium (International) on Combustion, Academic Press, p. 770.

Ferri, A., 1973. "Mixing controlled supersonic combustion," *Annual Review of Fluid Mech.* **5**.

Hooper, W. G., 1989. "Mach 5 and Mach 8 hypersonic test facility," AIAA Paper 89-2297, 25th Joint Propulsion Conference, July.

Jachimowski, Casimir J., 1992. "An analysis of combustion studies in shock expansion tunnels and reflected shock tunnels," NASA TP-3224, July.

Paull, A., Stalker, R., and Stringer, I., 1988. "Experiments on an expansion tube with a free piston driver," AIAA 15th Aerodynamic Testing Conference, San Diego, California, May 18-20.

Private Communication: Dr. Aaron Auslender, Lockheed Engineering and Sciences Company.

Private Communication: Facility data from Rogers (1990) were used to compute test section conditions with the NENZF computer code.

Puster, R. L., Reubush, D. E., and Kelly, H. N., 1987. "Modification to the Langley 8′ high temperature tunnel for hypersonic propulsion testing," AIAA Paper 87-1887, July.

Riggins, D. W., McClinton, C. R., Rogers, R. C., and Bittner, R. D., 1992. "A comparative study of scramjet injection strategies for high Mach number flows," AIAA Paper 92-3287, July.

Rogers, R. C. (editor), 1990. "Workshop on the application of pulse facilities to hypervelocity combustion simulation," NASP WP-1008. Eighth NASP Technology Symposium, Monterey, California, March 25-26.

Spalding, D. B., Launder, B. E., Morse, A. P., and Maples, G., 1974. "Combustion of hydrogen-air jets in local chemical equilibrium," NASA CR-2407.

Stalker, R. J., 1989. "Thermodynamics and wave processes in high Mach number propulsive ducts," AIAA Paper 89-0261, January.

Stalker, R. J. and Morgan, R. G., 1988. "Free-piston shock tunnel T4-initial operation and preliminary calibration," University of Queensland, Dept. of Mech. Eng. Research Rept., January. NASP CR-1023, August or NASA CR-181721, September.

Swithenbank, J., Eames, I., Chin, S., Ewan, B., Yang, Z., Cao, J., and Zhao, X., 1989. "Turbulent mixing in supersonic combustion systems," AIAA Paper 89-0260, January.

Tamagno, J., Bakos, R., Pulsonetti, M., and Erdos, J., 1990. "Hypervelocity real gas capabilities of GASL's expansion tube (HYPULSE) facility," AIAA Paper 90-1390, June.

RECENT EXPERIMENTS ON HYPERSONIC COMBUSTION IN AN EXPANSION TUBE TEST FACILITY

John I. Erdos

General Applied Science Laboratories, Inc. (GASL)
Ronkonkoma, New York 11779

ABSTRACT

The NASA LaRC expansion tube was recommissioned by GASL three years ago for the purpose of conducting hypersonic combustion experiments at Mach 17 total enthalpy in clean air in support of the NASP Technology Maturation Program. The first year of operation was essentially a learning period for development of the test techniques required by this new application of an expansion tube. During the past two years a series of test programs have been conducted which produced reassuring correlations of most experimental observations with CFD calculations, several perplexing data sets that seem at odds with prior observations, and some entirely unexpected difficulties in conducting one set of experiments. The latter are particularly relevant to this workshop since they involve reacting shear layers and detonations. We will briefly review the facility operation, test techniques and certain data sets for which corresponding CFD results are available to establish the basic capabilities and limitations of the facility and confidence in the data. In addition, these data will be used to illustrate some of the fundamental differences between subsonic, supersonic and hypersonic combustion. The focus of the remainder of the discussion will be on the reacting shear layer data and the detonation data. The challenges and opportunities which data from this type of test facility provide for mathematical and/or computational modeling of hypersonic combustion phenomena will hopefully be self-evident. However, we will also indicate a need for more detailed modeling of the facility itself to fully describe the test conditions under which the data is obtained.

J Buckmaster et al (eds), Combustion in High-Speed Flows, 53–91

LIST OF SYMBOLS

a	speed of sound
b	width of free shear layer
h	enthalpy
M	Mach number
M_c	convective Mach number
u	velocity
u_c	convective velocity
p	pressure
T	temperature
x	axial distance from model inlet

1. Description of the Test Facility

The experiments to be described in this presentation were performed over the past two years in the NASA Expansion Tube at GASL. This device is a large expansion tube (6-inch tube I.D., 120-ft overall length) which was originally constructed at NASA LaRC in the 1960's for real-gas aerothermal heating studies (Miller and Jones, 1983 and Trimpi, 1962). It was decommissioned in 1983, placed in storage, transferred to GASL (as part of our scramjet test complex), and recommissioned in 1989 as the HYPULSE facility (Tamagno et al., 1990). In essence, it consists of two shock tubes in tandem, followed by a free-jet test section/dump tank at the end of the second tube. The facility can be converted to an expansion tunnel by addition of a nozzle, but it is not currently configured in the tunnel mode. Figure 1 is a drawing of the facility as it existed at NASA, including a nozzle for tunnel operation. Figure 2 is a recent photo of the installation at GASL.

The operating principle of an expansion tube is, in simplest terms, use of an unsteady expansion process to perform additional work on a fluid which has been set into motion by passage of a shock wave. As in a steady expansion through a nozzle, the flow is accelerated, but unlike a steady expansion, its total enthalpy and total pressure are amplified by the unsteady wave process. In an expansion tube, the unsteady wave process is performed in a constant area duct. An expansion tunnel is created by addition of a diverging area nozzle in which a steady expansion occurs (without a converging area or plenum section typical of a wind tunnel or shock tunnel) (Trimpi and Callis, 1965). Although combination of a long, constant area,

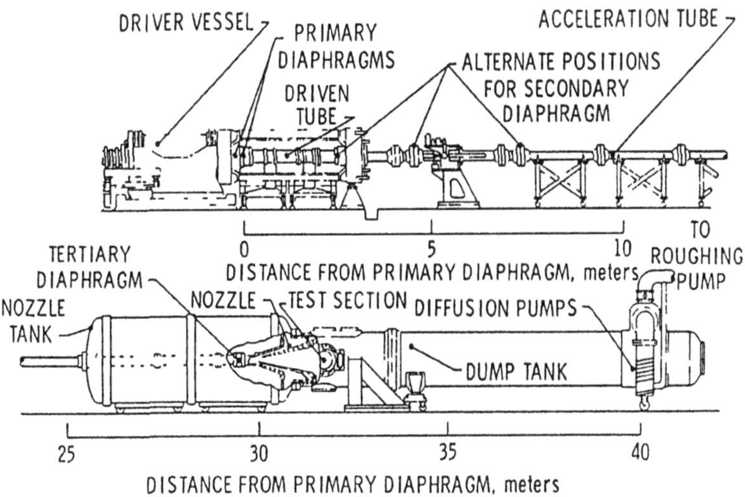

Figure 1. Drawing of Expansion Tunnel at NASA LaRC

Figure 2. HYPULSE Installation Photo at GASL (1990)

unsteady expansion process with a short, variable area, steady expansion makes analysis of an expansion tunnel tractable, an expansion tunnel nozzle is not necessarily short (due to the high initial Mach number,) and a practical design may have to combine both processes.

It is not the intent of the present paper to review the design or operating cycle of an expansion tube facility in detail. However, its unique characteristics bear so directly on the nature of the hypersonic combustion experiments that can be performed that a brief overview is considered imperative.

As stated, the facility is basically two shock tubes in tandem. The first shock tube is filled with the test gas (typically air) at relatively low (subatmospheric) pressure. A shock wave is driven into the test gas upon release of a light gas (typically helium) contained in the driver section at high pressure by rupturing a pair of metal diaphragms. The first shock tube is separated from the second shock tube (called the acceleration tube) by a thin secondary diaphragm (typically Mylar). The acceleration tube is filled with a fairly arbitrary choice of gas (typically nitrogen) at very low pressure (typically measured in microns of mercury). Ideally, the secondary diaphragm ruptures instantly upon impingement by the incident shock wave which has passed through the test gas. The driven section of the first shock tube effectively acts as the driver for the second (acceleration) tube. Upon encountering the very low pressure in the acceleration tube, the incident shock accelerates to higher speed, and a system of unsteady expansion waves is generated to balance the pressure between the shock and the test gas (air). The expansion waves propagate at velocity $(u - a)$ and therefore travel downstream, since the local Mach number in the test gas produced by the incident shock is supersonic. The test gas, travelling at velocity u, passes through the expansion waves and gains total enthalpy and total pressure, mainly in the form of kinetic energy (velocity), as a result of work performed on it by the unsteady expansion process. A region of constant velocity, static pressure and static temperature exists between the last expansion wave and the leading front between the air and the nitrogen. This region constitutes the test domain.

Note that the test gas is processed first by a downstream-travelling shock and then by a series of downstream-travelling expansion waves. At no point is the flow stagnated (with the possible exception of a momentary shock reflection off the secondary diaphragm, which is inconsistent with the calibration data obtained to date,) nor does the total pressure that is generated exist as a static pressure. The principal advantages of the expansion tube as compared to other hypersonic test facilities, especially shock tunnels, are three-fold:

- higher performance (total pressure and total temperature) for a given driver gas pressure and sound speed,
- absence of a sonic throat, with its high heat transfer rates, and
- the total temperature does not occur as a static thermodynamic state in the test gas, with the attendant dissociation of the gas.

The last two aspects are particularly important at hypersonic-hypervelocity test conditions. The problems of throat survival and/or contamination of the test gas by out-gassing, ablation or combustion of the throat material are avoided. The problem of recombination of dissociated species, produced in a plenum at the facility total temperature and total pressure, through the nozzle expansion is also avoided. Therefore the test gas reaches the test section in a more chemically pristine state than in a conventional wind tunnel or shock tunnel of comparable size and at comparable total pressure and total temperature. Furthermore, it is possible to conduct hypersonic flow experiments using pure oxygen (Spurk, 1970) and detonable mixtures (Srulijes et al., 1992) as test media. Thus, an expansion tube is an attractive device for hypersonic combustion research.

The trade-off to achieve these advantages is test time. Steady test time in the HYPULSE facility is typically in the 300 to 500 microsecond range. The ideal test time is diminished by boundary layer growth on the tube walls, which attenuates the shock speed and reduces the distance between the incident shock and the driver gas interface. However, the test period is always limited by the arrival of waves, either the unsteady expansion waves or the reflected waves from the intersection of the driver gas interface with the unsteady expansion waves. In any case, the arrival of waves is clearly identifiable, and always precedes the arrival of driver gas. Driver gas contamination of the test gas is sometimes difficult to identify in reflected shock tunnels.

The length of the test period determines the maximum model length in which steady flow can be fully established. Combustion models up to 1 meter in length, as depicted in Figure 3, have been successfully tested in HYPULSE.

2. Installation of Combustion Models in the Facility

In the current mode of operation, the test article (model) is placed in the free-jet at the exit of the acceleration tube. Since the local Mach number is about 8, the Mach wave angle is quite shallow and the test rhombus is relatively long. Aerodynamic models whose length is of the same order as the tube diameter can be tested to fairly large angles of attack (Miller et al., 1985).

The scramjet experiments performed in the facility to date are commonly referred to as "semi-direct connect" combustor tests. This means that the flow conditions produced by the facility are intended to represent the flow conditions which would exist at the entrance to the combustor section of a scramjet engine at a particular flight Mach number. However, the "semi" indicates that the combustor model is **not** directly connected to the facility itself. Rather, as depicted in Figure 3, it is mounted just downstream of the facility acceleration tube exit. In the experiments to date the model entrance area was sized to capture the inviscid "core flow," and allow the tube boundary layer flow to pass outside the model.

Accordingly, these experiments neglect the effects of the inlet boundary layer swallowed by a real combustor as well as the uncancelled inlet shock waves which may enter a real combustor. This was not intended to suggest that these effects are unimportant; indeed they may dominate the performance of real combustors at hypervelocity flight conditions. However, we have attempted to simplify and/or idealize the experiments to date in order to gain confidence in the test techniques, instrumentation, fidelity and repeatability of the data, and to focus initially on the fuel injector performance, per se.

Conduct of combustion experiments within the 300 to 500 microsecond test period requires a fuel (i.e. hydrogen) injection system that can establish steady fuel flow either just prior to arrival of the air or during establishment of steady air flow, and that can terminate fuel flow promptly following the test period. If the fuel flow begins too early, it can raise the pressure in the test section and the hydrogen can travel upstream into the acceleration tube, thereby altering the performance of the facility. A pair of Ludweig tubes operated by fast-acting valves is currently used as the fuel supply system. Fuel flow can be initiated about 2 to 3 milliseconds prior to shock arrival, timing the valve operation off of the incident shock, and fully

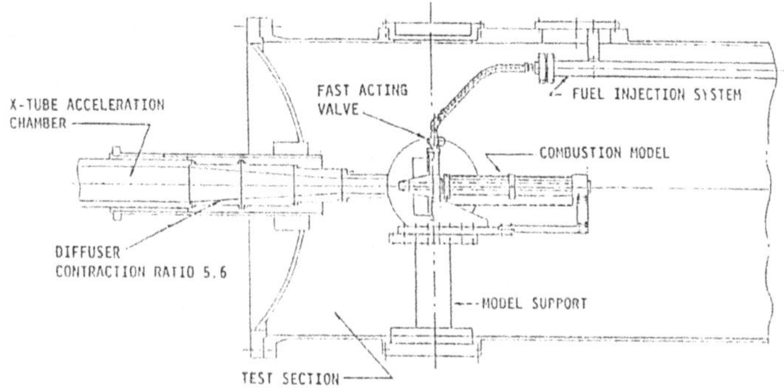

Figure 3. Typical Installation of a Combustion Model in the Test Section, also Showing Installation of Diffuser Section

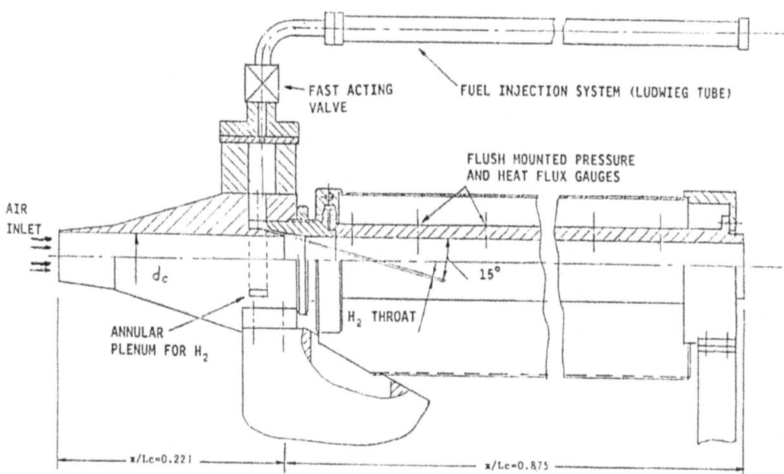

Figure 4. Schematic of the Axisymmetric Combustor Model

established fuel flow achieved virtually simultaneously with arrival of the test gas. Passage of the acceleration gas (nitrogen) through the model prior to arrival of the test gas assists in purging the excess hydrogen from the model and establishing a steady interaction of the fuel and test gas (e.g. air).

The pair of Ludweig tubes are sometimes used to independently supply injectors on two walls of the combustion channel, and sometimes used together to provide a higher flow rate (as in the shear layer experiments to be described.)

3. Facility Calibration

Following its recommissioning, the facility was recalibrated at GASL at the nominal 17,000 ft/s (5.2 km/s) test point in air at which it had been previously operated at NASA. It was subsequently recalibrated at 22,600 ft/s (6.9 km/s) in helium, 14,800 ft/s (4.5 km/s) in carbon dioxide, and 17,000 ft/s (5.2 km/s) in nitrogen, at which NASA had also operated the facility. In addition, GASL found two new operating points in air and in nitrogen at which calibration data was also taken, namely: 15,000 ft/s and 12,600 ft/s. Furthermore, we calibrated the facility at the 17,000 ft/s and 12,600 ft/s (nominal) conditions in pure oxygen.

At the nominal 17,000 ft/s operating point in air the static pressure is too low to produce autoignition. Therefore, a "diffuser" section (i.e., a converging duct) was added to the end of the facility for combustion experiments at this condition. The diffuser has an area ratio of 5.6 to 1, producing a 10 to 1 increase in pressure, albeit with an exit diameter of 2.5 inches. The diffuser was designed by the method-of-characteristics to produce an isentropic compression of the incoming flow to a uniform flow at the exit. The design and calibration of the diffuser is a separate topic (Bakos et al., 1992) into which we will not delve at this time. For the present purposes, it is sufficient to note that it functions adequately if not perfectly. The exiting flow apparently contains uncancelled waves which are too weak to be detected optically or by instream probes, but manifest themselves in a slightly higher pressure in the entrance section of a model than the wall static pressure measured near the diffuser exit station. Although we have frequently used the diffuser exit pressure as a reference value for nondimensionalizing the pressures within the combustor model, it should not be strictly interpreted as the freestream static pressure in the test section, esp. for the purpose of detailed comparisons with CFD solutions. Additional calibration measurements have recently been carried out to more accurately define the pressure field downstream of the diffuser.

Determination of the test section flow conditions (without the diffuser) is based on surveys of pitot pressure in the test section, measurements of wall static pressure along the tube up to the test section, and measurement of the shock speed along the tube up to the test section. Instream surveys of static pressure indicate that no measurable gradients exist across the tube exit. The test gas velocity is calculated from the shock speed, and the pitot-static pressure ratio is used to deduce all other flow properties assuming equilibrium chemical composition of the test gas. The procedure makes no a'priori assumption about the wave processes occurring in the facility. However, the resulting entropy of the test gas is generally much closer to that calculated behind the incident shock ahead of the secondary diaphragm than to that behind a reflection of the incident shock. In other words, the calibration data supports the notion of an instantaneous rupture of the secondary diaphragm without significant shock reflection. However, the occurrence of a complex, three-dimensional rupture mechanism resulting in some entropy generation cannot be ruled out.

The ratio of pitot pressure to driver gas initial pressure has been found to be sufficiently repeatable (less than a 4% standard deviation) in the calibration tests that pitot pressure has not been measured when a model is installed in the test section. The nominal test conditions at the 17,000 ft/s condition (simulated flight Mach number of 17) and the 12,600 ft/s condition (simulated flight Mach number of 13) are summarized in Table I.

With this background information in hand, we can proceed to a discussion of some typical test results.

4. Fuel Injector Performance at Flight Mach 13 and 17 Conditions

The first set of experiments on hypersonic combustion in the expansion tube were conducted at the Mach 17 flight condition using a constant area circular duct model with an annular slot injector (Anderson et al., 1990). The model details in the vicinity of the fuel injection station are shown in Figure 4. (The model installation

Table 1. Summary of HYPULSE Calibration Data[1]

Test Condition	Static Pressure[2] (psia)	Core Pitot Pressure[2] (psia)	Shock Velocity[2] (ft/s)	Mach Number	Velocity (ft/s)	Static Temperature (°R)
M17 (Air)	.26	20.3	17300	7.8	17300	2160
M17 (N_2)	.27	19.6	17200	7.5	17200	2230
M17 (O_2)	.27	22.1	17000	8.0	17000	2230
M15 (Air)[4]	.22	15.3	15440	7.3	15190	1890
M13.5 (Air-HP)[5]	3.40	74.0	13700	4.2	12660	4100
M13.5 (Air)	2.60	56.0	13680	4.1	12620	4230
M13.5 (O_2)	2.67	63.8	13490	4.3	12570	4220
Test Condition	**Total Enthalpy[3] (BTU/lbm)**	**Total Temperature (°R)**	**Total Pressure (psia)**	**Unit Reynolds x 10^{-5} $(ft)^{-1}$**	**c_p/c_v**	
M17 (Air)	6540	15100	23630	0.20	1.32	
M17 (N_2)	6500	15500	8340	0.20	1.33	
M17 (O_2)	6320	13000	57560	0.17	1.30	
M15 (Air)[4]	5090	12360	8660	0.16	1.33	
M13.5 (Air-HP)[5]	4360	10510	2130	0.60	1.29	
M13.5 (Air)	4380	10400	1490	0.44	1.29	
M13.5 (O_2)	4250	9130	3520	0.41	1.27	

(1) For nominal primary-diaphragm burst pressure of 5500 psia. M17 conditions uses room temperature helium driver gas. M13 uses a blend of helium and nitrogen.
(2) Measured quantities. All other parameters are calculated assuming local thermo-chemical equilibrium
(3) Referenced to 0°K
(4) M15 uses a mixture of 95% He and 5% N_2 by volume
(5) Nominal primary-diaphragm burst pressure of 7500 psia

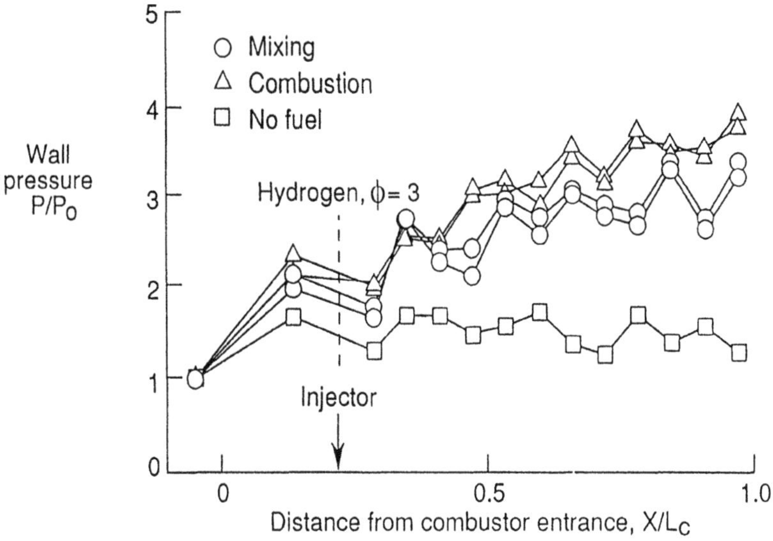

Figure 5. Wall Pressure Distributions in Axisymmetric Model at Mach 17 Test Condition

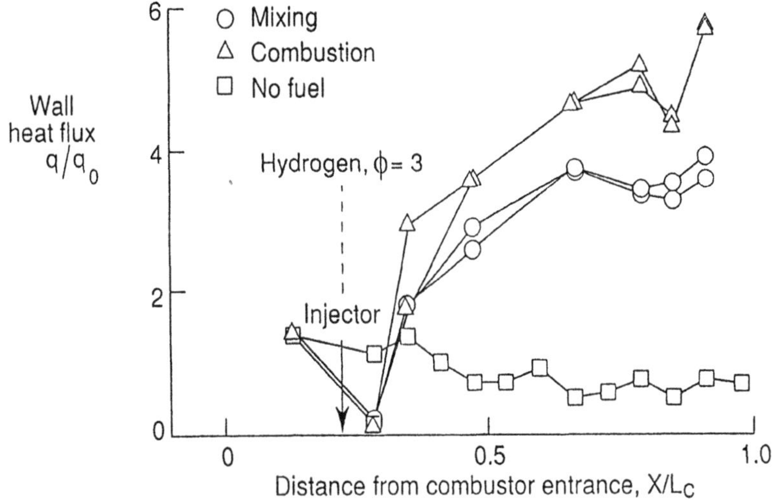

Figure 6. Wall Heat Transfer Rate Distributions in Axisymmetric Model at Mach 17 Condition

was previously depicted in Figure 3.) This configuration provided a truly two-dimensional internal flowfield which both minimized the required distribution of wall-mounted sensors and simplified analysis and interpretation of the data. However, it also presented a fuel injection mechanism that could not be expected to provide as rapid mixing as an array of discrete orifices. The configuration also precluded visualization of the flow in the model.

The test protocol followed a pattern set in earlier combustor tests in pulse facilities, namely: first determine the fuel-off characteristics (pressure rise due to viscous effects and shock waves, wall heat transfer rate, etc.) of the model ("tare" data), then determine the effects of inert fuel addition ("mixing" data), and finally determine the effects of heat addition ("combustion" data). Ideally, the "tare" data benchmarks the analysis method with respect to viscous effects not directly measured (e.g., skin friction drag). The "mixing" data then adds the effects of mass addition and thermal protection of the wall wetted by the layer of fuel. The difference between the "combustion" and "mixing" data then identifies the level of heat release, inferring at least the amount of fuel that has mixed with the air if not the combustion efficiency. The resulting data set is typified by Figures 5 and 6, which show measured axial distributions of wall pressure and heat transfer rate for the cylindrical model at Mach 17 flight conditions. Two runs are shown for each data set to indicate the repeatability of individual runs. In the case of the pressure data, the values shown at $x < 0$ are measured at the exit station of the diffuser, not in the model. The lines connecting the data points are to aid the reader in identifying the data type, and, as subsequent examples will indicate, should not be used to interpolate between data points.

Two observations can be made from Figures 5 and 6 that will carry over to most subsequent hypersonic combustion experiments: first, the pressure rise due to heat release is not very dramatic, and secondly, the augmentation of wall heat transfer rate is often a better indication of the occurrence of combustion than the pressure rise. There are several contributing factors to the first observation:

- The ratio of the heat of reaction of the hydrogen fuel to the sensible total enthalpy of the air (a Damkohler number) decreases rapidly with increasing flight speed, and the maximum attainable pressure rise decreases proportionately.

- The relative inefficiency of the mixing process for a slot injector restricts the degree to which reaction can proceed.
- The relatively low initial static pressure of these experiments can inhibit attainment of chemical equilibrium.
- The relatively high initial static temperature of these experiments can result in substantial dissociation of the expected combustion product, water vapor, even if chemical equilibrium is achieved, thereby reducing the exothermicity of the reaction process.

The first three factors are well known, and apply to the present experiments. The fourth is somewhat less obvious, but extremely important to both interpretation of these experimental observations and to the performance potential of hypersonic scramjets. The exothermic part of the combustion process is the final stage in which the hydroxyl radicals and atomic hydrogen combine to form water. Prior to reaching this stage the reactions are neutral or endothermic. Hence, no pressure rise due to combustion occurs until the last stage of reaction is at least entered, if not completed.

On the other hand, the wall heat transfer rate is less sensitive to the dissociation problem. This because the wall is cold (room temperature in these experiments) causing the dissociation products to thermally recombine and release their heat of reaction as they approach the wall. Furthermore, the rate of recombination is greatly accelerated at the wall by the catalytic efficiency of the metallic surface.

Therefore, the wall heat transfer rate data may be an effective discriminant between combustion "inefficiency" due to mixing and/or kinetic rate limitations and that due to dissociation of the combustion products. It is one of the most accurately measurable quantities in pulse facility experiments. Unfortunately, wall heat transfer rate is also one of the most challenging quantities to accurately model by CFD techniques, and therefore this data has not been extensively employed to date in analyses of hypersonic combustion phenomena.

Clearly, it would be desirable in these experiments, as well as in flight, to optimize the initial temperature and pressure in the combustor at a level where autoignition is rapid and reliable but dissociation of the combustion products is minimal. That is a design challenge for a flight vehicle. On the other hand, in ground test facilities excessive static temperature is invariably a consequence of attempting a flight simulation at the flight total enthalpy but with inadequate total pressure. In order to achieve adequate static pressure to sustain combustion, the facility must be run at a lower combustor Mach number than in flight and therefore at an excessively high static temperature. In the present experiments, this occurs through the isentropic compression process in the diffuser.

A further comparison of the heat transfer data in these experiments, Figure 7, for the fuel-off case with a simple flat-plate formula for laminar flow supports the contention that the flow is fully established throughout the model. Examination of individual data traces further corroborates this view. Fuel injection causes an immediate transition to turbulent flow downstream of the injection station, as can be seen in Figure 6, with a correspondingly more rapid establishment of the flow. Also evident in this figure is the "film cooling" effect of the fuel layer just downstream of the injection station.

The next set of experiments moved on to somewhat more realistic fuel injector configurations (Bakos et al., 1992). An ad-hoc injector was fabricated for the cylindrical model to improve mixing. It consisted of four diametrically opposed sonic orifices oriented at 30 degrees from the axis. This injector is depicted in Figure 8. As can be inferred from Figures 9 and 10, the mixing of fuel and air was substantially improved, resulting in a larger pressure rise due to combustion and a more rapid rise in wall heat transfer rate. Bear in mind, however, that these measurements were still made with the original wall sensor layout, which now produced values along a meridional line midway between a pair of fuel orifices.

The cylindrical model was also tested in the T4 tunnel at the University of Queensland, Australia (Morgan et al., 1991 and Bakos et al., 1992). The comparisons between data from a reflected shock tunnel and an expansion tube at the same nominal conditions are intended to elucidate facility-induced effects associated with the different chemical states of the air in the test section. We will not focus on this issue in the current discussion, other than to note that the second study devoted considerable attention to the most appropriate definition of comparable test conditions, which reduces the differences observed in the initial comparisons. It was also concluded that the high initial temperatures in these tests greatly mitigated the effects of facility-induced dissociation.

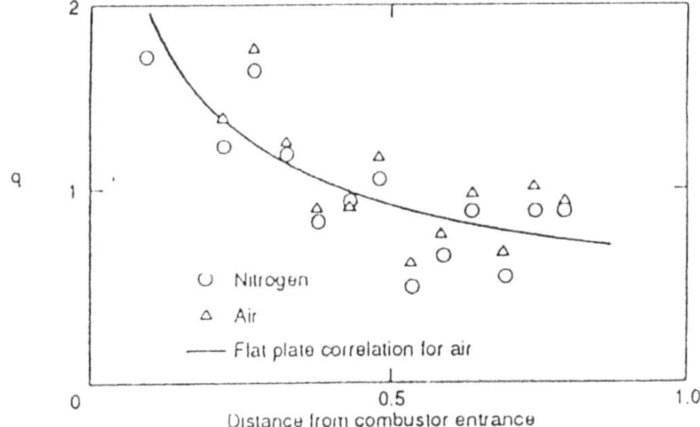

Figure 7. Comparison of Model Heat Transfer Rates Without Fuel to Compressible Flat Plate Theory

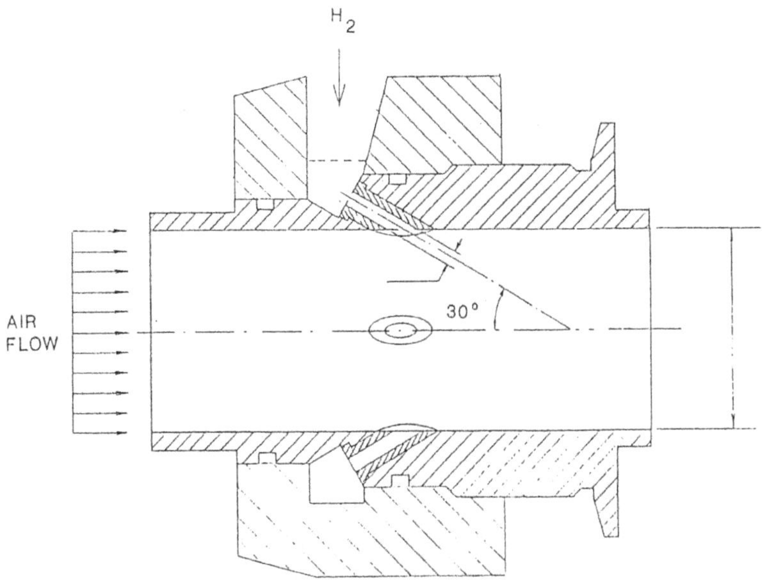

Figure 8. Schematic of the Flush-Wall Four Orifice Injector Section for the Axisymmetric Model

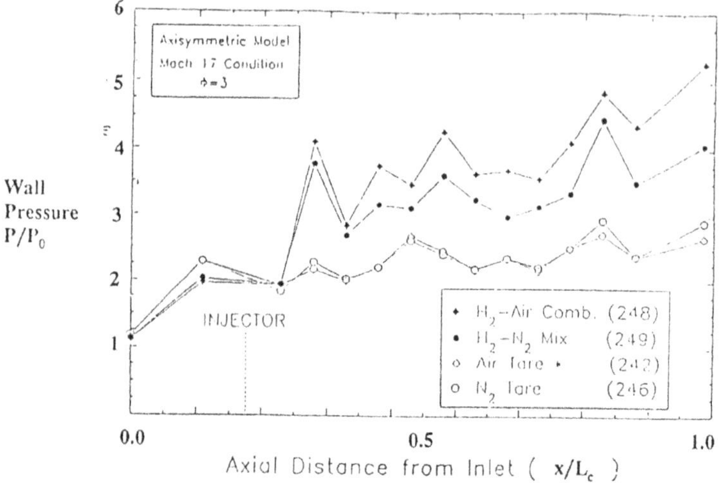

Figure 9. Wall Pressure Distributions in Circular Model With 4 Orifice Injector at Mach 17 Test Condition

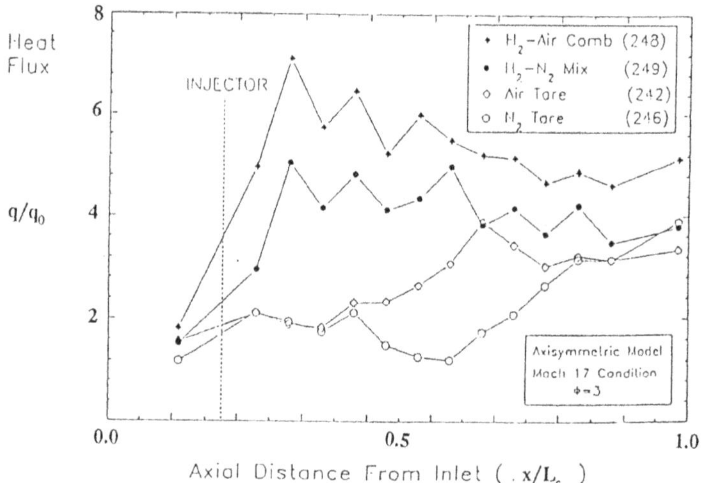

Figure 10. Wall Heat Transfer Rate Distribution in Circular Model With 4 Orifice Injector at Mach 17 Test Condition

All subsequent experiments have been conducted in a new combustor model having a rectangular cross section, 1 inch by 2 inches, as shown in Figure 11. One prominent feature of this model is the inclusion of sidewall windows permitting visualization of up to 12 inches of combustor length, beginning just upstream of the fuel injectors. On the other hand, the interior flowfield in this model is three-dimensional and the density of surface instrumentation needed to fully characterize it has increased correspondingly. Another feature of the new model is the ability to fuel the model from two independent supply systems for the upper and lower walls.

The initial experiments with the new model were performed at the nominal flight Mach 13 condition (12,600 ft/s). At this condition sufficient static pressure can be generated without use of the diffuser section, eliminating the performance of this device from interpretation of the combustor data. However, pitot surveys revealed a thicker boundary layer on the tube wall resulting in an inviscid core diameter no larger than that produced at the exit of the diffuser. These experiments also used an injector configuration similar to that in the previous tests, namely: sonic orifices aligned 30° from the wall. Three injector arrangements were examined: a single injector on one wall, two opposed injectors and four opposed injectors, all having the same total orifice area. The injectors, referred to as the single, dual and quad arrangements, are depicted in Figure 12. Each injector of the quad arrangement is a quarter-scale version of the single injector. A set of regression curves through the data sets, shown in Figure 13, (see Bakos et al. (1992) for the data) confirm that indeed the single and quad injector pressure data is correlated by the height of the injection cell for geometrically similar configurations.

In this same set of experiments, the use of pure oxygen, as opposed to air, as the oxidizer stream was tried for the first time. The conjecture was that if the combustion process was indeed rate limited by the low static pressure, then the collision frequency with oxygen (and hence the corresponding reaction rate constants) could be increased by about a factor of five by using pure oxygen in lieu of air. These tests were run at the highest hydrogen flow rate that had been previously run in air (an equivalence ratio of 2.7) to maximize the rate of mixing. The data and one-dimensional chemical kinetics calculations (which assume fully mixed flow), shown in Figure 14, indicate that the quad injector produces fully mixed flow within about 10 inches (20 gap heights) from the injector and that chemical

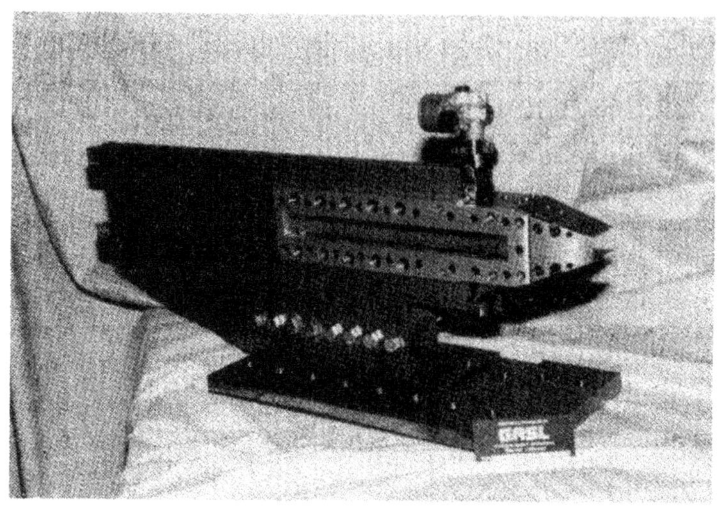

Figure 11. Photograph of the Rectangular Model

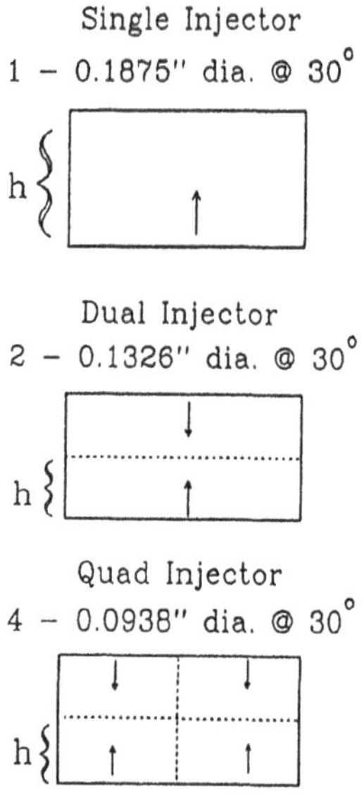

Figure 12. Schematic of the Single, Dual and Quad Injector Configurations

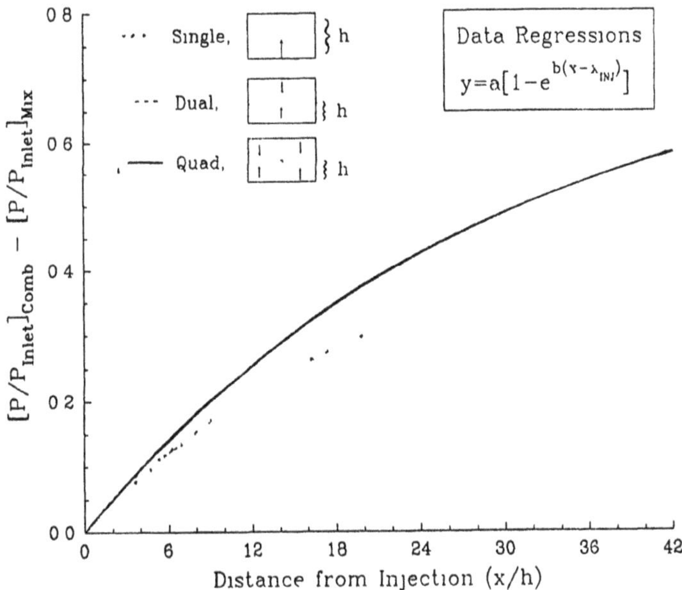

Figure 13. The Effect of Length-to-Gap Height Ratio on Combustion Pressure Rise

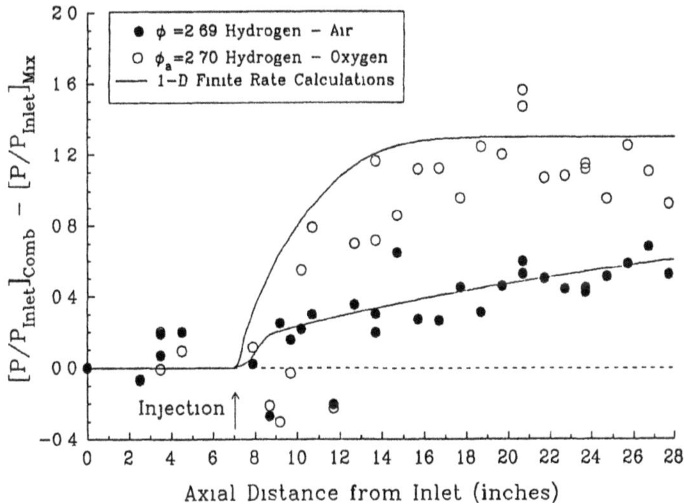

Figure 14. Comparison Between Combustion of Hydrogen in Air and in Oxygen

equilibrium is indeed achieved in pure oxygen, but the air case is rate-limited.

These injectors have also been tested at the flight Mach 17 condition (with the diffuser). An evaluation of the single injector data using a combination of a PNS code for the flow upstream of the injector, a FNS code in the region around the injector and the PNS code again further downstream of the injector, shows quite good agreement with the wall pressure data and moderately good agreement with the heat transfer data for a mixing (hydrogen injection into nitrogen) run (Bobskill et al., 1991). The solution was found to be grid-dependent, suggesting that further improvement in the heat transfer comparison could be achieved with further grid refinement. The FNS solution shows remarkably good agreement with a shadowgraph visualization of the injector region, including a small separated flow zone upstream of the injector. A further comparison with reacting flow data also shows good agreement with the pressure data but significant under-prediction of the heat transfer data. This may be due to a combination of grid-dependence and catalytic recombination effects at the wall. Overall, the evaluation further confirms the validity of the test technique and the data.

Interestingly, the CFD solution for this case produces a "mixing efficiency" which is about 90% at the combustor exit for both the combustion and mixing cases, and a combustion "efficiency" of about 75%. Nevertheless, a pressure rise due to combustion is not obvious in either the CFD solutions or the data. Only after one-dimensionalizing the CFD solution does a systematic difference in pressures between the combustion and mixing cases, consistent with a one-dimensional chemical kinetics calculation, become evident. This is because the overall pressure rise due to combustion, about 10%, is small compared to the local pressure variations due to the induced shock waves in the flow. The authors attribute the low pressure rise due to combustion to the large concentrations of atomic species and hydroxyl radical produced at the low static pressure and high static temperature prevailing at these test conditions, even at 75% combustion "efficiency." It should also be pointed out that the chemical kinetics model used in these calculations did not include nitrogen reactions. At these temperatures, the formation of nitric oxide is appreciable and may also contribute to low pressure rise.

The most recent work with this model has concentrated on development of optical diagnostics methods to quantitatively determine the extent of fuel-air mixing at hypervelocity flow conditions. Planar laser-induced fluorescence (PLIF) was used to visualize the hydroxyl radicals produced in the flame zone (Anderson et al., 1992). An example of a PLIF image and a corresponding laser schlieren image taken downstream of a single swept ramp fuel injector is shown in Figure 15. Unfortunately, the PLIF images could not, in this case, be made quantitative. Furthermore, even if they could, they would represent the instantaneous concentration of the hydroxyl radical in a turbulent field. While this is very useful in deducing the physics of the combustion processes, a large number of images must be acquired to obtain the mean values needed for comparison with the time-averaged values which most CFD solutions provide. The repetition rate of the high energy lasers required for PLIF imaging is too slow to permit more than one image to be obtained per run. Furthermore, the hydroxyl radical is not the most sensitive indicator of accomplished mixing. Therefore, PLIF imaging has not been pursued further.

A new technique, as yet unreported, has recently been developed to permit quantitative, cross-plane imaging of hydrogen concentration. The technique uses Mie scattering off of fine, submicron-sized silicon dioxide particles seeded into the fuel just prior to injection. The seed is created by burning a small amount of silane premixed with the hydrogen fuel (typically 3-5% silane in 97-95% hydrogen) with a stoichiometric proportion of oxygen in the plenum section of the fuel injectors. The combustion products of this process are solid silicon dioxide and water vapor. The hydrogen remains essentially inert, but is heated in the process to about 2700 R. The seeded hydrogen is then visualized by illuminating one (or more) cross-plane(s) downstream of the injector with a sheet (or sheets) of light from a laser. The resulting Mie scattering off the particles is recorded by a CCD camera. Since the intensity of scattered signal is directly proportional to the number density of particles, which in turn is directly proportional to the hydrogen concentration (in a pure mixing environment), the technique is capable of yielding quantitative data. In a combusting flow environment, quantitative interpretation is complicated by vaporization of the silicon dioxide within the flame zones, resulting in "holes" in the scattering image. Nevertheless, qualitative information on fuel dispersion is still provided.

An example of a Mie-scattering image obtained downstream of a pair of choked orifice fuel injectors at 90° to the surface is shown in

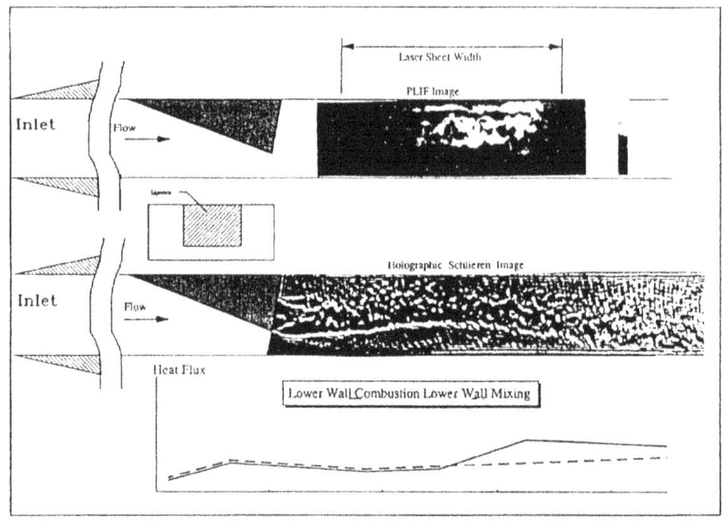

Figure 15. PLIF and Schlieren Images for Unswept Ramp Injector Model

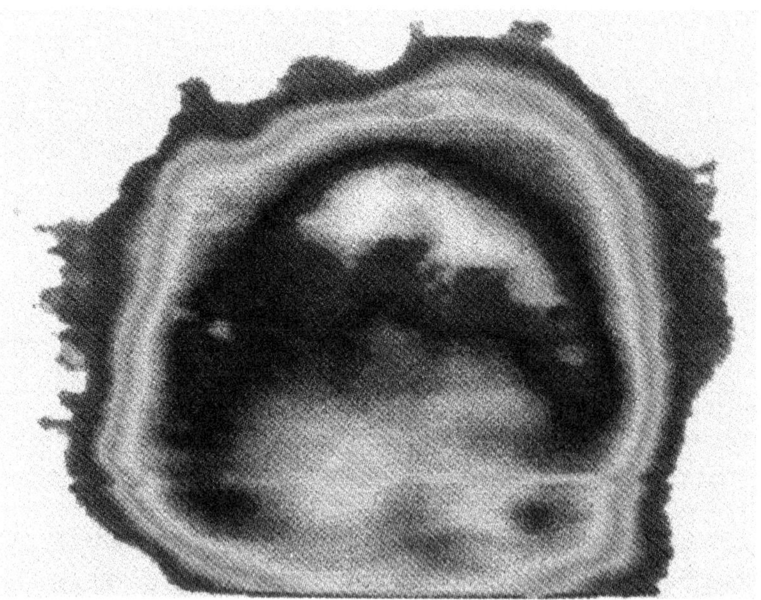

Figure 16. Example of Cross-Plane Imaging of a Fuel Plume Using Mie Scattering

Figure 16. This image was obtained in a small shock tunnel at GASL used for instrumentation development. Subsequent applications of the imaging technique in tests conducted in the HYPULSE facility have been quite successful.

5. Inert and Reacting Shear Layers at High Convective Velocity

The mixing of two parallel streams is a classical benchmark problem for verification of analytical and computational models. It is also a tractable starting point for comprehension of the physical processes by which mass, momentum, energy and chemical species are transported laterally by molecular diffusion. Consequently, it is the subject of considerable research, even though practical combustors seldom employ parallel stream fuel-oxidizer mixing due to its relative inefficiency as compared to three dimensional mixing.

A number of experimental studies have indicated a pronounced reduction in the rate of spreading of the mixing layer at the stream interface as compressibility of the streams increases. The implications of such a reduction in mixing could be quite adverse for diffusive-burning supersonic and hypersonic combustors. The available data tends to be correlated by a single-valued convective Mach number derived from the concept of a single relative speed with which organized structures should convect in the mixing layer (Bogdanoff, 1983 and Papamoschou and Roshko, 1991). However, recent observations of the speed with which organized structures actually moved revealed the existence of distinctly different convective speeds in each stream (Papamoschou, 1991). This has been attributed to the generation of shock waves within the structures as they move at locally supersonic speed. The effect of compressibility and, specifically of high convective velocity, therefore merits further evaluation, and the expansion tube offers a unique capability to achieve extreme conditions in terms of convective velocity.

Prior studies achieved large differences in convective Mach number mainly by varying the sound speed (molecular weight and isentropic exponent) rather than the velocities of the two streams. The study we carried out in the expansion tube (Erdos et al., 1992) employed a primary stream at 12,600 ft/s and we planned to vary the secondary stream from 2,100 ft/s to 7,830 ft/s using mixtures of nitrogen and hydrogen. In fact, tests were only run at the two limiting secondary stream velocities, obtained with room-temperature nitrogen and hydrogen, respectively.

Inert shear layer mixing data was obtained using nitrogen as the primary gas stream and both nitrogen and hydrogen as the secondary gas stream. These combinations yielded convective velocities of 4,000 and 10,000 ft/s, and convective Mach numbers of 2.8 and 0.8, respectively.

Reacting shear layer data was obtained using air and pure oxygen as the primary gas stream and hydrogen as the secondary stream. Thus the nominal convective velocity and Mach number were 10,000 ft/s and 0.8, respectively. The rate of reaction was varied by roughly a factor of five, corresponding to the increased partial pressure of oxygen, between the air stream and oxygen stream cases.

The model used in the prior fuel injection studies was modified as indicated in Figure 17. The front section was modified to have a 1/2 inch by 2 inch channel and a two-dimensional Mach 3 nozzle, having a 1/2 inch by 2 inch exit, was fit into the upper wall. The nozzle was designed by the method-of-characteristics to produce a parallel Mach 3 flow. Thus at the splitter plate formed by the trailing edge of the nozzle we had a two-dimensional shear layer formed at the interface between the two parallel streams. As will be evident from photos to be shown, the wall boundary layers were very thin. Nevertheless, with a 4:1 aspect ratio of each stream, two-dimensionality could not be preserved throughout the model, but within the optical window length the shear layer is believed to be essentially two-dimensional.

The test conditions and test parameters are summarized in Tables 2 and 3.

The nitrogen-nitrogen case is particularly interesting since it has the highest density ratio (24.0), highest mass flux ratio (4.0), highest convective Mach number (2.8) and lowest velocity ratio (0.168). Although the wall pressure and heat transfer traces give no reason to believe that the flow is not fully established in the test period, it must be pointed out that due to the relatively low sound speed in cold nitrogen, the secondary flow had to be allowed additional start-up time in advance of the primary stream arrival, as compared to hydrogen. Accordingly, if flow establishment were an issue, it should be more so for this case than for the hydrogen cases to follow.

Figure 18 is an infinite fringe interferogram for this case. As observed in prior experiments, the growth rate into the higher speed stream is slower than the rate into the lower speed stream. In this

Stream/location	Gas	Velocity (ft/s)	Static Pressure (psia)	Static Temperature (°R)	Mach Number	Unit Reynolds Number (ft^{-1})	Total Pressure (psia)	Total Temperature (°R)	Total Enthalpy* (Btu/lbm)
Primary / test section	N_2	12,600.	2.57	4,070.	4.12	0.45×10^6	787.	11,932.	4,298.
	Air	12,600.	2.61	4,230.	4.10	0.44×10^6	1,490.	10,400.	4,379.
	O_2	12,570.	2.67	4,220.	4.35	0.41×10^6	3,522.	9,133.	4,247.
Primary / at splitter**	N_2	12,492.	3.10	4,400.	3.93	0.48×10^6			
	Air	12,500.	3.10	4,397.	3.99	0.49×10^6	-	-	-
	O_2	12,485.	3.10	4,318.	4.25	0.46×10^6			
Secondary / at splitter	N_2	2,101.	3.10	183.	3.12	14.5×10^6	136.	536.	133.
	H_2	7,833.	3.10	186.	3.09	6.7×10^6	130.	536.	1,832.

* Referenced to zero at 0°R

** Assumes matched static pressure with secondary stream

Table 2. Nominal Test Conditions for Shear Layer Mixing Tests

$( )_2$ = Secondary, $( )_1$ = Primary

Case	Secondary / Primary Gases	ρ_2/ρ_1*	u_2/u_1	$\frac{\rho_2 u_2}{\rho_1 u_1}$	$\frac{\rho_2 u_2^2}{\rho_1 u_1^2}$	$\frac{H_2}{H_1}$	$\frac{\rho_2 u_2 H_2}{\rho_1 u_1 H_1}$	u_c (ft/s)	M_{c_1}	M_{c_2}	M_R
1	N_2/N_2	24.04	0.168	4.043	0.679	0.031	0.125	4,033.	2.764	2.867	5.565
2	H_2/N_2	1.689	0.627	1.059	0.664	0.427	0.426	9,989.	0.818	0.848	1.663
3	H_2/Air	1.637	0.627	1.026	0.643	0.418	0.429	9,989.	0.817	0.848	1.663
4	H_2/O_2	1.451	0.627	0.910	0.570	0.431	0.393	10,042.	0.838	0.869	1.705

* At matched pressures

Table 3. Summary of Test Parameters for Shear Layer Mixing Tests

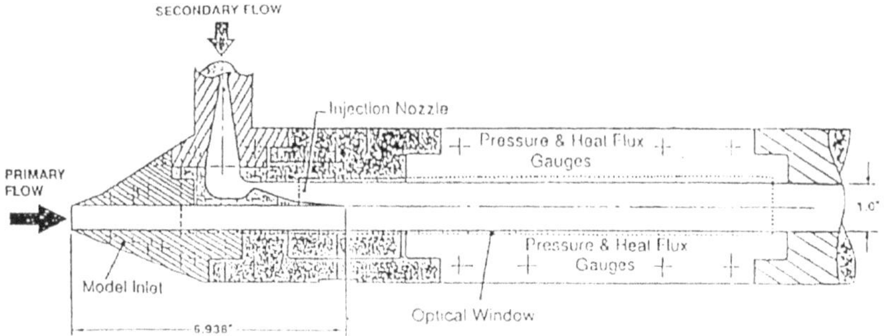

Figure 17. Side View of Model Details Near Window

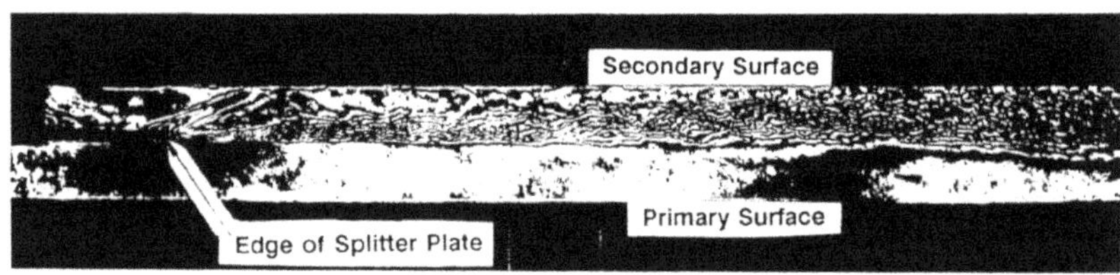

Figure 18. Infinite Fringe Interferogram for N_2/N_2 Case

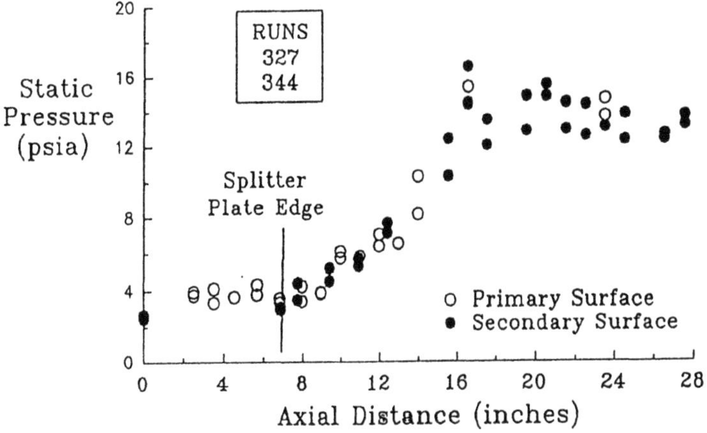

Figure 19. Surface Pressure Distribution for N_2/N_2 Case

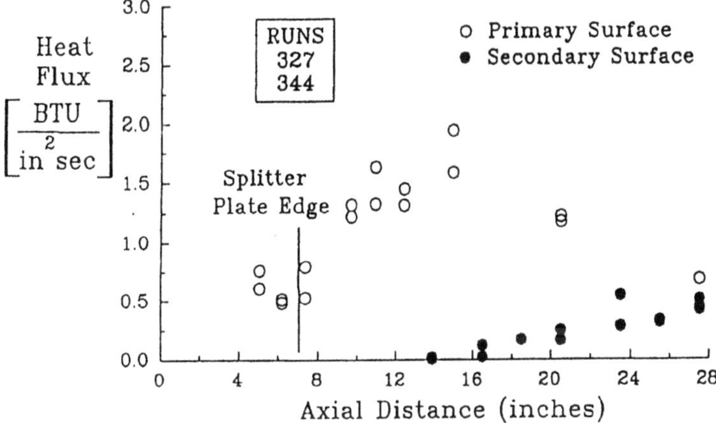

Figure 20. Surface Heat Flux Distribution for N_2/N_2 Case

case the spreading rate, db/dx, is virtually zero on the high speed side over the first 6 inches of length and about 0.07 on the low speed side. The equivalent incompressible growth rate from Papamouschou and Roshko's correlation is 0.438. Thus the normalized growth rate is about 0.16 at $M_c = 2.8$, which is very consistent with an extrapolation of the prior correlation by Papamouschou and Roshko to this convective Mach number and with a continued suppression of the growth rate. In other words, there are no surprises here, although the asymmetry of the shear layer is perhaps more extreme than previously observed.

The surface pressure and heat flux distributions for this case are shown in Figures 19 and 20. The gradually rising heat flux on the secondary surface downstream of $x = 16$ inches suggests that the high enthalpy of the primary stream first reaches the secondary wall at roughly $x = 16$ inches. However, this is considerably further downstream of where the visible mixing layer appears to reach the secondary wall in the interferogram, Figure 18, which represents the instantaneous refractive index (density) field. One is led to conclude that either there is a considerable difference between the instantaneous field captured by the interferogram and the mean field represented by the heat flux measurement, or mass and energy are being transferred laterally at very different rates.

Next consider the nitrogen-hydrogen case, which is also inert mixing. This case has a much lower density ratio (1.689), lower mass flux ratio (1.06), lower convective Mach number (0.84) and higher velocity ratio (0.627). In other words, compressibility effects should be smaller, although the convective velocity is higher (10,000 ft/s).

For this case, the shear layer was visualized by three different imaging modalities on three different runs. As evident from Figure 21, all three show the same general features of the shear layer, including a faster growth rate into the lower speed stream and spreading of the mixing layer to the secondary wall well within the field of view of the window. The visible spreading rate, normalized by the equivalent incompressible value, exceeds unity at $M_c = 0.84$ according to this data. This is in marked contrast to the prior data correlation which shows a very substantial suppression of the compressible growth rate at this convective Mach number. We can offer no explanation at this time.

The surface pressure and heat flux distributions for this case are shown in Figures 22 and 23. The heat flux data indicates that the

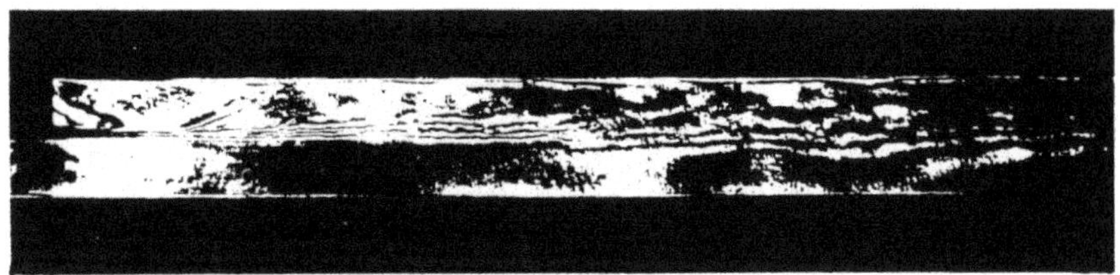

Infinite Fringe Interferogram for H_2/N_2 Case

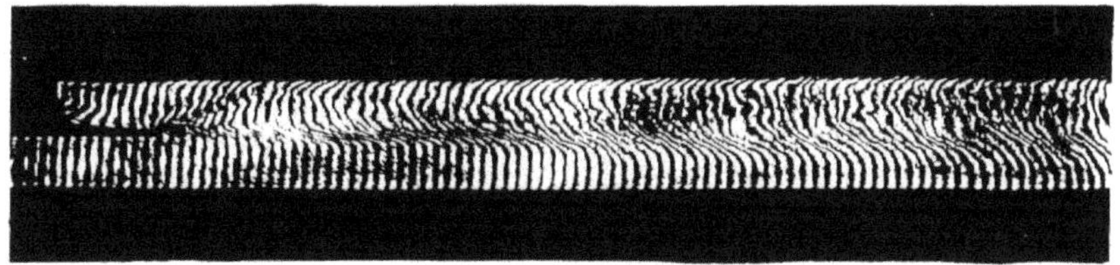

Finite Fringe Interferogram for H_2/N_2 Case

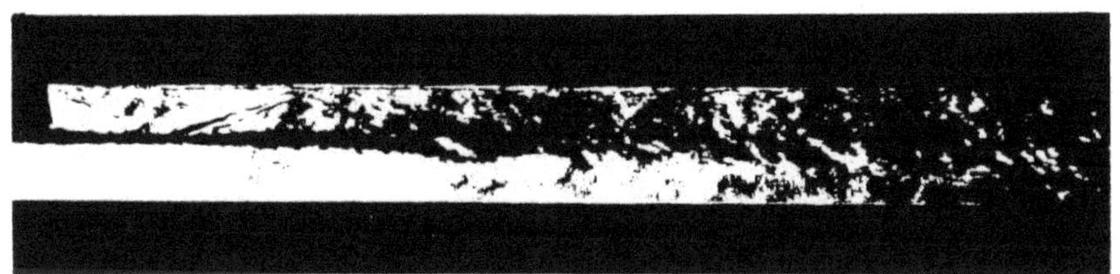

Laser Holographic Schlieren for H_2/N_2 Case

Figure 21. Visualization of H_2/N_2 Shear Layer by Three Imaging Modalities

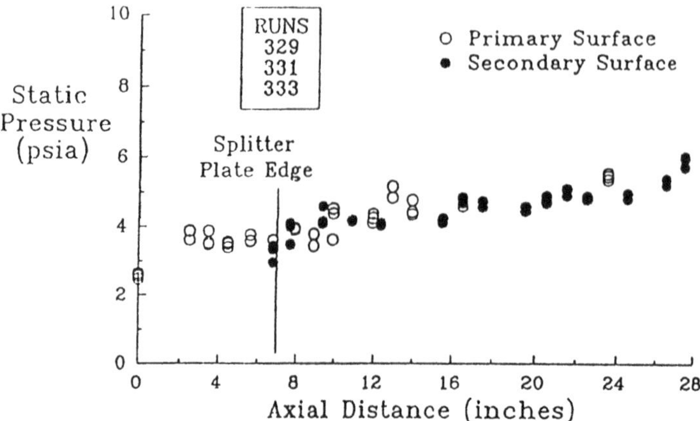

Figure 22. Surface Pressure Distribution for H_2/N_2 Case

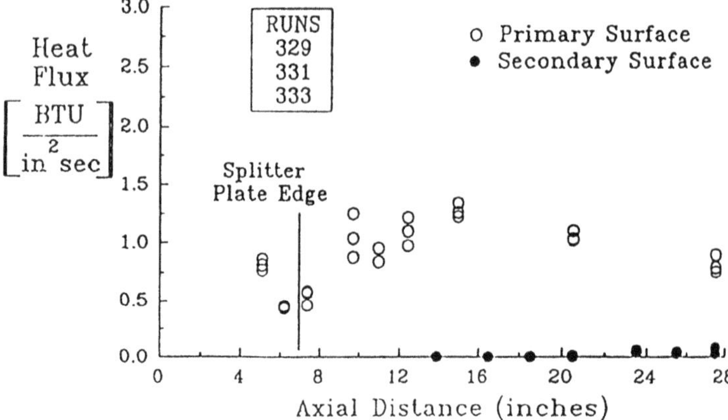

Figure 23. Surface Heat Flux Distribution for H_2/N_2 Case

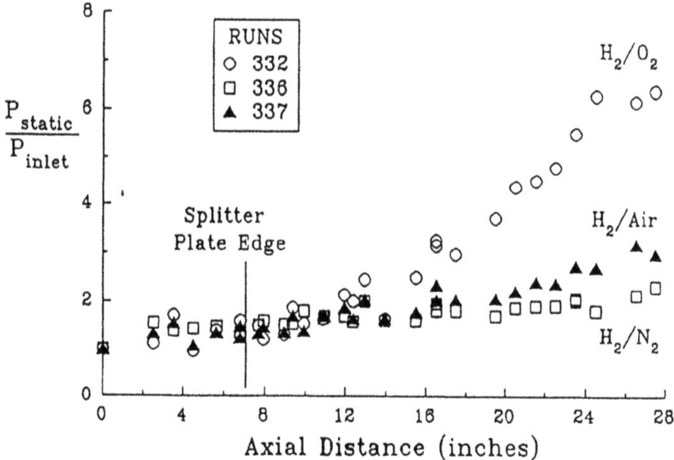

Figure 24. Summary of Wall Pressure Data Showing Effects of Heat Release

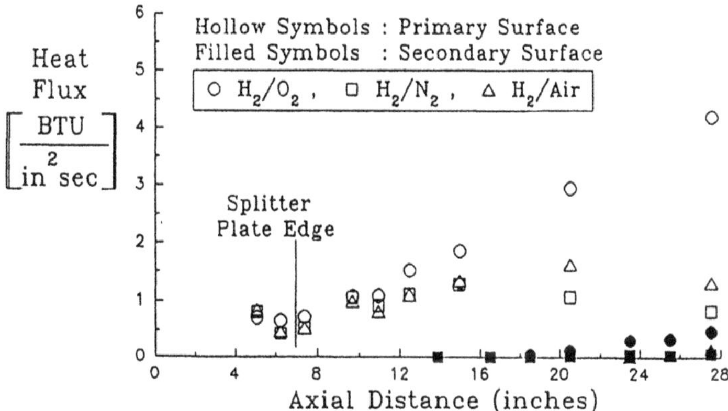

Figure 25. Summary of Heat Flux Data Showing Effects of Heat Release

secondary wall is effectively insulated from the primary stream by the cold hydrogen over virtually the entire model length. Again, the flow visualization, Figure 21, indicates that variations in density are reaching the secondary wall within the field-of-view of the window. Since this is seen in all three runs, a statistical aberration in the turbulent field is unlikely. The pressure field is almost constant in this case, therefore the refractive field variations must represent either temperature variations or species concentration variations, or both. Since the heat flux to the secondary wall is not varying, it is most probably species variations which we are seeing in the images. Hence, it appears that mass is being transferred laterally more rapidly than energy. The high heat capacity of the hydrogen is apparently sufficient to absorb the heat carried by the nitrogen and thereby thermally protect the secondary wall even though the hydrogen is being diluted with nitrogen. This scenario seems less likely in the prior nitrogen-nitrogen case, but it cannot be ruled out. Clearly, additional analysis of this data is called for, and may be very useful in establishing appropriate values for parameters such as the turbulent Schmidt number in very high speed flow.

Flow visualizations of the reacting shear layer cases (air-hydrogen and oxygen-hydrogen) were virtually indistinguishable from the inert case (nitrogen-hydrogen). However, the wall pressure and heat flux distributions, shown in Figures 24 and 25, clearly reveal that weak combustion occurred in air and vigorous combustion occurred in oxygen. The amount of heat produced in the mixing layer within the field of view of the window was apparently insufficient, even in the oxygen case, to produce a significant effect on the visible growth rate.

Although a large data base was not obtained in this study, the results indicate that a single-valued convective Mach number may not be an adequate correlating parameter for compressible turbulent shear layer growth. In view of the agreement found with the prior correlation at the high convective Mach condition (M_c = 2.8) and the lack of agreement at the transonic convective Mach condition (M_c = 0.84), it seems highly desirable to continue the originally planned parametric variation of convective Mach number by using various mixtures of nitrogen and hydrogen (and/or helium) as the secondary stream. Furthermore, the Mie scattering technique could offer an unambiguous determination of the extent to which mass is being transferred across the mixing layer.

6. Detonation Wave Experiments

The expansion tube offers a unique capability to conduct experiments on hypersonic premixed combustion and detonation waves by virtue of its ability to accelerate the test gas through a combination of shock and expansion processes. This has been recently demonstrated with methane-air mixtures (Srulijes et al., 1992). We attempted to extend their work to higher velocity and to the use of hydrogen-oxygen mixtures diluted in helium (Chinitz et al., 1992).

The concept of the experiment was to accelerate a premixed combustible hydrogen-oxygen-helium mixture to hypervelocity speed, and then allow it to flow over a two-dimensional wedge at conditions which would cause an oblique detonation wave to form. Ideally, the detonation wave would be formed after an induction distance roughly one-half the model dimension such that the change in shock shape could be clearly observed, as predicted by CFD calculations as shown in Figure 26.

In order to obtain this type of test condition, the following criteria were selected:

- stay at the lower end of the hypervelocity regime ($u > 10,000$ ft/s)
- use the minimum helium dilution
- stay well above the Chapman-Jouguet point.

The first two criteria were intended to maximize the ratio of heat of combustion to total enthalpy, so that the transition from shock wave to detonation wave could be readily apparent. The last criterion should insure stability of the oblique detonation wave.

The selected test section conditions were:

Stoichiometric hydrogen-oxygen ratio
30% helium dilution (by volume)
13,000 ft/s velocity
350°R static temperature
1 psia static pressure

Equilibrium reacting flow over a 30° wedge at these conditions would produce the following conditions behind an oblique detonation wave:

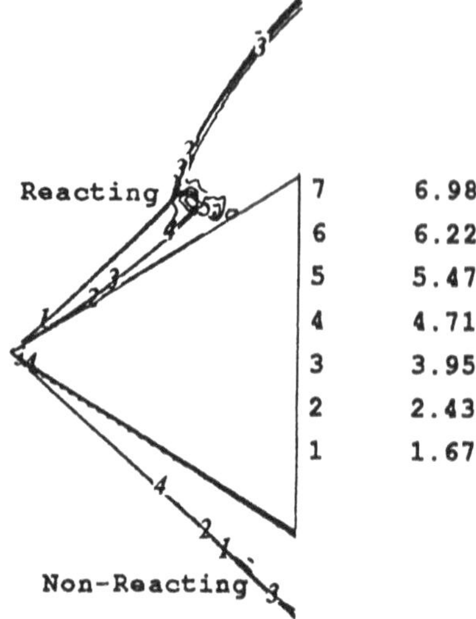

Figure 26. Computed Density Contours

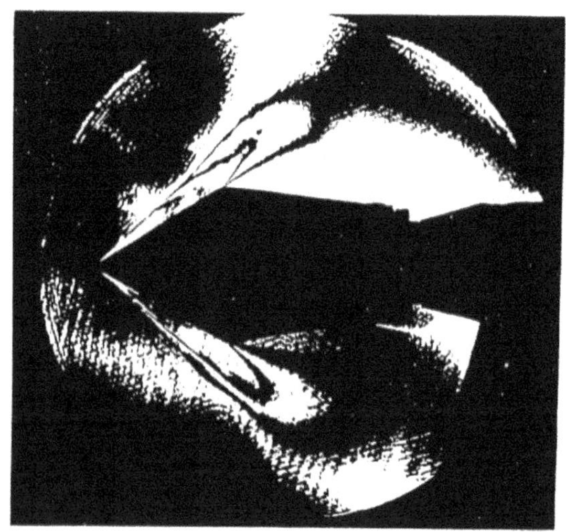

Figure 27. Laser Holographic Interferogram of Inert Test Gas Flow Over the Wedge Model

10,135 ft/s velocity
2273°R static temperature
31.4 psia static pressure

The finite rate CFD solution for this test condition is depicted on the upper half of Figure 26 and the non-reacting solution on the bottom half. An interferogram of the non-reacting case, obtained by substituting nitrogen for the oxygen in the test gas, is shown in Figure 27. Agreement with the CFD solution is considered quite satisfactory.

Unfortunately, no valid data for the reacting flow case has been obtained to date, due to a series of unanticipated difficulties in avoiding premature ignition of the mixture in either the shock tube or acceleration tube. A number of possible causes of premature ignition have been postulated, most associated in one way or the other with the secondary (Mylar) diaphragm. In all cases even a planar reflected shock, if it occurred due to failure of the diaphragm to rupture instantaneously, was theoretically incapable of causing autoignition. Combustion of the Mylar itself was also postulated as a contributing factor.

This study will resume shortly and focus on elimination of the secondary diaphragm as the cause of the observed premature ignition.

7. Facility Modeling Support

Although every attempt has been made to fully describe the test conditions achieved in the expansion tube (HYPULSE) facility within the constraints of available resources, it should be evident that the current calibration data will not fully satisfy the CFD validation community. Furthermore, in the premixed combustion/detonation wave experiments our incomplete knowledge of all details of the facility operation is painfully apparent. Additional work is planned on the premixed combustion/detonation wave problem, but it will be supported by more detailed CFD analysis, albeit one-dimensional, of the coupled unsteady wave and finite-rate chemical processes, using a recently developed code (Wilson, 1992).

It should be apparent that challenges lie in more accurate modeling of the facility operation itself, as well as the hypersonic combustion data it is uniquely able to generate. Particular attention needs to be devoted to the two-dimensional (axisymmetric), unsteady processes at play, especially boundary layer growth and transition on the tube walls and reflection of the incident (planar) shock off a weak, statically bulged, secondary diaphragm. The rupture process of the metal double-diaphragms and the resulting starting process of the primary shock wave are also inherently two- if not three-dimensional. The resulting disturbances are thought to contribute to the generation of multi-mode acoustic waves in the driver gas which propagate into the test gas under certain conditions and focus to a single dominant acoustic mode through the unsteady expansion wave system (Paull and Stalker, 1991). The rupture process of the secondary diaphragm is also thought to be capable of generating hydrodynamic instabilities under certain conditions (Trimpi, 1992).

The practical importance of modeling the facility is clearly secondary to that of modeling hypersonic combustion phenomena itself. However, as we progress beyond the phenomenological experiments to the benchmark quality experiments for CFD validation and to actual combustor development tests, CFD support of the facility performance will become increasingly more valuable if not essential.

8. Conclusions

The ability of the expansion tube (HYPULSE) to generate important data on hypersonic/hypervelocity fuel-air mixing and combustion has been demonstrated. Examples of two-dimensional (axisymmetric) and three-dimensional data taken at flight Mach 13 and 17 conditions have been presented.

The pressure rise due to combustion is often small, especially compared to pressure variations due to internal shock waves, due to the diminishing value of the ratio of heat released to air stream total enthalpy. This, in turn, is due to both the increasing total enthalpy of the air stream and to the reduced exothermicity of the combustion reactions at high temperature. However, the dissociated combustion products tend to recombine at a cold, metal wall resulting in augmented heat transfer rates due to combustion even when the pressure rise is suppressed. Thus, we are led to the current dilemma that the wall heat transfer rate is one of the most easily obtained and accurate hypersonic combustion diagnostics but the most difficult to accurately calculate by contemporary CFD methods.

On the other hand, glimpses have been offered herein of the sort of combustor flowfield diagnostics that can be obtained in the expansion tube, as well as other pulse-type facilities, using optical diagnostics and planar imaging. These clearly offer important opportunities for understanding the flow physics and for CFD validation.

Some data on inert and reacting two-dimensional shear layer mixing has also been obtained in the expansion tube at more extreme velocities than previously used in studies of compressibility effects. Some of the data is consistent with a prior correlation of compressible shear layer growth rate, others are notably at odds with the prior correlation. The data also suggests a much faster transfer of mass across the mixing layer than of energy. Additional analysis and further parametric variations of the test conditions are called for.

Finally, some recent attempts to obtain data on premixed combustion/detonation waves at hypervelocity conditions using hydrogen-oxygen mixtures did not meet with success to date. Prior experiments in another expansion tube facility using methane-air mixtures at lower speed were successful. This study should resume shortly.

The challenges and opportunities which data from this type of test facility provide for mathematical and/or computational modeling of hypersonic combustion phenomena are hopefully self-evident. However, a need for more detailed modeling of the facility itself to fully describe the test conditions under which the data is obtained also exists.

Acknowledgements

The author is pleased to acknowledge a number of people on whose work this paper is based and who contributed materially to its preparation. In the former category, particular thanks is due to Bob Bakos, Jose Tamagno, Richard Trucco, Sam Rizkalla, Wally Chinitz, John Calleja and Maria Pulsonetti, of GASL, Richard Morgan, Alan Paull and Ray Stalker, of the University of Queensland, Dave Swain and Larry Rubin, of Rocketdyne, George Havener and Joe O'Hare, of AEDC, and Griffin Anderson, Clay Rogers, Chuck McClinton and Ernie Mackley, of NASA Langley. In the latter category, special thanks is due to Jose Tamagno, Richard Trucco, Sam Rizkalla and Mary Ellen Kent.

References

Anderson, G., Kumar, A., and Erdos, J., 1990. "Progress in hypersonic combustion technology with computation and experiment," AIAA 90-5254.

Anderson, R. C., Trucco, R. E., Rubin, L. F., and Swain, D. M., 1992. "Visualization of hydrogen injection in a scramjet by simultaneous PLIF and laser holographic imaging," Proc. 1992 NASA Langley Measurement Technology Conference, NASA CP 3161.

Bakos, R., Morgan, R., and Tamagno, J., 1992. "Effects of oxygen dissociation on hypervelocity combustion experiments," AIAA 92-3964.

Bakos, R., Tamagno, J., Rizkalla, O., Pulsonetti, M., Chinitz, W., and Erdos, J., 1992. "Hypersonic mixing and combustion studies in the hypulse facility," *J. Prop. Power* **8**(4), pp. 900-906.

Bakos, R., Tamagno, J., Trucco, R., Rizkalla, O., Chinitz, W., and Erdos, J., 1992. "Mixing and combustion studies using discrete orifice injection at hypervelocity flight speeds," *J. Prop. Power* **8**(6), pp. 1260-1296.

Bobskill, G., Bittner, R., Riggins, D., and McClinton, C., "CFD evaluation of Mach 17 HYPULSE scramjet combustor data," AIAA 91-5093.

Bogdanoff, D. W., 1983. "Compressibility effects in turbulent shear layers," *AIAA J.* **21**(6), pp. 926-927.

Chinitz, W., Tamagno, J., Erdos, J., Singh, D., and Rogers, C., 1992. "Premixed shock-induced combustion studies in the HYPULSE facility," AIAA 92-3427.

Erdos, J., Tamagno, J., Bakos, R., and Trucco, R., 1992. "Experiments on shear layer mixing at hypervelocity conditions," AIAA 92-0628.

Miller, C. G. and Jones, J. J., 1983. "Development and performance of the NASA Langley Expansion Tube/Tunnel, A hypersonic-hypervelocity real-gas facility," 14th Intl. Symp. on Shock Tubes and Waves, August 15-18, Sydney, Australia.

Miller, C. G., Micol, J. R., and Gnoffo, P. A., 1985. "Laminar heat transfer distributions on biconics at incidence in hypersonic-hypervelocity flows," NASA TP-2213.

Morgan, R., Stalker, R., Bakos, R., Tamagno, J., and Erdos, J., 1991. "Scramjet testing - Ground facility comparisons," Paper No. 91-194(L), Proc. X ISABE Meeting, Nottingham UK.

Papamoschou, D., 1991. "Structure of the compressible turbulent shear layers," *AIAA J.* **21**(5), pp. 680-681.

Papamoschou, D. and Roshko, A., 1988. "The compressible turbulent shear layer: An experimental study," *J. Fluid Mech.* **197**, pp. 453-477.

Paull, A. and Stalker, R. J., 1991. "The effect of an acoustic wave as it traverses an unsteady expansion," *Physics of Fluids A.*

Spurk, J. H., 1970. "Experimental and numerical nonequilibrium flow studies," *AIAA J.* **8**(6), pp. 1039-1045.

Srulijes, J., Smeets, G., and Seiler, F., 1992. "Expansion tube experiments for the investigation of RAM accelerator related combustion and gasdynamic problems," AIAA 92-3246.

Tamagno, J., Bakos, R., Pulsonetti, M., and Erdos, J., 1990. "Hypervelocity real gas capabilities of GASL's expansion tube (HYPULSE) facility," AIAA 90-1390.

Trimpi, R. L., 1962. "Preliminary theoretical study of the expansion tube, A new device for producing high-enthalpy short-duration hypersonic gas flows," NASA TR-R-133.

Trimpi, R., 1992. NASA Langley Research Center, private communication.

Trimpi, R. L. and Callis, L. B., 1965. "A perfect gas analysis of the expansion tunnel, A modification of the expansion tube," NASA TR-R223.

Wilson, G., 1992. "Time-dependent quasi-one-dimensional simulations of high enthalpy pulse facilities," Proceedings AIAA 4th International Aerospace Planes Conference, Orlando, Florida.

SUPERSONIC COMBUSTION EXPERIMENTS IN FREE PISTON REFLECTED SHOCK TUNNELS

R. G. Morgan

Department of Mechanical Engineering
The University of Queensland
Queensland, Australia, 4072

ABSTRACT

Reflected shock tunnels have been used for a variety of experiments relating to scramjet combustion over the years. This paper describes three sets of experiments which were performed on T3 and T4, the free piston driven reflected shock tunnel at the Australian National University, and at the University of Queensland, and explains the methodology by which they were planned. The experiments include studies of thrust production in two dimensional and axisymmetric flows, and an investigation of the pressure/length scale interaction for scaling combustion processes. The experiments were all planned to be sensitive to areas of theoretical uncertainty, and to have sufficient accuracy and resolution to evaluate numerical and analytical simulations of the processes involved. The models used were deliberately kept simple, in order to minimize the number of interacting, coupled effects, and to facilitate data analysis and theoretical modeling.

1. Introduction

With the application of free piston driven shock tunnels to scramjet combustion phenomena (Stalker and Morgan, 1984), starting with the T3 facility at the Australian National University, it became possible to extend the range of ground facility simulation beyond the intermittent facility range of approximately 2 km/sec. Facility dimensions and test time limitations precluded the testing of full scale flight hardware, so a series of experiments were performed to gain an understanding of important processes relevant to scramjet operation.

For scramjet propulsion systems to be successfully implemented, a range of multidisciplinary skills must be brought to bear on an intricate Engineering problem. However, the dominating characteristic feature is the framework of supersonic and hypersonic ducted

J. Buckmaster et al. (eds.), Combustion in High-Speed Flows, 93–109.

TABLE 1

Thrust coefficients determined from measured nozzle pressure and predicted by SHARC

Equivalence ratio	Measured conical thrust coefficient	SHARC conical thrust coefficient	SHARC contoured thrust coefficient
0	1.49	1.39	1.55
0.93	1.47	1.46	1.55
1.91	1.50	1.43	1.50
3.05	1.41	1.41	1.47

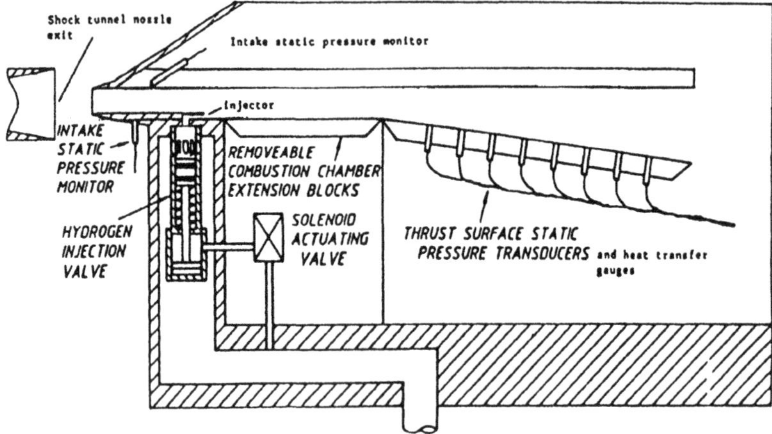

Fig. 1. Schematic Diagram of Experimental Apparatus.

and open flows against which the complimentary sciences must be interpreted. The approach taken in this laboratory has been to build upon a background of hypersonic gas dynamics to elucidate the influence of macroscopic fluid dynamic effects on a complex combustion and thrust producing process.

Experiment 1. Two dimensional thrust production

There is a considerable background of research into scramjet combustion chamber phenomena, directed particularly toward the important problem of fuel injection and mixing (Rogers, 1980). These studies have made it possible to estimate the combustion chamber length required for complete mixing and combustion. With this information , and the use of one dimensional theory to calculate the flow in the exhaust nozzle, the thrust performance of simple engine configurations can be ascertained.

However, such a procedure leads to an engine configuration which is not suited for practical use. Parametric studies (Martin, 1978) show that the overall mass of the engine is an important factor in limiting the performance of a scramjet flight vehicle, and therefore it is necessary to make it as short as possible. This implies that the thrust producing exhaust nozzle expansion may begin well upstream of the point where the fuel has mixed completely with the air in the combustion chamber, and therefore the thrust is produced through the interaction of a fuel jet with an expanding supersonic flow field. An experiment was configured to reproduce this situation in a shock tunnel, and to provide the data base necessary to validate analytical and numerical analyses of this situation, which is a fundamental step towards being able design a working engine.

The aim of the experiments was to investigate the thrust produced in a short , two dimensional nozzle by combustion of a hydrogen jet, and compare it, in the first instance, with the thrust predicted from equilibrium premixed one dimensional flow theory. This is a useful basis for comparison, as it forms the basis of much of the early analysis of scramjets, and because thermodynamic arguments establish that the thrust generated according to this theory represents an upper limit, against which the effectiveness of experimental configurations may be evaluated.

A schematic of the experimental apparatus is shown in Fig. 1. The experiments were performed in the free piston shock tunnel T3

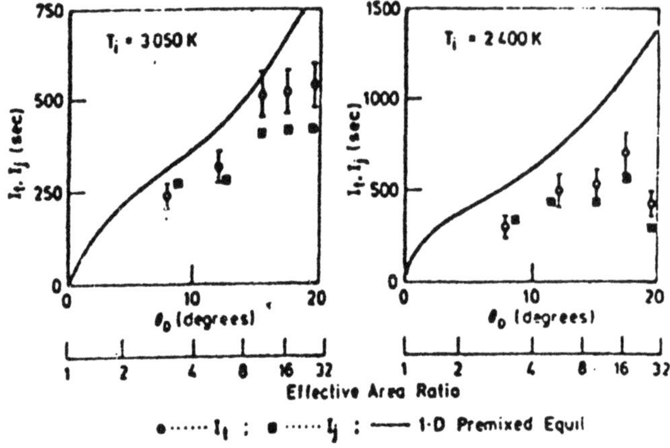

Fig. 2. Two dimensional scramjet performance (Stalker and Morgan, 1984).

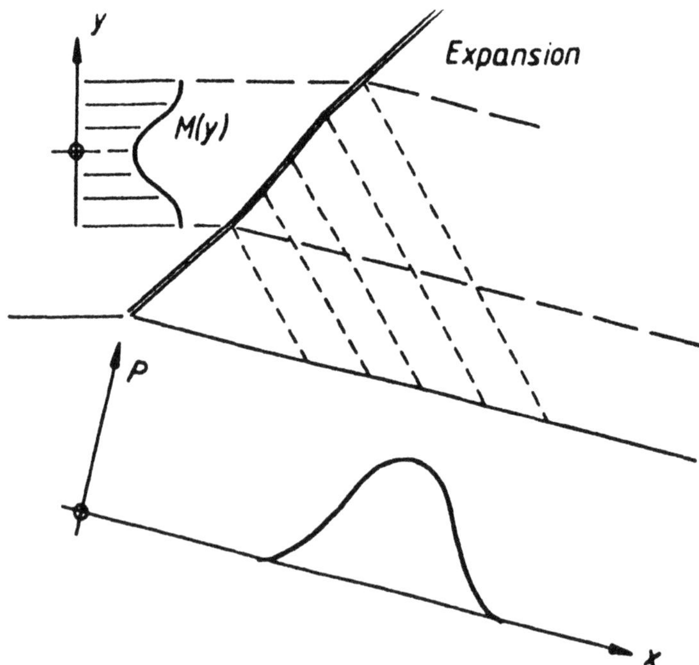

Fig. 3. Schematic of incremental expansion wave traversing a fuel jet (Stalker and Morgan, 1988).

(Stalker, 1972) at the Australian National University. The test flow from the supersonic nozzle passed directly into the model inlet. The model was two dimensional, with an internal width throughout of 50 mm, and consisted of a channel of constant height, 25mm, followed by an exhaust nozzle , in which the upper surface was fixed and parallel to the tunnel flow axis, whilst the lower surface was straight and could be set at any desired divergence angle up to 20 degrees. The lower surface was the only thrust surface, and was instrumented with Pcb Pietzotronic pressure transducers. Integration of the measured pressure along the surface provided an indication of the thrust developed, and formed the basis of comparisons with one dimensional predictions. The fuel injector strut was located midway between the upper and lower surfaces of the constant area section, and fully spanned its width. Hydrogen fuel, with room total temperature, was injected from both ends of the strut. The fuel was injected through a supersonic nozzle, with a profile consisting of two circular arcs, separated by 1.6mm at the throat. Equivalence ratio was controlled by adjustment of the total pressure of the fuel supply. The leading edge of the injector strut was a symmetric wedge of 20 degrees included angle, located upstream of the intake so that the leading edge shock/expansion system did not enter the duct, and confuse interpretation of the pressure measurements on the thrust surface.

A difficulty arises when a one dimensional analysis is used as a basis of comparison for experiments which have two or three dimensional flow characteristics. The flow in a short, two dimensional nozzle cannot be treated as one dimensional, even as an approximation, since the generation of thrust involves the interaction of supersonic expansion and compression waves within the nozzle (Stalker, 1989). Firstly, as the flow in the experimental situation is expanded by the propagation of a centered expansion fan across the duct, the flow at any axial station is not uniformly expanded, and the appropriate area ratio to use for comparison with one dimensional theory is not clear. The real flow situation also produces waves which detract from the thrust producing capacity of the nozzle. An approach is developed in Stalker and Morgan (1984), which creates an equivalent one dimensional flow, and is the basis of the comparisons presented here.

Sample data is shown in Fig. 2, and it can be seen that when the appropriate comparison is made, the experimental thrust is significantly less than the theoretical maxima. Subsequent theoretical

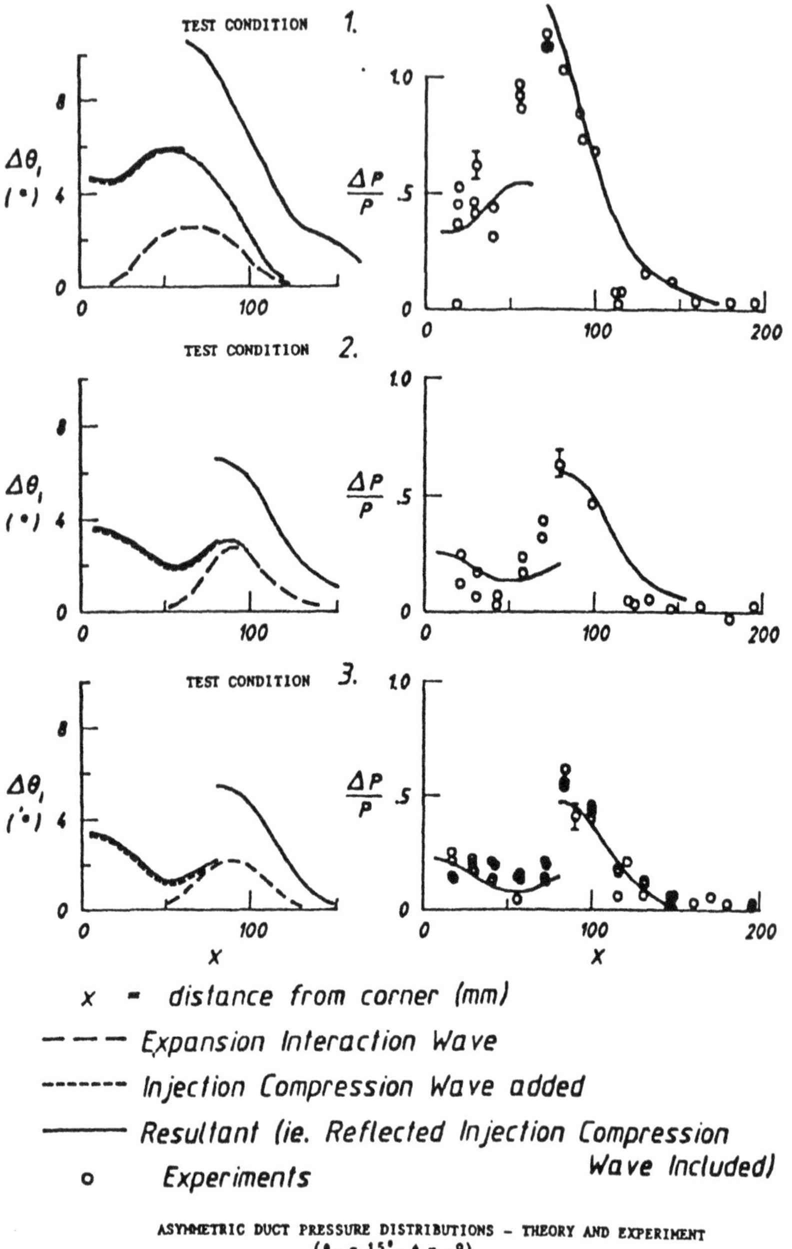

Fig. 4. Two dimensional scramjet thrust surface pressure distributions. (Stalker and Morgan, 1988).

analysis showed that the two dimensional nature of the wave system had a dominating influence on the way in which thrust was developed, and the distribution of thrust along the nozzle. In Fig. 3, from Stalker, Morgan, and Netterfield (1988), the primary wave structure created when incompletely mixed flow is expanded by a centered expansion fan is shown. It is noted that thrust increment above the fuel off reference condition may be due to two separate mechanisms. Firstly, combustion and mixing induced pressure rise in the combustion chamber leads to an overall increase in the static pressure level on the thrust surfaces, and a subsequent thrust enhancement. Secondly, when the expansion process has to be started before mixing and combustion are complete, the propagation of expansion waves through the region of non uniform flow causes waves to be reflected onto the thrust surface, which may cause additional thrust perturbations.

The nature of the processes may be quantified analytically, and Fig. 4 from Stalker, Morgan, and Netterfield (1988) shows a comparison of experimental and theoretical pressure profiles for a range of operating conditions. The good agreement between the two, and the difference of the profiles from one dimensional flow emphasizes the importance of experimental testing of this nature. A feature of these tests is that the theoretical approach was analytical, and based on assumptions made by the investigators as to the dominant mechanisms involved in developing thrust in a scramjet nozzle. The results indicate that these assumptions were essentially valid for the configuration and conditions studied. This approach lacks the precision of numerical simulation, but by its nature adds considerably to a physical understanding of the situation.

Experiment 2. Axisymetric thrust production

The two dimensional experiments discussed above are appropriate for investigating fundamental phenomena, because the simple geometry facilitates both experiment and analysis. However, for engineering scramjets to be viable as propulsion systems, they are expected to be based on three dimensional concepts. This complicates experimental thrust measurement, because a single pressure reading is no longer representative of the thrust surface static pressure at any axial station. Direct thrust measurement techniques for impulse facilities are now being developed to alleviate this problem (Paull,

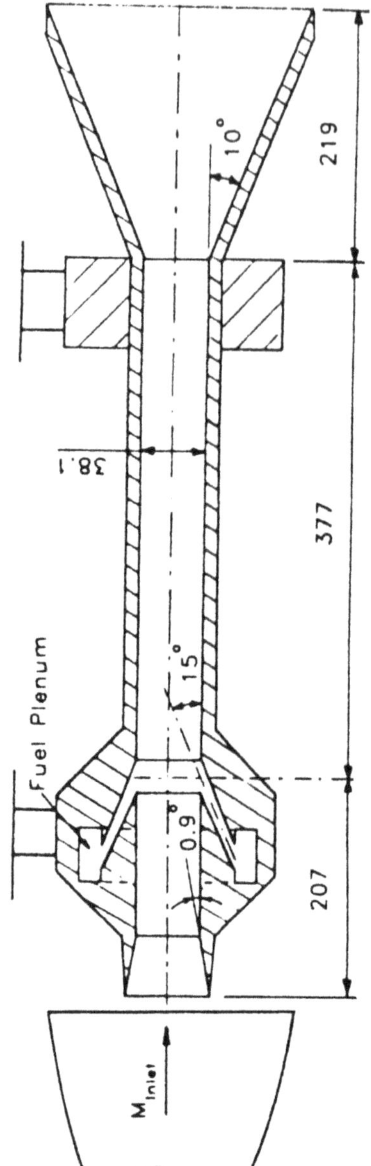

Fig. 5. Axisymmetric scramjet configuration with conical nozzle (Bakos and Morgan, 1992).

Mee, and Tuttle, 1993) and offer great promise for the evaluation of complex scramjet configurations.

The experiments described here feature an axisymmetric constant area duct with a conical thrust surface attached to the end of the combustion chamber. This arrangement contains some of the three dimensional flow characteristics, such as the rapid axial expansion of the gas, but retains the simplicity of a two dimensional flow for the purposes of analysis. The uniform angular distribution of pressure around the expansion nozzle allows the thrust to be estimated from single row of pressure transducers aligned with the flow axis. A schematic of the apparatus is shown in Fig. 5. The combustion chamber and injection system were constructed by the General Applied Science Laboratories Inc (GASL) for a cooperative research project to give comparisons of supersonic combustion in the HYPULSE expansion tube and the T4 reflected shock tunnel (Morgan et al., 1991). For these experiments the combustion chamber was truncated, and the conical nozzle was attached at a location where previous experiments had shown significant, but still incomplete, mixing and combustion had occurred.

Sample results at an equivalence ratio of 3.05 are shown in Fig. 6, taken from Bakos and Morgan (1992). Also shown in Fig. 6 are the results of predictions from the parabolic computer code SHARC, written by Brescianini (Brescianini and Morgan, 1992).

For the fuel off case the mean pressure level is well predicted, while the measured wave pattern created by the inlet appears to dampen more quickly than is predicted. The thrust surface pressure is of the correct magnitude, although the shape of the distribution is not picked up by the computation. For the fuel on case a large amplitude standing wave pattern has been created by the slightly transverse fuel injector, which is under expanded by a factor of approximately three at the injector lip. Agreement in the combustor is excellent, with the wave pattern and the combustion and mixing induced pressure rise well predicted. The calculated injection shock persists into the thrust nozzle as seen by the bump in the computed solution there, while no clearly defined perturbation appears in the data at this location.

The good agreement seen between the experiments and the SHARC predictions encourages further use of the code to reveal information which cannot be measured experimentally, or which can be found more accurately numerically. In Table 1 thrust coefficients

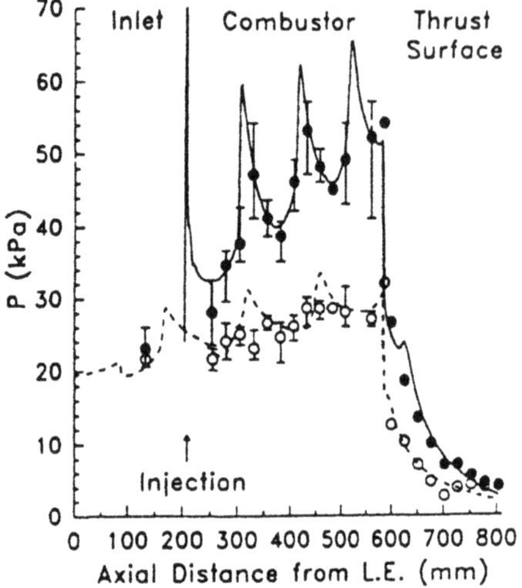

Fig. 6. Axisymmetric scramjet pressure profiles for $\phi = 3.05$. Fuel on and fuel off. Theory and experiment.

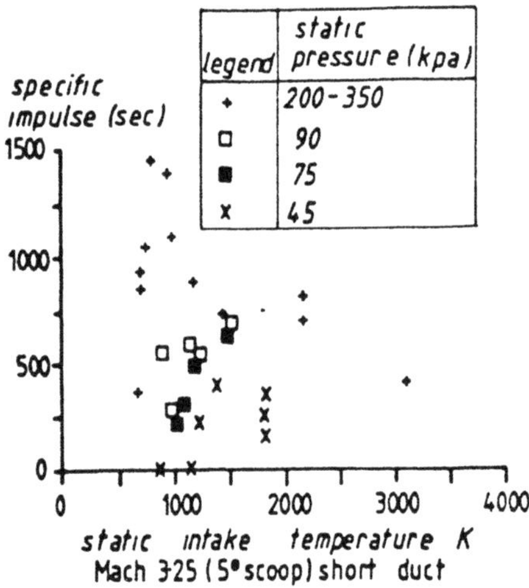

Fig. 7. Dependance of specific impulse on pressure, two dimensional (Morgan and Stalker, 1987).

are presented as estimated from the pressure distributions, and from SHARC computations. The conditions range from fuel off to an equivalence ratio of 3.05. The thrust coefficients from both experiment and theory show only slight dependence on equivalence ratio , while the thrust level itself varies considerably. Thus in this instance the major factor influencing thrust is the mean static pressure level at the combustor exit, while alteration of the Mach number distribution and subsequent nozzle wave pattern changes have not strongly influenced the thrust produced. This is consistent with fairly good mixing, producing a relatively uniform Mach number distribution which only interacts weakly with the wave propagation process. The addition of fuel was found to change the thrust coefficient by a maximum of 6% for both the measured and computed nozzle flows. This apparent insensitivity may be partially due to fact that disturbances in axisymmetric flow tend to focus on the centerline, and defocus at the walls, where the pressures are measured. Whilst only 6% of the thrust being dependent on Mach number distribution and the associated wave patterns may sound small compared to the two dimensional configuration discussed above, with the marginal thrust - drag balance expected in flight craft it could be a critical component.

Although only a slight dependence of thrust on wave pattern changes due to fuel injection has been demonstrated for this flow, it is of interest to determine how close the conical nozzle is to an optimum thrust contour for the given area ratio and combustor exit conditions. To investigate this, SHARC was run for the four fuel conditions with a uniform lateral pressure condition imposed in the combustor and nozzle. The nozzle thrust found is thus equivalent to a perfect, wave canceling expansion in a contoured nozzle of the same geometric area ratio. (Slight differences arise in the combustor exit conditions from the absence of waves this approach creates, but combustor exit stream thrust is found to agree within 1wave capturing solution for a sample calculation, and the comparisons presented are also believed to be similarly accurate.) The 'contoured' thrust coefficients thus calculated are also shown in Table 1, and are seen to fall within 11% of the conical values for all conditions. On this basis , the experimentally measured thrust coefficients are expected to fall within 11% of the value which could be achieved in a perfectly contoured nozzle.

To summarize these experiments, they used a simple configuration, with some three dimensional flow characteristics, which enabled

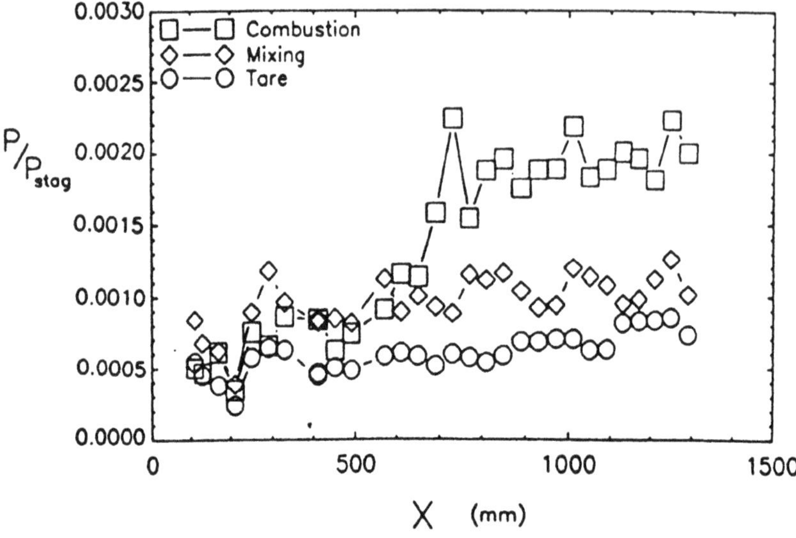

Fig. 8. Pressure distance profiles for large 2D scramjet model (Pulsonetti, 1992).

meaningful and useful comparisons to be made with a two dimensional nozzle, and with a CFD analysis. Agreement with the code (SHARC) was good which adds credibility to the experiments, and which strongly encourages further use of the same code for scramjet analysis, and for the planning and interpreting of experiments.

Experiment 3. Pressure-length scaling of supersonic combustion

By the very nature of their proposed flight paths, scramjet propulsion systems will have to operate over a wide range of static pressures. With the combustion chamber mass expected to have a strong influence on overall system viability (Martin, 1978) the coupling between pressure and the necessary combustor length required for mixing and combustion will be an important design parameter. Ground facility limitations also prescribe that testing will have to be performed on scale models, and understanding the so called pressure/length scale dependency will be an important step in making engineering use of shock tunnel experimental results.

A series of early experiments performed on the shock tunnel T3 at the Australian National University covered a range of operating pressures, and although it wasn't the primary objective of the tests, some pressure related effects were observed. In Fig. 7, from Morgan and Stalker (1987), sample data is shown illustrating the influence of pressure on specific impulse for a simple, centrally injected two dimensional combustor. For some conditions a simple correlation between pressure and length applied, with the pressure times length product acting as a single parameter to quantify the relative pressure rise which could be achieved in a given combustor geometry. However, given the complex nature of combustion and mixing in a scramjet, such a simple correlation would be surprising. Also, the precise test conditions of the experiments were not specifically targeted for this particular comparison. A number of different parameters, including pressure, were changed between tests, and the results, although interesting, cannot be considered definitive.

Further experimental work of this nature is required, and it can only be properly planned in the light of some understanding of the various processes involved. Pulsonetti (1992) has performed experiments in which the primary flow parameters, especially gas composition and static temperature are held constant as the pressure levels change. Test gas composition in a reflected shock tunnel depends primarily on the entropy of the flow in the shock tube stagnation region, due to chemical freezing in the nozzle (Harris, 1966), and this may only be conserved by keeping constant shock tube conditions. For instance, if the pressure was adjusted by changing all the shock tunnel operating pressure levels, then the influence of pressure on dissociation would cause the test gas composition to change. The approach taken by Pulsonetti is to control the static pressure level of the expanded test gas by means of a series of nozzles contoured to different Mach numbers, followed by shock recompression in the region of chemically frozen flow. The Mach number at which the recompression occurs is chosen to give the appropriate total pressure loss, and create the required static pressure at the planned Mach number.

The methodology of these ongoing experiments is to use a series of geometrically similar models, with the length scales adjusted to maintain at a constant value the product of static pressure with a characteristic length scale. Supersonic and hypersonic ducted flows are strongly influenced by viscous effects, but by maintaining the pressure length scale constant the Reynolds number is conserved, and viscous effects are decoupled from comparisons between the different models. Residual differences observed in the non dimensionalized pressure/axial distance profiles between the different models may therefore be attributed to the combustion and mixing processes departing from a pressure-length product scaling law.

Arguments presented in Pulsonetti (1992) indicate that, because the ignition process is controlled by binary reactions, the ignition delay time will be inversely proportional to duct static pressure. Maintaining the same relationship between duct residence time and ignition delay therefore also calls for conservation of the pressure-length product. Reaction times are more complicated, typically involving multiple reaction paths. If three body reactions dominate, then the product of pressure squared with a length scale may be shown to be inversely proportional to reaction time. Empirical and numerical studies have shown that pressure coupling to a slightly lower power may be more appropriate. The important implication, however, is that it is not possible to simultaneously satisfy the twin criteria for similarity of the pressure-length product and the product of pressure to a power other than one with length.

Another consideration must be the effect of pressure on heat release, once mixing and reaction have occurred. Increased pressure reduces dissociation of combustion products, which increases the combustion heat release, and therefore the non dimensionalized pressure rise. This process is not directly coupled with the length of the mixing/combustion layer, and quantifying its influence on scaling laws will be an important outcome of the experimental program.

Current work on compressible mixing layers (e.g. Papamoschou and Roshko (1988)) indicates that the mixing layer will propagate in a geometrically similar manner for all the model length scales, because the critical flow parameters of fuel and free stream Mach number, temperature and velocity are conserved, as are the density and total pressure ratios between the free stream and the injectant.

To summarize the methodology of the program, it is evident that the pressure-length scale product must be held constant when performing scaling experiments in order to conserve the macroscopic fluid dynamic features of the flow. Incidentally this also causes correct scaling of the ignition process. It is not yet clear how other combustion features will be altered by the changes in length scale and pressure, and it is the primary objective of the tests to quantify these unknown effects.

Figure 8 shows sample data from Pulsonetti (1992), for the largest model length scale. The important features to note are that the onset of combustion is clearly defined, and that a good indication is given of the post combustion pressure level. The elementary rule of an experimental investigation is seen to have been satisfied, in that the instrumentation can detect the phenomenon of interest with sufficient accuracy to make meaningful comparisons with theory. The current status of the work is that the methodology has been established, the experimental techniques are proven, and tests on the largest of three models are complete. The program is continuing with the testing of smaller models under identical flow conditions.

2. Conclusions

The usefulness of shock tunnels for the study of scramjet phenomena is demonstrated through a brief description of experiments performed on a range of topics of critical importance to understanding the principles involved, and to eventually developing flight worthy hardware. This utility is likely to increase in the future, with the development of new and larger facilities, and the development of new techniques, such as direct thrust measurement, and improved diagnostics. The importance of correctly targeting the experiments to examine effects which can be measured sufficiently accurately to comment on currently available numerical and analytical understanding of scramjet operation is observed.

Acknowledgements

The work reported here was not performed by the author, but by a large group of Research and academic staff, as evident from the reference list, and it has all been published previously. The work was funded by the Australian Research Grants Scheme, the NASA Langley Research Center (Grant NAGW 674, contract monitor R. C. Rogers), and the DEET scholarship scheme. Technical support from the Physics and Theoretical Physics Department of the Australian National University was fundamental to establishing the viability of the initial experiments on T3, without which there would have been no further progress.

References

Bakos, R. J. and Morgan, R. G., 1992. "Axisymetric scramjet thrust production," Proc. Australian Fluid Dynamics Conference, Hobart, Tasmania.

Brescianini, C. P. and Morgan, R. G., 1992. "An investigation of a wall injected scramjet using a shock tunnel," Paper No. AIAA 92-3965, 17th AIAA Aerospace Ground Testing Conference, Nashville, USA.

Harris, C. J., 1966. "Comment on 'non equilibrium effects on high-enthalpy expansion of air'," *AIAA J.* **4**(6).

Martin, J., 1978. *J. Spacecraft Rockets* **15**, pp. 259-260.

Morgan, R. G., Bakos, R. J., Tamagno, J., Stalker, R. J., and Erdos, J. I., 1991. "Scramjet testing, ground facility comparisons," Paper No. ISABE 91-194(L), 10th ISABE Symposium, Nottingham, U.K.

Morgan, R. G. and Stalker, R. J., 1987. "Pressure scaling effects in a scramjet combustion chamber," Paper No. ISABE 87-7080, 8th ISABE Symposium, Cincinnati, Ohio.

Papamoschou, D. and Roshko, A., "The compressible turbulent shear layer: An experimental study," *J. Fluid. Mech.* **197**, pp. 453-477.

Paull, A., Mee, D. J., and Tuttle, S., 1993. "Direct thrust measurement in impulse facilities," Private communication.

Pulsonetti, M. V., 1992. "Scaling and ignition effects in scramjets," Proc. Australian Fluid Dynamics Conference, Hobart, Tasmania.

Rogers, R. C., 1980. *AIAA J.* **18**, pp. 188-193.

Stalker, R. J., 1972. *Aeronautical Journal of the Royal Aero. Soc.* **76**, pp. 374-384.

Stalker, R. J., 1989. "Thermodynamics and wave processes in high Mach number propulsive ducts," Paper No. AIAA 89-0261, 27th Aerospace Sciences Meeting, Reno, Nevada.

Stalker, R. J. and Morgan, R. G., 1984. "Supersonic hydrogen combustion with a short thrust nozzle," *J. Comb. Flame* **57**(1), pp. 55-70.

Stalker, R. J., Morgan, R. G., and Netterfield, M. P., 1988. "Wave processes in scramjet thrust generation," *Combustion and Flame* **71**, pp. 63-77.

FUEL DISPERSION IN SUPERSONIC AIRSTREAMS

B. C. R. Ewan

The University of Sheffield
Department of Mechanical and Process Engineering
P. O. Box 600, Mappin Street
Sheffield S1 4DU

ABSTRACT

The work is concerned with the problems of achieving good fuel air mixing in supersonic combustion chambers within distances which minimize the frictional losses, and a number of the issues concerning the injection of fuel are discussed. The effective jet divergence angles arising from other work are reviewed and compared with present results for two types of injector. These are a wedge with trailing edge swept back fins and an acoustic reflector type.

As an alternative to these forms of fuel injection, a different strategy based on spray injection is explored. It is shown that this can give considerable improvements if certain injection conditions are met. These are principally that of increasing the axial injection velocity component of the fuel to minimize the Weber number of the droplets.

This enables the survival of larger droplets and permits their penetration into the central regions of the flow, which in turn permits wall injection with the potential minimization of intrusive shock losses.

1. Introduction

Up to the present, the launching of all space vehicles has been powered by chemical rockets. Although the technology and efficiency of chemical rocket propulsion systems have improved over the years, the launching costs still form a large proportion of the total space project cost, which very often constrain many space missions. This has been recognized and forms the basis for studies of more economical propulsion schemes, namely those of air breathing engines which use air as oxidant during the initial stage of space flight.

Many years of research has shown that the feasibility and indeed the efficiency of supersonic combustion largely depends on the effi-

J. Buckmaster et al. (eds.), Combustion in High-Speed Flows, 111–128.

ciency of fuel-air mixing, which in turn can rely on the turbulence generation under supersonic conditions. Acquisition of this fundamental knowledge about the turbulent mixing and its interaction with chemical kinetics has been crucial to the successful development and optimization of the scramjet.

Several contemporary studies have sought to elucidate some of the fundamental fluid dynamic processes occurring in shear layers in supersonic flow and these have shown that vortices arising in these layers are suppressed in their growth (Gilreath and Sullins, 1989) resulting in jet spreading angles of around one third those of subsonic flow. For single downstream facing hydrogen jets into Mach 2.5 flows (Drummond et al., 1989), typical spreading angles are around 5 deg. This has implications for the lengths of ducts which must be allowed to enable sufficient mixing, with the consequent implied penalty on frictional losses, or for the number of separate jets which must be included to achieve mixing within a short distance, with the consequent penalty through shock losses arising from the intrusion into the flow of the jet structures.

At Sheffield University, the research on scramjet engines has been active for many years with considerable success. The main research interest has been the behavior of fuel-air mixing under different supersonic flow conditions and hardware configurations.

2. Supersonic Mixing and Mixing Devices

The objective of our study is to investigate the effect of hardware variables and scale on the mixing process to help to find the optimum factor for mixing, heat release and combustion efficiency. An important facet of the study is the flow of mixing energy from the burner pressure drop via the shear layers through turbulence and dissipation to material mixed at the molecular level.

The relationship between the scale of mixing eddies and the scale of the inhomogeneities which are being mixed is also of importance since the combustion process is controlled by both chemical kinetics and mixing. Optimization of all combustion systems must take into account the requirement that the fuel be well mixed with air and this becomes more crucial and important in the case of supersonic combustion when the mean velocity through the combustor is only a few percent less than the flight velocity, and the time available for combustion is very short, typically only a tenth of a millisecond.

Mixing Power Concepts

Mixing power = $UA\Delta P = W\Delta P/\rho$

Mixing efficiency factor $\eta = \exp[-\lambda/\sigma]^2$

Turbulence time scale = $1.116k/\varepsilon$

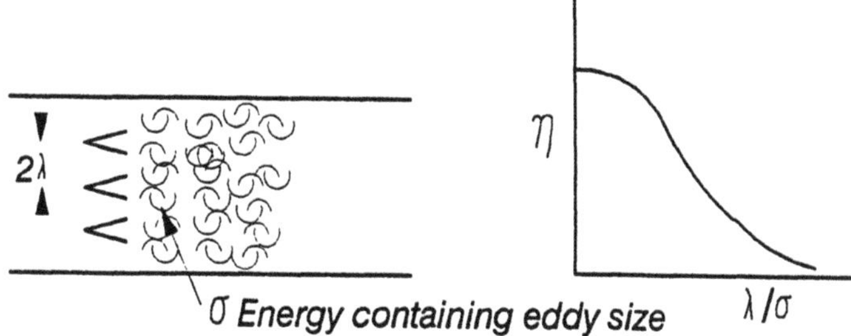

Figure 1.

Mixing Enhancement by Supersonic Flow Instability

This can be integrated with the Sheffield multi-vortex scramjet injector

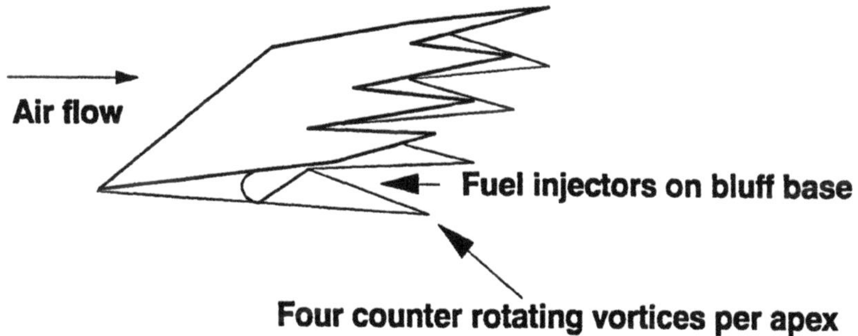

Figure 2.

ACOUSTIC REFLECTOR FOR ENHANCED JET MIXING

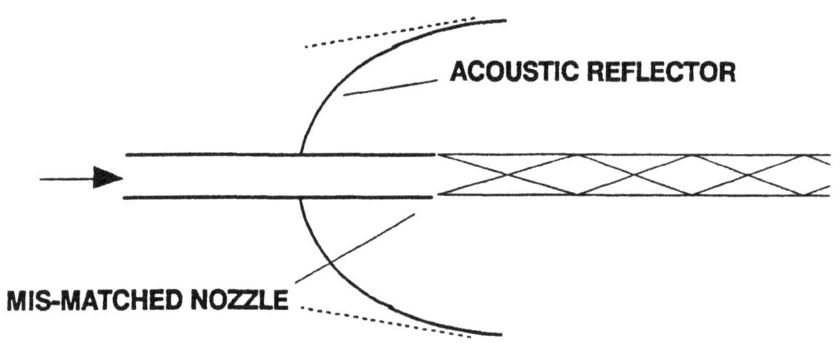

Figure 3.

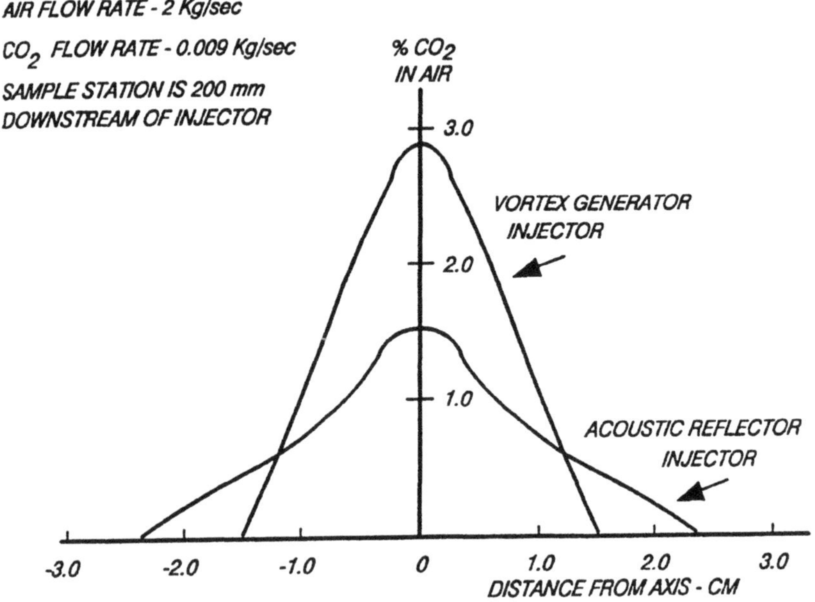

Figure 4.

To achieve high efficiency mixing within the short time duration requires a high efficiency turbulence generation which largely depends on hardware configuration.

The incorporation of a parameter to represent mixing efficiency in combustion systems has been discussed in detail by Swithenbank et al. (1989). The description of the key elements of this description are shown in Figure 1. The energy containing eddies have a scale of the order of the dimensions of the turbulence generating device σ. If the device is responsible for introducing fuel into a surrounding airstream, then the separation of such injection devices is a measure of the scale of the fuel inhomogeneities λ. The resulting mixing efficiency η is expected to follow a Gaussian function of the ratio λ/σ as shown. This has obvious implications for the scale and number of fuel injectors in the present area of interest and implies that either the devices can be small in number but relatively large to create a large turbulence scale, or smaller but more numerous. In either case there is an implication that shock losses due to intrusion into the flow will arise and minimization of these while maintaining the efficiency factor is a major consideration.

In our study, the fuel injector placed in the stream of supersonic flow plays the role of generating turbulence.

Two types of fuel injector have been designed for testing. One is a wedge type injector with a wedge angle of 12° and trailing edge swept back fins designed for vortex generation, whilst the other is the acoustically enhanced mixing device which has an acoustic reflector used to focus the disturbances from the first shock diamond onto the edge of the fuel injector nozzle.

These two types of injector are illustrated in Figures 2 and 3 respectively.

While the mechanism of turbulent mixing is being extensively studied with the aid of CFD code, experimental investigation is still needed to provide first hand information on turbulent and mixing behavior. The required information includes test section pressure, temperature, flow velocity, gas composition and turbulence structure in the form of flow visualization.

3. Initial Comparison Between Wedge and Acoustic Mixing

The two mixing devices described above have been compared outside of the shock tube environment in a supersonic airstream con-

GEOMETRY USED FOR HYDROGEN DROPLET INJECTION CALCULATION

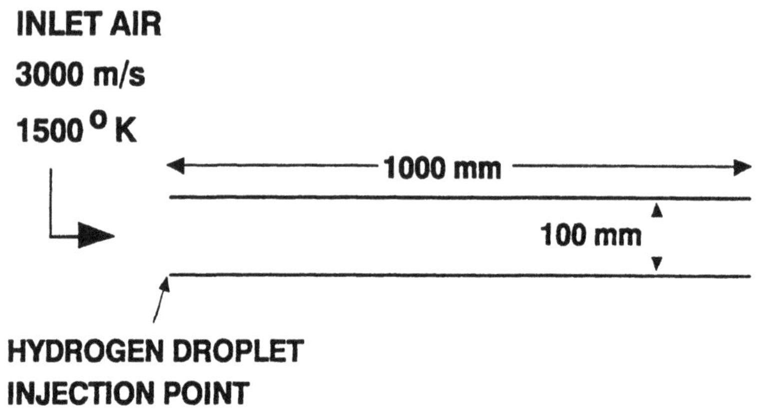

Figure 5.

Liquid Injection Systems

Drop breakup is governed by the ratio of disruptive aerodynamic force to the consolidating surface tension force.

$We = \rho\,(U_g - U_d)^2\, d\,/\,\sigma$ Critical We = 50

σ = 2E-3 N/m ρ = 0.07 kg/m³

For $(U_g - U_d)$ = 3000 m/s limiting diameter = 0.15 micron

For $(U_g - U_d)$ = 400 m/s limiting diameter = 10 micron

Figure 6.

tained within a duct of the same dimensions as the test section of the pulsed facility and using CO_2 injection.

The airstream mach number is 2. A single sampling station at 20cm downstream of the injector was used and a traverse across the flow was carried out using a transmitting laser beam and Rayleigh scattering method to measure concentrations.

The results from this comparison are shown in Figure 4. It can be seen that the jet spread included angle for the wedge device is around 8 deg, whereas for the acoustic reflector it is around 13 deg.

4. Consideration of an Alternative Mixing Strategy

The use of mixing devices in the flow to create turbulence inevitably raises concerns over the shock losses which can result and encourage the development of side wall injection strategies with minimum intrusion.

As part of the program of enquiry into fuel mixing in supersonic air streams, consideration has been given to an alternative dispersion scheme which adopts such a side wall injection scheme and uncouples the mixing process from the turbulent generation by using liquid injection as the means to propel the fuel into the body of the combustor duct.

Liquid hydrogen is presented in the form of a spray, and in this form is potentially attractive due to the fact that the momentum per unit volume is around 100 times larger than for gaseous fuel, and therefore may permit penetration from the side wall injector station to the centre of the combustor air stream.

As with perpendicular gaseous injection, such a station presents no obstruction in itself to the mainstream gas flow and has therefore the attraction of potentially lower shock pressure losses. The study has been a computational one, and to examine the injection process, a simple geometry has been chosen and is shown in Figure 5. It consists of a 2D duct 1000 mm long and 100 mm wide. At the inlet, gas enters at 3000m/s and 1500°K corresponding to Mach 4. Droplet injection takes place at the inlet plane and on the lower duct surface.

The main difficulty with liquid injection is likely to arise through the breakup processes in encountering the main stream. This is governed by the Weber numbers of the droplets, which are defined in Figure 6. For We less than 50, droplets are considered to be stable to further breakup. For the conditions in the chamber pertaining to

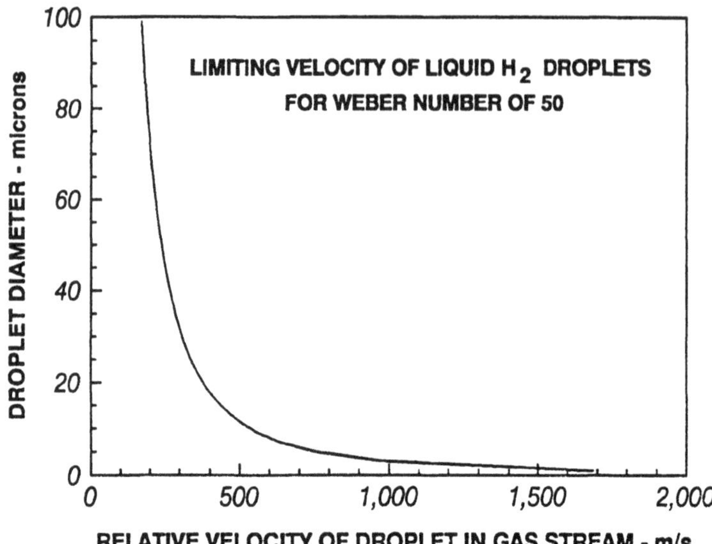

Figure 7.

DROPLET HEAT AND MASS TRANSFER

Heat balance for particle heating

$$m_p \, c_p \frac{d\,T_p}{d\,t} = h\,SA\,(T_\infty - T_p)$$

Heat transfer coefficient from

$$\frac{h\,D_p}{k_\infty} = 2.0 + 0.6\,Re_p^{0.5}\,Pr^{1/3}$$

$$Re = \frac{\rho\,D_p\,|\,V_p - V_\infty\,|}{\mu}$$

Droplet vaporisation rate

$$N = k_c\,(C_s - C_\infty)$$

Mass transfer coefficient from

$$\frac{k_c\,D_p}{D_{ab}} = 2.0 + 0.6\,Re_p^{0.5}\,Sc^{1/3}$$

Figure 8.

the likely flight regime, the gas density is around 0.07 Kg/m^3 and for liquid hydrogen $\sigma = 0.002$ N/m. Hence it can be seen that for low injection velocities perpendicular to the stream, $(U_g - U_d) = 3000$m/s and the maximum diameters which can survive are 0.15 micron. If however, mechanical means can be found to increase the injection velocity in the axial flow direction substantially, then the reduction in relative velocities will enable larger droplets to survive. For example, for $(U_g - U_d) = 400$m/s, the critical velocity becomes 10 micron. More generally, the limiting diameter versus relative velocity is given in Figure 7.

As an investigation of the likely benefits of the approach, it is proposed that droplets are injected with an axial velocity equal to the free gas stream velocity and a small transverse velocity. Although this is an ambitious demand in engineering terms, it is shown that the resulting benefits are considerable in terms of the resulting mixing lengths, and therefore merit a study of the means by which such injections may be achieved.

Since the duct length/height ratio is 10/1, and it is only required that droplets travel laterally to around 1/2 the duct height, then transverse velocities need only be around 1/20th of the axial value. For all cases, the transverse component at injection is taken as 200 m/s. This gives Weber limited diameters as high as 60 microns.

5. Integration with Fluid Dynamics

To assess the processes of heat, mass and momentum exchange, use has been made of the Fluent code to study the behavior of droplet trajectories and evaporated fuel.

The Fluent code has been described at length elsewhere and only the aspects particularly relevant to the current problem are detailed. The code enables particles and droplets to be injected from any location within the solution space. A large number of injections can take place simultaneously, and particular characteristics of mass flow rate and axial and transverse velocity can be assigned to each injection subgroup. In each subgroup, sizes and mass rates can be specified, or alternatively, a complete group can be defined by an overall mass flow and size distribution, e.g., Rosin Rammler.

Solution of the flow field is carried out with full two way coupling between droplets and airstream, with appropriate laws for drag and convective heat and mass transfer between phases.

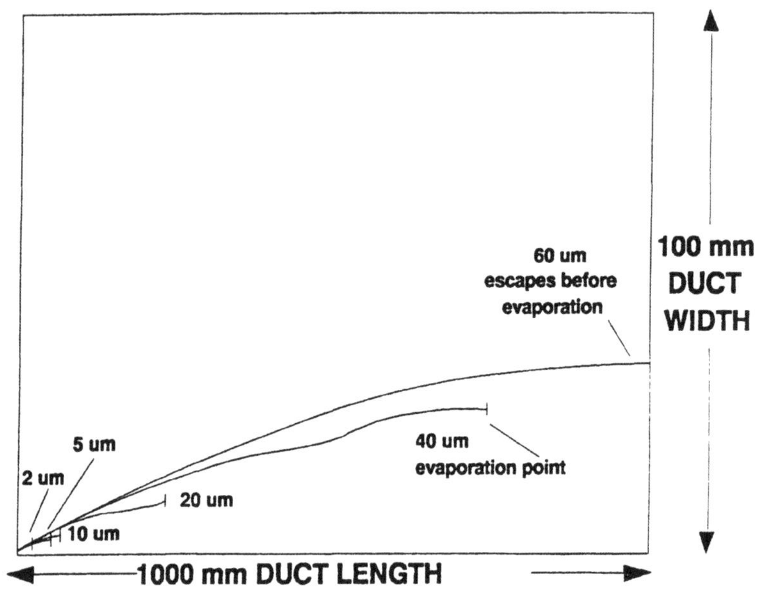

Figure 9.

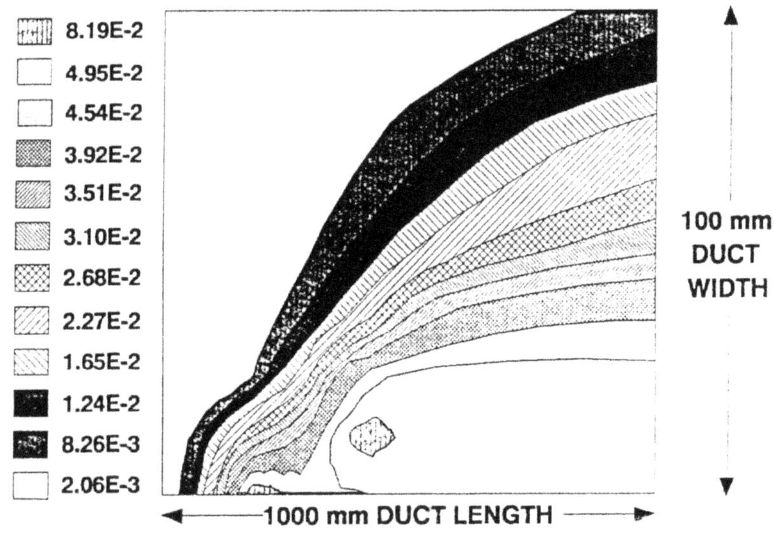

Figure 10.

These relations are summarized in Figure 8. In addition, a droplet heating scheme is employed, which recognizes boiling, and incorporates a four parameter vapor pressure curve vs temperature. Results of the calculation provide trajectories of each size group and indicate where complete evaporation has taken place. Mass fraction of evaporated fuel is provided for each cell and can be presented as contours or raster plots. Some properties of liquid hydrogen are collected together in Table 1 below.

Table 1. Some properties of liquid hydrogen.

Freezing point °K	13.8
Boiling point °K	20.3
Solid density Kg/m^3	86.5
Liquid density Kg/m^3	70.8
Liquid surface tension N/m	0.002
Viscosity of liquid g/cmsec	0.000132

6. Results

To identify the behavior of different sizes, separate calculations are performed for a range monosize droplet injections. Monosize groups range from 2 micron up to 60 micron and for each case injected mass flow rates are such as to provide a stoichiometric mixture with the incoming air stream if completely mixed. Trajectories are shown for each case and have been collected together in Figure 9. In this figure and those to follow, the transverse direction has been magnified for presentation.

It can be seen that for the 60 micron droplet case, some liquid material is still present at the end of the 1000mm duct and escapes. For the other droplets, complete evaporation is indicated by the vertical bar, and occurs progressively earlier as the droplet size decreases. Eventually, for the 2 and 5 micron sizes, evaporation occurs almost immediately, reducing the possibility for such sizes to contribute significantly to the centre of the duct.

The mass fractions of H_2 vapor are shown in Figures 10 to 15 for the same cases. The good penetration of the 60 micron droplets is reflected in relatively high concentrations even beyond the centerline

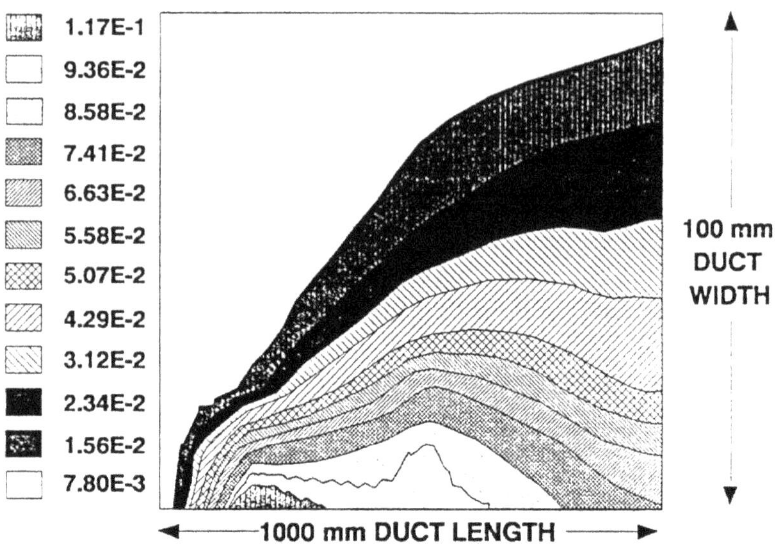

Figure 11.

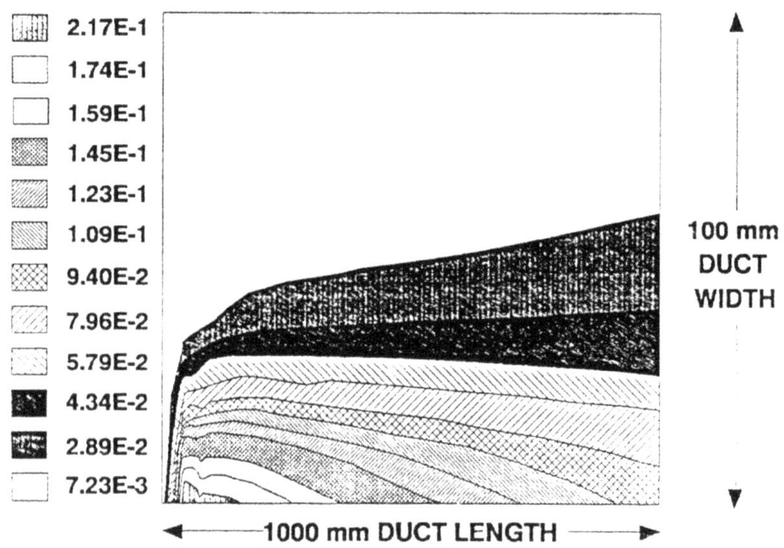

Figure 12.

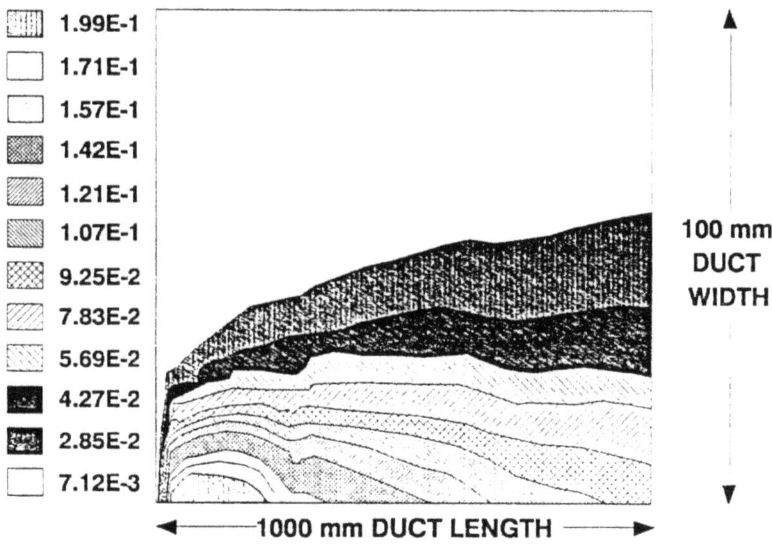

Figure 13.

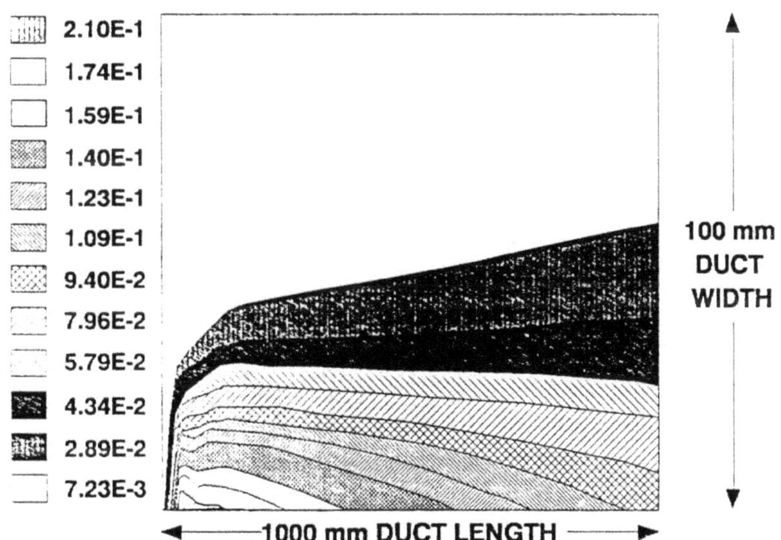

Figure 14.

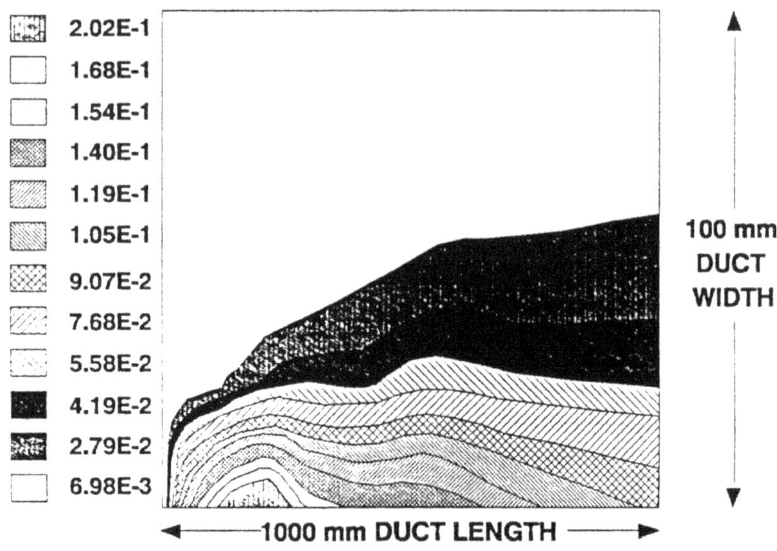

Figure 15.

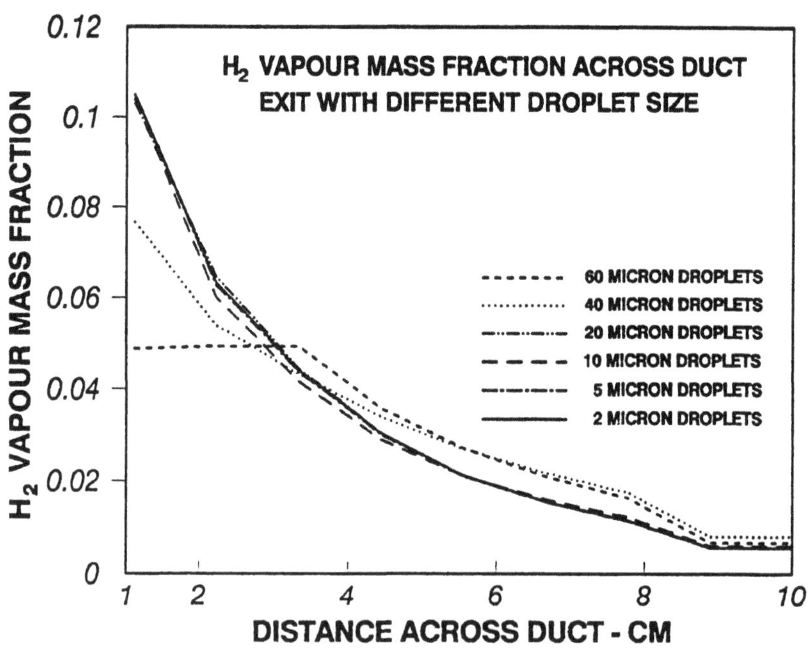

Figure 16.

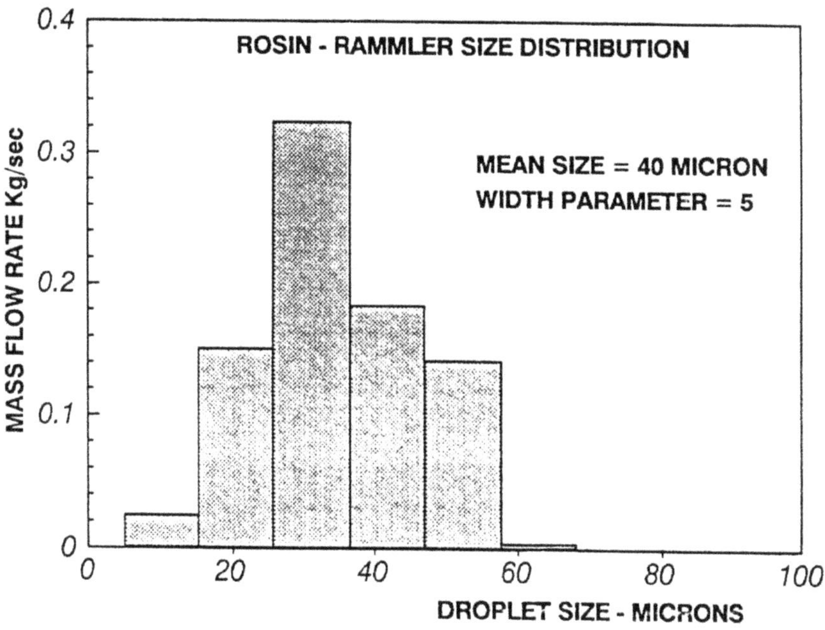

Figure 17.

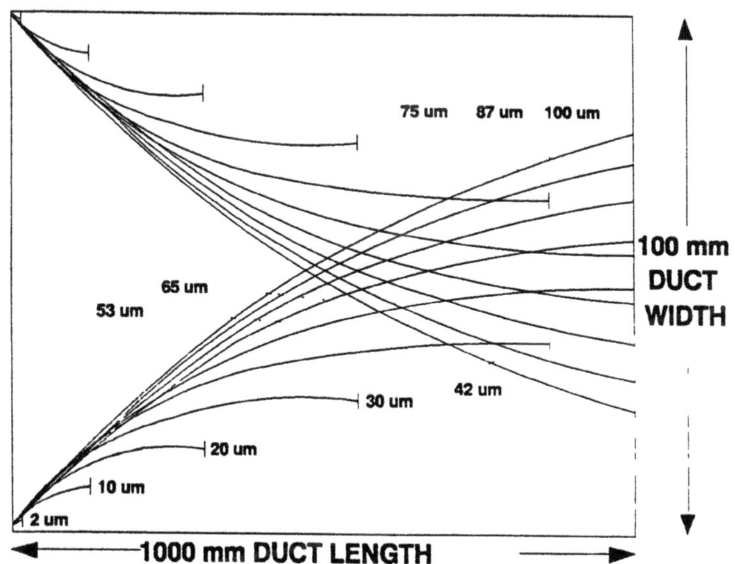

Figure 18.

MASS FRACTION CONTOURS OF HYDROGEN VAPOUR IN AIR
FOR UPPER AND LOWER WALL INJECTION WITH DISTRIBUTION
4.64E-2
3.72E-2
3.41E-2
3.10E-2
2.63E-2
2.32E-2
2.01E-2
1.70E-2
1.30E-2
9.3E-3
6.19E-3
3.10E-3
100 mm
DUCT
WIDTH
1000 mm DUCT LENGTH

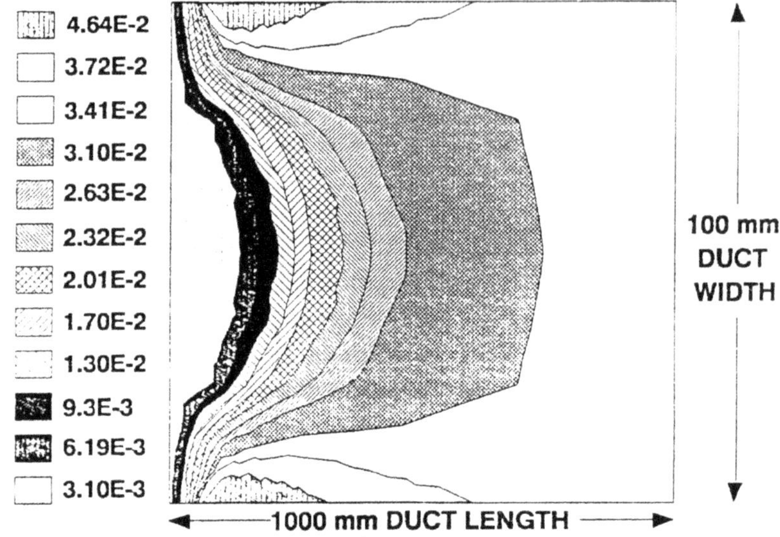

Figure 19.

at the exit plane and this is up to 0.8% at the upper wall. As expected, as one decreases in injected drop size, the mass fractions at the exit plane become more concentrated towards the lower face from which the injection arises. Below 20 micron however, the differences become less pronounced and appear to be accounted for principally by diffusion from the near lower wall region. The results for these test cases at the exit plane of the duct are collected together in Figure 16, which shows the variation of evaporated hydrogen mass fraction across the width.

These initial results are useful in that they indicate that the necessary dispersion can arise providing that a sufficient proportion of the fuel mass is supplied as droplets in the 40 - 60 micron range, in such a way that they can survive without breakup.

7. Effect of Size Distribution

In any real injection situation, a size distribution will be present, and the possibility exists for tailoring this to a suitable shape. For the purpose of illustration, an example of a Rosin Rammler size distribution has been computed. The function is given by

$$W = 1 - EXP\left[-\left(\frac{X}{\overline{X}}\right)^n\right] \tag{1}$$

where W is the fraction of weight below X.

For the example chosen $\overline{X} = 40$ micron and the width parameter $n = 5$. This distribution is shown in Figure 17 as the mass flow rate in each size group. Since it is intended in a real injection geometry to provide the load between upper as well as lower injection sites, it is expected that this may result in the mass fraction in the centre duct region receiving complementary contributions from each wall injection resulting in a more uniform profile.

The trajectory pattern for this injection case is shown in Figure 18, and the resulting H_2 vapor mass fraction is shown in Figure 19. This shows the exit plane profile completely uniform at 3.4% and shows complete mixing.

This is precisely the desired result, and it can further be seen that uniformity is achieved before the end of the duct corresponding to a length of around 7.5 duct heights.

This form of fuel dispersion relies on the possibility of being able to inject the fuel with very high axial velocities corresponding to the free stream velocity.

The manner of achieving this has not yet been established. However it should be noted that high velocity and supersonic liquid jets are used in other technologies such as rock cutting. The high pressure pumps needed for this purpose are already available and have operating pressures routinely of 4000 bar. It has been established that more specialized liquid pumps are available with operating pressures up to 15,000 bar, whereas the kinetic energy of a liquid hydrogen jet at 3000 m/s corresponds to a pressure drop of 3000 bar. It would appear from these feasibility calculations that this form of fuel injection can solve many of the problems which persist with gaseous injection and turbulent mixing.

References

Drummond, J. P., Carpenter, M. H., Riggins, D. W., and Adams, M. S., 1989. "Mixing enhancement in a supersonic combustor," AIAA 25th Joint Propulsion Conference, Monterey.

Gilreath, H. E. and Sullins, G. A., 1989. "Investigation of the mixing of parallel supersonic streams," Paper 7060, 9th ISABE Symposium, Athens.

Swithenbank, J., Eames, I., Chin, S., Ewan, B., Yang, Z., Cao, J., and Zhao, X., 1989. "Turbulence mixing in supersonic combustion systems," AIAA Paper 89-0260.

REACTING FREE SHEAR LAYERS

REACTING COMPRESSIBLE MIXING LAYERS: STRUCTURE AND STABILITY

C.E. Grosch[1]

Old Dominion University
Norfolk, Virginia 23529

ABSTRACT

Understanding the structure and the stability characteristics of a reacting compressible mixing layer is of fundamental importance. Also, this flow can regarded as the simplest relevant model of the combustion process in the scramjet. The theory describing the structure and stability of this flow is reviewed. This includes the structure of the mean flow and the combustion model, both of which determine the stability characteristics. Among the subjects included in the review of the stability characteristics are: the eigenvalue spectrum, convective Mach number, growth rates, and the transition from convective to absolute instability. Comparisions to experimental and numerical simulation results are made where possible.

1. Introduction

Understanding the stability characteristics of reacting compressible free shear flows is of fundamental importance (Jackson, 1992) and may have possible usefulness in the development of the scramjet engine (Beach, 1992). As discussed by Drummond and Mukunda (1988), the scramjet combustor flow is complex but spatially developing and reacting compressible mixing layers of fuel and oxidizer provide the simplest relevant model. In modeling this flow it is assumed that initially the fuel and oxidizer (and any non-reactive components) are in two unmixed co-flowing streams. Mixing of the two gases takes place in the shear layer between the streams and combustion occurs when there is both sufficient fuel and oxidizer present at the same

[1]Supported in part by the Air Force Office of Scientific Research under Contract No. 91-0250 and in part by the National Aeronautics and Space Administration under NASA Contract No. NAS1-18605 while in residence at the Institute for Computer Applications in Science and Engineering (ICASE), NASA Langley Research Center, Hampton,VA 23665.

J. Buckmaster et al. (eds.), Combustion in High-Speed Flows, 131–190.

point. The residence time of the fuel and oxidizer in the combustion chamber can be very short; therefore, it is extremely important that a high mixing rate be achieved so that complete combustion is attained before the fuel is convected out of the engine. Mixing is enhanced when the basic flow is unstable. Therefore knowledge of the flow's stability characteristics may improve the understanding of the mixing process in the scramjet and is also of fundamental interest in its own right.

It has been realized for some time that the problem of a very short residence time is compounded by the experimental observations that the mixing rates of shear layers decrease as the Mach number increases from zero (e.g., Brown and Roshko, 1974; Chinzei, Masuya, Komuro, Murakami, and Kudou, 1986; Papamoschou and Roshko, 1986, 1988; and Clemens, 1992). Numerical simulations of nonreacting compressible mixing layers (e.g., Guirguis, 1988; Lele, 1989; and Sandham and Reynolds, 1990) as well as reacting compressible mixing layers (Drummond and Mukunda, 1988; Drummond, Carpenter, Riggins, and Adams, 1989 for example) have also shown the same effect. It can be surmised that the initial growth of the mixing layer, formed by two turbulent boundary layers coming off a splitter plate, is entirely governed by entrainment due to the turbulent flow already present in the boundary layers. The scale of this motion is that of the turbulent boundary layers on the splitter plate. Further downstream inviscid Kelvin-Helmholtz instabilities, with scales much greater than that of the original boundary layer, appear. These instability waves lead to the large "rollers" observed at low Mach numbers. These, in turn, entrain fluid, pair, and evolve toward the large scale structures which are observed further downstream. Observation and simulations suggest that, while the initial entrainment continues, the long wavelength Kelvin-Helmholtz instability is suppressed at higher Mach numbers. Because of this increased stability, natural transition to turbulence may be delayed and occur at downstream distances which are larger than practical combustor lengths resulting in incomplete combustion.

The dynamical processes governing reacting compressible flows are very complex, involving strong interaction between the chemical and fluid dynamical effects. Apart from from a full numerical simulation of the chemistry and compressible fluid dynamics, including possibly turbulence modeling or large eddy simulation, any investigation of these flows must involve considerable modeling. This will, of necessity, rely heavily on the success achieved in modeling compressible non-reacting shear flows.

There is a very extensive literature dealing with stability and related topics for incompressible mixing layers (see Ho and Huerre (1984) for a comprehensive review) but, until quite recently, comparatively little dealing with even the non-reacting compressible flow. Earlier work on this problem has been reviewed by Jackson and Grosch (1989) as part of a comprehensive study of the stability of the non-reacting two dimensional compressible mixing layer. This study was extended to include the effect of variation in the thermodynamic properties on the stability characteristics (Jackson and Grosch, 1991) as well as the effect of three dimensionality of the mean flow due to skewing of the fast and slow streams (Grosch and Jackson, 1991a).

The structure and stability of the reacting flow is, of course, determined in part by the dynamics of the underlying non-reacting flow, in part by the reaction dynamics, and in part by the interaction of these two. In order to understand the dynamics of the reacting flow it is necessary to understand those of the non-reacting flow. Therefore in each section the results for the reacting flow are compared to those of the non-reacting flow.

In section 2 the structure of the mean flow is discussed. Section 3 contains a review of what is known of the stability of the compressible reacting mixing layer including, formulation of the problem (Section 3.1), the spectrum of the neutral waves (Section 3.2), the growth rates of unstable modes (Section 3.3), the convective Mach number (Section 3.4), and convective/absolute instabilities (Section 3.5). Finally, Section 4 contains concluding remarks.

2. Mean Flow: Ignition and Structure

A schematic of a reacting compressible mixing layer formed behind a splitter plate is shown in Figure 1. This shows the three regimes of ignition, deflagration, and diffusion flame (Linan and Crespo, 1976) which are expected to exist in this flow. The ignition regime is a region where the combustible gases mix until, at some finite distance downstream of the plate, a thermal explosion occurs and the gas is ignited. The second regime is the deflagration region. After ignition, a pair of well-defined deflagration waves (or "premixed flamelets") emerge according to classical thermal explosion theory. One of the flamelets is fuel-rich and the other is fuel-lean. There is

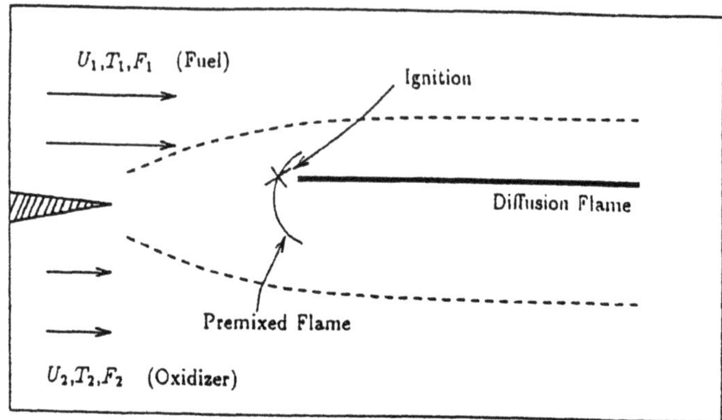

Figure 1. Schematic showing the reacting mixing layer, with the adiabatic flame temperature greater than T_1 and T_2.

excess fuel behind the fuel-rich flamelet and excess oxidizer behind the fuel-lean flamelet. Concentration gradients behind the flamelets drive the unburnt fuel and oxidizer towards the diffusion flame where they are consumed. These waves penetrate the mixing layer until all of the deficient reactant is consumed. Just downstream of the deflagration wave, a diffusion flame regime exists where the mixing process is governed by diffusion in the direction normal to the flame.

Theoretical analysis of the ignition and structure of the reacting compressible mixing layer has been confined to steady two dimensional flows with a zero pressure gradient. Further assuming that the Reynolds number is large but that the flow can be modeled by a laminar flow, the basic equations are the equation of state for a perfect gas, the compressible boundary layer equations with a heat source term in the energy equation modeling the reaction, and the species mass fraction conservation equations.

$$1 = \rho T, \tag{1}$$

$$(\rho U)_x + (\rho V)_y = 0, \tag{2}$$

$$\rho(UU_x + VU_y) = (\mu U_y)_y, \tag{3}$$

$$\rho(UT_x + VT_y) = Pr^{-1}(\mu T_y)_y + (\gamma - 1)M^2 \mu U_y^2 + \beta\,\Omega, \tag{4}$$

$$\rho(UF_{j,x} + VF_{j,y}) = Sc_j^{-1}(\mu F_{j,y})_y - \Omega_j. \tag{5}$$

These equations are nondimensionalized by the freestream values $T(\infty)$, $\rho(\infty)$, $U(\infty)$, $F_1(\infty)$ for the temperature, density, velocities and mass fractions, respectively, with lengths referred to some characteristic length scale of the flow. The x axis is along the direction of flow; the y axis is normal to the flow; U and V are the velocity components in the x and y directions, respectively; ρ is the density; T is the temperature; and $\{F_j\}$ are the mass fractions. The viscosity μ is assumed to be a function of temperature. The nondimensional parameters are the Prandtl number Pr, the Schmidt number $Sc_j = Pr\,Le_j$ for species j where Le_j is the Lewis number for species j, the Mach number $M = U(\infty)/a(\infty)$, with $a(\infty)$ the sound speed at $+\infty$, γ the specific-heats ratio, and β the heat release per unit mass fraction of the reactant. Finally, Ω is the reaction source term and $\{\Omega_j\}$ the species consumption rates. In the calculations of the mean flow and the stability calculations various models for the thermodynamic properties of the mean flow were used: (a) the Tanh model - the hyperbolic tangent profile for the mean speed and the Crocco relation for the mean temperature, with the Chapman viscosity-temperature relation and a Prandtl number of one; (b) the Lock model - the Lock profile for the mean speed and the Crocco relation for the mean temperature, with the Chapman viscosity-temperature relation and a Prandtl number of one; and (c) the Sutherland model - the similarity solution for the coupled velocity and temperature equations using the Sutherland viscosity temperature relation and arbitrary but constant Prandtl number.

A key element in the calculation is the specification of the combustion model, ie the form of Ω. Grosch and Jackson (1991b) assumed a one step irreversible reaction of the Arrhenius type between the fuel, F, (with mass fraction F_1) and the oxidizer, O, (with mass fraction F_2).

$$F + O \rightarrow P \tag{6}$$

Taking

$$\Omega = D\,\rho\,F_1\,F_2\,e^{-Ze/T}. \tag{7}$$

and

$$\Omega_1 = \Omega_2 = \Omega \tag{8}$$

with D the Damkohler number and the Zeldovich number $Ze = E/RT_\infty$ where E is the dimensional activation energy and R the universal gas constant. It is believed that this model will yield correct qualitative results but that there can be quantitative differences with the results of calculations using more accurate rate equation models. In any compressible flow calculations it is also necessary to specify the property variations with temperature and pressure. Because these authors used a simplified combustion model they also chose to use simple property variations: the Prandtl number constant and equal to one and a linear variation of the viscosity coefficient with temperature.

An appropriate set of initial (in x) and boundary (in y) conditions for the flow configuration of Figure 1 consistent with these equations is:

$$T = U = F_1 = 1, \quad F_2 = 0, \tag{9}$$

for $x = 0,\ y > 0$, and $x > 0,\ y \rightarrow \infty$.

$$T = \beta_T,\ U = \beta_U < 1,\ F_1 = 0,\ F_2 = \frac{F_2(-\infty)}{F_1(\infty)} \equiv \phi^{-1}, \tag{10}$$

for $x = 0$, $y < 0$ and $x > 0$, $y \to -\infty$. Here ϕ is the equivalence ratio. If $\phi = 1$, the mixture is stoichiometric, if $\phi > 1$ it is fuel rich, and if $\phi < 1$ it is fuel lean. If β_T is less than one, the gas in the slow stream is relatively cold compared to that in the fast stream, and if β_T is greater than one it is relatively hot.

Grosch and Jackson (1991b) solved these equations by (1) numerically marching downstream in x (note that the equations are parabolic in x) and (2) by using a combination of large activation energy asymptotics and numerics to analyze the ignition and diffusion flame regimes. The numerical solution of these equations was facilitated by first transforming them into the incompressible form by means of the Howarth-Dorodnitzyn transformation

$$Y = \int_0^y \rho \, dy, \qquad \hat{V} = \rho V + U \int_0^y \rho_x \, dy. \tag{11}$$

Solutions to the transformed equations were found by marching in x subject to the initial and boundary conditions (9 and 10) and it was found that the velocity profile, U, attained a self-similar form at a small value of x (Drummond and Mukunda, (1988) found that the solution of the Navier-Stokes equations also became self similar at small x). Because the solutions were self-similar, the independent variable Y was transformed to the similarity variable for the chemically frozen heat conduction problem

$$\eta = \frac{Y}{2\sqrt{x}}, \tag{12}$$

and all of the numerical calculations were carried out in the (x, η) variables.

The asymptotic analysis in the ignition regime was complicated by the necessity to consider a number of different cases depending on the magnitude of the dimensionless speed, β_U, and temperature, β_T, of the slow stream, and the Mach number, M. The cases analyzed were: (a) $|1 - \beta_T| \ll 1$, $|1 - \beta_U| \ll 1$, and $M = O(1)$, which is the case of ignition with nearly equal free stream temperatures and speeds (previously studied by Jackson and Hussaini (1988)); (b) $|1 - \beta_T| \ll 1$, $\beta_U < 1$, and $M \ll 1$, corresponding to ignition in a shear flow at nearly equal free stream temperatures at small Mach number; and (c) $\beta_T > 0$, $0 \leq \beta_U < 1$, and $M > 0$.

The results are presented and compared in Figures 2 to 7. Figure 2 is a plot of the maximum temperature in the shear layer as a

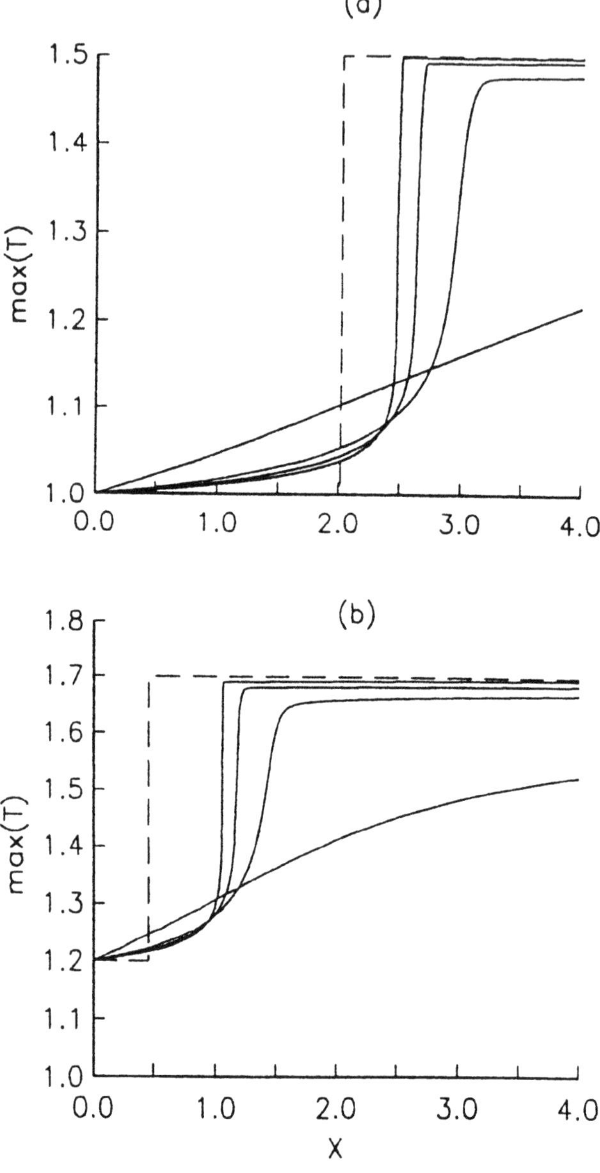

Figure 2. Plot of the maximum temperature in η versus x for $\beta_U = 0$, $\beta_T = \beta = \phi = 1$ with (a) $M = 0$, and (b) $M = 2$. In both graphs the Zeldovich number increases from right to left with $Ze = 10, 30, 40$, and 50. The dashed line corresponds to the solution in the infinite Zeldovich limit.

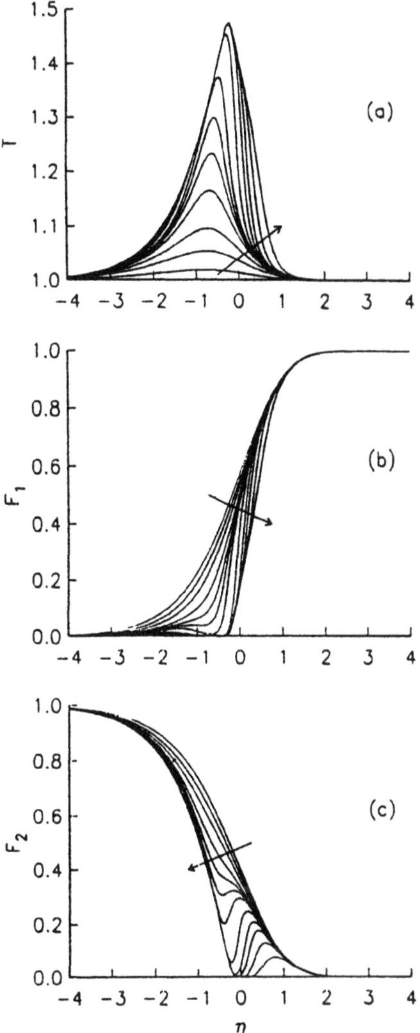

Figure 3. Plot of (a) temperature, (b) F_1 mass fraction, (c) F_2 mass fraction versus η for $M = 0$, $\beta_U = 0$, $\beta_T = \beta = \phi = 1$, $Ze = 30$. Here, the arrow denotes increasing x, with $x = 1, 2, 2.5, 2.8, 2.9, 2.95, 3, 3.1, 3.2, 3.3, 3.5, 4$.

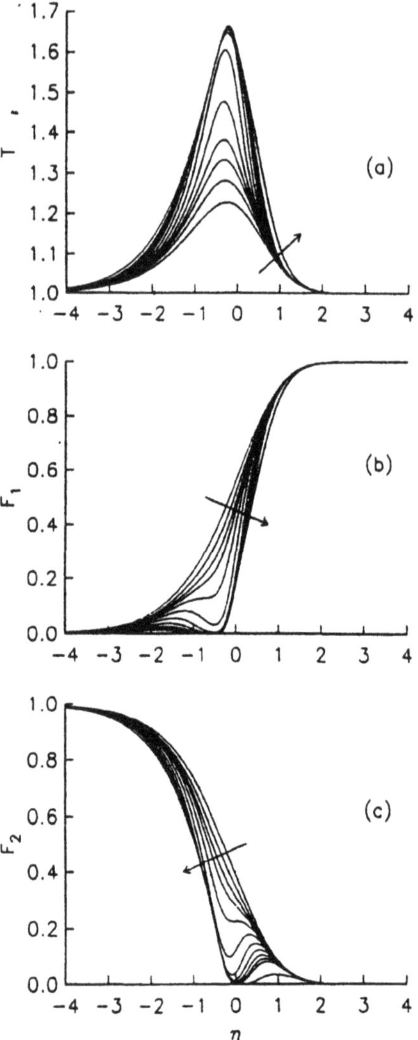

Figure 4. Plot of (a) temperature, (b) F_1 mass fraction, (c) F_2 mass fraction versus η for $M = 2$, $\beta_U = 0$, $\beta_T = \beta = \phi = 1$, $Ze = 30$. Here, the arrow denotes increasing x, with $x = 0.5$, 1, 1.2, 1.3, 1.4, 1.5, 1.6, 1.7, 1.8, 1.9, 2, 3.

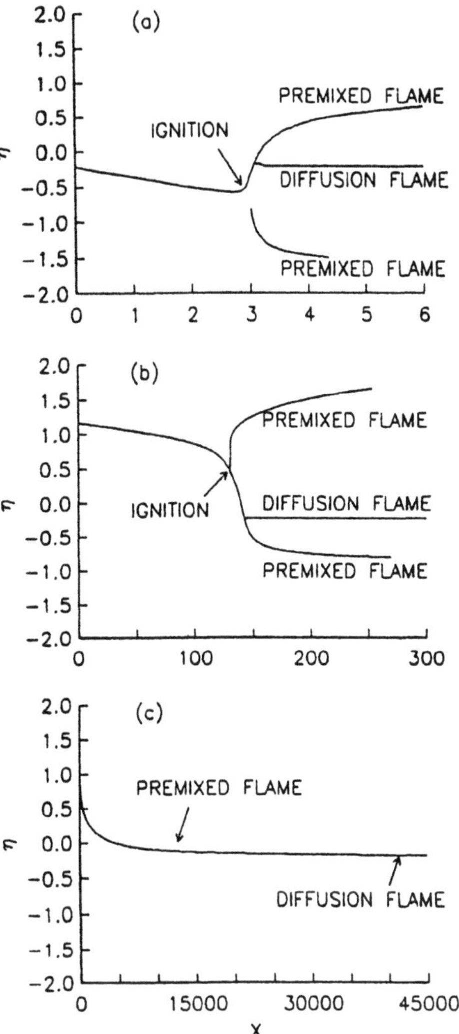

Figure 5. Plot of the loci in the (x, η) plane of the maxima of Ω for $M = \beta_U = 0$, $\phi = 1$, $Ze = 30$, and (a) $\beta_T = \beta = 1$; (b) $\beta_T = 0.5$, $\beta = 1.5$; and (c) $\beta_T = 0.5$, $\beta = 0.4$.

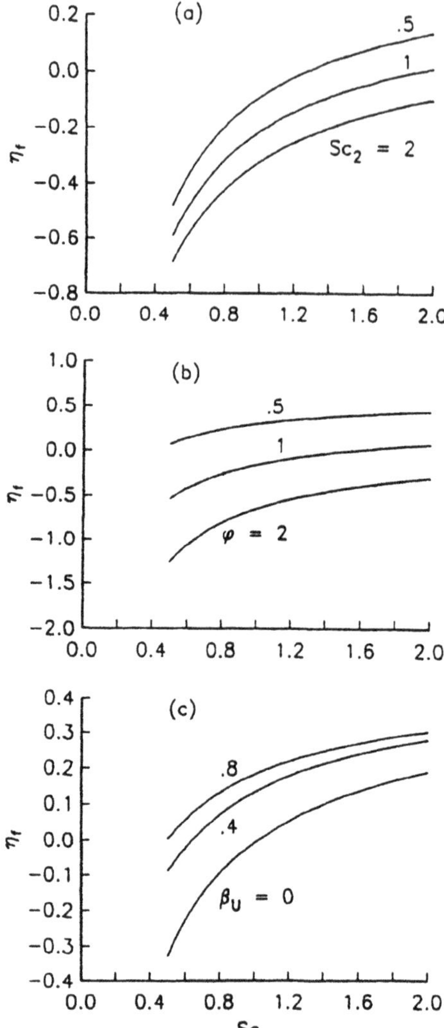

Figure 6. Plot of the flame location η_f versus Sc_1: (a) $\beta_U = 0$, $\phi = 1$, $Sc_2 = .5, 1, 2$; (b) $\beta_U = 0$, $\phi = .5, 1, 2$, $Sc_2 = .7$; (c) $\beta_U = 0, .4, .8$, $\phi = .8$, $Sc_2 = .7$.

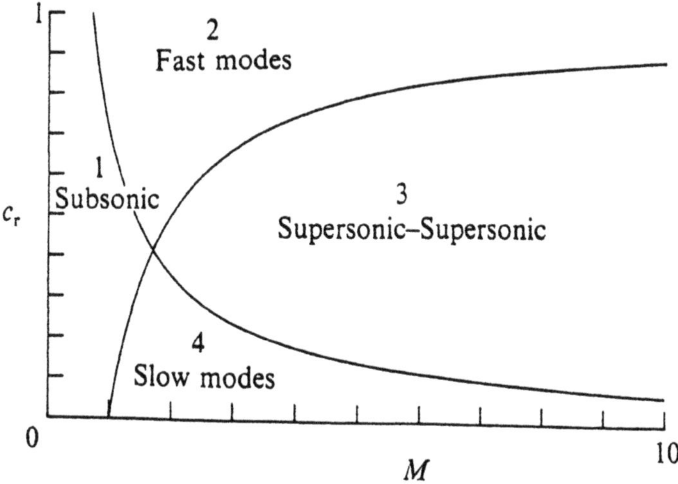

Figure 7. Plots of the sonic speeds $c_{\pm}$ versus Mach number for $\beta_T = 0.5$, $\beta_U = 0$, and $\theta = 0^o$.

function of the downstream position at $M = 0$ and $M = 2$ for various values of Zeldovich number (Grosch and Jackson, 1991b). At $x = 0$ the temperature is that of the inert solution and for large x the temperature is that of the diffusion flame. It is clear that for the smaller value value of the Zeldovich number ($Ze = 10$) a well defined ignition point does not occur; instead at both $M = 0$ and $M = 2$ there is a smooth and gradual transition from the inert solution at $x = 0$ to a diffusion flame. At the larger values of Ze there is a rapid transition from the inert solution to the diffusion flame solution, with the curves steepening as the Zeldovich number is increased. As the Zeldovich number increases, the solution approaches the infinite Zeldovich solution (dashed line) obtained from the asymptotic analysis for the ignition point and the diffusion flame. From this figure it is apparent that the ignition regime also exists in supersonic flows. As the Mach number increases there is a corresponding increase in the inert temperature at $x = 0$ due to viscous heating, thus lessening the relative effect of combustion on the overall temperature field. Also note that ignition occurs at a smaller value of x for the Mach 2 case as compared to that at zero Mach number.

Figures 3 and 4 show the corresponding temperature and mass fraction profiles as a function of position in the shear layer at various x locations for Mach numbers of 0 and 2, respectively. The rapid rise in the temperature (Figures 3a and 4a) over a narrow range of x indicates ignition. The temperature peak shifts and the profile is asymmetric due to the asymmetry in the velocity profile. The mass fraction profiles show that there is a diffusion of F_1 from the $\eta \geq 0$ region into the $\eta \leq 0$ region with $F_1 = 0$ only at $\eta = -\infty$. At larger x there is a small secondary maximum in the F_1 distribution in $-3 < \eta < -1$ showing the presence of a premixed flamelet in this region. As first pointed out by Linan and Crespo (1976), these arise because the mixture is not stoichiometric in the premixed region. One of the reactants is consumed locally, leaving behind an excess of the other reactant. These premixed flamelets are quite weak in that the temperature rise associated with them is small. The distribution of F_2 is the converse of that of F_1.

The existence of the premixed flamelets and the diffusion flame is shown quite clearly in Figure 5a, which shows the loci in the (x, η) plane of the maxima of Ω (the chemical production term) for equal free stream temperatures ($\beta_T = 1.0$). The adiabatic diffusion flame temperature is 1.5 and is greater than the free stream temperature.

As shown in this figure, the position of the maximum decreases from about $\eta = -0.2$ at $x = 0$ to nearly $\eta = -0.6$ at $x \approx 2.9$. The maximum value of Ω increases along this curve. At $x \approx 2.9$ ignition occurs and two maxima appear giving rise to the premixed flamelets. Beyond the ignition point the premixed flamelets move outwards in the shear layer until all of the deficient reactant is consumed. The appearance of the third maxima just behind the ignition point marks the appearance of the diffusion flame. As x is further increased, the diffusion flame becomes dominant and, as $x \to \infty$, the diffusion flame thins and approaches a flame sheet characterized by local chemical equilibrium and described by the asymptotics. Figure 5b shows similar results for the same values of the parameters except that $\beta_T = 0.5$ and $\beta = 1.5$. This case corresponds to unequal freestream temperatures. The adiabatic flame temperature is again 1.5 and is larger than either freestream temperature. The ignition point has moved into the region of higher freestream temperature and the location of the diffusion flame is unchanged while that of the premixed flamelets has changed. Figure 5c shows results for $\beta_T = 0.5$ and $\beta = 0.4$. The adiabatic flame temperature in this case is 0.95 and is smaller than the freestream temperature at $+\infty$. In contrast to the two previous figures, there is no well defined ignition point; the premixed flame merges smoothly into the diffusion flame whose location is unchanged. In addition, the premixed flamelets are absent. Finally, it should be noted that the authors stated that the behaviour shown in Figure 5 also was found for Mach numbers greater than zero. Detailed predictions of location of the ignition point as a function of the flow parameters were also made via the large Ze asymptotic analysis. The results showed that the ignition point moves toward the origin as the Mach number is increased, in agreement with the numerical results.

From both the numerical and asymptotic results it is apparent that the ignition regime exists in supersonic as well as subsonic and incompressible flows. As the Mach number increases there is a corresponding increase in the inert temperature at $x = 0$ due to viscous heating, thus lessening the relative effect of combustion on the overall temperature field. Also note that ignition occurs at a smaller value of x for the Mach 2 case as compared to that at zero Mach number.

The diffusion flame regime was analyzed by considering the limit of infinite Damkohler number. Grosch and Jackson showed that solutions for the mean flow could be found in terms of certain integrals of the velocity profile. In the case of $Pr = Le = 1$ this solution reduced to the flame sheet solution (Jackson and Grosch, 1990b). When these parameters are not unity it was necessary to evaluate the results numerically. The location of the flame sheet as a function of Sc_1 for various values of β_U, the slow stream nondimensional speed, ϕ, the equivalence ratio, and Sc_2 is shown in Figure 6. As Sc_2 or ϕ is increased the flame sheet moves into the slower moving stream, but as β_U is increased it moves into the faster stream.

3. Stability of the Reacting Compressible Mixing Layer

Rapidly growing broadband instabilities will generally enhance mixing and thus promote rapid and complete combustion. The current state of knowledge of the stability of the reacting compressible mixing layer is reviewed here. The topics reviewed include: the spectrum of the neutral modes, the growth rates of the unstable modes, and the convective Mach number. Finally, the recent results on the transition from convective to absolute instability in this flow is reviewed.

3.1 Formulation of the Stability Problem

It is generally agreed that the stability of free shear layers, both incompressible and compressible, is dominated by inviscid dynamics. Thus the governing equations for the stability problem are the compressible Euler equations. In the reacting case the source terms for the temperature and mass fractions must also be included if the combustion is modeled by finite rate chemistry. In the flame sheet limit the perturbation does not affect the heat release in the sheet, it merely wrinkles the sheet. Therefore, the only effect the reaction has on the flow stability is through the change in the mean temperature distribution from that of the non-reacting flow. With finite rate chemistry, the perturbations not only wrinkle the combustion zone but also change the rate of heat release in the reaction through changes in the temperature and mass fraction distribution within the combustion zone. The change in heat release then effects the temperature and mass fraction distributions. This, in turn, affects the stability of the flow.

To date there have been very few studies of this problem (Jackson and Grosch, 1990b; Hu, Jackson, Lasseigne and Grosch, 1993;

Shin and Ferziger, 1990 and 1991; and Planche and Reynolds, 1991). The starting point for all of these studies is the compressible Euler equations with the source terms those for a one step irreversible reaction of the Arrhenius type (equations (6) - (8)). All of the published studies make the parallel flow assumption in the derivation of the stability equations. However, it is straightforward to apply a weakly nonparallel approach to this stability problem. In this approach one obtains the parallel flow equations as the first approximation and also finds the next order correction to the growth rate.

The equation for the amplitude of the pressure perturbation is

$$\Pi'' - [2U'/(U-c) + (1-K_1)\,T'/T]\,\Pi'$$

$$- \alpha^2\, T\,[T - K_2 M^2 \cos^2\theta\,(U-c)^2]\,\Pi = 0, \tag{13}$$

where

$$K_1 = J_3/J_1, \tag{14}$$

$$K_2 = \gamma - (\gamma - 1)(J_2/J_1), \tag{15}$$

and

$$J_1 = 1 + i\frac{(\beta\, Q_3 - Q_1 - Q_2)T}{\alpha \cos\theta\,(U-c)}, \tag{16}$$

$$J_2 = 1 - i\frac{(Q_1 + Q_2)T}{\alpha \cos\theta\,(U-c)}, \tag{17}$$

$$J_3 = 1 - i\frac{(Q_1 H_1' + Q_2 H_2')T}{\alpha \cos\theta\, T'\,(U-c)}, \tag{18}$$

$$Q_1 = \frac{\partial \Omega}{\partial F_1}, \quad Q_2 = \frac{\partial \Omega}{\partial F_2}, \quad Q_3 = \frac{\partial \Omega}{\partial T}, \tag{19}$$

and

$$H_j = T + \beta\, F_j, \tag{20}$$

with the primes indicating differentiation with respect to the similarity variable η. Here α is the wavenumber, θ is the direction of propagation of the disturbance wave in the $(x-z)$ plane, $c = \omega/\alpha$ is the complex phase speed and ω is the frequency. For spatial theory, ω is required to be real and solutions are sought for which α is complex. For temporal theory, α is assumed to be real and solutions are sought for which ω is complex. The amplification rates of the disturbances are then $-\alpha_i$ or ω_i, respectively. The disturbances are two dimensional for $\theta = 0^o$ and otherwise oblique.

If $\beta = 0$ it is easily seen that $K_1 = K_2 = 1$ and equation (13) reduces to the compressible Rayleigh equation governing the stability of the non-reacting flow (Jackson and Grosch, 1989). If a flame sheet model is used instead of a finite rate chemistry model the reaction is confined to a sheet of zero thickness and $\beta = 0$ outside of the sheet. Thus the stability equation is again (13) with $K_1 = K_2 = 1$ and, of course, the appropriate velocity and temperature distributions for the flame sheet model (Jackson and Grosch, 1990b; Hu, Jackson, Lasseigne, and Grosch, 1993).

The boundary conditions for Π are obtained by considering the limiting form of (13) as $\eta \to \pm\infty$ which gives

$$\Pi \to e^{(\pm \Delta_{\pm} \eta)}, \tag{21}$$

where

$$\Delta_{+}^{2} = \alpha^2[1 - M^2 cos^2\theta \, (1-c)^2], \tag{22}$$

$$\Delta_{-}^{2} = \alpha^2 \beta_T[\beta_T - M^2 cos^2\theta \, (\beta_U - c)^2]. \tag{23}$$

The values of the phase speed for which $\Delta_{\pm}^{2}$ vanishes are

$$c_{+} = 1 - \frac{1}{M cos\theta}, \tag{24}$$

$$c_{-} = \beta_U + \frac{\sqrt{\beta_T}}{M cos\theta}, \tag{25}$$

where c_{+} is the phase speed of a sonic disturbance in the fast stream and c_{-} is the phase speed of a sonic disturbance in the slow stream. For

$$M cos\theta = M_* = \frac{1+\sqrt{\beta_T}}{1-\beta_U} \tag{26}$$

$c_{\pm}$ are equal.

The nature of the disturbances for the stability problem can be illustrated by Figure 7 (Jackson and Grosch, 1989), which is a plot of $c_{\pm}$ versus M for $\beta_T = 0.5$, $\beta_U = 0$, and $\theta = 0^o$. These curves divide the phase speed-Mach number plane into four regions. If a neutral disturbance exists with a Mach number and phase speed in region 1, it is subsonic at both boundaries, and is classified as a subsonic neutral mode. In region 3, the neutral disturbance is supersonic at both boundaries, and is classified as a supersonic-supersonic neutral mode. In region 2, the neutral disturbance is subsonic in the fast stream and supersonic in the slow stream, and is classified as a fast neutral mode. Finally, in region 4, the neutral disturbance is supersonic in the fast stream and subsonic in the slow stream, and is classified as a slow neutral mode. For oblique modes ($\theta \neq 0^o$) the four regions still exist and only the boundaries, as defined by the $c_\pm$ curves in the phase speed - Mach number plane, are changed from those of the two dimensional modes (Grosch and Jackson, 1991a). Finally, it is important to note that the sonic speeds are independent of the reaction since the far field is chemically frozen. Thus the classification scheme does not depend on the reaction model used.

Because of causality there can be no incoming waves for an unbounded domain, unless the flow is being driven by an external source. Assuming that this is not the case, boundary conditions for an unbounded domain require that the far field solution be outgoing waves. The appropriate boundary condition for outgoing waves in the fast stream is,

$$\Pi \to e^{(-\Delta_+ \eta)}, \tag{27}$$

if $c_r > c_+$, and

$$\Pi \to e^{(-i\eta\sqrt{-\Delta_+^2})}, \tag{28}$$

if $c_r < c_+$. For the slow stream the appropriate boundary condition for outgoing waves is,

$$\Pi \to e^{(\Delta_- \eta)}, \tag{29}$$

if $c_r < c_-$, and

$$\Pi \to e^{(-i\eta\sqrt{-\Delta_-^2})}, \tag{30}$$

if $c_r > c_-$.

For a bounded domain the boundary condition is that the pressure gradient normal to the boundary is zero. Thus, on the boundaries

$$\hat{n} \cdot \nabla\Pi = 0, \tag{31}$$

with $\hat{n}$ the unit normal to the boundary.

3.2 The Spectrum of Neutral Waves

Both the non-reacting and reacting mixing layers have a complicated eigenvalue spectrum. The first step in finding and analyzing this spectrum is to find the neutral modes. For the subsonic modes, which lie in region 1 of the $c_r - M$ diagram, a theorem of Lees and Lin (1946) can be used. This result is derived from consideration of the equation governing the normal velocity perturbation $\hat{v}$. The disturbance equation for the normal velocity component is (Jackson, 1992)

$$(\xi\, e^H\, \hat{v}')' - e^H(q+\alpha^2)\,\hat{v} = 0, \tag{32}$$

where

$$\xi = T^{-1}(T - K_2\, M^2\,(U-c)^2)^{-1}, \tag{33}$$

$$H = -\int (1-K_1)\,(T'/T)\,d\eta, \tag{34}$$

and

$$q = [\xi\,(U' + (U-c)\,(1-K_1)\,(T'/T))]'/(U-c). \tag{35}$$

Note that (32) has a singularity at $U = c$. Define

$$S(\eta) \equiv \frac{d}{d\eta}(T^{-2}\frac{dU}{d\eta}). \tag{36}$$

Let $\tilde{c} = U(\eta_c)$, where η_c is a root of $S(\eta)$. If $\tilde{c}$ lies in region 1 of the $c_r - M$ diagram (Figure 7), then (Lees and Lin, 1946) $\tilde{c} \equiv c_N$ is the phase speed of a neutral mode provided that $\alpha \neq 0$. The corresponding neutral wave number and frequency must be determined numerically. These modes are the regular subsonic neutral modes.

In addition to the neutral modes with $\alpha_N \neq 0$ there may exist neutral modes having zero wavenumber. The phase speed of such modes do not satisfy (36) but can be found by an asymptotic analysis of (13) in the limit $\alpha \to 0$ (Hu, Jackson, Lasseigne, and Grosch, 1993). The result of this analysis is, for $M = \beta_U = 0$,

$$c_N = \frac{1 + i\,e^{B/2}}{1 + e^B}, \tag{37}$$

with

$$B = \int_{-\infty}^{\infty} \frac{(F_1+F_2)T' + \beta(F_1F_2)'}{(F_1+F_2)T - \beta(Ze-1)F_1F_2T'}\,d\eta. \tag{38}$$

In the nonreactive case ($\beta = 0$) this reduces to

$$c_N = \frac{\beta_T + i\sqrt{\beta_T}}{\beta_T + 1}, \tag{39}$$

which shows that the neutral phase speed is complex for $\alpha_N = 0$.

If $\tilde{c}$ lies in regions 2, 3, or 4 of the $c_r - M$ diagram, then $\tilde{c}$ does *not* correspond to the phase speed of a true neutral mode. The phase speed of the neutral modes in these regions must, in general, be found numerically. One exception is the supersonic-supersonic neutral modes with $\alpha = 0$ (Grosch, Jackson, Klein, Majda, and Papageorgiou, 1991). In this case an expansion of the solution in powers of α, along the lines previously used by Drazin and Howard (1962) and Blumen, Drazin and Billings (1975) in related studies, yields an eigenvalue relation which is analytically tractable.

The leading-order term in the expansion is independent of the detailed form of U and T, and only depends on the basic flow characteristics at infinity. This is to be expected from physical arguments because the wavelength of the instability in the limit $\alpha \to 0$ is much larger than the length scale over which the undisturbed flow is non-uniform. Setting the leading-order term in the expansion to zero yields an equation for $c \equiv c_N$:

$$\beta_T[M^2(\beta_U - c_N)^2 - \beta_T](1 - c_N)^4 = [M^2(1 - c_N)^2 - 1](\beta_U - c_N)^4. \quad (40)$$

This equation is identical to (5.3a) of Miles (1958) if his result is expressed in the notation used here. Miles showed (in this notation) that:

[1] A single real root of (40) exists for

$$M \geq M_* \equiv (1 + \sqrt{\beta_T})/(1 - \beta_U), \quad (41)$$

with phase speed

$$c_N = (\beta_U + \sqrt{\beta_T})/(1 + \sqrt{\beta_T}). \quad (42)$$

This is classified as a constant speed supersonic-supersonic neutral mode lying in region 3 of the $c_r - M$ plane. It is independent of Mach number and corresponds to the phase speed at which the sonic speeds in the two streams are equal. In this regime there is also a pair of complex conjugate eigenvalues corresponding to one unstable and one stable eigenmode. The associated instability is analogous to the classical Kelvin-Helmholtz instability for subsonic vortex sheets (Artola and Majda, 1987). This instability disappears as the Mach number increases.

[2] A double root first appears at

$$M_{CR} = (1 + \beta_T^{1/3})^{3/2}/(1 - \beta_U), \quad (43)$$

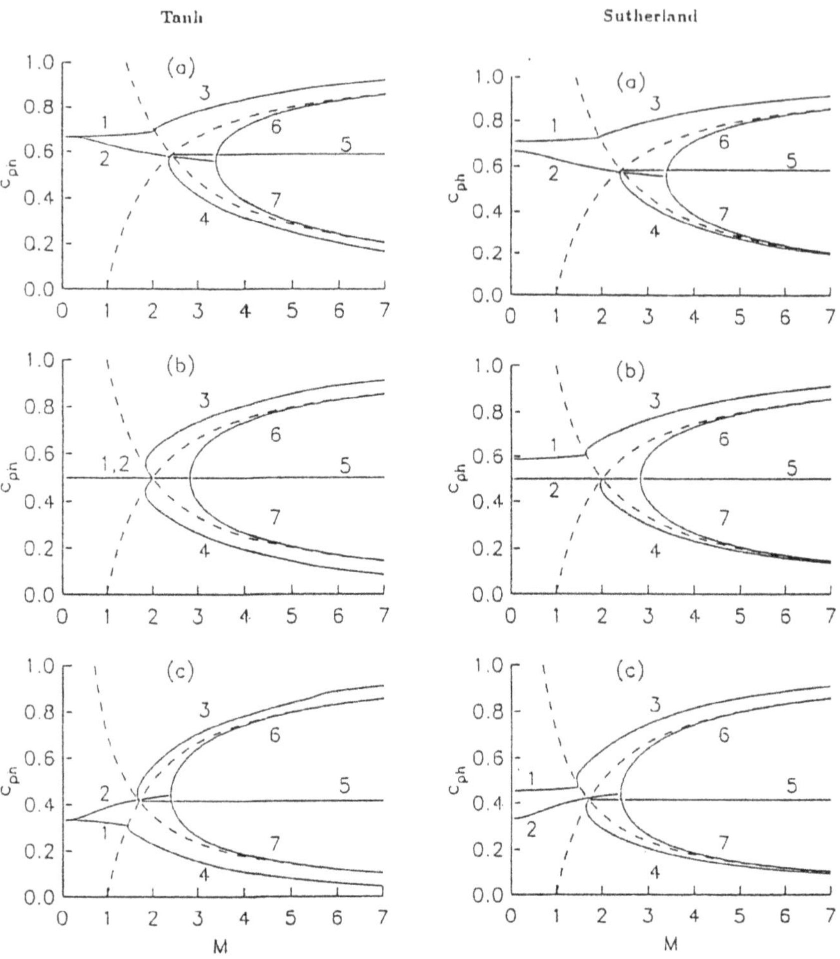

Figure 8. Plots of the neutral phase speeds as a function of Mach number for the Tanh model (left hand column) and Sutherland model (right hand column). For both $\beta_U = 0$, $\theta = 0^o$ and (a) $\beta_T = 2.0$, (b) $\beta_T = 1.0$, and (c) $\beta_T = 0.5$. The neutral mode classification is: (1) subsonic, $\alpha_N \neq 0$; (2) subsonic, $\alpha_N = 0$; (3) fast supersonic, $\alpha_N \neq 0$; (4) slow supersonic, $\alpha_N \neq 0$; (5) constant speed supersonic-supersonic, $\alpha_N = 0$; (6) fast supersonic-supersonic, $\alpha_N = 0$; and (7) slow supersonic-supersonic, $\alpha_N = 0$. The sonic curves are shown as dashed.

with phase speed

$$c_N = (\beta_U + \beta_T^{1/3})/(1 + \beta_T^{1/3}). \tag{44}$$

There are three distinct real roots for $M > M_{CR}$. One of these is the phase speed of the constant speed supersonic-supersonic neutral mode while the other two roots must be found numerically from (40). For the special case of $\beta_T = 1$, these roots are given by

$$c_N = \frac{1+\beta_U}{2} \pm \frac{1}{2M}[M^2(1-\beta_U)^2+4-4(M^2(1-\beta_U)^2+1)^{1/2}]^{1/2}. \tag{45}$$

The root which corresponds to the (+/−) sign is classified as a fast/slow supersonic-supersonic neutral mode. Note that all three of these neutral modes lie in region 3.

The neutral phase speeds given above are exact for $\alpha = 0$. In order to obtain the higher order corrections for $\alpha \neq 0$ the value of c must also be expanded in powers of α. When this was done it was found that the overall growth rate was $O(\alpha^2)$. (Balsa and Goldstein (1990) also found, numerically, the $O(\alpha^2)$ growth rate for these modes.) It was also found that the growth rate at $O(\alpha^2)$ becomes singular at M_{CR}. This singular behaviour was studied by expansions about the singular value of M. A connection between the regimes $M_* < M < M_{CR}$ and $M > M_{CR}$ was found and yielded the transition from a stable/unstable pair of eigenmodes plus a supersonic neutral mode for $M < M_{CR}$ to three supersonic neutral modes for $M > M_{CR}$.

The asymptotic expansion for the supersonic-supersonic modes gave the neutral curves in region 3. The other regions of the $c_r - M$ diagram were also investigated numerically using all three of the thermodynamic models (Grosch, et al, 1991). Representative results are presented in Figure 8 for the case of the Tanh and Sutherland profiles. Note the qualitatively similar results of these models. Plots of the phase speed c_N of the neutral modes as a function of the Mach number are shown in this figure.

In order to understand the variation of phase speed of the neutral modes with β_T and M it is important to recall (Jackson and Grosch, 1991) that, for each thermodynamic model, there exists a transition value of β_T, denoted by $\hat{\beta}_T$. For $\beta_T = \hat{\beta}_T$, the neutral mode phase speed is independent of M in region 1. For $\beta_T > (<)\ \hat{\beta}_T$, c_N is a monotonically increasing (decreasing) function of M in region 1.

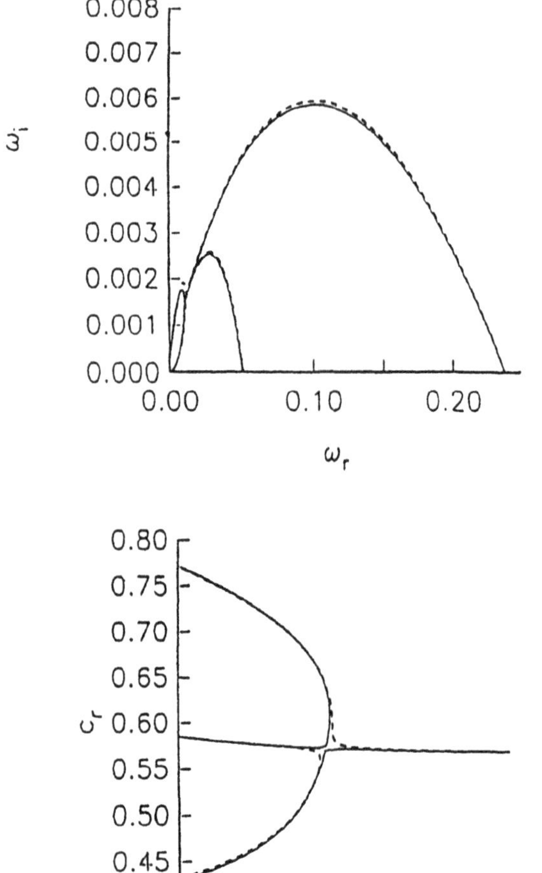

Figure 9. Plot of (a) ω_i versus ω_r and (b) c_r versus c_i with $\beta_U = 0$, $\beta_T = 2.0$ and (dashed) $M = 2.86$ and (solid) $M = 2.88$.

The value of $\hat{\beta_T}$ is 1.0, 0.57753, and 0.445 for the Tanh, Lock, and Sutherland ($Pr = 1$) profiles, respectively. For all β_T there are two subsonic neutral modes, labeled in this figure by 1 and 2. Mode 1 is that whose phase speed is found from the regularity condition (equation (36), above) and mode 2 is found numerically as $(\alpha, \omega) \to 0$ with $c \neq 0$. Because of the symmetry of the Tanh velocity and temperature profiles, the phase speeds of these two modes are identical at $M = 0$ for any β_T and also for all Mach numbers in region 1 when $\beta_T = 1$. In Figure 8a, $\beta_T = 2$ and is greater than the transition value for both models. Thus as the Mach number is increased, mode 1 is transformed into a fast supersonic mode (labeled 3 in Figure 8a) and mode 2 is transformed into a supersonic-supersonic mode. Mode 4 is the slow supersonic mode which appears at $M = M_*$. Mode 5 is the constant speed supersonic-supersonic mode which also appears at $M = M_*$. Modes 6 and 7 are the fast and slow supersonic-supersonic modes which exist for $M > M_{CR}$. Note that, with $\beta > \hat{\beta_T}$, mode 2 merges with mode 7, the slow supersonic-supersonic mode, at an M near M_{CR}. The phase speed curves of Figure 8b show the symmetry due to the Tanh profiles when $\beta_T = \hat{\beta_T} = 1$. Modes 1 and 2 coincide in region 1 as do modes 2 and 5 for $M_* \leq M \leq M_{CR}$ in region 3. The phase speeds of modes 1 and 2 for the Sutherland model do not coincide in region 1. The results shown in Figure 8c are for $\beta_T = 0.5 < \hat{\beta_T}$ for the Tanh model but $> \hat{\beta_T}$ for the Sutherland model. Because of this, mode 1 merges, with increasing Mach number, with mode 4, the slow supersonic mode, at an M near M_{CR} for the Tanh model. However, for the Sutherland model, mode 1 still merges with mode 3, the fast supersonic mode as the Mach number increases.

Coalescence and switching of a pair of unstable modes was found by Grosch, et al (1992) at other values of the parameters and for the other thermodynamic models. This phenomena is not a feature solely due to the symmetry of the profiles of the Tanh model at $\beta_T = 1$. An example of coalescence and mode switching, with $\beta_T = 2$, is shown in Figure 9, for the Tanh model. This is a plot of ω_i versus ω_r (9a) and c_r versus c_i (9b) at two Mach numbers which closely bracket the mode switch. At $M = 2.86$ (dashed curve) the fast supersonic mode has its neutral value on curve 2 of Figure 9 and its growth rate goes to zero linearly with α. The slow supersonic mode has its neutral point on curve 5, the constant speed supersonic-supersonic neutral mode. Its growth rate goes to zero quadratically with α. At $M = 2.88$ (solid

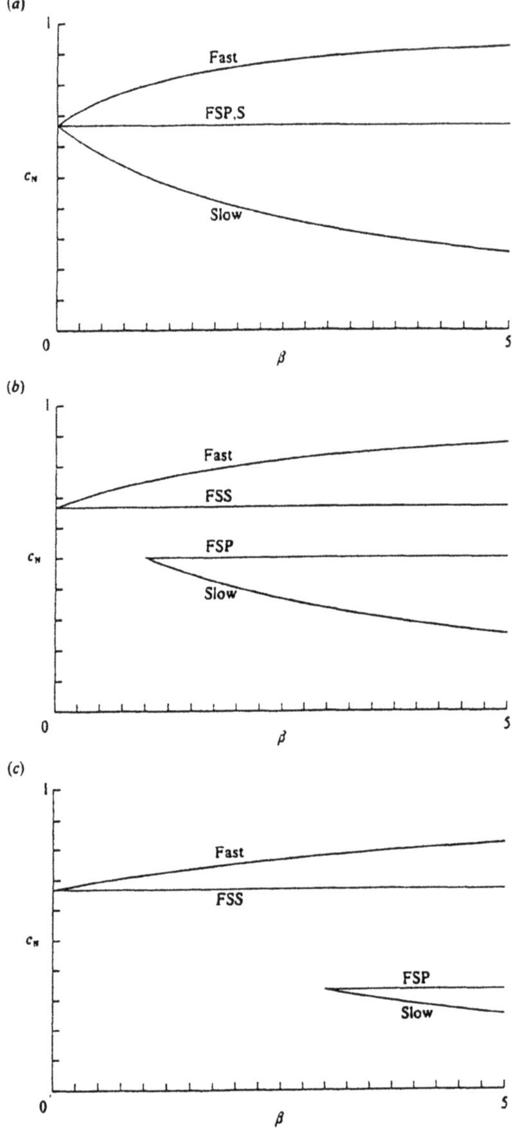

Figure 10. (a) Plot of the neutral phase speeds c_N versus β for $\beta_T = 2$, $\phi = 0.5$, and $M = 0$. (b) Plot of the neutral phase speeds c_N versus β for $\beta_T = 2$, $\phi = 1.0$, and $M = 0$. (c) Plot of the neutral phase speeds c_N versus β for $\beta_T = 2$, $\phi = 2.0$, and $M = 0$.

curve) these two modes have switched. The switching is most clear in Figure 9b.

The addition of the reaction further increases the complexity of the spectrum of the disturbances. Consider first the flame sheet model. Because T' is discontinuous at the flame sheet for nonzero β, $S(\eta)$ will also be discontinuous at this point. It was found (Jackson and Grosch, 1990b) that S can have a single root, two roots one of which corresponds to η positive and the other negative, or two roots one of which is a one-sided zero. The roots of S, which corresponds to phase speeds that are subsonic at both boundaries, are the phase speeds of subsonic neutral modes. The one-sided zero of S may or may not yield a phase speed of a neutral mode. Finally, for non-zero β, there can also be singular neutral modes whose phase speeds are not given by roots of S and are subsonic at the boundaries. If the phase speed corresponding to a zero of S is supersonic at either or both boundaries it may or may not be that of a neutral mode. This can only be determined numerically.

In order to illustrate the complexity, results for a typical case obtained using the flame sheet model with a Tanh profile (Jackson and Grosch, 1990) are shown. The phase speeds of the neutral modes for $M = 0$ and $\beta_T = 2.0$ are plotted as a function of β with $\phi =$ 0.5, 1.0, 2.0 in Figure 10. There are both fast and slow subsonic neutral modes. It was shown that, for the Tanh model with $M = 0$, fast waves only exist for

$$\beta \geq 1 - \beta_T \phi \tag{46}$$

with corresponding neutral phase speed

$$c_N = \frac{\beta + \beta_T \phi}{\beta + (1 + \beta_T)\phi}, \tag{47}$$

while slow waves only exist for

$$\beta \geq \beta_T \phi - 1 \tag{48}$$

with corresponding neutral phase speed

$$c_N = \frac{\beta_T}{1 + \beta + \beta_T}. \tag{49}$$

In addition, there are both fast and slow singular subsonic neutral modes, adjacent to their corresponding regular neutral modes. The

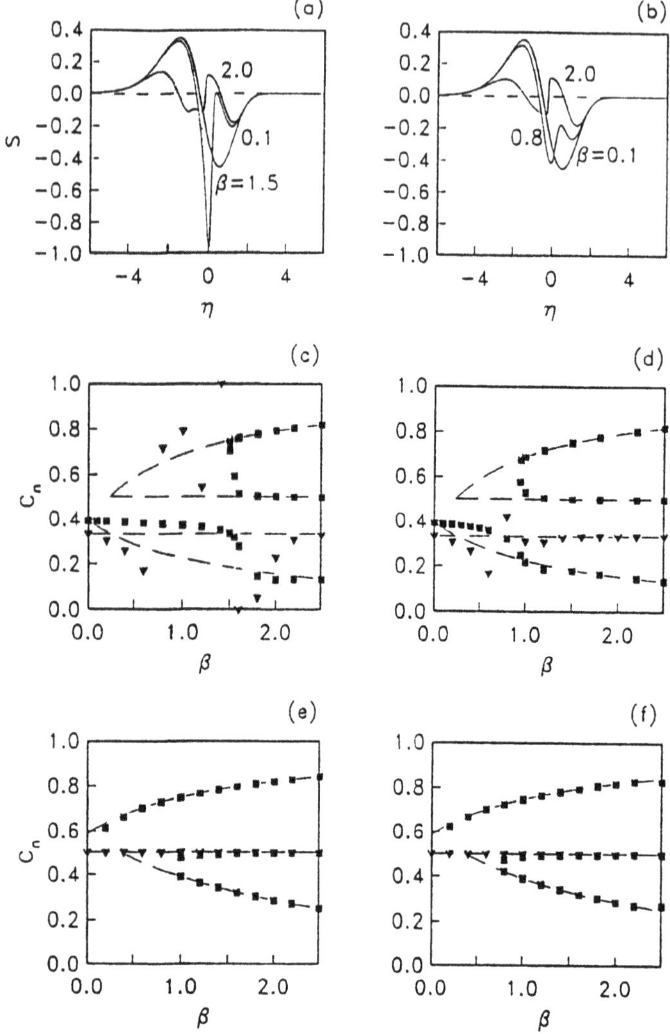

Figure 11. Plot of S versus η for various values of β, $\beta_U = 0$, $\beta_T = 0.5$, $\phi = 1$, $Ze = 20$, $M = 0$ at (a) $x = 3$ and (b) $x = 10$. Plot of neutral phase speeds versus β with $\beta_T = 0.5$ at (c) $x = 3$ and (d) $x = 10$, and with $\beta_T = 1$ at (e) $x = 3$ and (f) $x = 10$ for same parameters as (a,b). Here – – – – – in (c-f) denotes the neutral phase speeds obtained from the flame sheet model. The disturbances are two dimensional with $\theta = 0°$. In all cases the one step irreversible Arhenius reaction was used.

phase speeds of these singular modes are independent of β. These modes are labeled FSP (Flame Sheet Primary) and FSS (Flame Sheet Secondary) in the figure. These singular modes have both zero wavenumber and frequency.

An important question is the extent to which this mode structure is an artifact of (a) the Tanh model and (b) the flame sheet model. At least a partial answer was provided by Hu, Jackson, Lasseigne, and Grosch (1993). They compared the phase speeds of the neutral modes at $M = 0$ using the finite rate model discussed in Section 2 and a flame sheet model, both with $Pr = 1$ and a linear viscosity-temperature relation (the Lock model). The phase speed of the singular modes with zero wavenumber was found by using equations (38) or (39).

Figures 11a,b are plots of S versus η for various values of the heat release parameter, β, using the finite rate model. The slow stream has a speed $\beta_U = 0$ and temperature $\beta_T = 0.5$ and the equivalence ratio $\phi = 1$. The results shown in Figure 11a were obtained using the temperature distribution upstream of ignition while those of Figure 11b were obtained using the temperature distribution far downstream of ignition where the temperature and mass fraction concentrations are close to those of the flame sheet. The rate of heat release has a significant effect not only on the number of roots of S, but also on their values. When the heat release parameter is small ($\beta = 0.1$), there is a single root of S with η_c close to zero at both locations. With $\beta = 1.5$ there are three roots of S at the downstream location, one at $\eta \approx -0.5$ and a double root close to $\eta = 0$. A further increase in β to 2 results in a shift of the first root to more negative values of η and a splitting of the double root into two distinct roots, one close to zero and the other near $\eta = 1$. Qualitatively similar behavior is shown in Figure 11b at the downstream location.

The corresponding neutral phase speeds (indicated by boxes) obtained from the roots of S are shown in Figure 11c and 11d for the upstream and downstream locations, respectively, both for $\beta_T = 0.5$ and 11e and f, again upstream and downstream of ignition, with $\beta_T = 1$. These are shown as functions of the heat release parameter β. The real part of the neutral phase speeds for the $\alpha_N = 0$ mode, found from (38), are shown in these figures by inverted triangles. The flame sheet model results are shown as dashed lines in these figures.

In the nonreactive case ($\beta = 0$), there are two neutral modes with different phase speeds which coincide at $\beta_T = 0.57753$ (Jackson and Grosch, 1990b). With $\beta_T = 0.5$ these are slow neutral modes. One of these neutral modes has a phase speed determined by a root of S and the other member of this pair has a phase speed determined by (39). With $\beta_T = 1$, these neutral modes are fast modes since they have phase speeds greater than 0.5. Again, one of these has a phase speed determined by a root of S and the other member of this pair has a phase speed determined by (39).

When heat release is included ($\beta > 0$) and the flame sheet model is used (denoted by the dashed lines) there are, in general, four neutral modes: two are found from the Lees and Lin condition, called modes 1 and 2; one is found from the zero wavenumber asymptotics, called mode 3; and the remaining one, mode 4, is a mode with phase speed $c_N = U(\eta_f)$ where η_f is the location of the flame sheet. Mode 1 is a slow mode for $\beta_T < 0.57753$ and its phase speed is a decreasing function of β (Figures 11c,d); while for $\beta_T > 0.57753$, mode 1 is a fast mode whose phase speed is an increasing function of β (Figures 11e,f). Mode 2 only exists for $\beta > 0$ and shows the opposite behavior of mode 1. The third neutral mode, that with $\alpha_N = 0$, exists at $\beta = 0$ and has a phase speed which is constant for all values of the heat release parameter, β. Finally, the fourth neutral mode appears at the same value of β as the second mode, has a phase speed which is equal to $U(\eta_f)$ and is independent of β. When both streams have the same temperature, $\beta_T = 1$, the phase speeds of the third and fourth modes are equal. These neutral curves separate stable from unstable regions with an unstable region lying between modes 1 and 3 (called the slow branch) and another between modes 2 and 4 (called the fast branch).

As with the flame sheet model, there are also four neutral modes when using the finite rate chemistry model. The phase speeds of modes 1, 2 and 4 are determined from the Lees and Lin condition, and the third neutral mode is again the zero wavenumber mode with phase speed determined from (38). The reason the fourth mode of the flame sheet model is not determined from (36) is that the Lees and Lin condition fails to hold because S is discontinuous and the derivatives of the eigenfunctions become discontinuous at the flame sheet position. For the finite rate chemistry model, the phase speed of the fourth neutral mode approaches that given by mode 4 of the flame sheet model, i.e., $c_N \rightarrow U(\eta_f)$, as x increases. The phase speeds of the neutral modes 1, 2, and 4 are indicated by boxes in Figures 11c-e, and the phase speed of the third neutral mode is indicated by inverted triangles. Unlike the flame sheet model, the neutral phase speeds for modes 3 and 4 are functions of the heat release parameter β and the downstream position x.

The value of the phase speeds of all four neutral modes will depend critically on whether the x location is upstream or downstream of ignition. In the region of ignition, the temperature and mass fraction fields vary rapidly with position and consequently the parallel flow approximation no longer holds. If x is sufficiently downstream of the ignition point, neutral modes 2 and 4 are present. At $x = 3$ and with $\beta_T = 0.5$ (Figure 11c) the phase speeds of neutral mode three show large variations between $0.5 < \beta < 2$. This is to be expected because ignition occurs in this region and the parallel flow assumption fails. As β is increased past 2, the phase speeds of all four neutral modes approach the phase speeds predicted by the flame sheet model. Similar behavior is shown in Figure 11d at $x = 10$. The variations in the real part of the phase speeds of the $\alpha_N = 0$ neutral mode appear smaller than at $x = 3$ which is consistent since the source term is proportional to $x\,\beta$ and thus at larger x the ignition region extends over a smaller range of β.

These results indicate that the flame sheet is a good approximation to the the flow and combustion field resulting from using finite rate chemistry to calculate the flame, as least as far as is required for stability analysis. However, the finite rate chemistry model used in these calculations is the one step irreversible model given by equations (6) - (8). It is also important to determine how sensitive the results of the stability calculations are to the details of finite rate chemistry model. It appears that almost nothing has been done to address this problem.

The only results currently available are those of Hu (1992). He used the Birkan and Law (1988) three step chain reaction model

$$F + R_1 \rightarrow 2R_2 \tag{50}$$

$$O + R_2 \rightarrow 2R_1 \tag{51}$$

$$R_1 + R_2 + M \rightarrow 2P + M \tag{52}$$

where F, O, P and M are the fuel, oxidizer, product, an inert third body and R_1 and R_2 are radicals. The first two reactions are irreversible, thermoneutral, high activation energy branching reactions. The third reaction is a highly exothermic, zero activation energy, three body termination reaction. Hu repeated the calculations of

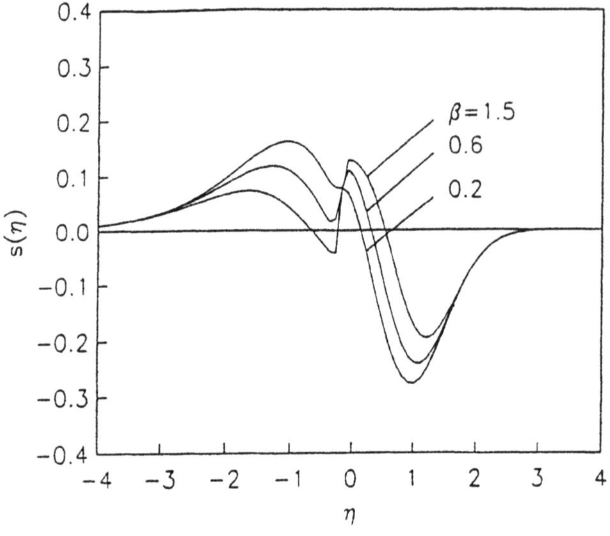

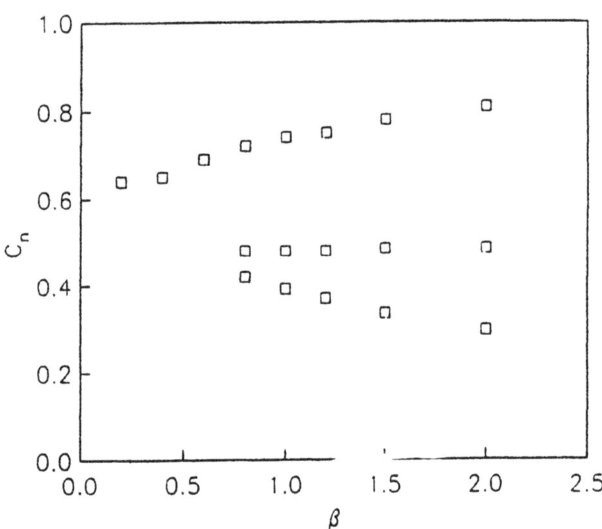

Figure 12. Plot of S versus η for various values of β, $\beta_U = 0$, $\beta_T = 1.0$, $\phi = 1$, $Ze = 20$, $M = 0$ at $x = 10$. Plot of neutral phase speeds versus β with $\beta_T = 1.0$ at $x = 10$, and with the other parameters unchanged. The disturbances are two dimensional with $\theta = 0^o$. In this case the Birkan and Law (1988) reaction model was used.

Grosch and Jackson (1991b) for the mean velocity, temperature, and mass fraction distributions with the Birkan-Law reaction replacing the one step irreversible reaction (equation (6)).

This required having four equations for the mass fraction distributions, $\{F_j\}$ $j = 1, 2, 3, 4$, one each for F, O, R_1 and R_2, respectively. The reaction rates, w_j, for equations (50) - (52) were modeled by

$$w_1 = D_1\, F_1\, F_3\, e^{Ze(1-1/T)}, \tag{53}$$

$$w_2 = D_2\, F_2\, F_4\, e^{Ze(1-1/T)}, \tag{54}$$

$$w_3 = D_3\, F_3\, F_4. \tag{55}$$

Then in equations (4) and (5)

$$\Omega = w_3, \tag{56}$$

$$\Omega_1 = -w_1, \tag{57}$$

$$\Omega_2 = -w_2, \tag{58}$$

$$\Omega_3 = -w_1 + 2w_2 - w_3, \tag{59}$$

and

$$\Omega_4 = 2w_1 - w_2 - w_3. \tag{60}$$

Hu then used the calculated velocity and temperature distributions to calculate $S(\eta)$ (equation (36)) in order to find its roots and thus the phase speeds of the subsonic neutral modes as a function of β. His results for $M = 0$, $\beta_U = 0$ and $\beta_T = 0.5$ are shown in Figure 12, with 12a being a plot of $S(\eta)$ for selected values of β and 12b being a plot of the phase speeds of the neutral modes c_N as a function of β. These results should be compared to those of Figures 11b and 11f which are the equivalent results for the same values of the parameters using the one step irreversible reaction. The general behavior of the $S(\eta)$ curves in Figure 12a is similar to that of Figure 11b. The agreement between the $c_N s$ of 11f and 12b is extremely good. On the basis of this very limited evidence, it appears that the spectrum of the neutral modes may not be very sensitive to the details of the chemistry model.

Figure 13 is a plot of the c_N as a function of the Mach number for $\beta_T = 2$, $\phi = 1$ and various values of $\beta > 0$ using the flame sheet model. The structure of the neutral mode spectrum at $M = 0$ is here extended into the range of non-zero Mach numbers. There is an

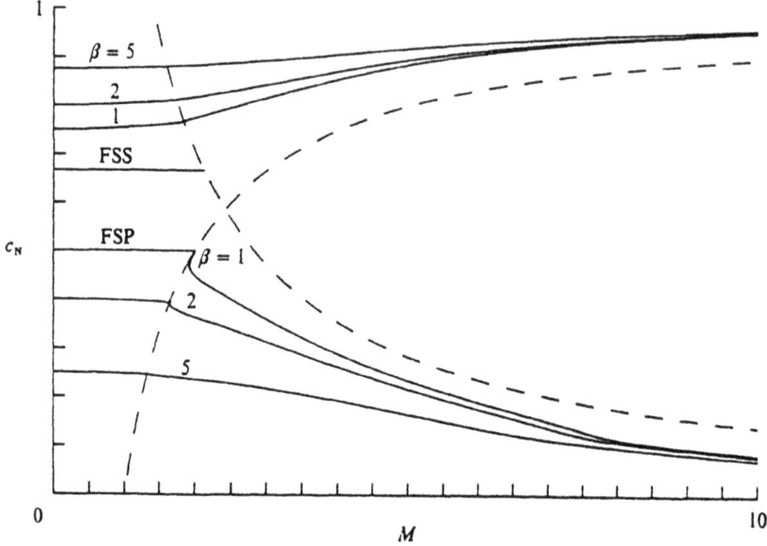

Figure 13. Plot of the neutral phase speeds (solid) and sonic speeds (dashed) versus Mach number for $\beta_T = 2$, $\beta = 1, 2, 5$, and $\phi = 1$.

increase in the phase speed of the fast modes as β is increased. The phase speed of the slow supersonic modes of region 4 decreases with increasing β and, for $\beta > 1$, a singular subsonic neutral mode with $c_N = 0.5$ appears in region 1. Correspondingly, for $\beta > 0$, there also exists a singular subsonic neutral mode in region 1 associated with the fast modes. Again the curves for the fast and slow neutral modes are each asymptotic to a single curve for large M. As in other cases, an increase in β causes an increase in the range of the phase speeds of the unstable waves and hence an increase in the dispersion. Jackson and Grosch (1990b) also reported calculations of c_N as a function of the Mach number with fixed β_T and β for various values of ϕ. It was reported that, as the equivalence ratio ϕ increased, the phase speed of the slow mode was unchanged, consistent with the Mach zero results (see equation (39)). The only effect was a change in the critical value of the Mach number below which this neutral mode did not exist; the smaller the value of ϕ the larger was the value of the critical Mach number. The phase speed of the fast neutral modes was reported to decrease with increasing ϕ. Finally, it should be noted in Figure 13 that, at larger Mach numbers, the neutral modes have quite different phase speeds. Consequently the associated unstable waves will also have much different phase speeds. They will appear as fast and slow unstable waves.

3.3 Growth Rates

The spectrum of the neutral modes has been found to be rather complex and that of the unstable modes is, of course, equally complex. Perhaps the most striking feature is the existence, at supersonic Mach numbers, of two bands of unstable modes; the fast and slow unstable modes. For the non-reacting flow it has been shown (Jackson and Grosch, 1991) that the growth rates are not very sensitive to the detailed shape of the mean velocity and temperature profiles. This is shown by the results presented in Figures 14 and 15.

These figures show the variation of the growth rate with the frequency of the disturbance, for both the fast and slow unstable supersonic modes with $\beta_T = 2.0$ and M = 2.5 and M = 5.0 for the three thermodynamic models: Tanh, Lock and Sutherland. In both cases the slow unstable supersonic modes exist in a very narrow range of frequencies compared to that of the unstable fast supersonic modes. The shape of the $-\alpha_i$ versus ω curves is similar for all of the models

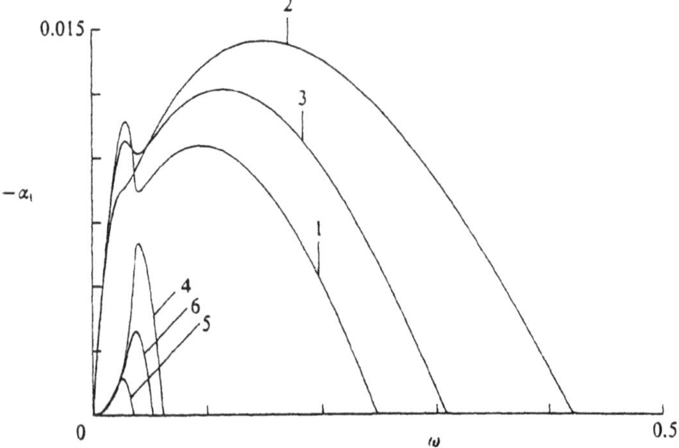

Figure 14. Plot of growth rates $-\alpha_i$ of the fast and slow two dimensional modes versus frequency for $\beta_T = 2$ and M = 2.5; fast modes: (1) Tanh, (2) Lock, (3) Sutherland; slow modes: (4) Tanh, (5) Lock, (6) Sutherland.

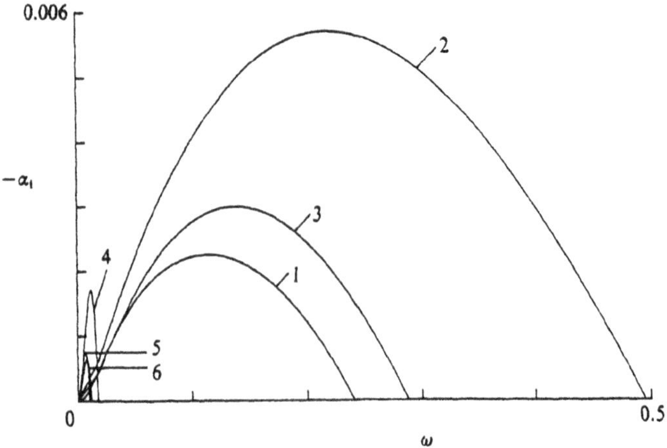

Figure 15. Plot of growth rates $-\alpha_i$ of the fast and slow two dimensional modes versus frequency for $\beta_T = 2$ and M = 5.0; fast modes: (1) Tanh, (2) Lock, (3) Sutherland; slow modes: (4) Tanh, (5) Lock, (6) Sutherland.

with the maximum growth rate of the fast modes being substantially greater than that of the slow modes. This is a result of β_T of 2.0 being greater than the critical value, $\hat{\beta_T}$, as discussed in Section 3.2. The double maximum in the growth rate curves for the fast modes at $M = 2.5$ is quite characteristic. It occurs because there is a mode switch, as discussed in the previous Section, at an near value of β_T.

Figure 16 shows the maximum growth rates versus Mach number with $\beta_T = 2.0$, again for the non-reacting flow. The general variation is similar for all of the thermodynamic models. The maximum growth rate is largest at Mach zero and decreases by a factor of 5 to 10 as the Mach number increases from zero to M_* and approaches a limiting value as the Mach number is further increased. The second group of unstable modes, the slow supersonic modes, appear just below M_*. The growth rate of the most unstable of these modes first increases over a small range of Mach numbers and then decreases, approaching a limiting value at larger values of the Mach number. These results are typical of those found at other values of β_T (Jackson and Grosch, 1990b).

The effect on the stability of the non-reacting compressible mixing layer of a skewing of the streams at $\pm\infty$ has also been investigated (Grosch and Jackson, 1991a). The mean flow at $+\infty$ has a magnitude of 1.0 and is at an angle ψ with respect to that at $-\infty$, which has a magnitude β_U. These parameters can be restricted to $0^o \leq \psi \leq 90^o$ and $0 < \beta_U \leq \cos\psi$. The direction of propagation of the disturbance, θ, was also taken to be non-zero.

The theorems of Rayleigh and Howard provide bounds on the phase speed and/or growth rates of temporally growing disturbances in unstable, inviscid, incompressible shear flows (see Drazin and Reid (1984) for a comprehensive review). Some of these results have been extended by Chimonas (1970) so as to include compressibility. Similar results were also obtained by Djorddjevic and Redekopp (1988). Chimonas' results were further extended by Grosch and Jackson to include crossflow, with the extension applying to flows in a channel with boundaries at a finite distance or in an infinite domain provided the disturbances are subsonic, and hence decay, at $\pm\infty$. The Lees and Lin condition for the existence of a regular subsonic neutral mode was extended to the case of crossflow. The definition of the convective Mach number (Jackson and Grosch, 1990a) was also generalized. Finally it was shown that, at zero Mach number, a generalization of Squire's theorem could be found.

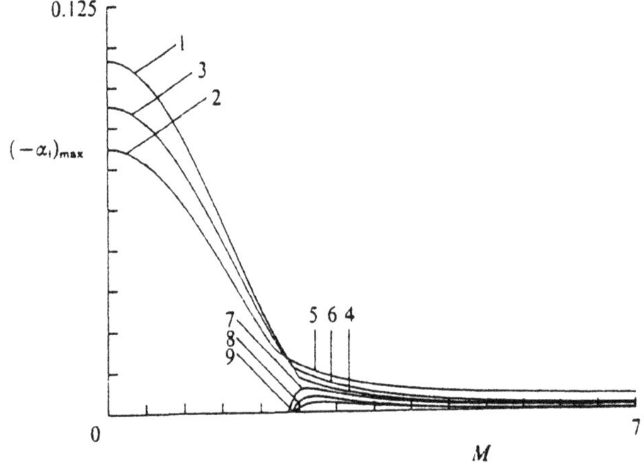

Figure 16. Plot of maximum growth rates of the two dimensional modes versus Mach number for $\beta_T = 2.0$; subsonic modes: (1) Tanh, (2) Lock, (3) Sutherland; fast modes: (4) Tanh, (5) Lock, (6) Sutherland; slow modes: (7) Tanh, (8) Lock, (9) Sutherland.

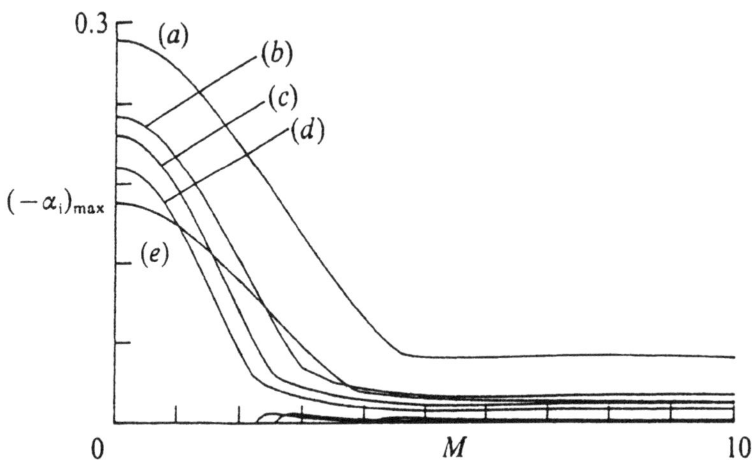

Figure 17. Plot of the maximum growth rates of the subsonic modes and their slow supersonic continuation and some fast supersonic modes versus Mach number for $\beta_T = 0.5$, $\beta_U = 0.25$, $\psi = 10^o$, and (a) $\theta = 80^o$, (b) $\theta = 60^o$, (c) $\theta = 45^o$, (d) $\theta = 0^o$, (e) $\theta = -45^o$.

Results of calculations of the growth rates showed that a non-zero crossflow ($\psi > 0$) combined with an obliquely propagating disturbance ($\theta > 0$) could have an appreciable effect. A typical result is shown in Figure 17 where ψ is taken to be 10° and maximum growth rates are shown for several values of θ. A decrease in θ from 80° results in a decrease in the maximum value of the growth rate for low Mach numbers. For all values of θ the trend in the maximum growth rate with Mach number is similar. For the subsonic mode and its slow supersonic continuation the growth rates decrease by a factor of five to ten with Mach number and then level off. The growth rates of the fast supersonic modes, which appear at M_*, increase slightly and then decrease with increasing Mach number. The growth rate for $\theta = 80^o$ is always greater than the growth rates of the other cases. At $M = 2$, the growth rate of the $\theta = 80^o$ case is comparable to the growth rate of the $\theta = 0^o$ case at $M = 0$.

It was further found, using a flame sheet model, that the addition of combustion had important, and complex, effects on the flow stability (Jackson and Grosch, 1990b). A set of typical results is shown in Figure 18. The maximum growth rate as a function of M with $\beta_T = 2$, $\phi = 1$ and various values of the heat release parameter, β is plotted in Figure 18a. Figure 18b contains similar results, but with β fixed at a value of 2 and $\phi = 0.5, 1.0, 2.0$. The results in Figure 18a show that the maximum growth rates of the slow modes are strictly increasing and those of the fast modes are strictly decreasing with increasing β at zero Mach number. The results of Figure 18b show that increasing ϕ yields an increased in the growth rates of the fast modes while decreasing those of the slow modes. However, the growth rates of both are independent of ϕ at higher Mach number.

Finally, as will be discussed in more detail in Section 3.5, Jackson and Grosch (1990b) found that the addition of heat from the reaction could cause a transition from convective to absolute instability without any backflow.

3.4 The Convective Mach Number

In order to correlate experimental results, a number of experimentalists have used a heuristically defined "convective Mach number". This idea, first introduced by Bogdanoff (1983) for compressible flows, has permeated much of the experimental work on non-reacting compressible mixing layers. The convective Mach num-

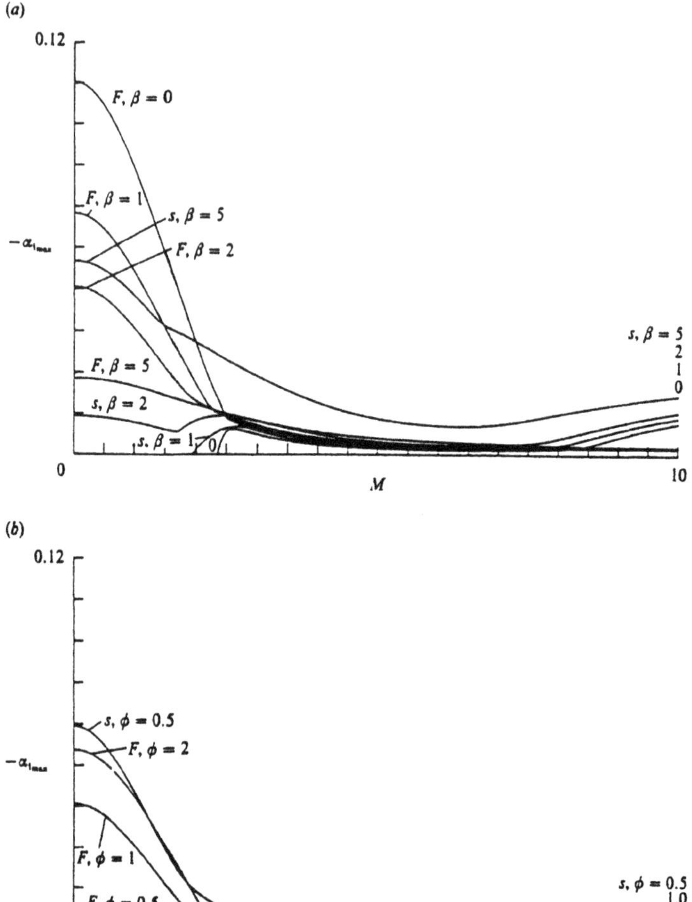

Figure 18. (a) Plot of the maximum growth rates of the fast and slow modes versus Mach number for $\beta_T = 2$, $\beta = 0, 1, 2, 5$, and $\phi = 1$. (b) Plot of the maximum growth rates of the fast and slow modes versus Mach number for $\beta_T = 2$, $\beta = 2$, and $\phi = 0.5, 1.0, 2.0$. In both, the flame sheet model was used.

ber, M_c, has been used to correlate the reduction in growth rate (Papamoschou and Roshko, 1986, 1988) and the reduction in the Reynolds stresses (Elliott and Samimy, 1990; Samimy and Elliot, 1990; Goebel and Dutton, 1991) which are observed to occur in this flow as the Mach number is increased. Several different definitions of a convective Mach number have been advanced. One approach is to define a Mach number in a moving frame of reference which is fixed either to the large scale structures of the mixing layer (Bogdanoff, 1983; Papamoschou and Roshko, 1986, 1988) or to the most unstable wave (Zhuang, Kubota and Dimotakis, 1988). These definitions were for unbounded flows. Tam and Hu (1988, 1991) have applied the same ideas to the non-reacting mixing layer in a channel.

Jackson and Grosch (1990a) defined a convective Mach number in terms of a generalization of the M_* defined in equation (26). They argued that M_* marked marked the transition between two different flow regimes: if $M < M_*$ it is possible to have disturbances which are subsonic in both streams but if $M > M_*$ only disturbances which are supersonic in either or both streams are possible. They derived the equations of linear stability theory for the flow of multispecies streams of gas and showed that the $c_\pm$ can be rewritten in terms of the mixture quantities as

$$c_+ = 1 - 1/M, \tag{61}$$

$$c_- = \beta_U + 1/M\sqrt{\beta_\rho/\beta_\gamma}, \tag{62}$$

where M is the mixture Mach number at $\eta = +\infty$ and β_ρ and β_γ are defined in terms of the mixture quantities at $\pm\infty$ by

$$\beta_\rho = \rho(-\infty)/\rho(\infty), \tag{63}$$

and

$$\beta_\gamma = \gamma(-\infty)/\gamma(\infty). \tag{64}$$

Thus Jackson and Grosch (1990a) defined a convective Mach number for a multi-species gas as

$$M_c = M(1 - \beta_U)/(1 + \sqrt{\beta_\gamma/\beta_\rho}). \tag{65}$$

The other definitions of the the convective Mach number for a multi-species gas are (in the notation used here): that of Bogdanoff (1983)

$$M_c = M(1 - \beta_U)/(1 + 1/\sqrt{\beta_\rho})\beta_\gamma^{1/4}, \tag{66}$$

Papamoschou and Roshko (1986, 1988)

$$M_c = M(1 - \beta_U)/(1 + 1/\sqrt{\beta_\rho}), \tag{67}$$

and Zhuang, Kubota, and Dimotakis (1988)

$$M_c = M(1 - c_{ph}), \tag{68}$$

where c_{ph} is the phase speed of the most unstable mode from linear stability theory. For a single species gas $\beta_\gamma = 1$ and the definition of the convective Mach number of Jackson and Grosch, Bogdanoff, and Papamoschou and Roshko are equal. For most multi-species gases β_γ is close to one and all three definitions yield values of the convective Mach number which are nearly equal. These definitions are for two dimensional disturbances propagating in the direction of the mean flow. If the disturbances are oblique waves propagating at an angle θ to the mean flow direction, the definition of M_c can be generalized by replacing the free stream Mach number M by the effective Mach number $M\ cos\theta$.

The convective Mach number definitions of Bogdanoff and Papamoschou and Roshko are based on the speed of the large scale structures and that Zhuang, Kubota and Dimotakis is based on the speed of the most unstable wave. Both concepts refer to events within the mixing layer. The reaction may be expected to change the speed of both the large scale structures and the most unstable wave. Thus the definitions of M_c should be different for a reacting as compared to a non-reacting flow. The definition of Jackson and Grosch is based on the sonic speeds in the far field. The far field is chemically frozen so their definition of M_c is independent of the reaction.

Jackson and Grosch (1990b) plotted R, the maximum growth rate at an M_c (normalized by that at $M_c = 0$) versus M_c as obtained from the results of spatial stability calculations for a non-reacting compressible mixing layer using the Tanh, Lock and Sutherland models. In addition, the results included the data for three values of the dimensionless temperature of the slow stream, $\beta_U = 0.5,\ 1,\ 2$. They showed, that with these scalings, the data collapsed onto essentially a single curve for $M_c < 1$, and a narrow band for $M_c > 1$ with additional unstable modes appear around $M_c = 1$. This result is similar to that obtained by Ragab and Wu (1988) using a single thermodynamic model but not including the second supersonic modes and using Bogdanoff's heuristic definition of the convective Mach number.

Although the convective Mach number does correlate the results of stability calculations, its main use has been in correlating experimental results and, in this role, there have been certain problems associated with its use for supersonic mixing layers. In the interpretation of experimental results two convective Mach numbers are computed, one using the difference in speed between the fast stream and the speed of the large scale structures (the "fast" one) and the other using the difference in speed between the slow stream and the speed of the large scale structures (the "slow" one). Assuming that there is a stagnation point between each pair of structures, an argument based on isentropic pressure matching leads to the result that the two convective Mach numbers should be equal, or nearly so.

Figure 19 contains curves of the normalized maximum growth rate versus the effective convective Mach number for the Tanh model with $\beta_T = 0.5$ and $\beta_U = 0$ and for the instability waves traveling at an angle θ to the mean flow direction. These curves are based on the numerical calculations of Jackson and Grosch (1989). Results are shown for $\theta = 0^o, 20^o, 40^o, 60^o$. For angles of propagation greater than 60° it was found (Jackson and Grosch, 1989) that the maximum growth rate begins to decrease with increasing angles of propagation. The data points shown in this figure are taken from the experimental results of Papamoschou and Roshko (1988), Samimy and Elliot (1990), Clemens and Mungal (1990), and Hall, Dimotakis and Rosemann (1991). The trends in both the theoretical and experimental results are roughly similar although there is a great deal of spread in the data and no single theoretical curve "fits" all of the data.

The results of some recent experiments have also cast doubt on the relevance of the concept of a convective Mach number. The measurements of Papamoschou (1989) yielded convection speeds which were very close to one or the other of the free stream speeds. Thus the "fast" and "slow" convective Mach numbers were very different. In another experiment (Hall, Dimotakis and Rosemann, 1991) large scale coherent structures were in general *not* seen in Schlieren photographs. However the presence of such structures was inferred from the existence of traveling shock and expansion waves in the low speed side of the flow. The inferred convection speeds of these structures were reported to be much higher than would be predicted for the "fast" convective Mach number. Samimy, Reeder, and Elliot (1992) reported measurements of the convective speed, U_c, of individual structures in compressible mixing layers for two cases. In both cases the fast stream was supersonic and the slow stream subsonic. These speeds were obtained from the signals of a pair of pressure probes in the layer. Their paper contains histograms of the measured U_c at various positions in the mixing layer for two cases. For case 1 the theoretical value of U_c was 352 m/sec and for case 2 it was 428 m/sec, both obtained from Bogdanoff's definition of the convective Mach number. In both cases the results of the measurements showed that there was not a single value of U_c but rather a wide range of values. On the centerline the mean of the distribution was close to the theoretical value but the range was appreciable, about $\pm 40\%$ of the mean. Off the centerline, the mean was 5% to 10% lower than on the centerline and the range was equally large (about 40%).

These experimental results suggest that the convective Mach number can be regarded as an indication of the importance of compressibility effects but may not be very useful beyond that. It is suggested that the observation of convection speeds very close to one or the other of the free stream speeds (Papamoschou, 1989) may be a reflection of the existence of instability waves with both fast and slow phase speeds. Similarly, the observations of Hall, Dimotakis and Rosemann (1991) could be a nonlinear form of the fast instability waves which are subsonic in the fast stream and are supersonic in the slow stream. Linear stability theory predicts that the disturbances are constant amplitude outgoing waves in the slow stream and the nonlinear form of these might be the shock-rarefaction wave pattern observed. It would be desirable to examine the data in the light of this theory. Numerical experiments could probably provide a definitive test of this hypothesis.

3.5 Convective/Absolute Instabilities

In the stability problem, the eigenvalue is a zero of the characteristic equation relating the wavenumber α and the frequency ω at fixed Mach number. Since $\alpha(\omega)$ has a square root branch point singularity at a zero of the complex group velocity $d\omega/d\alpha$ (Briggs, 1964; Gaster, 1968), transition from convective to absolute instability occurs when the zero lies on the real ω axis. Therefore Jackson and Grosch (1990b), using a flame sheet model, and Hu, Jackson, Lasseigne, and Grosch (1993), using finite rate chemistry, choose ω to be

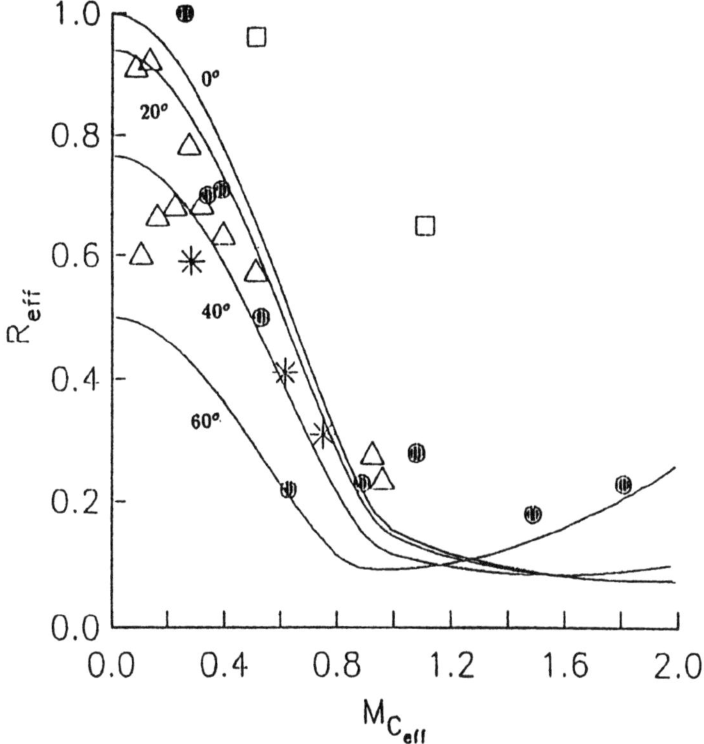

Figure 19. Plots of the normalized maximum growth rate versus the effective convective Mach number for the Tanh model with $\beta_T = 0.5$ and $\beta_U = 0$ and with $\theta = 0°, 20°, 40°, 60°$. The experimental results are the data of Papamoschou and Roshko (1988) •; Samimy and Elliot (1990) □; Clemens and Mungal (1990) ∗; and Hall, Dimotakis and Rosemann (1991) Δ.

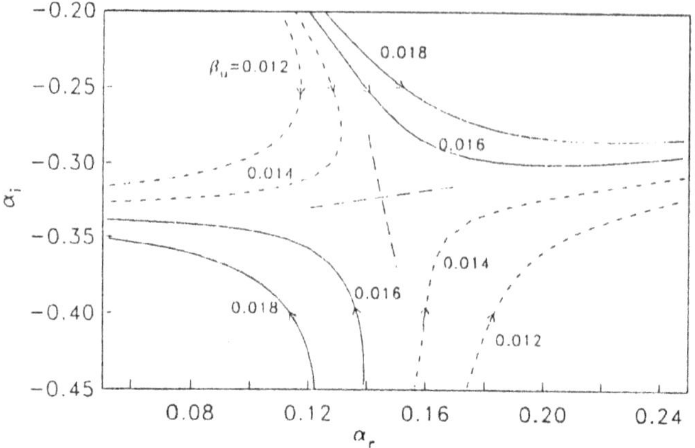

Figure 20. Plot of α_i versus α_r as ω was varied for various values of β_U at $x = 10$ showing the saddle point. Here $\beta = 2$, $\phi = 1$, $\beta_T = 0.5$, $Ze = 20$ and $M = 0$. The mean flow was calculated using the Lock model and the one step irreversible reaction. The disturbances are two dimensional with $\theta = 0^o$.

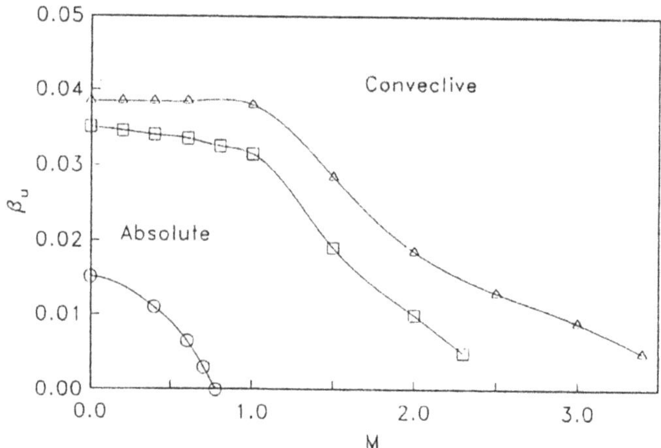

Figure 21. Transition value of β_U from absolute to convective instability for the flame sheet model as a function of M with $\phi = 1$. Results are shown for: ○ $\beta_T = 0.5$ and $\beta = 2$; □ $\beta_T = 0.15$ and $\beta = 2$; and △ $\beta_T = 0.15$ and $\beta = 4$. The disturbances are two dimensional with $\theta = 0^o$.

real, α to be complex, and carried out a numerical search for a zero of $d\omega/d\alpha$. It was shown by Jackson and Grosch (1990b) that the fast branch was convectively unstable while the slow branch undergoes a transition from convective to absolute instability.

Qualitatively similar behavior was found for both the flame sheet and finite rate models; in particular (Hu, et al, 1993) a plot of α_i versus α_r as the real frequency ω varies continuously has (Figure 20) a saddle point for the speed of the slow stream, β_U, between 0.014 and 0.016 showing the presence of a square root branch point singularity due to a transition from convective to absolute instability. Figure 21 (Hu,et al, 1993) shows the effect of varying the temperature at $-\infty$, β_T, and the heat release parameter, β, on the boundary between the regions of convective and absolute instability in the $\beta_U - M$ plane. With β fixed, decreasing β_T, that is cooling the flow at $-\infty$, results in an increase in the range of β_U for which the flow is absolutely unstable. Similarly, increasing the heat release parameter, β, with fixed temperature at $-\infty$ also increases the range of β_U over which the flow is absolutely unstable. Although the range of β_U over which the flow is absolutely unstable is largest for subsonic flow, sufficient cooling at $-\infty$ and/or heat release can cause an absolute instability in supersonic mixing layers.

Similar calculations were made to determine how obliquely traveling disturbances affects the convective/absolute instability transition. The results presented in Figure 22 (taken from those of Hu, et al, 1993) show that increasing the angle of propagation with respect to the mean flow direction, θ, increases the range of β_U over which the reacting flow is absolutely unstable, at least over the parameter ranges examined in their study. This effect seems to be a purely kinematic one in that the wave propagating at an oblique angle relative to the mean flow direction "sees" a flow with a lower Mach number. This is evident from the scaling to an effective Mach number, $M\ cos\theta$, which collapses all of the curves for the oblique disturbances onto essentially a single curve, corresponding that for $\theta = 0^o$, see Figure 23.

The results of Hu, et al (1993) show that, whatever the values of the other parameters, the reacting mixing layer will be convectively, rather than absolutely, unstable at sufficiently large Mach number unless there is a backflow. It is easily shown that, for any β and ϕ, if M is large enough the temperature distribution in the layer will be approximated by that of a non-reacting flow and this requires a

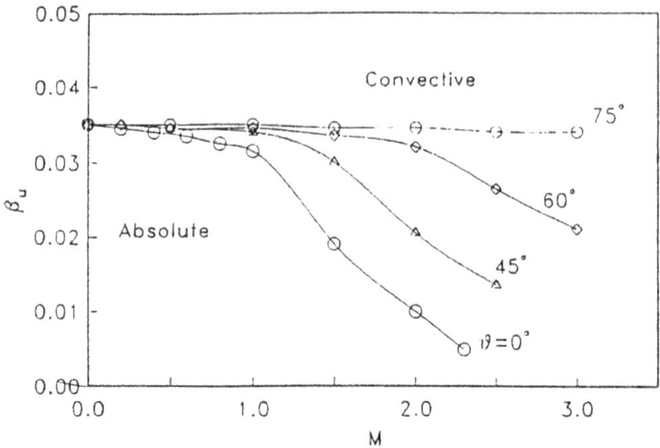

Figure 22. Transition value of β_U from absolute to convective instability for the flame sheet model as a function of M for $\phi = 1$, $\beta = 2$, $\beta_T = 0.15$ for two dimensional and oblique disturbances with $\theta = 0^o$, 45°, 60°, and 75°.

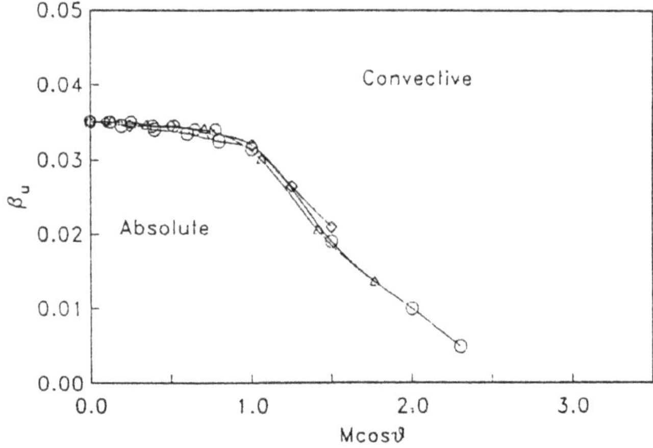

Figure 23. Transition value of β_U from absolute to convective instability for the flame sheet model as a function of $M\ cos\,\theta$ for $\phi = 1, \beta = 2, \beta_T = 0.15$ for two dimensional and oblique disturbances with $\theta = 0^o$, 45°, 60°, and 75°.

negative β_U for an absolute instability. Thus large Mach numbers cause the flow to be convectively, rather than absolutely, unstable. It was also found that cooling the slow stream (decreasing β_T) and increasing the heat release (increasing β) both caused an increase in the range of β_U and Mach number over which an absolute instability existed. These results suggested that it was the magnitude of the temperature gradient induced by the flame which must be large for the absolute instability to occur.

Hu, et al (1993) also used a complementary approach to investigate the transition from convective to absolute instability, namely to examine the response, $I(x,t)$, of the flow to an impulse in space and time (see Huerre and Monkewitz 1985, and the references cited therein). The impulse generates a wave packet in the (x,t) plane with the real part of I the wave packet and its absolute value the envelope. An asymptotic expansion of the impulse response for large time can be determined by the method of steepest descent (Gaster, 1981, 1982) The leading term in the expansion is

$$I = [\frac{2\pi}{(d^2\omega/d\alpha^2)\,t}]^{1/2}_{\alpha=\alpha^*} \; e^{\Sigma t} \; [1+O(t^{-1})], \tag{69}$$

where

$$\Sigma = i(\alpha^*\frac{x}{t} - \omega(\alpha^*)). \tag{70}$$

The value of α^* was found from the requirement that the rays in the wave packet had constant real values of the group velocity, C_g. This gave

$$C_g = [\frac{d\omega}{d\alpha}]_{\alpha=\alpha^*} = \frac{x}{t}. \tag{71}$$

Sets of $\{\alpha^*, \omega(\alpha^*)\}$ pairs which satisfy this equation were then found.

Two, generally distinct, wave packets were found by Hu, et al: the first was made up of the unstable modes of the slow branch, which are absolutely unstable in certain parameter ranges, and the second was made up of the unstable modes of the fast branch which are always convectively unstable. The real part of Σ, the temporal growth rate along the rays, as found by Hu, et al is plotted in Figure 24 for both the fast and slow unstable branches at $M = 0$ for various values of β. As β increases, the maximum of the real part of Σ for the slow branch decreases and the range of x/t for which the real part of Σ is positive decreases. For the fast branch, the maximum growth rate increases by a small amount, and the range of unstable

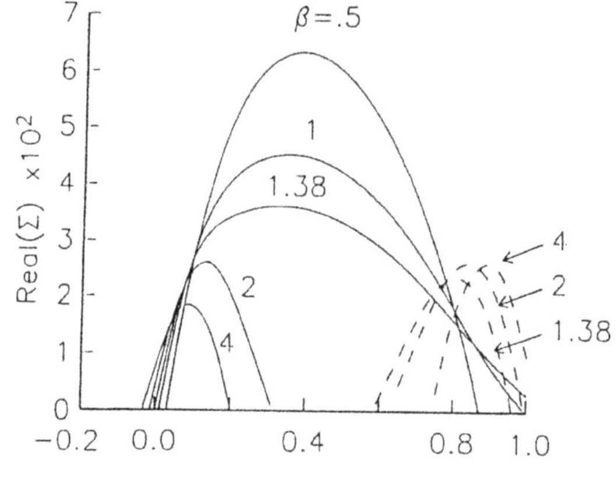

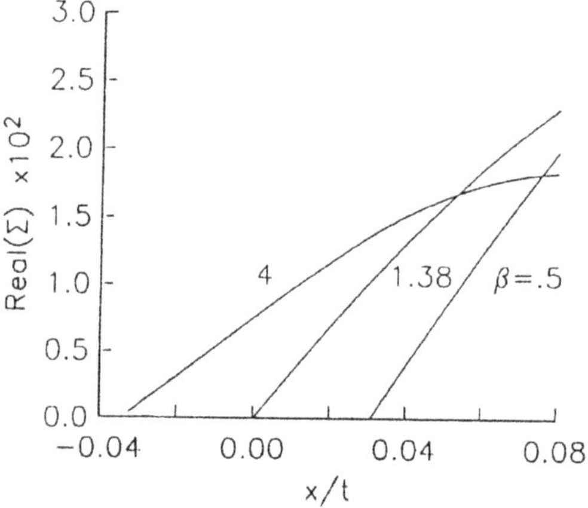

Figure 24. (a) Plot of the real part of Σ as a function of x/t for various values of β. (b) Enlargement of (a) in the region $-0.04 \leq x/t \leq 0.08$. Here $M = 0$, $\beta_T = 0.5$, $\beta_U = 0$ and $\phi = 1$. The disturbances are two dimensional with $\theta = 0^o$.

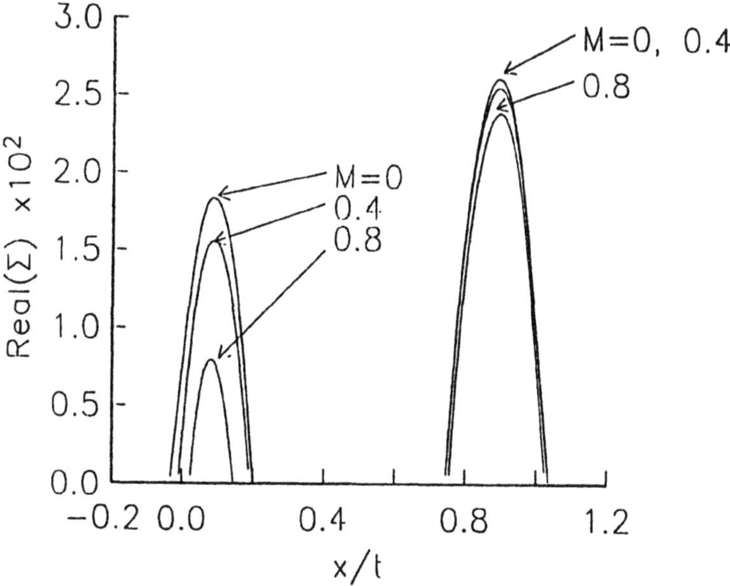

Figure 25. Plot of the real part of Σ as a function of x/t for various Mach numbers. Here $\beta = 4$, $\beta_T = 0.5$, $\beta_U = 0$ and $\phi = 1$. The disturbances are two dimensional with $\theta = 0°$.

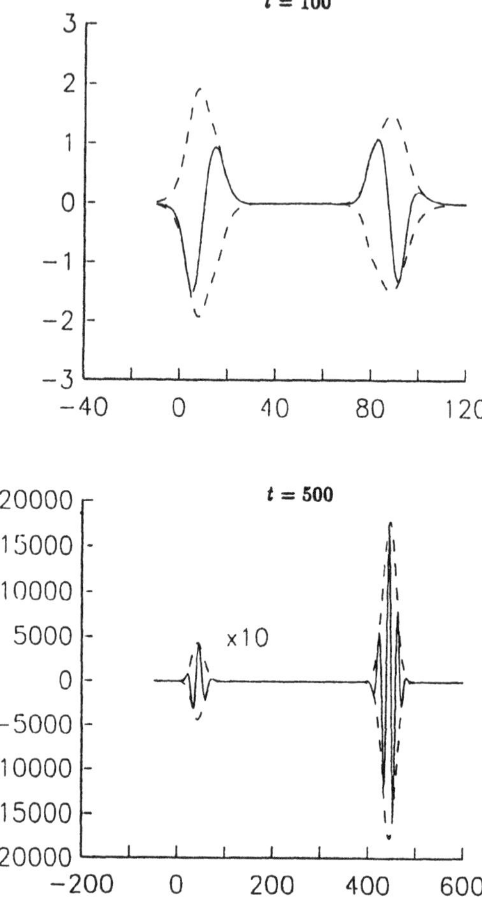

Figure 26. Plot of the wave packets and envelopes for the fast and slow modes as a function of x at (a) $t = 100$ and (b) $t = 500$. The slow packet is absolutely unstable and the fast packet convectively unstable. Here $\beta = 4$, $M = 0.4$, $\beta_T = 0.5$, $\beta_U = 0$ and $\phi = 1$. The disturbances are two dimensional with $\theta = 0^o$.

frequencies increases. Figure 24b is an enlargement of Figure 24a near x/t = 0. The real part of Σ goes to zero at x/t slightly above 0.03 for $\beta = 0.5$. For $\beta = 1.38$, it is zero at x/t = 0, and for $\beta = 4$, it is zero at $x/t \approx -0.035$. The fact that the real part of Σ is positive for a range of negative values of x/t shows that the wave packet is traveling *both* upstream and downstream and therefore that the flow is absolutely unstable. It is important to note that the growth rates in the region of $x/t < 0$ are small compared to the maximum growth rate. Consequently the upstream propagating portion of the wave packet grows slowly compared to downstream propagating part.

The effect of increasing the Mach number on the temporal growth rate along the rays is shown in Figure 25 where the variation of Σ with x/t for $\beta = 4$ and various Mach numbers is shown. The temporal growth rates for the fast branch are only slightly effected by the change in M from 0 to 0.8 with a small decrease in the maximum and the range of x/t over which it is positive. There is a much greater effect on the slow branch. The peak value decreases by more than a factor of 2 as M increases to 0.8 and the range of x/t over which this branch is unstable decreases. For $M = 0.8$ the slow branch only has a positive growth rate for $x/t > 0$, indicating that there is no absolute instability at this Mach number.

Typical wave packets, as computed by Hu, et al, resulting from the impulse with $M = 0.4$ and $\beta = 4$ are shown at (a) $t = 100$ and (b) $t = 500$ in Figure 26. In each figure there are a pair of wave packets: one is a fast packet containing the unstable modes of the fast branch and the other a slow packet containing the unstable mode of the slow branch. As the pair evolves in time, they move apart because of the substantial differences in their group velocities. At $t = 100$ (26a) the slow packet is somewhat larger than the fast packet and clearly exists in a region of $x < 0$, showing the absolute instability. At $t = 500$ (26b) both packets have grown, spread, and moved apart. The notation, $X10$, close to the slow packet means that the amplitude of the slow packet, but not that of the fast packet, has been multiplied by a factor of 10 in order that it be visible on this scale. In 26b the fast packet is much larger than the slow packet because of its greater growth rate. The slow wave packet extends into the region $x < 0$, but because of the scaling, it is difficult to see this on the figure. As time increases, the slow packet will continue to grow, but at a much slower rate than the fast packet, and spread both upstream and downstream. However the upstream propagation is very slow. These results are for $\beta_T = 0.5$. Hu, et al state that similar results were also found at other values of β_T.

These wave packet calculations showed that when the reacting shear layer is absolutely unstable it is *weakly* unstable. That is, with increasing Mach number from zero and a fixed rate of heat release, the absolute instability becomes progressively weaker in that the range of negative x/t over which the growth rate is non-negative grows smaller and the growth rate in this region and the speed of the upstream traveling waves also becomes smaller. Thus a wave packet will grow and spread throughout the entire domain, but it may take a long time for this to happen.

4. Concluding Remarks

The results of the numerics and asymptotics show that the reacting compressible mixing layer has a rather complex structure. In particular, the ignition point and the location of the diffusion flame in the layer are sensitive to the parameters of the flow. All of these results (numeric as well as asymptotic) were found while using the simplifying assumptions of $Pr = 1$ and a linear viscosity-temperature relation. It is highly desirable to determine to what extent these results are dependent on the simplifying assumptions for the flow properties. Would the use of $Pr = 0.72$ and a Sutherland viscosity law, say, result in *qualitative* changes in the results or merely *quantitative* ones? It is certainely expected that there would be quantitative changes in the results and it is desirable to find out how large these would be.

The structure of the reacting compressible mixing layer has been studied using a very simple model for the combustion process: fuel and oxidizer undergoing a one step irreversible reaction. It is clear from calculations using more complex, and realistic, combustion models (see for example the calculations of Birkan and Law (1988) for counterflow flames) that these models can yield a richer and more complex structure than that found for the one step irreversible model. It is desirable to re-do the analysis of the structure of the reacting compressible shear layer using at least one, and possibly more, of the more realistic combustion models. Candidate models include the two step reversible model of Rogers and Chinitz (1983), the three step irreversible model of Birkan and Law (1988), and the three and two step models of Balakrishnan (1992).

The experimental results cited above suggest that the convective Mach number can be regarded as an indication of the importance of compressibility effects but may not be very useful beyond that. Carefully designed numerical experiments may be able to shed additional light on the role of the convective Mach number. In particular, they could be used to look for "structures" with convection speeds very close to one or the other of the free stream speeds, as observed by Papamoschou (1989). One could test the hypothesis that they are a reflection of the existence of instability waves with both fast and slow phase speeds. Again, are Hall, Dimotakis and Rosemann (1991) observing a nonlinear form of fast instability waves which are subsonic in the fast stream and supersonic in the slow stream? Linear stability theory predicts that there are disturbances which are constant amplitude outgoing waves in the slow stream. The nonlinear form of these might be the shock-rarefaction wave pattern observed. Finally, numerical experiments might be used to examine the nonlinear form of the absolutely unstable waves predicted using linear theory. Very careful calculations would be required in order to minimize the effects of numerical dissipation. The role of physical damping on the convective/absolute transition also needs to be investigated.

Most of the theoretical studies of structure and stability have dealt with compressible mixing layers. While this is a useful model, these studies should be extended to other flow configurations, perhaps compressible jets.

References

Artola, M. and Majda, A.J. 1987 "Nonlinear Development of Instabilities in Supersonic Vortex Sheets". Physica D, 28, 253-281.

Balakrishnan, G. 1992 "Studies of Hydrogen-Air Diffusion Flames and of Compressibility Effects Related to High-Speed Propulsion". PhD Dissertation, University of California, San Diego.

Balsa, T.F. and Goldstein, M.E. 1990 "On the Instabilities of Supersonic Mixing Layers: A High Mach Number Asymptotic Theory". J. Fluid Mech., 216, 585-611.

Beach, H.L. 1992 "Supersonic Combustion Status and Issues". In Major Research Topics in Combustion, M.Y. Hussaini, A. Kumar, R.G. Voigt, (eds.) Springer-Verlag, 1-20.

Birkan, M.A. and Law, C.K. 1988 "Asymptotic Structure and Extinction of Diffusion Flames with Chain Mechanism". Twenty-Second Symposium (International) on Combustion, The Combustion Institute, 127-146.

Blumen, W., Drazin, P.G. and Billings, D.F. 1975 "Shear Layer Instability of an Inviscid Compressible Fluid". Part 2, J. Fluid Mech., 71, 305-316.

Bogdanoff, D.W. 1983 "Compressibility Effects in Turbulent Shear Layers". AIAA Journal, 21, 926-927.

Briggs, R.J. 1964 Electron-Stream Interaction with Plasmas, Research Monograph No. 29, MIT Press, Cambridge, Mass.

Brown, G.L. and Roshko, A. 1974 "On Density Effects and Large Structure in Turbulent Mixing Layers". J. Fluid Mech., 64, 775-816.

Chimonas, G. 1970 "The Extension of the Miles-Howard Theorem to Compressible Fluids". J. Fluid Mech., 43, 833-836.

Chinzei, N., Masuya, G., Komuro, T., Murakami, A. and Kudou, D. 1986 "Spreading of Two-Stream Supersonic Turbulent Mixing Layers". Phys. Fluids, 29, 1345-1347.

Clemens, N.T. and Mungal, M.G. 1992 "Two- and Three-Dimensional Effects in the Supersonic Mixing Layer". AIAA Paper 90-1978.

Clemens, N.T. 1992 "An Experimental Investigation of Scalar Mixing in Supersonic Turbulent Shear Layers". HTGL Report T-274, Mechanical Engineering Department, Stanford University.

Djordjevic, V.D. and Redekopp, L.G. 1988 "Linear Stability Analysis of Nonhomentropic, Inviscid, Compressible Flows". Physics of Fluids, 31, 3239-3245.

Drazin, P.G. and Howard, L.N. 1962 "Shear Layer Instability of an Inviscid Compressible Fluid. Part 2". J. Fluid Mech., 71, 305-316.

Drazin, P.G. and Reid, W.H. 1984 Hydrodynamic Stability, Cambridge University Press, Cambridge.

Drummond, J.P. and Mukunda, H.S. 1988 "A Numerical Study of Mixing Enhancement in Supersonic Reacting Flow Fields". AIAA Paper 88-3260.

Drummond, J.P., Carpenter, M.H., Riggins, D.W. and Adams, M.S. 1989 "Mixing Enhancement in a Supersonic Combustor". AIAA Paper 89-2794.

Elliott, G.S. and Samimy, M. 1990 "Compressibility Effects in Free Shear Layers". Physics of Fluids A, 2, 1231.

Gaster, M. 1968 "Growth of Disturbances in Both Space and Time". Physics of Fluids, 11, 723-727.

Gaster, M. 1981 "Propagation of Linear Wave Packets in Laminar Boundary Layers". AIAA Journal, 19, 419-423.

Gaster, M. 1982 "Estimates of the Errors Incurred in Various Asymptotic Representations of Wave Packets". J. Fluid Mech., 121, 365-377.

Goebel, S.G. and Dutton, J.C. 1991 "Experimental Study of Compressible Turbulent Mixing Layers". AIAA Journal, 29, 538-546.

Grosch, C.E. and Jackson, T.L. 1991a "Inviscid Spatial Stability of a Three Dimensional Compressible Mixing Layer". J. Fluid Mech., 231, 35-50.

Grosch, C.E. and Jackson, T.L. 1991b "Ignition and Structure of a Laminar Diffusion Flame in a Compressible Mixing Layer with Finite Rate Chemistry". Physics of Fluids A, 3, 3087-3097.

Grosch, C.E., Jackson, T.L., Klein, R., Majda, A. and Papageorgiou, D.T. 1991 "The Inviscid Discrete Eigenvalue Spectrum of the Compressible Mixing Layer". Unpublished manuscript.

Guirguis, R.H. 1988 "Mixing Enhancement in Supersonic Shear Layers: III. Effect of Convective Mach Number". AIAA Paper 88-0701.

Hall, J.L., Dimotakis, P.E. and Rosemann, H. 1991 "Experiments in Non-Reacting Compressible Shear Layers". AIAA Paper 91-0629.

Ho, C. M. and Huerre, P. 1984 Perturbed Free Shear Layers. Annual Review of Fluid Mechanics. 16, 365-424.

Hu, F.Q. 1992 Personel Communication.

Hu, F.Q., Jackson, T.L., Lasseigne, G.L. and Grosch, C.E. 1993 "Absolute/Convective Instabilities and Their Associated Wave Packets in a Compressible Reacting Mixing Layer". Physics of Fluids A, in press.

Huerre, P. and Monkewitz, P.A. 1990 Local and Global Instabilities in Spatially Developing Flows. Annual Review of Fluid Mechanics. 22, 473-537.

Jackson, T.L. 1992 "Stability of Laminar Diffusion Flames in Compressible Mixing Layers. In Major Research Topics in Combustion, M.Y. Hussaini, A. Kumar, R.G. Voigt, (eds.) Springer-Verlag, 131-161.

Jackson, T.L. and Grosch, C.E. 1989 "Inviscid Spatial Stability of a Compressible Mixing Layer". J. Fluid Mech., 208, 609-637.

Jackson, T.L. and Grosch, C.E. 1990a "Absolute/Convective Instabilities and the Convective Mach Number in a Compressible Mixing Layer". Physics of Fluids A, 2, 949-954.

Jackson, T.L. and Grosch, C.E. 1990b "Inviscid Spatial Stability of a Compressible Mixing Layer. Part 2. The Flame Sheet Model". J. Fluid Mech., 217, 391-420.

Jackson, T.L. and Grosch, C.E. 1991 "Inviscid Spatial Stability of a Compressible Mixing Layer. Part 3. Effect of Thermodynamics". J. Fluid Mech., 224, 159-175.

Jackson, T.L. and Hussaini, M.Y. 1988 "An Asymptotic Analysis of Supersonic Reacting Mixing Layers." Comb. Sci. Tech. 57, 129-140.

Lees, L. and Lin, C.C. 1946 "Investigation of the Stability of the Laminar Boundary Layer in a Compressible Fluid". NACA Tech. Note 1115.

Lele, S.K. 1989 "Direct Numerical Simulation of Compressible Free Shear Layer Flows". AIAA 89-0374.

Linan, A. and Crespo, A. 1976 "An Asymptotic Analysis of Unsteady Diffusion Flames for Large Activation Energies". Comb. Sci. Tech., 14, 95-117.

Miles, J. W. 1958 "On the Disturbed Motion of a Plane Vortex Sheet". J. Fluid Mech., 4, 538-552.

Papamoschou, D. and Roshko, A. 1986 "Observations of Supersonic Free-Shear Layers". AIAA Paper 86-0162.

Papamoschou, D. and Roshko, A. 1988 "The Compressible Turbulent Shear Layer: An Experimental Study". J. Fluid Mech., 197, 453-477.

Papamoschou, D. 1989 "Structure of the Compressible Turbulent Shear Layer". AIAA Paper 89-0126.

Planche, O.H. and Reynolds, W.C. 1991 "Compressibility Effect on the Supersonic Reacting Mixing Layer". AIAA Paper 91-0739.

Ragab, S.A. and Wu, J.L. 1988 "Instabilities in the Free Shear Layer Formed by Two Supersonic Streams". AIAA Paper 88-0038.

Rogers, R.C. and Chinitz, W. 1983 "Using a Global Hydrogen-Air Combustion Model in Turbulent Reacting Flow Calculations". AIAA Journal, 21, 586-592.

Samimy, M. and Elliott, G.S. 1990 "Effects of Compressibility on the Characteristics of Free Shear Layers". AIAA Journal, 28, 439-445.

Samimy, M., Reeder, M.F. and Elliott, G.S. 1992 "Compressibility Effects on Large Structures in Free Shear Flows." Physics of Fluids A, 4, 1251-1258.

Sandham, N. and Reynolds, W. 1989 "The Compressible Mixing Layer: Linear Theory and Direct Simulation". AIAA Paper 89-0371.

Shin, D. and Ferziger, J. 1990 "Linear Stability of the Reacting Mixing Layer". AIAA Paper 90-0268.

Shin, D.S. and Ferziger, J.H. 1991 "Stability of Compressible Reacting Mixing Layer". AIAA Paper 91-0372.

Tam, C.K.W. and Hu, F.Q. 1988 "Instabilities of Supersonic Mixing Layers Inside a Rectangular Channel". AIAA Paper 88-3675.

Tam, C.K.W. and Hu, F.Q. 1991 "Resonant Instability of Ducted Free Supersonic Mixing Layers Induced by Periodic Mach Waves". J. Fluid Mech., 229, 65-85.

Zhuang, M., Kubota, T. and Dimotakis, P.E. 1988 "On the Instability of Inviscid, Compressible Free Shear Layers". AIAA Paper 88-3538.

SUPPRESSION AND ENHANCEMENT OF MIXING IN HIGH-SPEED REACTING FLOW FIELDS

J. Philip Drummond

NASA Langley Research Center
Hampton, Virginia 23681

Peyman Givi[1]

State University of New York
Buffalo, New York 14260

ABSTRACT

Work is underway at the NASA Langley Research Center to develop a hydrogen-fueled supersonic combustion ramjet, or scramjet, that is capable of propelling a vehicle at hypersonic speeds in the atmosphere. Recent research has been directed toward the optimization of the scramjet combustor and, in particular, the efficiency of fuel-air mixing and reaction taking place in the engine. With increasing Mach number, the degree of fuel-air mixing through natural convective and diffusive processes is significantly reduced leading to an overall decrease in combustion efficiency and thrust. Even though the combustor flow field is quite complex, it can be viewed as a collection of spatially developing and reacting supersonic mixing layers or jets from fuel injectors mixing with air, one of which serves as an excellent physical model for the overall flow field. This work is focused on understanding the mechanisms of mixing (or lack thereof) and on the development of techniques for its enhancement in compressible turbulent reacting flows. Results generated by direct numerical simulations (DNS) are first used to demonstrate the mechanisms for reduced mixing in shear layers. To counter the effects of suppressed mixing, several mixing enhancement techniques are then discussed. The most successful approaches involve longitudinal vorticity induced into the flow field. Several means for inducing vorticity are studied and assessed.

[1]The work at SUNY-Buffalo is sponsored by NASA Langley Research Center under Grant NAG-11122, and by the Office of Naval Research under Grant N00014-90-J-4013.

J. Buckmaster et al. (eds.), Combustion in High-Speed Flows, 191–229.

1. Introduction

Research has been underway for a number of years, both in the United States and abroad, to develop advanced aerospace propulsion systems for use late in this century and beyond. One program is now underway at the NASA Langley Research Center to develop a hydrogen-fueled supersonic combustion ramjet (scramjet) that is capable of propelling a vehicle at hypersonic speeds in the atmosphere. A part of that research has been directed toward the optimization of the scramjet combustor and, in particular, the efficiency of fuel-air mixing and reaction taking place in the engine. In the very high-speed vehicle configurations currently being considered, achieving a high combustor efficiency becomes particularly difficult. With increasing combustor Mach number, the degree of fuel-air mixing that can be achieved through natural convective and diffusive processes is reduced leading to an overall decrease in combustion efficiency and thrust.

Because of these difficulties, attention has now turned to the development of techniques for enhancing the rate of fuel-air mixing in the combustor. In an early study of high-speed mixing, Brown and Roshko (1974) show that the spreading rate of a supersonic mixing layer decreases with increasing Mach number, exhibiting a factor of three decrease in spread rate as compared with an incompressible mixing layer with the same density ratio. They conclude that the reduced spread rate is primarily due to compressibility. Papamoschou and Roshko (1986) and Papamoschou and Roshko (1988) also observe that the spreading rate of compressible mixing layers is significantly reduced over that of incompressible layers. To characterize the structure of the flow quantitatively, they define a convective Mach number (Bogdanoff, 1983). The reduction in mixing layer spreading rate (by approximately a factor of three or four) is shown in these experiments to correlate well with increasing convective Mach number. The results of linear stability analyses (Ragab and Wu, 1988; Ragab and Wu, 1989; Jackson and Grosch, 1989) also show that the decreased spreading rate of the mixing layer correlates well with the convective Mach number.

Faced with this challenge, several techniques have been developed for enhancing the mixing rates in supersonic mixing layers and jets. Guirguis *et al.* (1987) show that the spreading rate of a confined mixing layer can be improved if the pressure of the two streams is different. Encouraged by this result, Guirguis (1988) employed a bluff body at the base of the splitter plate separating the two streams. It is shown that the body produces an instability further upstream in the layer and results in a more rapid rate of spread. Kumar *et al.* (1989) discuss a number of mixing problems that may exist in scramjet combustors. Several techniques for enhancing turbulence and mixing in combustor flow fields are suggested, and one enhancement technique that employs an oscillating shock is studied numerically. Drummond and Mukunda (1988) have studied fuel-air mixing and reaction in a supersonic mixing layer and have applied several techniques for enhancing mixing and combustion in the layer. They show that when the mixing layer, with its large gradients in velocity and species, is processed through a shock with strong curvature, vorticity is produced. The vorticity then interacts with the layer and results in a significant increase of the degree of mixing and reaction. Drummond *et al.* (1989) and Drummond *et al.* (1991) continued this investigation further by studying fuel-air mixing in a supersonic combustor. They describe a technique using swept-wedge fuel injectors (Northam *et al.*, 1989) to enhance the mixing processes and overall combustion efficiency in the flow. The swept-wedge injectors introduce streamwise vorticity in the inlet air passing over them, and that air then entrains fuel being injected from the base of the strut. Fuel-air mixing efficiency is shown to be significantly improved by the fuel-jet-air interaction. Marble *et al.* (1987) and Marble *et al.* (1990) employ a planar oblique shock to enhance the mixing between a co-flowing circular helium or hydrogen jet and air. They show that when the jet is processed by the oblique shock, a strong vorticity component is induced at the interface between the low density jet and the relatively high density airstream by the pressure gradient of the shock. Vorticity is generated when the density and pressure gradients are not aligned. The induced vorticity in the fuel jet provides a significant degree of mixing enhancement.

With the brief literature survey presented above, our hope in this article is to describe several numerical experiments on fuel-air mixing and reaction in mixing layers and jets. The initial studies involve simulations of mixing layers conducted to improve the understanding of mechanisms contributing to reduced mixing at high Mach numbers. The latter simulations involve studies of configurations designed to improve the degree of mixing and reaction in such flows.

2. Theory

The flow field considered in this study is described by the two-dimensional (2D) or three-dimensional (3D) Navier-Stokes, energy, and species continuity equations governing multiple species fluid undergoing chemical reaction (Drummond, 1988; Carpenter, 1989), (Drummond, 1991). The finite-rate chemical reaction of gaseous hydrogen and air is modeled with either a three-species, one-reaction model or a seven-species, seven-reaction model. The coefficients governing the diffusion of momentum, energy, and mass are determined from models based on kinetic theory (Drummond, 1988). Sutherland's law is employed to compute the individual species viscosity; the mixture viscosity is evaluated by the Wilke's law. An alternate form of Sutherland's law is also used to compute the individual species thermal conductivity. The mixture thermal conductivity is then determined by the Wassilewa's formula. The Chapman and Cowling law is used to determine the binary diffusion coefficients which describe the diffusion of each species into the remaining species. Knowing the diffusion coefficients, the diffusion velocities of each species are determined by solving the multicomponent diffusion equation (Drummond, 1988). Alternately, in some simulations the calculation of diffusion velocities is simplified by assuming only binary diffusion and applying Fick's law.

Once the thermodynamic properties, chemical production rates, and diffusion coefficients have been computed, the governing equations are solved with the 2D or 3D SPARK computer code using Carpenter's convective fourth-order symmetric predictor-corrector compact algorithm (Carpenter, 1989). The algorithm is constructed on a compact three by three stencil which provides high-order accuracy while allowing boundary conditions to be specified to fourth-order accuracy in a straightforward manner. Details of the algorithm are given by Carpenter (1989).

3. Results

With the development of the theory and the solution procedure described above, several temporally developing mixing layer flows are studied to explore the phenomenon of reduced mixing with increasing Mach number. These results are summarized in the next subsection. Following these studies, two strategies for enhancing the mixing in high Mach number flow fields are examined to determine their effectiveness for enhancing fuel-air mixing. These strategies are discussed in sections 3.2 and 3.3.

3.1. Temporally developing mixing layers

The results obtained by direct numerical simulations (DNS) have been very useful in portraying the problem of mixing in high-speed turbulent combustion. A reasonably updated review of the state of progress on DNS of shear flows is provided in the proceedings of the first ICASE Combustion Workshop (Givi and Riley, 1992). Since then, DNS have been widely utilized for the analysis of high-speed flows in both temporally developing and spatially developing mixing layers (Soetrisno *et al.*, 1988; Lele, 1989; Sandham and Reynolds, 1989; Sekar and Mukunda, 1990; Givi *et al.*, 1991; Grinstein and Kailasanath, 1991; Steinberger, 1992; Mukunda *et al.*, 1992; Planche and Reynolds, 1992; Steinberger *et al.*, 1993). To demonstrate the problems discussed above, it is useful to consider some of the results of these simulations. Here we discuss the results by Givi *et al.* (1991) and Steinberger (1992) of a temporally developing reacting mixing layer since these results contain all the information pertinent to this article.

The configuration of a temporally developing mixing layer is shown in Fig. 1. In this configuration the flow on the top stream is toward the right. The stream on the bottom side of the layer flows to the left with the same speed as that on the top stream. The justifications for temporal simulations are provided in several previous contributions (see Oran and Boris (1987) and Givi (1989) for reviews). The reacting species are introduced into the layer at the free streams. The chemical reaction occurring within the flow is idealized to a simple irreversible second-order form of $A + B \rightarrow Products + Heat$. Reactant A is introduced on the top stream and reactant B on the bottom stream. Calculations are performed with different values of the convective Mach number (M_c) and the heat release parameter (Ce) to assess the influence of these parameters on the structure of the layer (see Givi *et al.* (1991) for a definition of the non-dimensional parameters). In this assessment all of the other non-dimensional parameters are kept constant to isolate the effects of compressibility and exothermicity.

The influences of compressibility are captured by examining the effects of the convective Mach number on the rate of chemical product

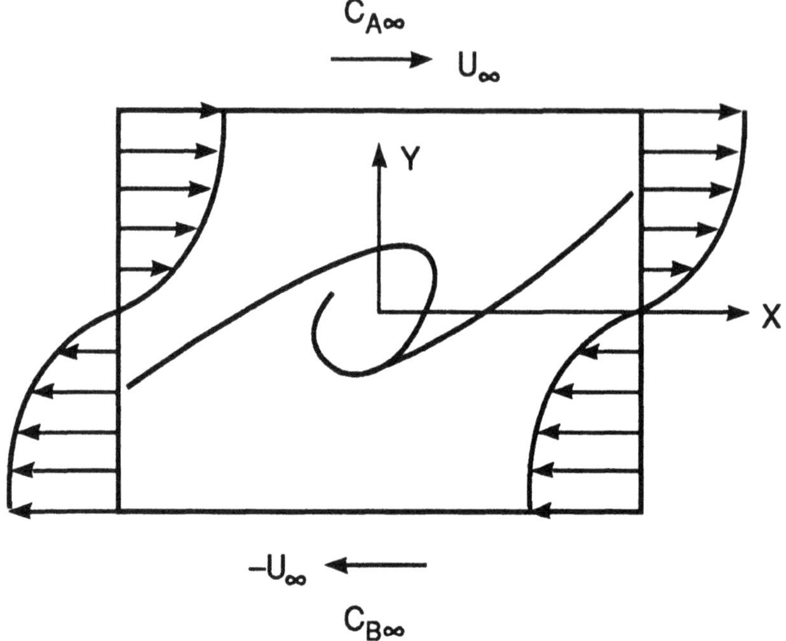

Figure 1. Schematic diagram of a temporally evolving mixing layer.

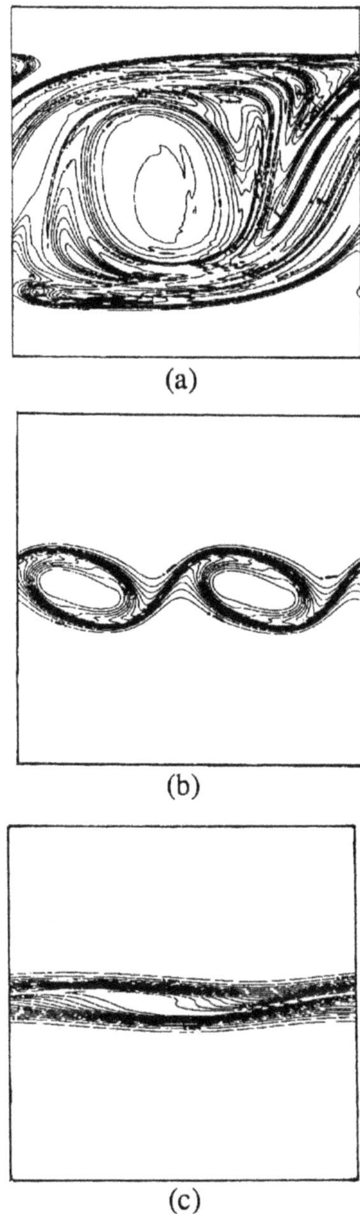

Figure 2. Plots of product mass fraction contours at three convective Mach numbers (M_c). (a) $M_c = 0.2$, (b) $M_c = 0.8$, (c) $M_c = 1.2$.

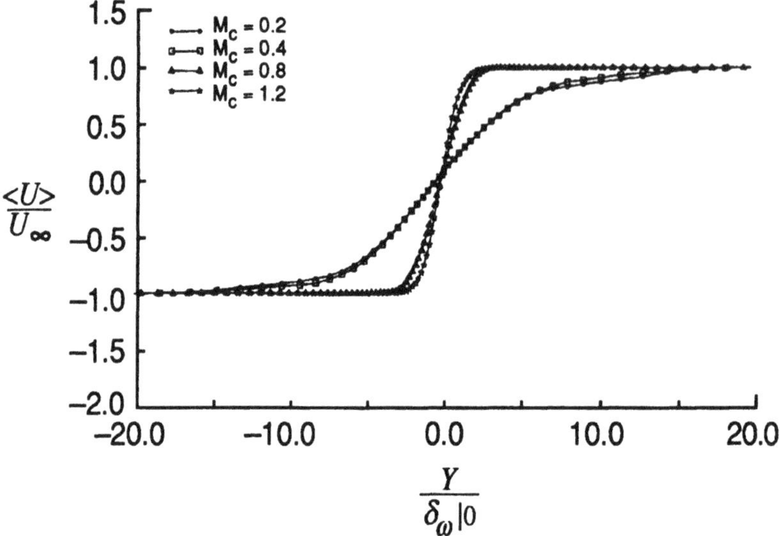

Figure 3. Profiles of normalized mean velocity $\frac{<U>}{U_\infty}$ versus the normalized cross-stream direction $\frac{Y}{\delta_\omega|0}$ for different values of the convective Mach number. U_∞ represents the free-stream velocity and $\delta_\omega|_0$ denotes the vorticity thickness at the initial time.

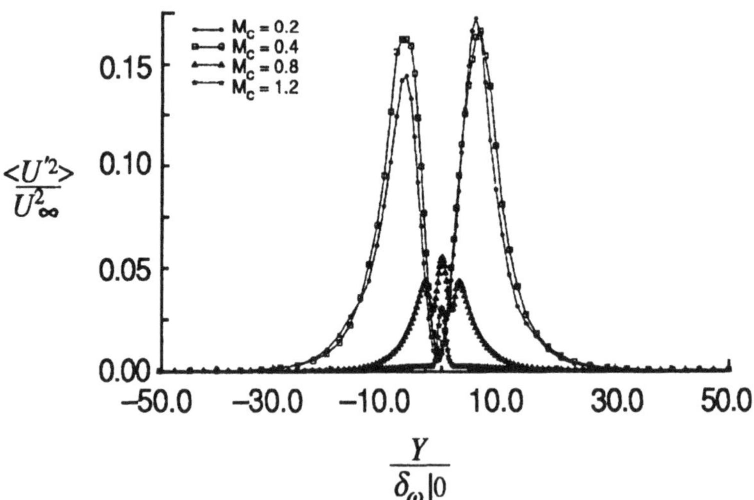

Figure 4. Profiles of normalized mean square velocity $\frac{<U'^2>}{U^2_\infty}$ versus $\frac{Y}{\delta_\omega|0}$ for different values of the convective Mach number.

formation. Figure 2 presents the plot of the product mass fraction contours for different values of the convective Mach numbers (keeping heat release rate at $Ce = 0$). This figure shows a reverse relation between the magnitude of the convective Mach number and the extent of large scale mixing and chemical product formation. As M_c increases it takes longer for background perturbations to grow, and the layer becomes more sluggish in responding to such perturbations. The trend is enhanced as the Mach number is increased; and at the largest Mach number considered, the rate of the layer's growth and the amount of products formed are the smallest.

The response of the shear layer to increased compressibility is further appraised by examining the statistical and the integral properties of the flow. In Figs. 3 and 4, the cross stream variations of the mean and the mean square of the streamwise velocity are shown. The most significant feature displayed in Fig. 3 is the steepness of the mean velocity profiles at high Mach numbers. In view of the contour plots of the product mass fraction, this is to be expected, and the increase in the velocity steepness (caused by the reduced growth rate) implies a reduced rate of mixing and, thus, decreased product formations. This trend can also be described by examining Fig. 4. Note the double hump characteristics of the mean square velocity profile at low Mach numbers. Also note that as the magnitude of the convective Mach number is increased, the amplitude of the fluctuations decreases, and this amplitude becomes very small at $M_c = 0.8$ and $M_c = 1.2$.

Another interesting characteristic of the increased compressibility is captured by examining the plots of pressure contours at high convective Mach numbers as shown in Fig. 5. The pressure response in Fig. 5 shows the regions of pressure maxima and minima at the braids and the cores of the vortices. At higher convective Mach numbers it is observed that the increased compressibility results in steepness of the gradients of instantaneous pressure and the formation of "eddy shocklets." These shocklets are initiated at the shear zone of the layer and extend to the outer region of the flow near the boundaries. A rationale for the formation of these shocklets is provided by noting the increased compressibility within the domain at high convective Mach numbers. In these cases, the layer is dominated by regions of supersonic and subsonic flows; and in order for the flow to adapt to high pressures at the braids, it must go through a shocklet to make the proper adjustment. Also, it is noted that the

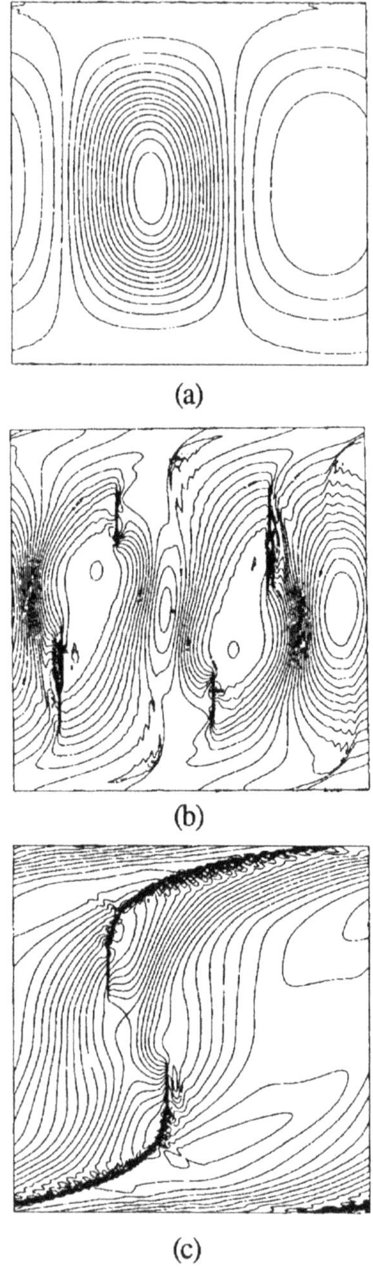

Figure 5. Plots of pressure contours. (a) $M_c = 0.4$, (b) $M_c = 0.8$, (c) $M_c = 1.2$.

currents do not necessarily have to be supersonic at the free streams, and compression occurs within the flow as a result of the formation of large scale structures. This point is demonstrated by examining the contour plots of the instantaneous Mach numbers in Fig. 6. It is shown in this figure that for the case of $M_c = 0.8$ the flow at the interior is characterized by localized regions of supersonic ($Ma > 1$) and subsonic ($Ma < 1$) flows. The adjustment from supersonic to subsonic conditions is provided by the formation of eddy shocklets. The strength of these shocklets becomes stronger as the convective Mach number is increased (*i.e.*, as the effects of compressibility become more pronounced).

The results of the simulations presented here are consistent with those of experimental measurements of Elliott and Samimy (1990) in that as the compressibility increases, the magnitudes of turbulence fluctuations decrease. The results are also in agreement with laboratory data of Hall (1991) in that mixing is reduced at higher compressibility. However, the conclusions drawn here are not in accord with those of Dutton *et al.* (1990), Clemens *et al.* (1991), and Clemens (1991) who suggest higher mixing at elevated compressibility levels. This issue is the subject of current investigations. Also, it has been suggested (Menon and Fernando, 1990; Sandham and Reynolds, 1989) that eddy shocklets form only in 2D simulations. However, the results of recent simulations by Lee *et al.* (1991) and Miller *et al.* (1993) indicate that such shocklets do indeed occur in 3D, both in isotropic and in shear flows.

The influence of the heat release on the structure of the reacting layer is assessed by examining the amount of normalized total product mass fraction shown in Fig. 7. In these simulations two chemistry models are considered; a constant rate kinetics model and an Arrhenius prototype. Figure 7 shows that at the initial stages of the layer development, the effect of heat release is a somewhat enhanced product formation, whereas at intermediate and final stages a reverse scenario holds. At early stages, the effect of heat release is to expand the fluid at the cores of the layer. Therefore, a mixing zone is expected and, thus, a higher amount of product is formed. However, as the extent of heat release increases and the layer thickens, the rate of growth of the instability modes becomes subdued, postponing the rate of formation of large scale vortices. After the initial stages, the non-heat releasing simulations predict a sharp increase in the product formation; and as the magnitude of the heat release is increased,

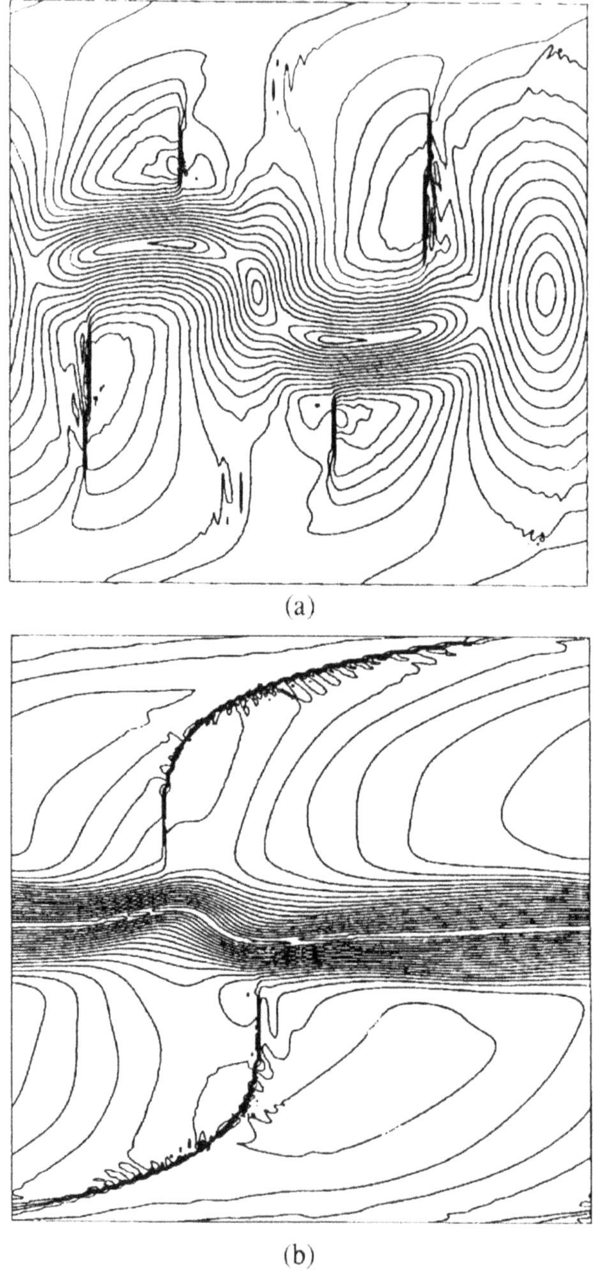

Figure 6. Plots of Mach number contours. (a) $M_c = 0.8$, (b) $M_c = 1.2$.

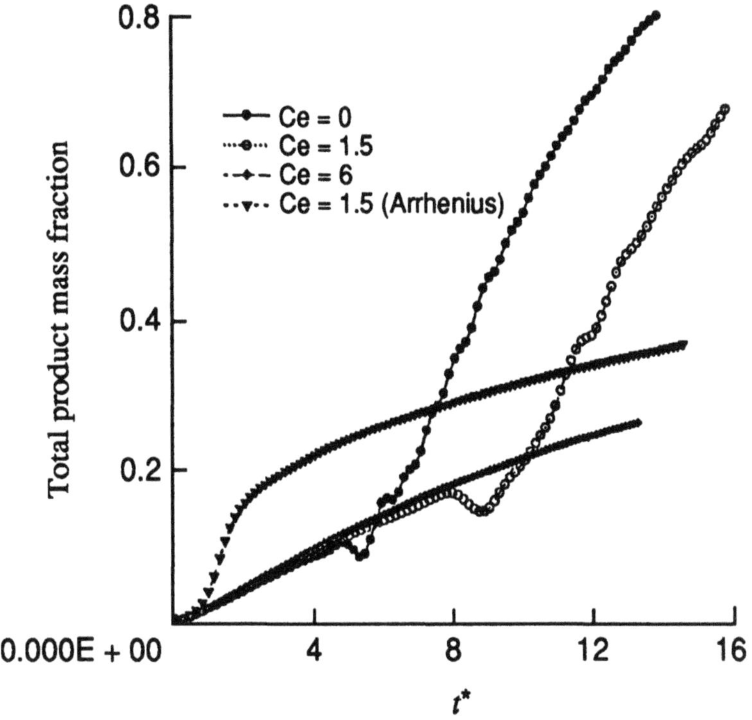

Figure 7. Normalized total product mass fraction versus normalized time (t^*) for different values of the heat release parameter. $t^* = tU_\infty/L$, where L is the size of the computational box.

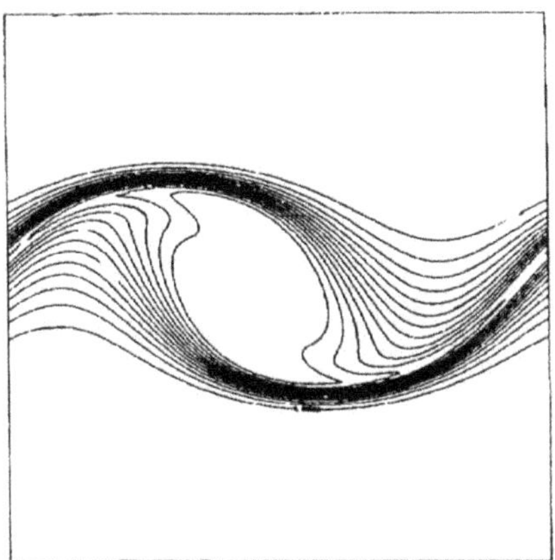

Figure 8. Plots of product mass fraction contours for $M_c = 0.2$, and $Ce = 1.5$.

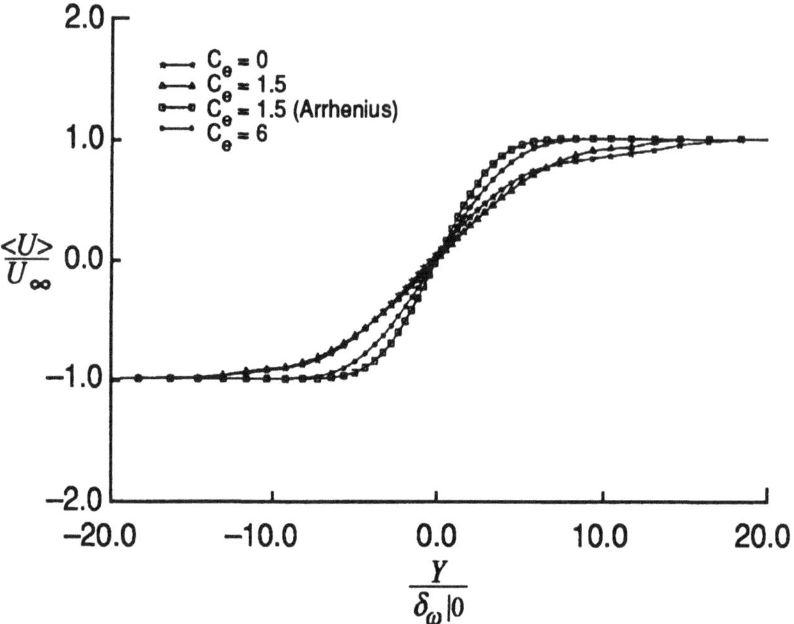

Figure 9. Profiles of normalized mean velocity $\frac{\langle U\rangle}{U_\infty}$ versus $\frac{Y}{\delta_\omega|0}$ for different values of the heat release parameter.

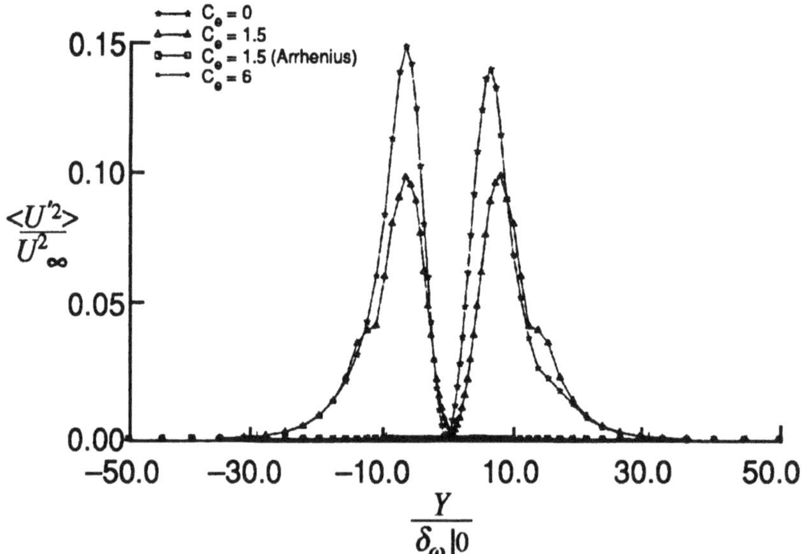

Figure 10. Profiles of normalized mean square velocity $\frac{\langle U'^2\rangle}{U^2_\infty}$ versus $\frac{Y}{\delta_\omega|0}$ for different values of the heat release parameter.

the time at which such structures are formed is delayed. The lowest rate of product formation is for $Ce = 6$ simulations in which the only mechanism of mixing is through diffusion. This reduction in product formation is also evidenced by a comparison between the contour plots of the product mass fraction with heat release (Fig. 8) and those without heat release (Fig. 2(a)). Further influences of heat release become evident by examining its effect on statistical quantities. In Fig. 9 the normalized profiles of mean streamwise velocity component are presented. This figure shows that heat liberation results in a steeper gradient of the velocity and, therefore, less mixing. This has a substantial influence on the two-dimensional turbulence transport, as indicated by the cross stream variations of the mean square velocity presented in Fig. 10. It is shown in this figure that as exothermicity becomes dominant, the amplitude of the fluctuation decreases. For the most significant heat release cases ($Ce = 6$ and the Arrhenius model), the amplitude of the mean square velocity is very close to zero, indicating virtually no turbulence fluctuations.

The conclusion drawn here in regard to mixing reduction caused by exothermicity is consistent with those of laboratory experiments (Hermanson and Dimotakis, 1989), inviscid linear stability analyses (Jackson and Grosch, 1990; Jackson, 1992) and previous DNS results based on low Mach number approximations (McMurtry *et al.*, 1989). However, it has recently been suggested by Steinberger *et al.* (1993) and Miller *et al.* (1993) that in flames where chemistry is described by an Arrhenius kinetics model, the effect of heat release is to increase the rate of product formation. This is due to the increase in the magnitude of the temperature due to heat release which is not considered in the experiments. Based on this observation, it is recommended to further assess the effects of exothermicity by means of laboratory measurements. These measurements must involve a reacting system whereby the rate of reaction conversion is temperature dependent and in which the large scale mixing intensity is not significantly affected by the heat release.

3.2. Mixing enhancement using swept wedges

A number of approaches have been suggested for enhancing the mixing of high-speed fuel-air flows. Several of these approaches are discussed in the Introduction. A particularly attractive option has been suggested by Northam *et al.* (1989) in their experimental study

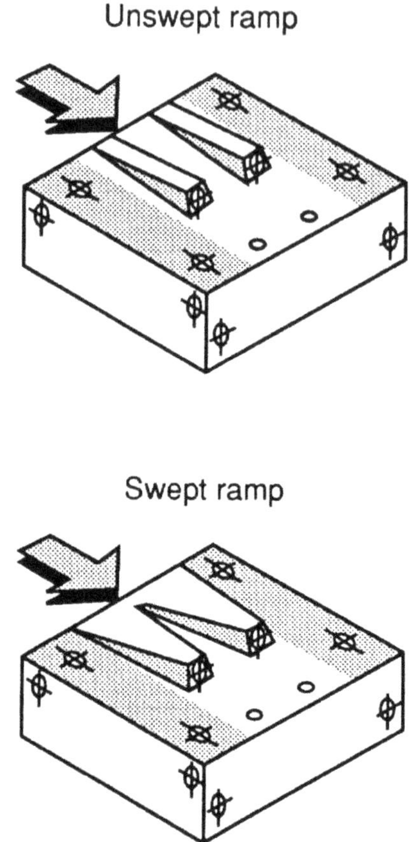

Figure 11. Swept and unswept ramp fuel-injector configurations.

of wall-mounted parallel injector ramps used to enhance the relatively slow mixing of fuel and air normally associated with parallel fuel injection. Parallel injection may be useful at high speeds to extract energy from hydrogen that has been used to cool the engine and the airframe of a hypersonic cruise vehicle. The ramp injector configurations are intended to induce vortical flow and local recirculation regions similar to the rearward-facing step that has been used for flame holding in reacting supersonic flow.

It is instructive to study some aspects of these experiments here. Two ramp configurations are considered in the experiment of Northam *et al.* (1989) as shown in Fig. 11. In both configurations, hydrogen gas is injected at Mach 1.7 from conical nozzles in the base of the two ramps which are inclined at 10.3 degrees to the combustor wall. The injector diameters are 0.762 cm. The sidewalls of the unswept ramps are aligned with Mach 2 streamwise airflow from a combustion facility, whereas the swept ramps are swept at an angle of 80 degrees. Each ramp is 7 cm long and ends in a nearly square base, 1.52 cm on a side. Both ramp designs are chosen to induce vorticity to enhance mixing and base flow recirculation to provide flame holding. The swept ramp injector, because of its delta shape, is intended to induce higher levels of vorticity and, therefore, higher levels of mixing. Hydrogen injection occurs at a streamwise velocity of $1,747$ m/s, a transverse velocity of 308 m/s, and a static temperature and pressure of 187 K and $325,200$ Pa, respectively. The facility air crosses the leading edge of the wedges at a streamwise velocity of $1,300$ m/s, a static temperature of $1,023$ K, and a static pressure of $102,000$ Pa. The air is vitiated following heating by a burner with oxygen, nitrogen, and water mass fractions of 0.2551, 0.5533, and 0.1818, respectively. The overall fuel-air equivalence ratio is 0.6. Both the unswept and swept parallel injector ramps are studied computationally. Only fuel-air mixing is considered. The facility test section surrounding the ramps and considered in the computation is 13.97 cm long and 3.86 cm high. Symmetry planes are chosen to pass transversely through each fuel injector to define the spanwise computational boundaries.

Results from the computational study for both the unswept and swept injector ramps are shown in Figs. 12-17. Figures 12 and 13 show the cross-stream velocity vectors for the unswept and swept cases at two downstream planes ($x = 6.6$ and 13.2 cm) oriented perpendicular to the test section walls. Part (a) of the figures displays

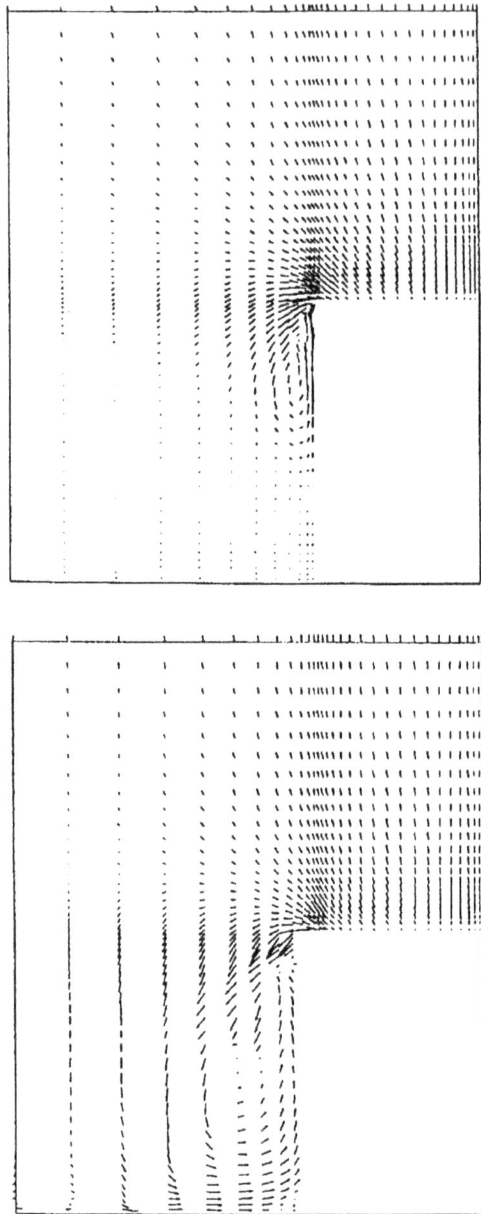

Figure 12. Cross-stream velocity vectors for (a) unswept, and (b) swept, ramp at $x = 6.60$ cm.

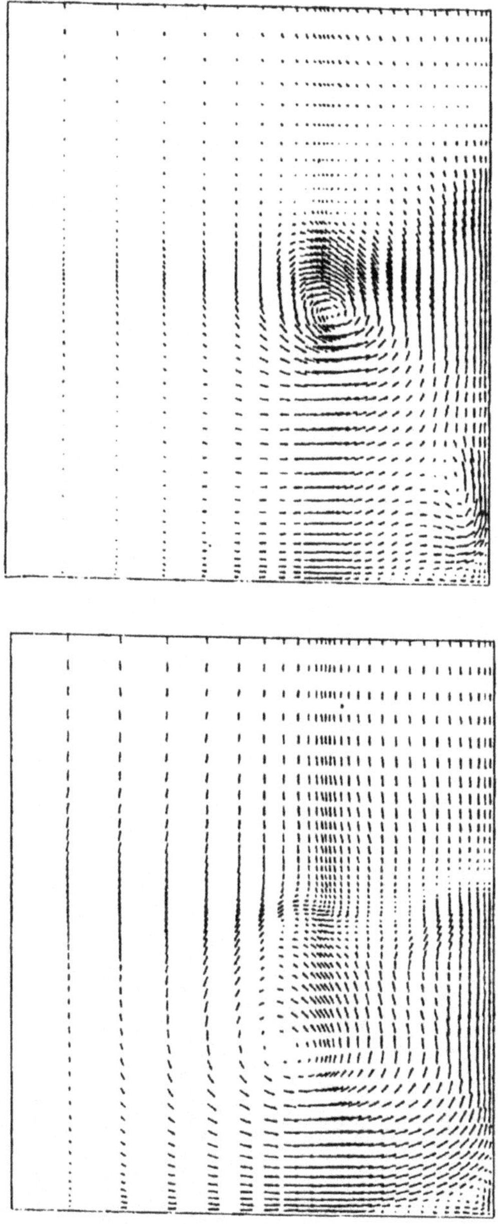

Figure 13. Cross-stream velocity vectors for (a) unswept, and (b) swept, ramp at $x = 13.2$ cm.

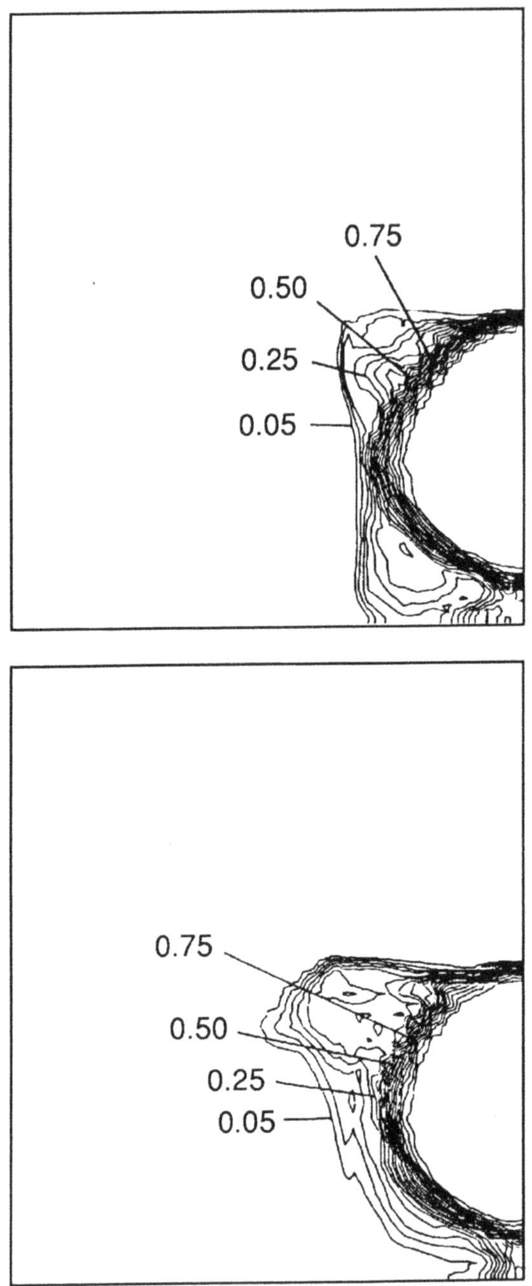

Figure 14. Cross-stream hydrogen mass fraction contours for (a) unswept, and (b) swept, ramp at $x = 7.30$ cm.

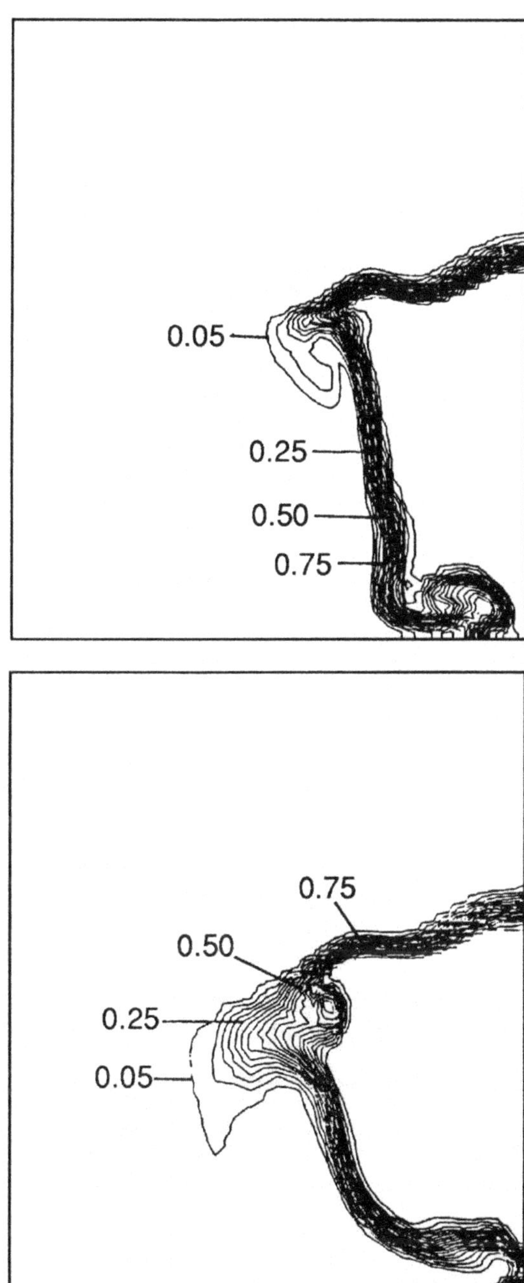

Figure 15. Cross-stream hydrogen mass fraction contours for (a) unswept, and (b) swept, ramp at $x = 9.60$ cm.

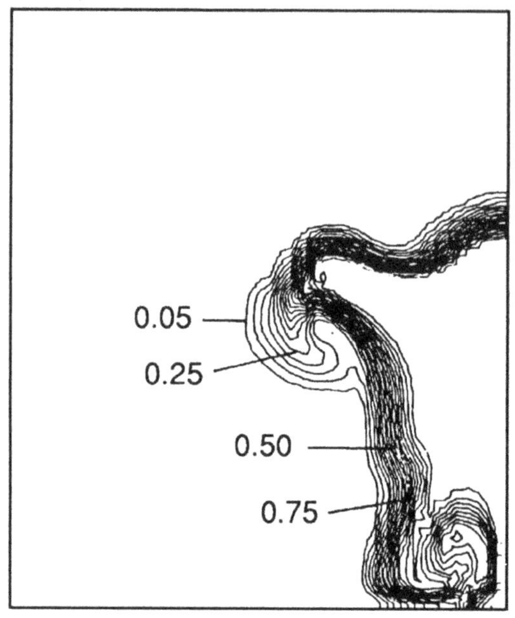

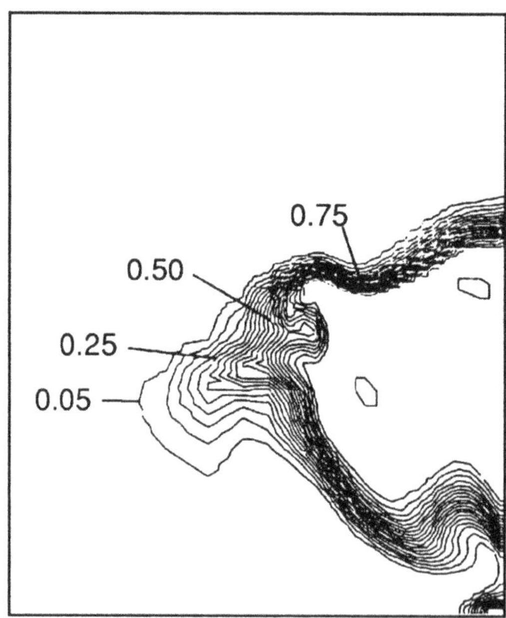

Figure 16. Cross-stream hydrogen mass fraction contours for (a) unswept, and (b) swept, ramp at $x = 11.3$ cm.

the unswept ramp results and part (b) shows the swept ramp results. The planar cut extends from the lower to the upper wall of the test section, and it slices through the center of the right fuel jet. The left boundary is located halfway between the two ramps. At the $x = 6.6$ cm station, which lies just ahead of the end of the ramps, a streamwise vortex has formed at the edge of each ramp. The vortex formed by the swept ramp is considerably larger, however, and it persists well into the flow above the ramp and to the ramp centerline. At the $x = 13.2$ cm station, located 6.2 cm beyond the end of the ramps, the swept ramp vortex has significantly grown and has moved well toward the jet centerline. The swept ramp vortex has now interacted with the hydrogen fuel jet, enhancing its penetration into the airstream. There is pronounced fuel-air mixing enhancement as the vortex spreads across the test section, convecting hydrogen fuel into the airstream. Some enhancement is also provided by the unswept ramp, but it is not nearly as pronounced as that provided by the swept ramp.

The transport of hydrogen fuel into the airstream can be observed more clearly by studying the location of hydrogen mass fraction contours in several test section cross planes, plotted with increasing streamwise distance. Figures 14-17 show the hydrogen mass fraction contours at four successive downstream planes ($x = 7.3$, 9.6, 11.3, and 13.2 cm), again oriented perpendicular to the test section walls. As before, part (a) of the figures displays the unswept ramp results, and part (b) displays the swept ramp results. The results in Fig. 14 occur 0.3 cm downstream of the end of the ramp. With the swept ramp, the larger streamwise vortex has already begun to sweep the hydrogen fuel across into the airstream and away from the lower wall. The smaller streamwise vortex of the unswept ramp also begins to transport hydrogen away from the jet, but not nearly as much as does the swept ramp. As a result, more hydrogen is transported toward the lower wall boundary layer in the unswept case. The same trends continue at the $x = 9.6$ cm station as shown in Fig. 15. At $x = 11.3$ cm, as shown in Fig. 16, the swept ramp enhancer has lifted the fuel jet almost completely off the lower wall. Significant amounts of hydrogen have also been carried across the test section. On the other hand, the unswept ramp enhancer still allows a large amount of hydrogen to be transported along the lower wall, and the spanwise transport is not nearly as great. At $x = 13.2$ cm, the final streamwise station shown in Fig. 17, the spanwise spread of the

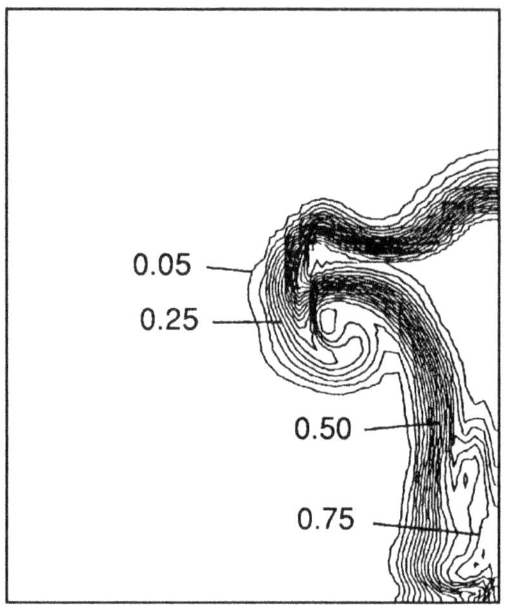

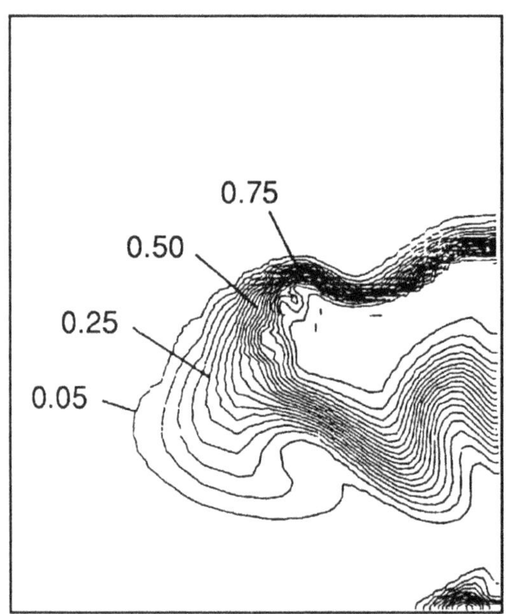

Figure 17. Cross-stream hydrogen mass fraction contours for (a) unswept, and (b) swept, ramp at $x = 13.2$ cm.

fuel jet enhanced by the swept ramp is 46 percent greater than the spanwise spread due to the unswept ramp. In addition, the swept enhancer has resulted in the fuel jet being transported completely off the lower wall. Finally, an eddy of hydrogen has broken completely away from the primary hydrogen jet, increasing the fuel-air interfacial area even further. Clearly then, the swept ramp enhancer significantly increases the overall spread and mixing of the hydrogen fuel jets.

3.3. Mixing enhancement using shocks

Following the analysis of swept wedge injectors, a study of the parallel fuel jet configuration described in the introduction is conducted. As noted before, fuel injected parallel to inlet air entering a combustor is normally assumed to mix relatively slowly with that air. Therefore, to employ parallel injection, it is quite important to enhance mixing of parallel fuel jets and air to the greatest extent possible.

The configuration used for the study of enhanced mixing of parallel fuel jets and air is shown in Fig. 18. It consists of a parallelepiped 6 cm long with a square cross-section 2 cm on a side. A circular hydrogen jet with a 2 mm diameter is injected into the domain from the left face. The hydrogen gas is introduced with a streamwise exponential velocity profile with a peak centerline value of 2,883 m/s, a temperature of 1,000 K, and a pressure of 101,325 Pa (1 atm.), resulting in a peak hydrogen Mach number of 1.2. Air, co-flowing with the hydrogen, is also introduced from the left face at a velocity of 1,270 m/s, a temperature of 1,000 K, and a pressure of 101,325 Pa, resulting in an air Mach number of 2. An oblique shock is introduced across the flow from the lower wall, by a 10 degree wedge also shown in Fig. 18. In the computations, the shock is produced by specifying the appropriate jump conditions for a 10 degree turning angle along the lower boundary where the shock enters the domain.

To establish a baseline for mixing and chemical reaction, calculations are first carried out without the shock. These calculations are conducted for 4 ms in time until a pseudo-steady state is reached following 85 computational sweeps of the flow field. Results for this computation are presented in Figs. 19-22. Figure 19 shows the streamwise development of the hydrogen jet along its centerline in the $x-z$ plane. Values of the hydrogen mass fraction, shown as contours

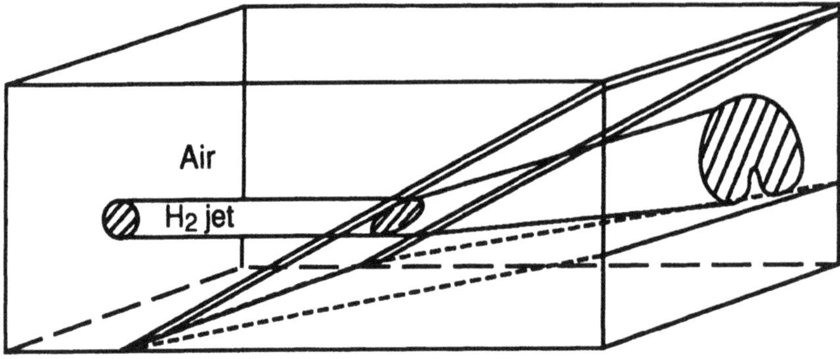

Figure 18. Schematic of shocked, parallel hydrogen fuel jet in air.

in the figure, are defined in the legend. The hydrogen jet develops very slowly with only a small degree of mixing. The cross- stream hydrogen mass fraction distribution at the 6 cm station is shown in Fig. 20. It is also clear from this figure that very little mixing of the hydrogen and air has occurred at the end of the domain, with peak values of hydrogen mass fraction as high as 0.56 still persisting in the flow. Figures 21 and 22 show the water mass fraction resulting from chemical reaction in the $x - z$ and $y - z$ planes, respectively. Due to poor fuel-air mixing, reaction occurs only on the edge of the hydrogen jet, and peak values of water mass fraction of only 0.008 are achieved in the outflow cross-plane at $x = 6$ cm. Combustion efficiency for this case rises to only 0.4 percent at the 6 cm station. Combustion efficiency is defined as the ratio of hydrogen in water to the total hydrogen, integrated over each cross-plane. Therefore, credit in efficiency is taken only for exothermically produced final product water, and not for the remaining product species.

To enhance the degree of mixing and combustion of the hydrogen jet and air, the flow is then processed through the 10 degree shock. It was earlier noted that the shock causes the hydrogen jet to split into a vortex pair and spread quickly downstream. The vortices convect hydrogen away from the jet centerline in a spanwise and transverse direction, entraining and mixing the hydrogen with the surrounding airstream. Reacting results for the shocked jet are given in Figs. 23-29. Figure 23 shows the streamwise development of the hydrogen jet along its centerline in the $x - z$ plane. The jet passes through the shock at $x = 1.1$ cm and flows downstream at an angle of 10 degrees to the original horizontal path. Due to jet mixing and initial chemical reaction, no hydrogen mass fraction contour greater than 0.09 exists beyond the 2 cm station. The water mass fraction distribution resulting from reaction is shown in Fig. 24. Water production begins a short distance downstream of the shock. Peak water production at each station occurs downstream along the stoichiometric line roughly located 75 percent across the water profile. This location is coincident with the lower hydrogen concentration lying between and above the stable hydrogen vortex pair. However, water production is still significant above and below this line as indicated in Fig. 24. The streamwise temperature distribution in the $x - z$ plane is given in Fig. 25. Consistent with the previous results, maximum temperatures occur along the stoichiometric line, with a peak temperature of $2,105$ K at and beyond 4.8 cm.

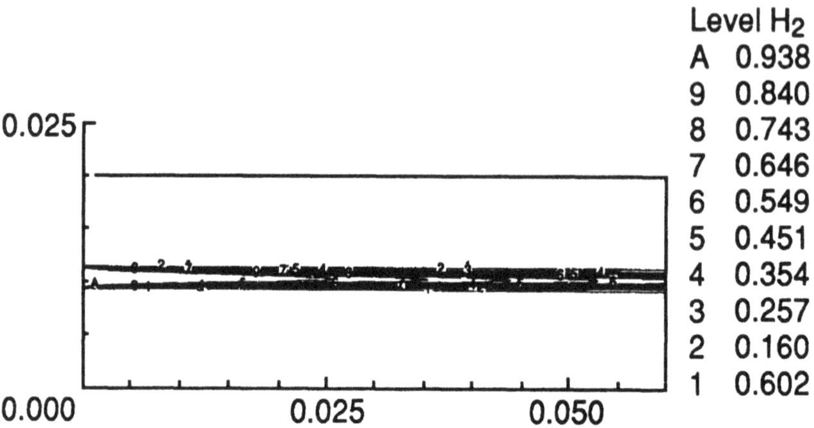

Figure 19. Hydrogen mass fraction of reacting, unshocked jet in $x-z$ plane at $y = 1$ cm.

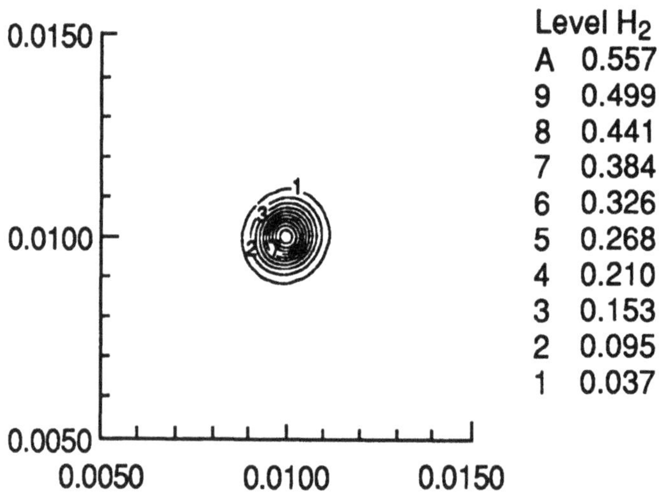

Figure 20. Hydrogen mass fraction of reacting, unshocked jet in $y-z$ plane at $x = 6$ cm.

The vorticity field with chemical reaction in the $y-z$ cross-plane at the $x = 6$ cm station is shown in Fig. 26. Two streamwise vortices have formed in the hydrogen jet, with the left vortex containing positive and the right vortex containing negative components of vorticity when viewed from the outflow of the domain. This vortex structure distorts the initial circular cross-section of the hydrogen jet, entraining fuel and air and enhancing mixing. The jet distortion can be seen in Fig. 27 which shows the hydrogen species mass fractions at the same station displayed in the previous figure. Hydrogen is concentrated toward the interior of each vortex with peak values of around 0.012. Hydrogen is stretched away from the upper portion of the jet, however, and the mass fraction is most greatly reduced in that region. This region of reduced concentration favors the highest initial degree of combustion since the fuel-air ratio is nearest to stoichiometric conditions.

Figure 28 shows the resulting water mass fraction distribution in the $y-z$ plane at the $x = 6$ cm station. Combustion begins in the stoichiometric region at the top of the vortices and along the outer edge of the remainder of the vortices. At $x = 6$ cm, the flame has propagated into the interior of the vortex structure such that significant reaction is occurring near the center of each vortex. The peak water mass fraction of 0.2 occurs at this location. As shown in Fig. 29, there is also a significant temperature rise near the top and near the center of the vortices due to reaction. A peak temperature of 2, 158 K occurs at this location. It is quite interesting to compare the resulting combustion efficiency for the shocked reacting case with the unshocked reacting jet case. Recall that in the unshocked case, the combustion efficiency at $x = 6$ cm is only 0.4 percent whereas in the shocked case, a combustion efficiency of 72 percent is achieved.

4. Concluding Remarks

In high-speed airbreathing propulsion systems, the extent of fuel-air mixing is significantly reduced with increasing Mach number. Direct numerical simulations of reacting mixing layer flows indicate that there is a reduction in turbulence levels with both increased compressibility due to an increase in either Mach number or heat release. To counter the effects of suppressed mixing and reaction, two mixing enhancement techniques have been developed. The first one involves the use of swept wedges placed in the airstream to introduce lon-

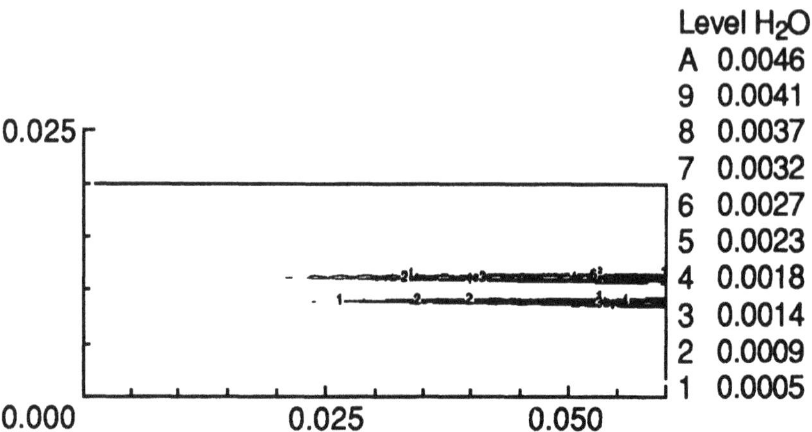

Figure 21. Water mass fraction of reacting, unshocked jet in $x - z$ plane at $y = 1$ cm.

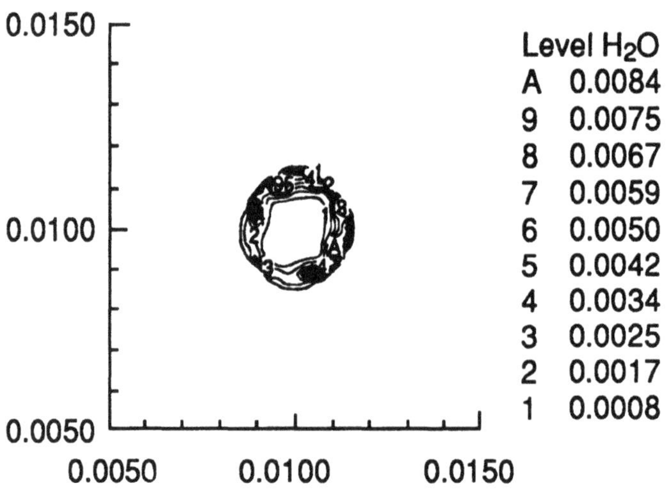

Figure 22. Water mass fraction of reacting, unshocked jet in $y - z$ plane at $x = 6$ cm.

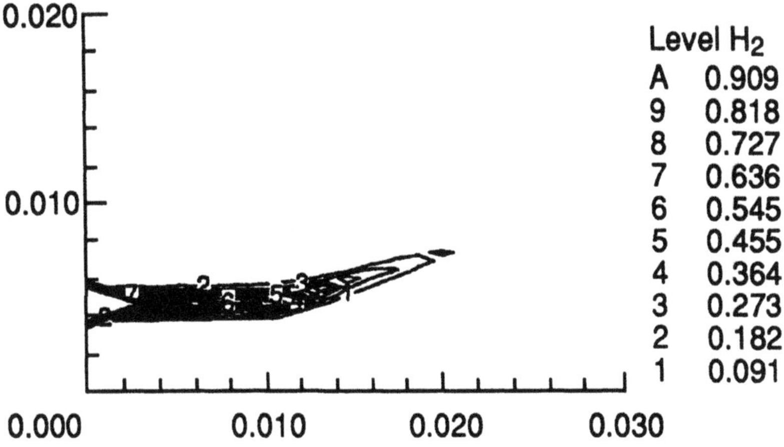

Figure 23. Hydrogen mass fraction of reacting, shocked jet in $x-z$ plane at $y = 1$ cm.

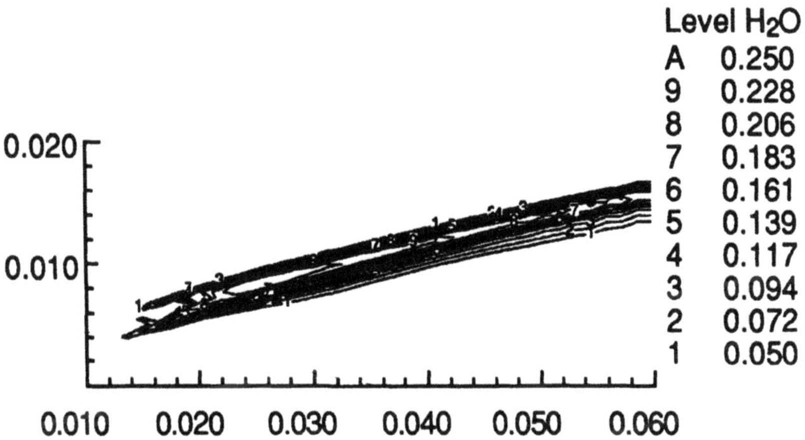

Figure 24. Water mass fraction of reacting, shocked jet in $x-z$ plane at $y = 1$ cm.

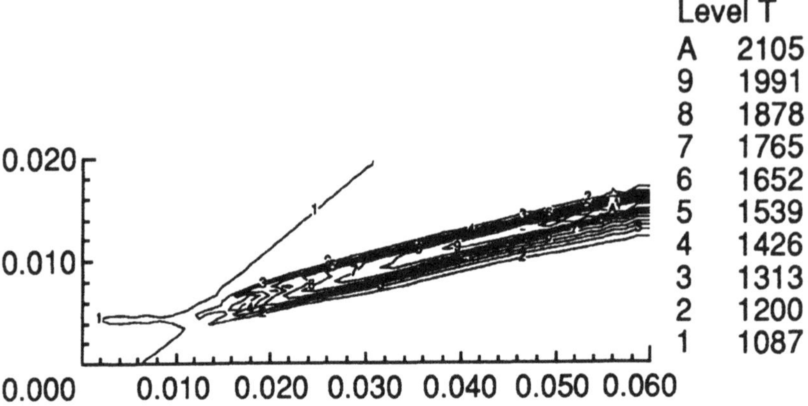

Figure 25. Temperature of reacting, shocked jet in $x - z$ plane at $y = 1$ cm.

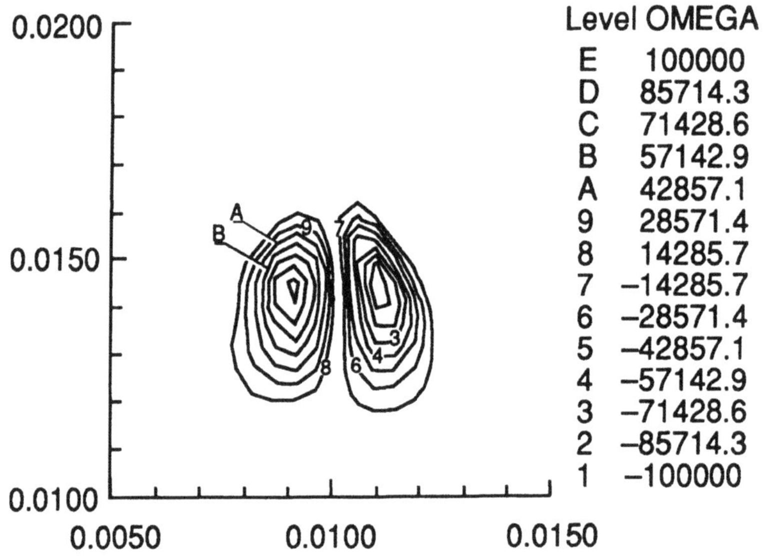

Figure 26. Streamwise vorticity of reacting, shocked jet in $y - z$ plane at $x = 6$ cm.

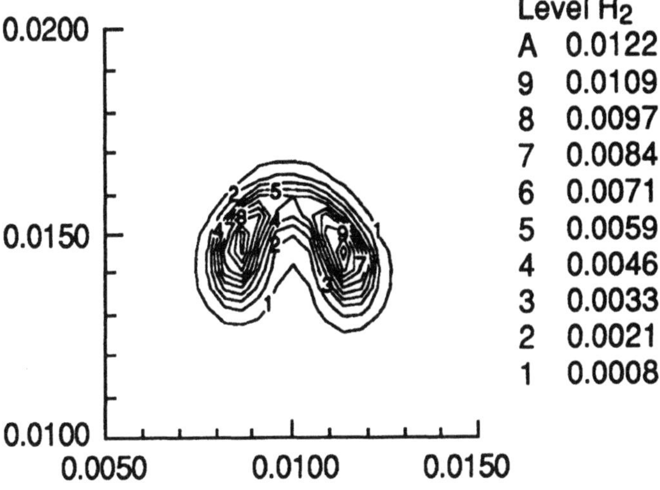

Figure 27. Hydrogen mass fraction of reacting, shocked jet in $y-z$ plane at $x = 6$ cm.

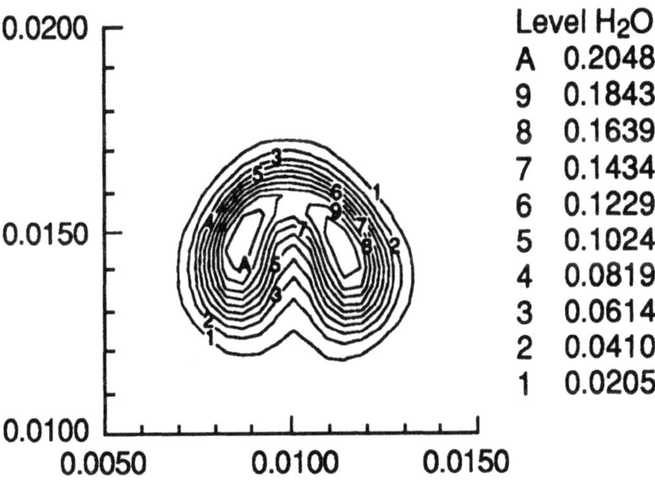

Figure 28. Water mass fraction of reacting, shocked jet in $y-z$ plane at $x = 6$ cm.

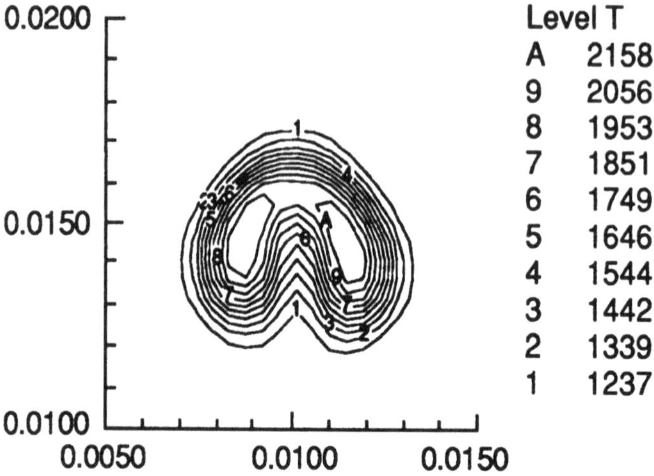

Figure 29. Temperature of reacting, shocked jet in $y-z$ plane at $x = 6$ cm.

gitudinal vorticity leading to large scale mixing enhancement. The second technique utilizes the interaction of a shock with the large density gradient existing between a hydrogen fuel jet and the surrounding airstream to introduce streamwise vorticity and mixing. Both of these approaches have proven effective in providing mixing enhancement mechanisms in nonpremixed high-speed reacting flows.

References

Bogdanoff, D. W., 1983. "Compressibility effects in turbulent shear layers," *AIAA J.* **21**, 926–927.

Brown, G. L. and Roshko, A., 1974. "On density effects and large structure in turbulent mixing layers," *J. Fluid Mech.* **64**, 775–816.

Carpenter, M. H., 1989. "Three-dimensional computations of cross-flow injection and combustion in a supersonic flow," AIAA Paper 89-1870.

Clemens, N. T., Paul, P. H., Mungal, M. G., and Hanson, R. K., 1991. "Scalar mixing in the supersonic shear layer," AIAA Paper 91-1720.

Clemens, N. T., 1991. "An experimental investigation of scalar mixing in supersonic turbulent shear layers," Stanford University Report HTGL T-274, Department of Mechanical Engineering, Stanford University, Stanford, CA.

Drummond, J. P. and Mukunda, H. S., 1988. "A numerical study of mixing enhancement in supersonic reacting flow fields," AIAA Paper 88-3260.

Drummond, J. P., Carpenter, M. H., Riggins, D. W., and Adams, M. S., 1989. "Mixing enhancement in a supersonic combustor," AIAA Paper 89-2794.

Drummond, J. P., Carpenter, M. H., and Riggins, D. W., 1991. "Mixing and mixing enhancement in supersonic reacting flow fields," in Murthy, S. N. B. and Curran, E. T., editors, *High Speed Propulsion Systems*, volume 137 of *AIAA Progress Series*, chapter 7, pages 383–455, American Institute of Aeronautics and Astronautics.

Drummond, J. P., 1988. "Two-dimensional numerical simulation of a supersonic, chemically reacting mixing layer," NASA TM 4055.

Drummond, J. P., 1991. "Supersonic reacting internal flow fields," in Oran, E. S. and Boris, J. P., editors, *Numerical Approaches to Combustion Modeling*, volume 135 of *Progress in Astronautics and Aeronautics*, chapter 12, pages 365–420, AIAA Publishing Co., Washington, D.C.

Dutton, J. C., Burr, R. F., Goebel, S. G., and Messersmith, N. L., 1990. "Compressibility and mixing in turbulent free shear layers," in *12th Symposium on Turbulence*, Rolla, MO.

Elliott, G. S. and Samimy, M., 1990. "Compressibility effects in free shear layers," AIAA Paper 90-0705.

Givi, P. and Riley, J. J., 1992. "Some current issues in the analysis of reacting shear layers: Computational challenges," in Hussaini, M. Y., Kumar, A., and Voigt, R. G., editors, *Major Research Topics in Combustion*, pages 588–650, Springer-Verlag.

Givi, P., Madnia, C. K., Steinberger, C. J., Carpenter, M. H., and Drummond, J. P., 1991. "Effects of compressibility and heat release in a high speed reacting mixing layer," *Combust. Sci. and Tech.* **78**, 33–68.

Givi, P., 1989. "Model free simulations of turbulent reactive flows," *Prog. Energy Combust. Sci.* **15**, 1–107.

Grinstein, F. F. and Kailasanath, K., 1991. "Chemical energy release, spanwise excitation, and dynamics of transitional, reactive, free shear flows," AIAA Paper 91-0247.

Guirguis, R. H., Grinstein, F. F., Young, T. R., Oran, E. S., Kailasanath, K., and Boris, J. P., 1987. "Mixing enhancement in supersonic shear layers," AIAA Paper 87-0373.

Guirguis, R. H., 1988. "Mixing enhancement in supersonic shear layers: III. effect of convective Mach number," AIAA Paper 88-0701.

Hall, J. L., 1991. *An Experimental Investigation of Structure, Mixing and Combustion in Compressible Turbulent Shear Layers*, Ph.D. Thesis, California Institute of Technology, Pasadena, CA.

Hermanson, J. C. and Dimotakis, P. E., 1989. "Effects of heat release in a turbulent, reacting shear layer," *J. Fluid. Mech.* **199**, 333–375.

Jackson, T. L. and Grosch, C. E., 1989. "Inviscid spatial stability of a compressible mixing layer," *J. Fluid Mech.* **208**, 609–637.

Jackson, T. L. and Grosch, C. E., 1990. "Inviscid spatial stability of a compressible mixing layer. Part 2. The flame sheet model," *J. Fluid Mech.* **217**, 391–420.

Jackson, T. L., 1992. "A review of spatial stability analysis of compressible reacting mixing layers," in Hussaini, M. Y., Kumar, A., and Voigt, R. G., editors, *Major Research Topics in Combustion*, pages 131–161, Springer-Verlag.

Kumar, A., Bushnell, D. M., and Hussaini, M. Y., 1989. "A mixing augmentation technique for hypervelocity scramjets," *J. Propulsion and Power* **5**, 514–522.

Lee, S., Lele, S. K., and Moin, P., 1991. "Eddy shocklets in decaying compressible turbulence," *Phys. Fluids A* **3**, 657–664.

Lele, S. K., 1989. "Direct numerical simulation of compressible free shear flows," AIAA Paper 89-0374.

Marble, F. E., Hendricks, G. J., and Zukoski, E. E., 1987. "Progress toward shock enhancement of supersonic combustion processes," AIAA Paper 87-1880.

Marble, F. E., Zukoski, E. E., Jacobs, J. W., Hendricks, G. J., and Waitz, I. A., 1990. "Shock enhancement of and control of hypersonic mixing and combustion," AIAA Paper 90-1981.

McMurtry, P. A., Riley, J. J., and Metcalfe, R. W., 1989. "Effects of heat release on the large scale structures in a turbulent reacting mixing layer," *J. Fluid Mech.* **199**, 297–332.

Menon, S. and Fernando, E., 1990. "A numerical study of mixing and chemical heat release in supersonic mixing layers," AIAA Paper 90-0152.

Miller, R. S., Madnia, C. K., and Givi, P., 1993. "The structure of a reacting turbulent mixing layer," Submitted for publication.

Mukunda, H. S., Sekar, B., Carpenter, M. H., Drummond, J. P., and Kumar, A., 1992. "Direct simulation of high-speed mixing layers," NASA TP 3186.

Northam, G. B., Greenberg, I., and Byington, C. S., 1989. "Evaluation of parallel injector configurations for supersonic combustion," AIAA Paper 89-2525.

Oran, E. S. and Boris, J. P., 1987. *Numerical Simulations of Reactive Flows*, Elsevier Publishing Company, Washington, D.C.

Papamoschou, D. and Roshko, A., 1986. "Observations of supersonic free shear layers," AIAA Paper 86-0162.

Papamoschou, D. and Roshko, A., 1988. "The compressible turbulent shear layer: An experimental study," *J. Fluid Mech.* **197**, 453–477.

Planche, O. H. and Reynolds, W. C., 1992. "A numerical investigation of the compressible reacting mixing layers," Report No. TF-56, Stanford University, Department of Mechanical Engineering, Thermosciences Division, Stanford, CA.

Ragab, S. A. and Wu, J. L., 1988. "Instabilities in the free shear layer formed by two supersonic streams," AIAA Paper 88-0038.

Ragab, S. A. and Wu, J. L., 1989. "Linear instabilities in two dimensional compressible mixing layers," *Phys. Fluids A* **1**, 957–966.

Sandham, N. D. and Reynolds, W. C., 1989. "A numerical investigation of the compressible mixing layer," Report No. TF-45, Stanford University, Department of Mechanical Engineering, Thermosciences Division, Stanford, CA.

Sekar, B. and Mukunda, H. S., 1990. "A computational study of direct numerical simulation of high speed mixing layers without and with chemical heat release," in *Proceedings of 23rd Symp. (Int.) on Combustion*, pages 707–713, The Combustion Institute, Pittsburgh, PA.

Soetrisno, M., Eberhardt, D. S., Riley, J. J., and McMurtry, P. A., 1988. "A study of inviscid, supersonic mixing layers using a second-order TVD scheme," AIAA Paper 88-3676-CP.

Steinberger, C. J., Vidoni, T. J., and Givi, P., 1993. "The compositional structure and the effects of exothermicity in a non-premixed planar jet flame," *Combust. Flame*, in press.

Steinberger, C. J., 1992. "Model free simulations of a high speed reacting mixing layer," AIAA Paper 92-0257.

MODELING TURBULENT SCALAR MIXING WITH MAPPING CLOSURE METHODS

Sharath S. Girimaji[1]

Analytical Services and Materials Inc.
Hampton, Virginia 23666

ABSTRACT

The Mapping closure concept was first introduced by Chen, Chen, and Kraichnan (1989) to model turbulent scalar mixing. Since then it has been modified and extended to multiscalar mixing. In this paper, the time-evolving reference-field version of the mapping closure method (Girimaji 1992a, 1993) is presented. In this approach, the physical scalar field $\phi(\mathbf{y},t)$ ($\mathbf{y}$ - space coordinate, t - time) is mapped using a mapping function, $X(t)$, on to a reference field $\theta(\mathbf{y},t)$ which could be time evolving. The mapping function evolves in time such that the implied evolution of the ϕ field is a model of the true evolution. Two-species mixing and multispecies mixing are considered. Model calculations are compared with data from the direct numerical simulations of stochastic diffusion.

1. Introduction

In many combustion problems of practical importance, the flow field is turbulent. Turbulent combustion can be broadly classified into two categories depending upon the initial mixedness of the fuel and oxidant: premixed and nonpremixed combustion. In turbulent premixed combustion, the fuel and oxidant are initially well mixed and the flame propagates through the mixture as chemical reaction takes place. For a review of the problems associated with this branch of combustion and the models available to solve them, the reader is referred to Pope (1987). In turbulent nonpremixed combustion, the fuel and oxidant are initially segregated and chemical reactions occur as they mix. The rate of reaction is controlled by the rate of mixing. A recent review of the research issues in this type of combustion

[1]This research was supported by the Theoretical Flow Physics Branch, Fluid Mechanics Division, NASA Langley Research Center, Hampton, VA 23681, under contract No. NAS1-18599.

J. Buckmaster et al. (eds.), Combustion in High-Speed Flows, 231–252.

is provided in Bilger (1989). One of the most important processes that requires modeling is scalar mixing. Due to the nonlinear nature of the chemical reaction rate (in terms of scalar concentrations and temperature), an accurate probabilistic description of the scalar field is required (Pope 1985). In other words, for accurate calculation of turbulent combustion, the joint probability density function (pdf) of the scalars and temperature is needed. Much effort has been directed recently towards modeling the scalar pdf evolution in turbulent flows. In most cases, for the sake of simplicity, consideration is restricted to passive scalar mixing in constant density isotropic turbulence. At this stage of development of the subject, the inclusion of variable density effects, inhomogeniety, and chemical reaction renders the modeling intractable.

Of the many turbulent scalar mixing models available, the mapping closure model introduced by Chen, Chen, and Kraichnan (1989) (henceforth referred to as CCK) comes closest to capturing some of the details of scalar pdf evolution observed in the direct numerical simulations (DNS) of Eswaran and Pope (1988). The mapping closure approach was further developed by Gao (1991) and Pope (1991). A different mapping closure methodology which utilizes a time-evolving reference field was developed by Girimaji (1992a). The new mapping closure method allows the reference field to be (i) time evolving, and (ii) non-Gaussian. The time-evolving method has also been extended to multiscalar mixing by Girimaji (1993). In this paper, we review the development of the time-evolving mapping closure model and its application to multiscalar mixing. The main purpose is to explain the concepts of the mapping closure method to the general combustion community without dwelling on the mathematical details of the approach.

The time-evolving mapping closure method for two-scalar mixing is developed and validated in Section 2. In Section 3, this mapping closure method is extended to multiple scalar mixing. The paper concludes in Section 4 with a discussion.

2. Two scalar mixing

2.1. Model development

As the simplest nontrivial case, we consider the passive mixing of two scalars in a constant-density isotropic turbulent field $\mathbf{U}(\mathbf{y}, t)$.

Let the entire field be composed of the two scalars, so that if ϕ is the mass fraction of one of the scalars, $(1-\phi)$ is the mass fraction of the other. The scalar field ϕ evolves from its initial state according to

$$\frac{\partial \phi}{\partial t}+\frac{\partial \phi U_i}{\partial y_i}=D\frac{\partial^2 \phi}{\partial y_i \partial y_i}, \tag{1}$$

where D represents the coefficient of Fickian diffusion. Here, and throughout the remainder of the paper, repeated subscripts imply summation.

The pdf of the mass fraction evolves in the isotropic field according to (Pope 1985)

$$\begin{aligned}\frac{\partial f(\psi)}{\partial t} &= -\frac{\partial}{\partial \psi}\{f(\psi)\Theta(\psi)\} \\ &= -\frac{\partial^2}{\partial \psi \partial \psi}\{f(\psi)\chi(\psi)\},\end{aligned} \tag{2}$$

where ψ represents the probability space value of ϕ. In the above equation, the conditional scalar diffusion $\Theta(\psi)$ is given by

$$\Theta(\psi)=D\langle\frac{\partial^2 \phi}{\partial y_i \partial y_i}|\phi=\psi\rangle, \tag{3}$$

where the notation $\langle P|R\rangle$ denotes the conditional expectation of P with respect to R. The conditional scalar dissipation $\chi(\psi)$ is given by

$$\chi(\psi)=D\langle\frac{\partial \phi}{\partial y_i}\frac{\partial \phi}{\partial y_i}|\phi=\psi\rangle. \tag{4}$$

The cumulative distribution function (cdf, $F(\psi)\equiv\int f(\psi)d\psi$) evolves by the equation (Pope 1985)

$$\frac{\partial F(\psi,t)}{\partial t}=-\Theta(\psi,t)\frac{\partial F(\psi,t)}{\partial \psi}. \tag{5}$$

The variance (σ_ϕ^2) of the mass fraction evolves according to

$$\frac{d\sigma_\phi^2}{dt}=-2\int \chi(\psi)d\psi=-2\epsilon_\phi, \tag{6}$$

where ϵ_ϕ is the mean scalar dissipation of the ϕ field.

For any arbitrary initial pdf, even for the simplest case of isotropic turbulence, neither χ nor Θ is closed in terms of the pdf, and a closure model is required to determine the pdf evolution. However, for certain specific initial conditions (pdf's), approximate descriptions of the subsequent scalar pdf evolutions are known from experiments and DNS. For instance, it is demonstrated in Girimaji (1991, 1992c) that if the initial pdf is double-delta (nonpremixed), the subsequent pdf evolution can be described approximately by a β pdf. It is also known from experiments that if the initial pdf is Gaussian, the subsequent evolution is nearly Gaussian. In the analysis that follows, this scalar field evolving in a known manner is called the *reference scalar field* θ. The scalar field ϕ evolving from an arbitrary initial condition in an unknown manner is called the *physical scalar field.* The objective of the mapping closure method is to extract information about the unknown evolution of the scalar ϕ from the known evolution of the scalar field θ, using an appropriate *mapping function* $X(\theta, t)$.

As in the case of previous mapping closure models, it is assumed here that although the evolution of the scalar pdf is unknown, the evolution of the variance (and, hence, ϵ_ϕ) is known.

Reference field θ(y, t). Consider a notional scalar field $\theta(\mathbf{y}, t)$ residing in the same physical space as ϕ, evolving according to the same instantaneous evolution equation (1), and subject to the same velocity field $\mathbf{U}(\mathbf{y}, t)$. The probability space variable, the pdf, and the cdf of θ are η, $b(\eta)$, and $B(\eta)$, respectively. The cdf evolves according to

$$\frac{\partial B(\eta, t)}{\partial t} = -\Theta(\eta, t)\frac{\partial B(\eta, t)}{\partial \eta}. \tag{7}$$

The choice of the initial θ-field is such that an approximate description of the subsequent evolution is known.

Mapping function $X(\eta, t)$ The mapping function $X(\theta, t)$ describes a one-to-one transformation between the physical and the reference fields:

$$\theta \overset{X(\theta,t)}{\rightarrow} \phi. \tag{8}$$

The corresponding mapping in the probability space is

$$\eta \overset{X(\eta,t)}{\rightarrow} \psi. \tag{9}$$

The mapping function is defined in terms of cdf's of ϕ and θ:

$$F(X(\eta, t), t) = B(\eta, t). \tag{10}$$

Such a mapping function exists and is a non-decreasing function of η at any given time, leading to (differentiating equation 10 with respect to X):

$$f(X(\eta,t),t) = b(\eta,t)\frac{\partial \eta}{\partial X}. \tag{11}$$

Differentiating equation (10) with respect to time we get

$$\frac{\partial F(X(\eta,t),t)}{\partial t} = -f(X(\eta,t)=\psi,t)\frac{\partial X(\eta,t)}{\partial t} + \frac{\partial B(\eta,t)}{\partial t}. \tag{12}$$

At any instant in time, the unknown Laplacian of ϕ can be written in terms of the derivatives of the θ field using the mapping relationship (equation 8):

$$\frac{\partial^2 \phi}{\partial y_i \partial y_i} = \frac{\partial^2 \theta}{\partial y_i \partial y_i}\frac{\partial X}{\partial \theta} + \frac{\partial \theta}{\partial y_i}\frac{\partial \theta}{\partial y_i}\frac{\partial^2 X}{\partial \theta \partial \theta}. \tag{13}$$

Taking the conditional expectation of the above equation and using the notation that prime denotes differentiation with respect to η we get

$$\Theta(\psi,t) = X'\Theta(\eta,t) + X''\chi(\eta,t). \tag{14}$$

In the above equation, $\chi(\eta,t)$ and $\Theta(\eta,t)$ are the conditional scalar dissipation and diffusion of the θ field.

Evolution of the mapping function. Substituting the expression (equation 14) for conditional scalar diffusion of the ϕ field into the cdf evolution equation (equation 5) we get

$$\frac{\partial F(\psi,t)}{\partial t} = -[X'\Theta(\eta,t) + X''\chi(\eta,t)]\frac{\partial F}{\partial \psi}. \tag{15}$$

Comparison of this result (equation 15) with equation (12) leads to

$$\dot{X}(\eta,t) = \chi(\eta,t)X'' + \frac{1}{(\partial F/\partial \psi)}[\dot{B}(\eta,t) + X'\frac{\partial F}{\partial \psi}\Theta(\eta,t)], \tag{16}$$

where overdot represents differentiation with respect to time. From equation (11) it is easily seen that

$$\frac{\partial F}{\partial \psi} = \frac{\partial B}{\partial \eta}\frac{\partial \eta}{\partial X} = \frac{1}{X'}\frac{\partial B}{\partial \eta}. \tag{17}$$

The quantity in the square brackets in equation (16) can then be written as

$$\begin{aligned} \dot{B}(\eta,t) + X'\frac{\partial F}{\partial \psi}\Theta(\eta,t) &= \dot{B}(\eta,t) + \Theta(\eta,t)\frac{\partial B}{\partial \eta} \\ &= 0. \end{aligned} \tag{18}$$

The last equality follows from equation (7).

Therefore, the evolution equation of the mapping function is

$$\dot{X}(\eta,t) = \chi(\eta,t)X'', \tag{19}$$

where $\chi(\eta,t)$ is the conditional scalar dissipation of the reference scalar field $\theta(\mathbf{y},\mathrm{t})$.

The mapping closure model. Given the initial cdf's of ϕ and θ, the initial mapping function is determined from equation (10). Knowing $\chi(\eta,t)$, the mapping function evolution is calculated from equation (19). The mapping function at any time t is substituted back into equation (10) to obtain the cdf of ϕ. The pdf is obtained easily by differentiating the cdf.

The conditional scalar dissipation $\chi(\eta,t)$ depends on the choice of the reference scalar field and the mean scalar dissipation ϵ_ϕ. For Gaussian and β reference fields, the relationship between $\chi(\eta,t)$ and $\epsilon_\phi(t)$ is provided in Girimaji (1992b). The mapping closure model with the Gaussian reference field (MCMG) has received a good deal of attention in the past (Gao 1991, Pope 1991). In this work, attention is restricted to the time-evolving mapping closure with β reference field (MCMB). It is shown in Girimaji (1992a) that MCMB may be more accurate than MCMG during the final stages of mixing.

2.2. The β reference field

The β-pdf of the reference scalar field θ is given by

$$b(\eta,t) = \eta^{\beta_1-1}(1-\eta)^{\beta_2-1}/N; \quad (0 \le \eta \le 1), \tag{20}$$

where the normalization factor N is

$$N = \int_0^1 \eta^{\beta_1-1}(1-\eta)^{\beta_2-1}d\eta = \frac{\Gamma(\beta_1)\Gamma(\beta_2)}{\Gamma(\beta_1+\beta_2)}, \tag{21}$$

and Γ represents the gamma function. The time-dependent parameters β_1 and β_2 of the model are related to the mean (μ) and the variance ($\sigma_\theta^2(t)$) of θ according to

$$\begin{aligned} \beta_1(t) &= \mu(\frac{\mu(1-\mu)}{\sigma_\theta^2(t)} - 1), \\ \beta_2(t) &= (1-\mu)(\frac{\mu(1-\mu)}{\sigma_\theta^2(t)} - 1). \end{aligned} \tag{22}$$

In a pure mixing problem, the mean remains constant, whereas the variance decays in time.

The conditional scalar dissipation corresponding to the β-pdf evolution is (Girimaji 1992b)

$$\frac{\chi(\eta,t)}{\epsilon_\theta} = -2\frac{\mu(1-\mu)}{\sigma_\theta^4}\frac{I(\eta)}{b(\eta)}, \tag{23}$$

where $I(\eta)$ is given by

$$I(\eta) = \int_0^\eta [\mu\{\ln\eta' - \langle\ln\theta\rangle\} + (1-\mu)\{\ln(1-\eta') - \langle\ln(1-\theta)\rangle\}]b(\eta')(\eta-\eta')d\eta'. \tag{24}$$

2.3. Mapping function

Consider the evolution of the mapping function in scaled time τ:

$$\tau = \int_0^t 2\epsilon_\theta(t)dt, \tag{25}$$

so that

$$\sigma_\theta^2(\tau) = 1 - \tau. \tag{26}$$

The relationship between the scaled time and the real time can be calculated by knowing the mean scalar dissipation of the physical field. In this case the mean dissipations of the two fields are related according to

$$\epsilon_\theta(t) = \epsilon_\phi(t)/Q_2(t), \tag{27}$$

where

$$Q_2(t) = -2\frac{\mu(1-\mu)}{\sigma_\theta^4}\int\frac{\partial X}{\partial\eta}\frac{\partial X}{\partial\eta}I(\eta,t)d\eta. \tag{28}$$

The mapping-function evolution equation (19) in scaled time is

$$\frac{\partial X(\eta,\tau)}{\partial \tau} = \frac{1}{2}[\frac{\chi(\eta,t)}{\epsilon_\theta}]\frac{\partial^2 X}{\partial\eta\partial\eta}. \tag{29}$$

The quantity in the square brackets is modeled with equation (23). By definition, ψ_{min} is mapped on to $\eta = 0$, and ψ_{max}, to $\eta = 1$. So the boundary conditions for the above evolution equation are (from equation 10)

$$X(0,\tau) = 0; \quad X(1,\tau) = 1. \tag{30}$$

The mean and the initial variance of the reference field are chosen to be those of the initial physical scalar field. The initial mapping, $X(\eta, 0)$, is then obtained using equation (10), with $B(\eta, 0)$ being the cdf corresponding to the β pdf.

In scaled time, the β pdf parameters evolve according to

$$\begin{aligned}\frac{d\beta_1(\tau)}{d\tau} &= \frac{\mu^2(1-\mu)}{(\sigma_0^2-\tau)^2} \\ \frac{d\beta_2(\tau)}{d\tau} &= \frac{\mu(1-\mu)^2}{(\sigma_0^2-\tau)^2}.\end{aligned} \tag{31}$$

Equations (31), (23), and (29) are then solved knowing the initial mapping.

Range of validity of the model. The range of validity of the mapping closure model is discussed in detail in Girimaji (1992a). It is shown that the mapping closure model is formally invalid during the final stages of mixing. This is due to the fact that the transformation from the physical to reference field ceases to be one-to-one during the final stages. This leads to the mapping closure model being inaccurate during the final stages of the mixing. The original model of CCK, Pope's (1991) version of the mapping closure model, and the present model all suffer from this shortcoming. In this paper, we shall not dwell on the shortcomoings of the model, but rather present the comparison between the model and data.

2.4. Comparison with data

In this subsection, we test MCMB against numerical data. For this purpose, rather than DNS data, we use stochastic diffusion or heat conduction simulation (HCS) data of Girimaji (1992c). In the HCS, the scalar field evolves from random initial conditions in the absence of a velocity field. It is shown in Girimaji (1992c) that validation of turbulent mixing models against such data is reasonable in appropriately normalized time.

In mixing problems of practical importance, the initial scalar pdf is arbitrary in shape. It is not possible to evaluate the performance of the model for all arbitrary shapes. In Girimaji (1992a) the model is tested closely against many initial pdf's. Of those, we present the comparison between the model and HCS data of scalar mixing from the quadruple-δ initial pdf:

$$f(\psi,0) = 0.2\delta(\psi)+0.3\delta(\psi-0.25)+0.3\delta(\psi-0.75)+0.2\delta(\psi-1). \quad (32)$$

(The mean and the initial variance are 0.5 and 0.1375, respectively.) In Figure 1, the cdf of the scalar calculated from the model is compared against HCS data at various stages of mixing characterized by the scalar variance. Also plotted for comparison is the unmodified β pdf at the various variances. At the early stages of mixing, the HCS pdf exhibits steep changes due to the intermediate δ-functions. The MCMB appears to capture the steep changes in the cdf quite well. However, the unmodified β cdf is quite inaccurate. Clearly, the mapping is essential for capturing the early-time behavior. With mixing, the intermediate δ-functions dissipate quickly, and during the latter stages, the cdf appears quite smooth. The MCMB captures these changes in the cdf quite adequately. During the final stages the β-cdf is fairly close to the HCS cdf. At these values of the variance, the β pdf is close to Gaussian, indicating that the asymptotic shape of the HCS cdf is close to Gaussian even for the quadruple-δ initial condition.

3. Multiscalar mixing

Most turbulent mixing problems of practical interest in turbulent combustion involve multiple scalar mixing. Pope (1991) and Gao (1991b) have extended the original mapping closure methodology (of CCK) to the mixing of multiple scalars. The application of these models depends on the ordering of the various scalars involved in the mixing process. This, as the authors themselves point out, is a clear violation of the linearity and independence principles outlined by Pope (1983). In this Section, we present a new multiscalar mixing model based on the time-evolving mapping closure methodology presented in the previous Section. The present multiscalar model makes the simplification that the conditional diffusion of a particular scalar depends only upon the value of that scalar and is independent of the composition of the remainder of the mixture. This simplification is described in detail and justified elsewhere (Girimaji 1992c). The ensuing model satisfies the linearity and independence requirements of Pope (1983). In the latter part of this Section, the model is compared against numerical data. Owing to the lack of detailed multiscalar mixing data from either DNS or experiments, the model is validated, again, by comparison against heat conduction simulations (HCS) data.

We consider the mixing of $(n+1)$ scalars (of mass fraction ϕ_1, $\phi_2, \cdots, \phi_{n+1}$) in a constant-density homogeneous velocity field $\mathbf{U}(\mathbf{y}, t)$. The definition of mass fraction is such that

$$0 \leq \phi_k \leq 1, \text{ and } \sum_{k=1}^{n+1} \phi_k = 1. \tag{33}$$

We consider the case when molecular diffusion is governed by Fick's law. In such a case, each mass fraction evolves according to

$$\frac{\partial \phi_k}{\partial t} + \frac{\partial \phi_k U_i}{\partial y_i} = D_k \frac{\partial^2 \phi_k}{\partial y_i \partial y_i}, \tag{34}$$

where D_k is the coefficient of Fickian diffusion of species k.

In homogeneous turbulence, the evolution of the joint probability density function (jpdf) of the scalar composition vector is governed by (Pope 1985)

$$\frac{\partial f(\underline{\psi})}{\partial t} = -\sum_{k=1}^{n} \frac{\partial \Theta_k(\underline{\psi}) f(\underline{\psi})}{\partial \psi_k}, \tag{35}$$

where f represents the jpdf, $\underline{\psi}$ is the sample space variable of the composition vector $\underline{\phi}$, and Θ_k is the conditional diffusion of the turbulent scalar given by

$$\Theta_k(\underline{\psi}) = D_k \langle \frac{\partial^2 \phi_k}{\partial y_i \partial y_i} | \underline{\phi} = \underline{\psi} \rangle. \tag{36}$$

It is important to note that the conditional scalar dissipation in the multiscalar case is conditioned upon the composition of the entire mixture rather than the mass fraction of the scalar in question only, as was the case in the previous Section. However, it is argued in Girimaji (1993) that within the framework of Fickian diffusion it is reasonable to replace this conditioning upon the entire composition vector by that on the particular scalar in consideration only:

$$\begin{aligned}\Theta_k(\underline{\psi}) &= D_k\langle\frac{\partial^2\phi_k}{\partial y_i\partial y_i}|\underline{\phi}=\underline{\psi}\rangle \\ &\approx D_k\langle\frac{\partial^2\phi_k}{\partial y_i\partial y_i}|\phi_k=\psi_k\rangle.\end{aligned} \tag{37}$$

We first develop the model for three-scalar mixing. Subject to the simplification, the jpdf evolution equation is

$$\frac{\partial f(\psi_1,\psi_2)}{\partial t} = -\frac{\partial[\Theta_1(\psi_1)f(\psi_1,\psi_2)]}{\partial\psi_1} - \frac{\partial[\Theta_2(\psi_2)f(\psi_1,\psi_2)]}{\partial\psi_2}. \tag{38}$$

The joint cumulative distribution function (jcdf), F, is defined as

$$F(\psi_1,\psi_2) = \int_0^{\psi_1}\int_0^{\psi_2} f(\gamma_1,\gamma_2)\mathrm{d}\gamma_1\mathrm{d}\gamma_2. \tag{39}$$

The evolution equation of the jcdf is obtained by integrating equation (38):

$$\frac{\partial F(\psi_1,\psi_2)}{\partial t} = -\Theta_1(\psi_1)\int_0^{\psi_2} f(\psi_1,\gamma_2)d\gamma_2 - \Theta_2(\psi_2)\int_0^{\psi_1} f(\gamma_1,\psi_2)d\gamma_1. \tag{40}$$

In deriving the above expression, the following identities have been used (from Girimaji 1992b):

$$\Theta_1(\psi_1=0)=0;\;\;\Theta_2(\psi_2=0)=0. \tag{41}$$

Equation (40) can be simplified further:

$$\begin{aligned}\int_0^{\psi_2} f(\psi_1,\gamma_2)d\gamma_2 &= \int_0^{\psi_2}\frac{\partial}{\partial\gamma_2}[\frac{\partial F(\psi_1,\gamma_2)}{\partial\psi_1}]d\gamma_2 \\ &= \frac{\partial F(\psi_1,\psi_2)}{\partial\psi_1} - \frac{\partial F(\psi_1,0)}{\partial\psi_1}.\end{aligned} \tag{42}$$

By definition, for continuous random variables ϕ_1 and ϕ_2,

$$F(\psi_1,0) = Prob\{\phi_1<\psi_1,\phi_2<0\} = 0 \;\;\text{for all } \psi_1. \tag{43}$$

Therefore,

$$\int_0^{\psi_2} f(\psi_1, \gamma_2) d\gamma_2 = \frac{\partial F(\psi_1, \psi_2)}{\partial \psi_1}. \tag{44}$$

Similarly,

$$\int_0^{\psi_1} f(\gamma_1, \psi_2) d\gamma_1 = \frac{\partial F(\psi_1, \psi_2)}{\partial \psi_2}. \tag{45}$$

Equation (40) can then be rewritten as

$$\frac{\partial F(\psi_1, \psi_2)}{\partial t} = -\Theta_1(\psi_1)\frac{\partial F(\psi_1, \psi_2)}{\partial \psi_1} - \Theta_2(\psi_2)\frac{\partial F(\psi_1, \psi_2)}{\partial \psi_2}. \tag{46}$$

The simplified jcdf evolution equation (46) is hyperbolic in nature. This can be solved by the method of characteristics (Snedden 1984). It can be shown using this method that the jcdf value is conserved in the (ψ_1, ψ_2, t) - space along lines called the characteristic lines. The coordinates of the constant-jcdf line originating (at time $t = 0$) from ψ_1^0 and ψ_2^0 evolve according to

$$\begin{aligned} \psi_1^*(\psi_1^0, t) &= \psi_1^0 + \int_0^t \Theta_1[\psi_1^*(\psi_1^0, s)] ds, \\ \psi_2^*(\psi_2^0, t) &= \psi_2^0 + \int_0^t \Theta_2[\psi_2^*(\psi_2^0, s)] ds. \end{aligned} \tag{47}$$

Along these lines we have

$$F[\psi_1^*(\psi_1^0, t), \psi_2^*(\psi_2^0, t); t] = \text{Constant} = F[\psi_1^0, \psi_2^0; 0]. \tag{48}$$

Determination of jcdf from mcdf's. The evolution of the jcdf can be determined knowing the evolution of the marginal cumulative distribution functions (mcdf) of the participating scalars. The mcdf of the scalar k is defined as

$$F_k(\psi_k, t) = Prob\{\phi_k(t) \leq \psi_k\}. \tag{49}$$

The mcdf evolves according to

$$\frac{\partial F_k(\psi_k; t)}{\partial t} = -\Theta_k(\psi_k)\frac{\partial F_k(\psi_k; t)}{\partial \psi_k}. \tag{50}$$

This equation is again hyperbolic. In the (ψ_k, t) - plane, the mcdf value is conserved along the lines $[\psi_k^*; t]$:

$$F_k[\psi_k^*(\psi_k^0, t); t] = F_k[\psi_k^0; 0], \tag{51}$$

where the coordinate evolves according to

$$\psi_k^*(\psi_k^0, t) = \psi_k^0 + \int_0^t \Theta_k[\psi_k^*(\psi_k^0, s)]ds. \tag{52}$$

Equations (47) and (52) are identical. The implication is the following. Let $[\psi_1^*(\psi_1^0, t); t]$ be a constant-mcdf (of species 1) line on the (ψ_1, t) plane. Similarly, $[\psi_2^*(\psi_2^0, t); t]$ is a constant-mcdf line on the (ψ_2, t) plane. Then, according to equations (47) and (52), the jcdf is conserved along the line $[\psi_1^*, \psi_2^*; t]$ in the (ψ_1, ψ_2, t)-space. Therefore, if the constant-mcdf lines of the participating scalars are known, then the constant-jcdf line can be easily identified. We use the time-evolving MCMB presented in the previous Section to determine the evolution of the constant-mcdf lines. As mentioned in the previous Section, the model for the mcdf is formally invalid during the final stages of mixing. Hence, the model for the jcdf cannot be expected to be accurate at the final stages of mixing.

The above analysis is easily extended to any number of scalars. Further, the model does not depend on the ordering of the species, unlike the previous multiscalar mapping closure models. As is the case with all mapping closure models, the mean scalar dissipation of each scalar needs to be specified from an external source.

3.1. Multiscalar model implementation

The calculations from the model are compared against HCS data for the case of stochastic mixing of three nonpremixed scalars. With initially segregated species, the initial ($t = 0$) joint and marginal pdf's are composed of dirac delta functions. As a result, the initial jcdf and mcdf's contain discontinuities. This leads to difficulties in the numerical implementation of the procedure, described below, to calculate the jcdf. To circumvent this problem, we specify the jcdf and mcdf's at a time, $t_r (> 0)$, called the reference time. Knowing the jcdf and mcdf's at the reference time, the procedure described below can be used to calculate the jcdf at earlier ($0 < t \leq t_r$) and later ($t > t_r$) times. For problems with more benign initial conditions, the initial time ($t = 0$) can be taken as the reference time, and the analysis below would still be valid.

For the case of initially segregated mixing, the time-evolving MCMB of the previous Section reduces to the basic β-pdf model of Girimaji (1991). In that case, the mcdf of scalar k is given by

$$F_k(\psi_k;t) = I(\beta_{1k}(t), \beta_{2k}(t); \psi_k), \tag{53}$$

where the incomplete beta function I is given by

$$I(a,b;x) = \frac{\Gamma(a+b)}{\Gamma(a)\Gamma(b)} \int_0^x y^{a-1}(1-y)^{b-1} dy. \tag{54}$$

The parameters β_{1k} and β_{2k} are functions of the mean and variance of the scalar k as given in equation (22).

The MCMB model is implemented as follows. Let t_r represent the reference time and the index r denote quantities at that time. From the HCS data at t_r (slightly larger than zero), the jcdf at the reference time is specified at a set of J points, $(\psi_1^{rj}, \psi_2^{rj}; t_r)$, $j = 1, J$. The mean scalar-dissipation evolution from HCS is used to determine the scalar variance at any time t. The parameters of the β pdf at time t are then determined from equation (22). Next, the coordinates of each constant-jcdf line are calculated from equation (53):

$$\psi_k^*(\psi_k^r, t) = B[\beta_{1k}(t), \beta_{2k}(t); I(\beta_{1k}^r, \beta_{2k}^r; \psi_k^r)]. \tag{55}$$

The function B is the inverse-incomplete-β function defined such that

$$I[\beta_{1k}(t), \beta_{2k}(t); B = \psi_k^*(t)] = I[a_{1k}^r, b_{2k}^r; \psi_k^r]. \tag{56}$$

The jcdf is conserved along the line $[\psi_1^*(t), \psi_2^*(t); t]$:

$$F[\psi_1^{*j}(t), \psi_2^{*j}(t); t] = F[\psi_1^{rj}, \psi_2^{rj}; t_r], \quad \text{for all } j. \tag{57}$$

Thus, the jcdf at time t is determined. The various joint statistics are calculated knowing the jcdf.

3.2. Model vs. HCS comparison

We first present some results from the HCS calculations to demonstrate the complexity of multiscalar mixing. Five heat conduction simulations are performed on a 128^3 grid. For all the cases, the initial marginal pdf's are the same:

$$\begin{aligned} f_1(\psi_1;0) &= \frac{1}{2}[\delta(\psi_1) + \delta(\psi_1 - 1)], \\ f_2(\psi_2;0) &= 0.75\delta(\psi_2) + 0.25\delta(\psi_2 - 1), \\ f_3(\psi_3;0) &= 0.75\delta(\psi_3) + 0.25\delta(\psi_3 - 1). \end{aligned} \tag{58}$$

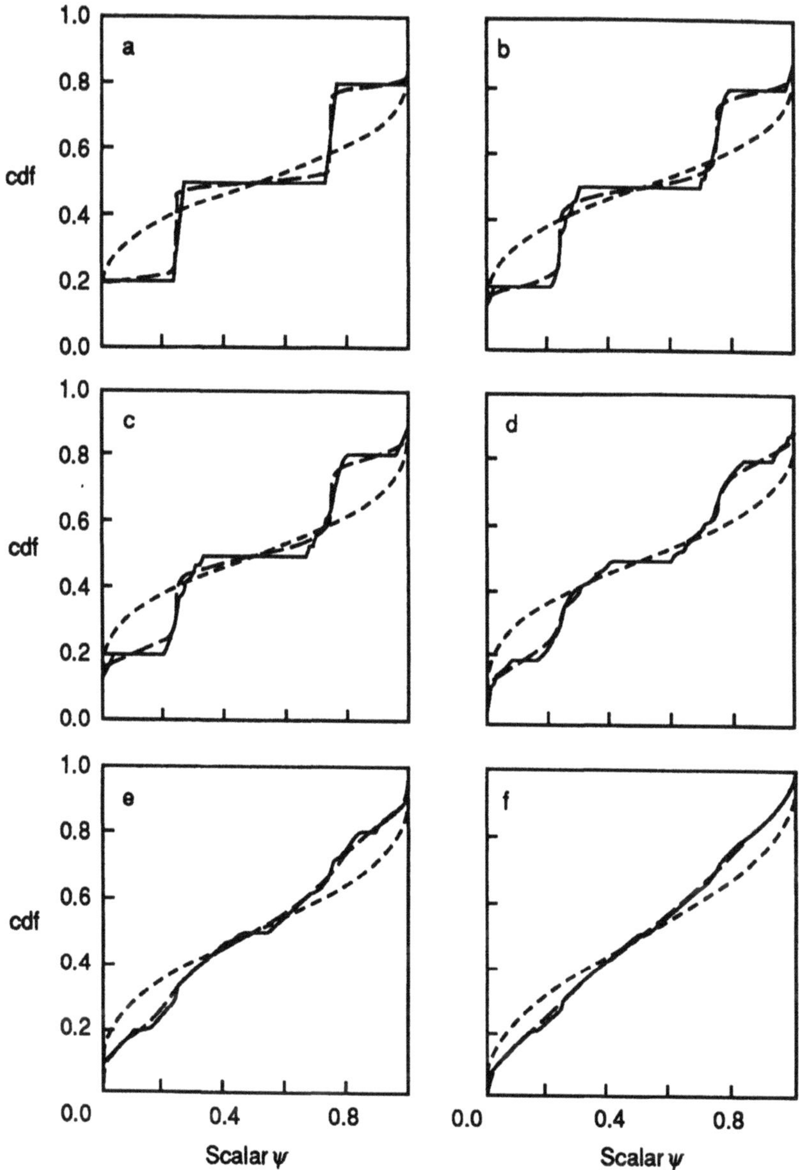

Figure 1: Evolution of the scalar cdf from the quadruple-δ initial condition. Solid line represents HCS data (initial wavenumber of 8); - - - represents the unmodified β-pdf model; – – – – – – represents the mapped β-pdf model. (a) $\sigma_\phi^2 = 0.133$; (b) $\sigma_\phi^2 = 0.132$; (c) $\sigma_\phi^2 = 0.124$; (d) $\sigma_\phi^2 = 0.117$; (e) $\sigma_\phi^2 = 0.107$; (f) $\sigma_\phi^2 = 0.057$.

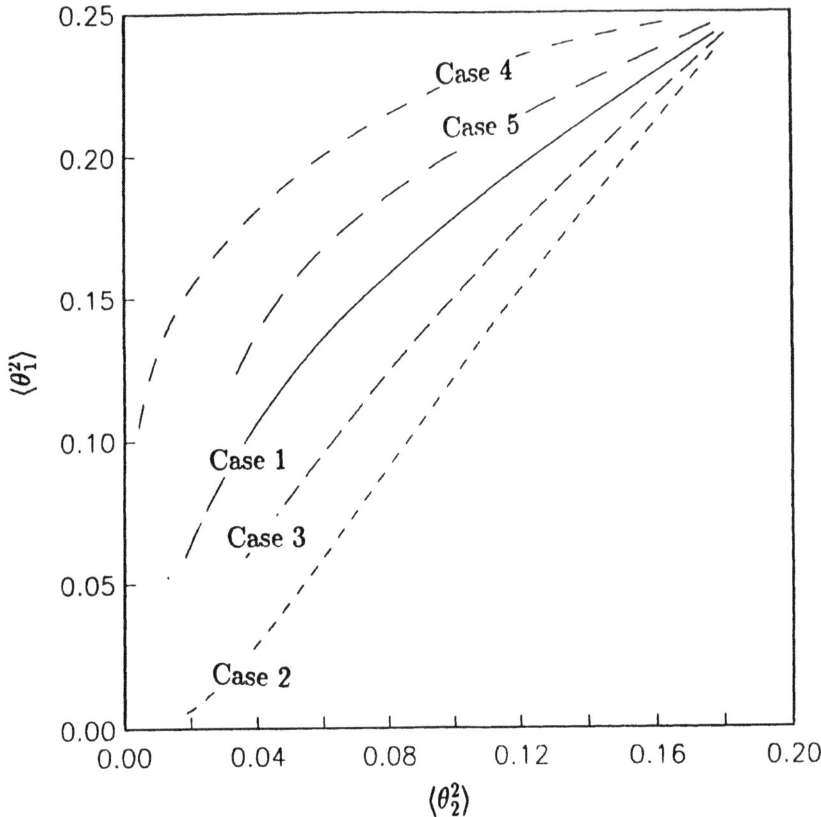

Figure 2: HCS data. $\langle\theta_2^2\rangle$ vs. $\langle\theta_1^2\rangle$ for various cases.

The joint pdf of the two independent mass fractions is

$$f(\psi_1,\psi_2;0) = 0.25\delta(\psi_1)\delta(\psi_2)+0.50\delta(\psi_1-1)\delta(\psi_2)+0.25\delta(\psi_1)\delta(\psi_2-1). \tag{59}$$

The parameters of the various simulations are provided in Table 1. The initial joint-scalar statistics are identical for all the cases. However, the subsequent evolutions are different, depending upon the length scales and the Fickian-diffusion coefficients. The variance of each scalar decays monotonically, going to zero at large times. Hence, the variance value is a good indicator of the degree of mixedness. In all the figures presented, the variance of scalar 2 is plotted on the x- axis, rather than time, as a measure of the mixedness. Also, throughout the remainder of the paper, angular brackets imply the spatial average value. The results are presented in terms of the fluctuating quantities:

$$\begin{aligned} \theta_1 &= \phi_1 - \mu_1, \\ \theta_2 &= \phi_2 - \mu_2, \\ \theta_3 &= \phi_3 - \mu_3. \end{aligned} \tag{60}$$

These fluctuations are related according to

$$\theta_1 + \theta_2 + \theta_3 = 0. \tag{61}$$

In Figure 2, $\langle\theta_1^2\rangle$ is plotted against $\langle\theta_2^2\rangle$. Although the $\langle\theta_1^2\rangle$ decays monotonically in all cases, the manners of evolution are quite different. For the cases considered, $\langle\theta_1^2\rangle$ decreases most rapidly in Case 2 and least rapidly in Case 4.

In Figure 3, the correlation $M1$ $(= \langle\theta_1\theta_2\rangle/\sqrt{\langle\theta_1^2\rangle\langle\theta_2^2\rangle})$ is plotted against $\langle\theta_2^2\rangle$ for various cases. For all the cases, the initial correlation is the same, approximately -0.58. Since the sum of the fluctuations is constrained to be zero (equation 61), the negative correlation is to be expected. With mixing, in some of the cases (Cases 1 and 5), the correlation values become more negative, approaching negative unity. So, in the limit of complete mixedness, the scalars 1 and 2 appear to be perfectly negatively correlated. In Cases 2 and 4, with mixing, the two scalars appear to get more and more uncorrelated, with the correlation value approaching zero. In Case 3, the correlation appears to hold steady at its initial value.

It is quite clear that the variances of the two scalars do not uniquely determine the correlation between the scalars. It is very

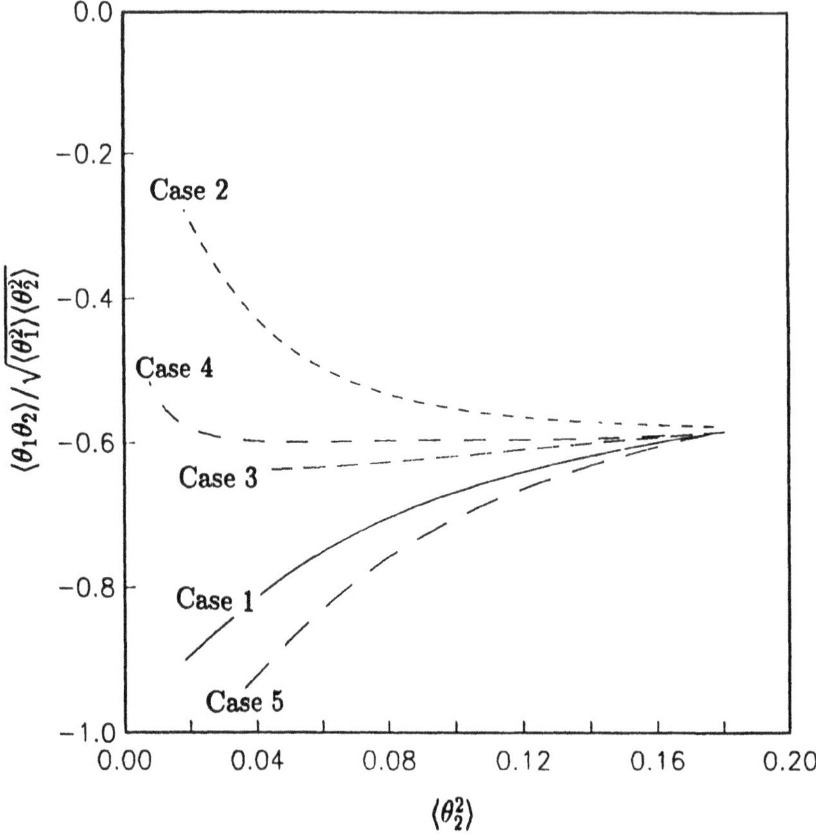

Figure 3: HCS data. $\langle\theta_2^2\rangle$ vs. correlation $M1$ for various cases.

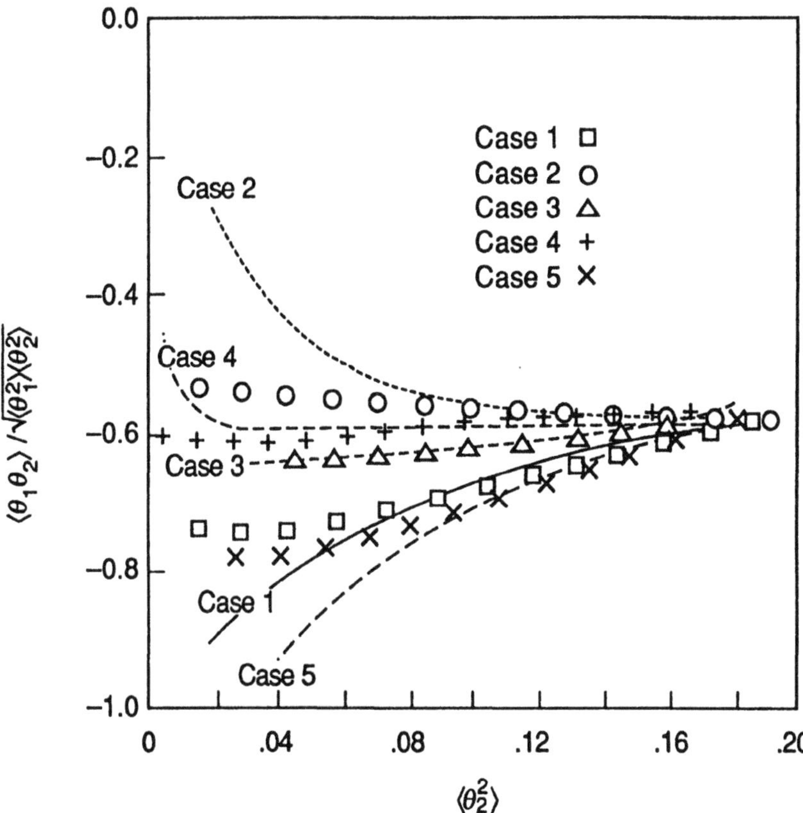

Figure 4: Model vs. HCS data. $\langle\theta_2^2\rangle$ vs. correlation $M1$ for various cases. Lines – HCS, symbols – model.

likely that the higher order joint statistics would also not be unique functions of the two variances. Given this complex behavior of the joint moments, can the present model – where the scalar-variance evolutions are the only inputs – be adequate?

In Figure 4, the correlations from MCMB are compared against HCS data. For all the five cases, in the early stages, the model performs well. It gives the increasing or decreasing trend as required. During the final stages the performance of the model is not as good. This is, as mentioned earlier, due to the failure of the MCMB for the mcdf at this stage of mixing. The formal failure of the model occurs for all the Cases at $\theta_2^2 \approx 0.8$. Clearly the agreement before this value of variance is good, and poor thereafter. More detailed comparison between the model and HCS data is performed in Girimaji (1993). It is demonstrated that the model captures the behavior of the higher order moments well also.

4. Conclusion

In this paper, the time evolving mapping closure method (Girimaji 1992a) and its application to multiscalar mixing (Girimaji 1993) are described. The model calculations with the β reference field are compared against stochastic mixing (HCS) data for the case of non-premixed two and multiple scalar mixing. The agreement is good except during the final stages of mixing, when the mapping transformation is formally invalid.

Despite the success, the mapping closure model suffers from inadequacies that must be overcome before it can be used in turbulent combustion calculations. The most important shortcoming is the fact that it does not address the issue of the scalar variance (or mean scalar dissipation) evolution. In some applications, this evolution may be more important than the details of the scalar pdf evolution. Further, the model is insensitive to the velocity field characteristics (Reynolds number, compressibility) and scalar field parameters (Prandtl and Schmidt numbers). The model permits these parameters to affect the scalar pdf evolution only via their effect on mean scalar dissipation. The extension of the model to multiscalar mixing, when the conditional dissipation simplification (equation 37) is invalid, is not straightforward. When mixing is accompanied by chemical reaction, calculations using the model could be computationally very expensive. Even for the case of two-scalar mixing, the model needs to be tested for a wider variety of initial conditions before it can be classified as a good mixing model. Research addressing these issues related to the mapping closure model should be useful and interesting.

Acknowledgements

The author would like to thank Dr. J. P. Drummond for his constant encouragement and support.

References

Bilger, R. W., 1980. *Turbulent flows with non-premixed reactants.* Turbulent reacting flows (ed. P. A. Libby and F. A. Williams), Springer - Verlag.

Chen, H., Chen, S., and Kraichnan, R. H., 1989. *Probability distribution of a stochastically advected scalar field.* Physical Review Letters, **63** (24), 2657 - 2660, 1989.

Eswaran, V., and Pope, S. B., 1988. *Direct Numerical Simulations of the turbulent mixing of a passive scalar.* Physics of Fluids, **31** (3), 506 - 520.

Gao, F., 1991a. *An analytical solution for the probability density function in homogeneous turbulence.* Phys. Fluids A **3** (4), 511 - 513.

Gao, F., 1991b. *Mapping closure for multispecies Fickian diffusion.* Phys. Fluids **3** (10), 2438 - 2444.

Girimaji, S. S., 1991. *Assumed β-pdf model for turbulent mixing: validation and extension to multiple scalar mixing.* Combust. Sci. and Tech. **78**, 4 - 6, 177 - 196.

Girimaji, S. S., 1992a. *A mapping closure for turbulent scalar mixing using a time-evolving reference field.* Phys. Fluids. A. **4**, 12, 2875 - 2886.

Girimaji, S. S., 1992b. *On the modeling of scalar diffusion in isotropic turbulence.* Phys. Fluids. A, **4** (11), 2529 - 2537.

Girimaji, S. S., 1992c. *Towards understanding turbulent scalar mixing.* NASA CR 4446.

Girimaji, S. S., 1993. *A study of multiscalar mixing.* Submitted to Phys. Fluids A.

Pope, S. B., 1983. *Consistent modeling of scalars in turbulent flows.* Phys. Fluids **26**, 404 - 408.

Pope, S. B., 1985. *PDF methods for turbulent reacting flows.* Prog. Energy Combust. Sci., **11**, 119 - 192.

Pope, S. B., 1987. *Turbulent premixed flames.* Ann. Rev. Fluid Mech. **19**, 237 - 270.

Pope, S. B., 1991. *Mapping closures for turbulent mixing and reaction.* Theo. and Comp. Fluid Dynamics, **2**, 255 - 270.

Sneddon, I. N., 1984. *Elements of partial differential equations.* McGraw- Hill International Book Company.

Table 1: Parameters of the various simulations.

Case	$p_{10} = 2\pi/l_{10}$	$p_{20} = 2\pi/l_{20}$	D_1	D_2	$\langle\theta_2^2\rangle$ *at* t_r
1	8	16	1.0	1.0	0.0877
2	16	8	1.0	1.0	0.0876
3	8	8	1.0	1.0	0.0750
4	8	8	0.5	4.0	0.0889
5	4	16	1.0	1.0	0.0967

FINITE-RATE CHEMISTRY EFFECTS IN SUBSONIC AND SUPERSONIC COMBUSTION*

R. W. Pitz, T. M. Brown, T. S. Cheng[1], *S. Nandula, and J. A. Wehrmeyer*[2]

Department of Mechanical Engineering
Vanderbilt University
Nashville, Tennessee 37235

O. Jarrett, Jr. and G. B. Northam

NASA Langley Research Center
Hampton, Virginia 23681-0001

J.-Y. Chen

Department of Mechanical Engineering
University of California
Berkeley, California 94720

ABSTRACT

Finite-rate chemistry effects are revealed in subsonic and supersonic lifted hydrogen flames using narrowband KrF excimer laser (248 nm) induced Raman scattering and fluorescence to measure simultaneously species concentrations (H_2, H_2O, N_2, O_2, OH) and temperature. The fluorescence measurement of OH concentration is quantitative since it is predissociative and insensitive to collisional quenching. Using these simultaneous measurements, the chemistry of turbulent reaction zones is compared to the limits of fast chemistry (equilibrium), slow chemistry (mixing without reaction), and 1D stretched laminar flamelets. The reaction zone chemistry in turbulent subsonic flames is found to be inconsistent with 1D stretched laminar diffusion flames.

*This work was supported by the NASA Langley Research Center (Grants NGT-50263 and NAG-1-770) and the National Science Foundation (PYI Grant CTS-8657130).

[1]Present Address: Department of Mechanical Engineering, Chung Hua Polytechnic Institute, Hsin Chu, Taiwan, 30067, R.O.C.

[2]Present Address: Department of Mechanical and Aerospace Engineering, University of Missouri-Columbia, Columbia, MO 65211.

J Buckmaster et al. (eds.), Combustion in High-Speed Flows, 253–273.

1. Introduction

Hypersonic propulsion research has been underway for a number of years to develop propulsion systems for hypersonic aircraft that are propelled by air-breathing supersonic combustion ramjet (scramjet) engines fueled with hydrogen (Northam and Anderson, 1986). Hydrogen is the preferred fuel because of its short ignition delay time, its higher energy per unit weight, and its better cooling ability. Although hydrogen-air kinetics are fast, the very short mixing times in scramjets are comparable to H_2-air reaction times. Incomplete combustion due to slow chemical kinetics can affect combustion efficiency dramatically. Excessive production of super-equilibrium free radical species due to slow three-body recombination can lead to reduced thrust. However, because of the difficulty in measuring high-speed, high-enthalpy reacting flows, experimental studies of finite-rate chemistry effects in supersonic reacting flows are lacking.

In supersonic reacting flows, measurement of chemical scalars such as temperature and species concentrations are difficult since these scalars usually have Kolmogorov time and length scales much less than $\sim$ 1 ms and $\sim$ 1 mm, respectively. In previous studies of supersonic reacting flows, temperature and concentration measurements were usually made with thermocouples and gas sampling probes (Ferri, Libby and Zakkay, 1964; Beach, 1972). Physical sampling probes lack the necessary spatial and temporal resolutions to measure the small scales of supersonic combustion. Moreover, these probes produce shocks that accelerate reactions approaching the probe and bias the concentration measurement.

Non-intrusive laser methods avoid physical contact with the flame and can measure temperature and species concentration with adequate temporal and spatial resolutions. Laser-based non-intrusive techniques have been applied to measure the state properties of gases in combustion environments for many years (Eckbreth, 1988). However, most of the measurements have been made in subsonic reacting flows. The first OH concentration measurement, using UV absorption, was performed in an exhaust nozzle that simulates Mach 6 flight with subsonic combustion by Lezberg and Franciscus (1963). Jarrett et al. (1988) applied coherent anti-Stokes Raman spectroscopy (CARS) and laser Doppler velocimetry (LDV) techniques separately in a supersonic jet flame and measured temperature, species concentrations (O_2, N_2), and velocity. However, the spatial resolution ($\sim$ 4 mm) of the CARS system was poor for analysis of supersonic reacting flows. Also, the two major species concentrations measured simultaneously (N_2, O_2) were insufficient to determine finite-rate chemistry effects.

Recently, KrF excimer laser-induced spontaneous Raman scattering and predissociative fluorescence techniques have been developed to measure simultaneously the temperature and species concentrations (O_2, N_2, H_2O, H_2, OH) in hydrogen-air flames with good spatial resolution ($\sim$ 0.4 mm) (Cheng et al., 1991; Cheng, Wehrmeyer and Pitz, 1992; Pitz, Nandula and Brown, 1992). The major species concentrations and temperature are measured by spontaneous vibrational Raman scattering. The OH concentration is measured by laser-induced predissociative fluorescence (LIPF) that is insensitive to collisional quenching. From these concentration measurements, the local fuel/air ratio (mixture fraction) can be calculated.

With this Raman/LIPF method, the instantaneous local state of mixing (mixture fraction) and chemistry can be determined in turbulent reacting flows. The simultaneous measurements of the species concentrations and temperature can be compared to the limits of frozen chemistry and equilibrium chemistry to determine the importance of finite-rate chemistry throughout a supersonic flame.

In addition, the chemistry measurements can be compared to models of finite-rate chemistry. Laminar flamelet theory has been used to predict finite-rate chemistry effects in turbulent diffusion flames (Peters and Williams, 1983; Peters, 1984). In this theory, the reaction zones are assumed to be identical to those found in 1D steady stretched laminar diffusion flames. The local rate of scalar dissipation in the turbulent flame is related to the rate of strain imposed on the 1D steady laminar diffusion flame. A probability density function (pdf) of the scalar dissipation rate and a library of strained laminar diffusion flames are used to predict ignition and extinction effects in turbulent flames (Peters, 1984).

In turbulent flames, reaction zones can be affected by unsteady rate of strain and distorted by turbulent eddies. Unsteady rate of strain and flame curvature may affect the flame chemistry and cause it to differ from that predicted by 1D steady flamelets. Flamelet theory assumes specific variations between the major species, minor species, and temperature within the stretched flames. In attached and lifted hydrocarbon-fueled diffusion flames, simultaneous measurements of major species, minor species and temperature have been compared to steady 1D strained laminar flames (Barlow et al., 1990; Stårner, Bilger and Barlow, 1992). In the attached hydrocarbon flames, super-flamelet values of OH and sub-flamelet values of H_2O are measured (Barlow et al., 1990). In the lifted hydrocarbon flames, the chemistry differed from stretched laminar diffusion flames and evidence of premixing was found (Stårner, Bilger and Barlow, 1992).

Swirl, used to enhance mixing in combustors, can lead to increased flame curvature and unsteady rate of strain. Finite-rate chemistry effects in swirl flames have only recently been studied (Pitz et al., 1991, 1992).

In this paper, the UV Raman/LIPF method is applied to measure finite-rate chemistry effects in subsonic and supersonic flames. Simultaneous measurements of major species, minor species, and temperature are made in the turbulent reaction zones. The chemistry in the supersonic flame is compared to the limits of frozen chemistry and equilibrium chemistry. Flamelet theory is evaluated in a subsonic lifted turbulent hydrogen jet diffusion flames with no swirl, weak swirl, and strong swirl. Joint scatter plots of species concentrations from turbulent reaction zones are compared to stretched laminar diffusion flame calculations.

2. UV Raman/LIPF Method

The UV Raman system (Fig. 1) was developed to provide improved spatial resolution ($0.75 \times 0.25 \times 0.4$ mm^3) and temporal resolution (20 ns) for supersonic flame measurements. Details of the Raman system have been reported previously (Cheng, Wehrmeyer and Pitz, 1992) and only a brief description is included in this paper. The Lambda-Physik EMG 160T MSC narrowband KrF excimer laser emits light that is tunable from 248 to 249 nm with a bandwidth of $\sim$ 0.001 nm and a pulse energy of $\sim$ 200 mJ. Light scattered by the 2000 mm focusing lens is measured by a photomultiplier tube (PMT) to provide a relative measure of laser pulse energy. The laser beam is focused toward either a flat-flame "Hencken" calibration burner or the turbulent jet flame. The fluorescence and Stokes vibrational Raman scattering signals are collected by a Cassegrain mirror and focused into a single-grating spectrometer. The anti-Stokes vibrational Raman scattering signal is collected by a separate Cassegrain mirror and focused into a double-grating monochromator. The different flu-

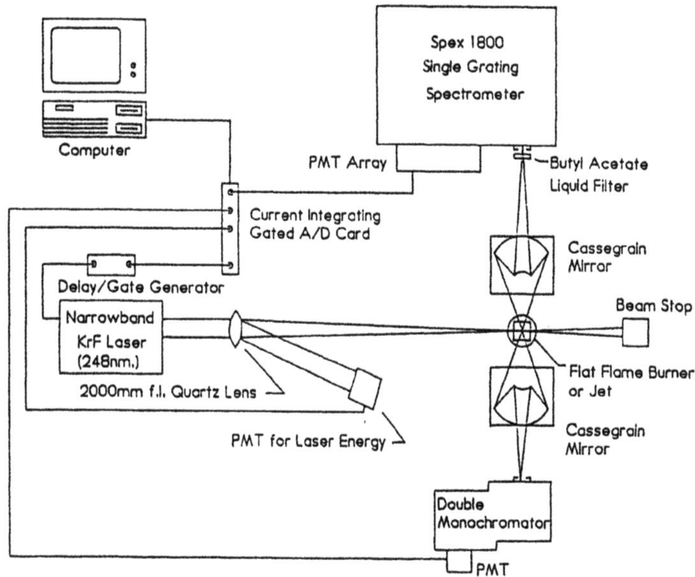

Fig. 1. Schematic of the UV Raman system.

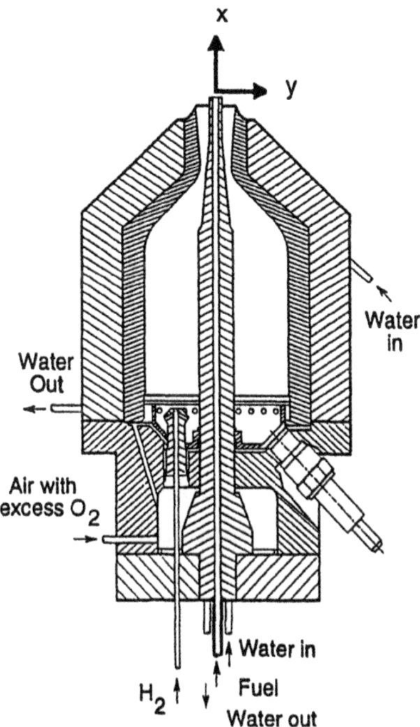

Fig. 2. Schematic of the supersonic burner.

orescence and Raman signals are measured by PMTs whose outputs are collected and measured by a gated A/D system connected to a personal computer for data reduction.

The Raman system is calibrated with the flat-flame "Hencken" burner operating at several known equivalence ratios from lean to rich. Calibration and data reduction procedures for the Raman scattering and LIPF measurements have been described previously (Cheng, Wehrmeyer and Pitz, 1992). The temperature is measured with vibrational Raman scattering of nitrogen by using the ratio of the integrated anti-Stokes signal at 235 nm to the integrated Stokes signal at 263 nm. For temperatures below 800 K, this method is not accurate and the ideal gas law equation is used to determine the temperatures by summing all the major species concentrations and assuming a local pressure.

The detailed background theory, calibration, and precision of the UV Raman system are described elsewhere (Wehrmeyer, Cheng and Pitz, 1992; Cheng, Wehrmeyer and Pitz, 1992). The major contributor to the uncertainties for both concentration and temperature measurements is shot noise originating from the photoelectron emission process in the PMT photocathode. For a single laser pulse, the experimental relative standard deviation for the N_2 concentration measurement is 2.7% at room temperature and 6.3% in a stoichiometric flame at 2300 K. The relative standard deviations in a stoichiometric flame for temperature and OH single-pulse measurements are 11.7% and 13.2%, respectively.

3. Supersonic Flame Results

3.1. Supersonic burner

A schematic diagram of the supersonic burner is shown in Fig. 2. This supersonic burner provides an annular, axisymmetric jet of hot, vitiated air at Mach 2. The annular air jet is concentric to the choked main H_2 fuel jet and the nozzle exit conditions are nearly perfectly expanded. The exit Reynolds number of the central jet is 15,600. Secondary hydrogen is injected into the combustion chamber directly through four injectors (only one of which is shown in the figure), and oxygen enriched air is distributed at the base of the heater by a liner. This hydrogen is burned with the oxygen enriched air, raising the stagnation temperature to the desired level, and the vitiated air is accelerated through a convergent-divergent nozzle, exiting into the atmosphere. The walls of the divergent portion of the convergent-divergent nozzle are conical (half angle of 4.3°). The combustion of the main hydrogen fuel and the vitiated air forms the supersonic flame for this work. The parameters and operating conditions of the supersonic burner are given in Table 1.

UV Raman measurements are made in the radial direction at the following downstream locations: x/D = 0.85, 10.8, 21.5, 32.3, 43.1, 64.7, and 86.1 where x is the downstream distance (see Fig. 2) and D is the inside diameter of the fuel jet (2.36 mm). For each downstream location, a radial scan took about 30 to 40 minutes. Data at each measurement location consists of either 500 or 2000 laser shots. Only a portion of the data is presented here. A complete presentation of the supersonic measurements are given elsewhere (Cheng et al., 1991).

3.2. Mean and rms profiles

Typical profiles of the mean and rms values of species mole fractions, temperature, and mixture fraction are shown in Fig. 3. The mixture fraction, f, is a conserved scalar that describes the state of mixing between fuel and oxidizer. The mixture fraction is calculated from the Raman scattering measurements of the major species concentrations for each laser shot as the mass of hydrogen originating from the fuel stream divided by the total mass (Cheng, Wehrmeyer and Pitz, 1992). The mean temperature profile (Fig. 3c) show higher temperatures at positive y than at negative y. For this axisymmetric jet flame, exact axisymmetry is not obtainable due to installation difficulties and the thermal expansion of the main fuel jet in the hot vitiated air. Hence, the center of the jet (y = 0) is chosen at the maximum mixture fraction and the minimum N_2 mole fraction, as shown by the dashed line in the profiles. This asymmetrical temperature profile may be due to a asymmetrical shock wave interaction that causes ignition at the positive side. This is consistent with the O_2, H_2O, H_2, and OH profiles. On the positive side, more O_2 and H_2 are consumed (Fig. 3e, 3k), and more H_2O and OH are produced (Fig. 3i, 3m). In the temperature rms profiles (Fig. 3d), the two outer peaks are not symmetric because the vitiated air stream is tilted slightly in the positive direction with respect to the centerline. At positive y, the fluctuating rms temperatures near the still air do

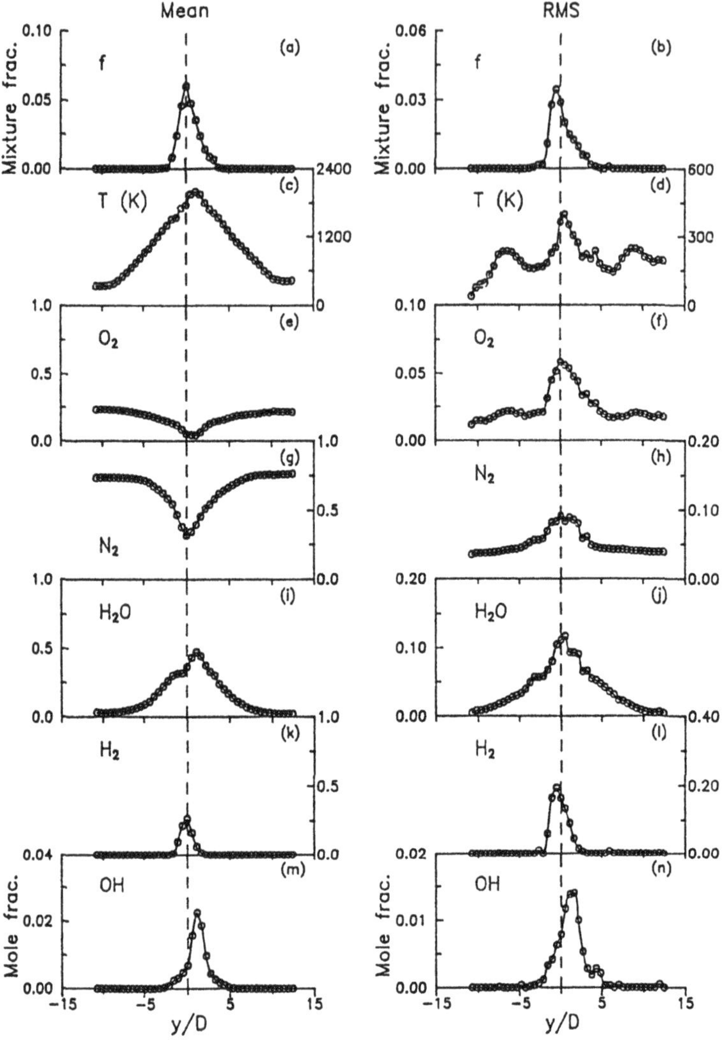

Fig. 3. Mean and rms profiles in the supersonic flame of mixture fraction, temperature, and species mole fractions (H_2, O_2, N_2, H_2O, OH) at $x/D = 32.3$.

not decrease as on the negative side. This is due to tilt in the vitiated air stream and travel limitations of the burner translation stage.

3.3. Finite-rate chemistry effects

To evaluate finite-rate chemistry effects in the flame, the instantaneous chemical measurements are compared to the limits of mixing without reaction and equilibrium in scatter plots (Figs. 4,5). The solid curves represent the adiabatic equilibrium condition, and the dashed curves are for mixing without reaction. At $x/D = 32.2$, the scatter plots for two relative radial positions ($y/D = -1.1$ and 1.1, see Fig. 3) are shown in Figs. 4 and 5. On the negative side, the temperature scatter plot (Fig. 4a) shows that reaction occurs at this point, but the flame temperature is far below the equilibrium curve. Although reaction takes place at this position, super-equilibrium OH is not observed, and most of the scatter points of OH are below the equilibrium curve (Fig. 4b). The scatter points of major species are between the equilibrium curve and the mixing without reaction curve (Fig. 4c, 4d). This suggests that flame extinction/ignition processes are occurring at this position. Another possible explanation is spatial averaging across the steep temperature and species concentration gradients in thin reaction zones.

On the positive side ($x/D = 32.3$, $y/D = 1.1$), the flame temperature (Fig. 5a) is higher than on the negative side due to more reaction. More reaction is possibly due to shock wave interactions on the positive side. The mean temperature is about 2000 K which is about 650 K less than the equilibrium temperature at the mean mixture fraction of 0.035 (Fig. 5a). This lower mean temperature may be due to flame extinction/ignition phenomena or spatial averaging of the thin reaction zones. Some scatter points above the equilibrium curve may be due to experimental uncertainty or differential diffusion. Scatter points near but below the equilibrium curve are caused by the slow three-body recombination of OH radicals which depress the flame temperature below its equilibrium value.

Super-equilibrium OH from slow three-body recombination reactions is apparent in Fig. 5b. The peak OH mole fraction is about 2.5 times the peak equilibrium value. This finding is similar to that in a subsonic lifted flame (Cheng, Wehrmeyer and Pitz, 1992). The sub-equilibrium OH could be due to ignition/extinction phenomena or spatial averaging. Except for N_2, the major species mole fractions

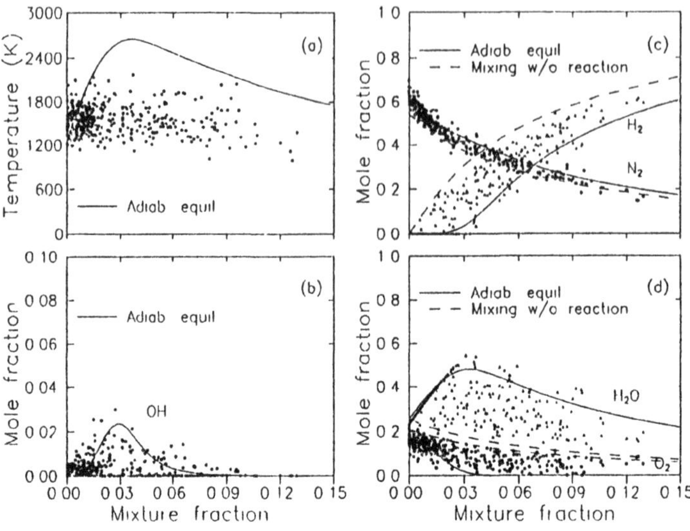

Fig. 4. Scatter plots in the supersonic flame of simultaneous measurements of temperature and species mole fractions (H_2, O_2, N_2, H_2O, OH) versus mixture fraction at $x/D = 32.3$, $y/D = -1.1$.

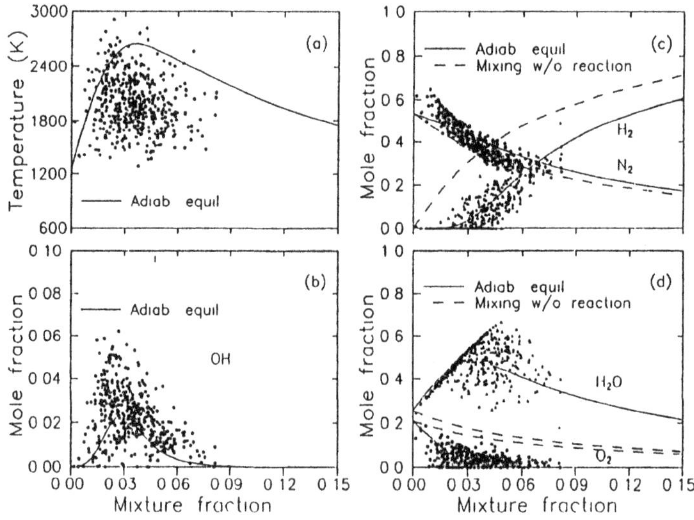

Fig. 5. Scatter plots in the supersonic flame of simultaneous measurements of temperature and species mole fractions (H_2, O_2, N_2, H_2O, OH) versus mixture fraction at $x/D = 32.3$, $y/D = 1.1$.

are on the equilibrium curves (Fig. 5c). Excess N_2 from the ambient air has diffused to this point and causes scatter points to deviate from the equilibrium curve. Comparing the major species concentration measurements to the subsonic flame measurements, higher fluctuations in the supersonic flame are observed. This is due to higher total enthalpy fluctuations in a supersonic flame.

4. Subsonic Flame Results

4.1. Subsonic burner

Subsonic lifted turbulent hydrogen jet diffusion flames with no swirl, weak swirl, and strong swirl are evaluated in the jet burner shown in Fig. 6. The fuel exits the jet into still atmospheric air. Hydrogen is supplied to the combustor either axially and/or tangentially with each of these flows being controlled separately. The nozzle of the burner is machined with a 5° converging angle to produce a plug flow exit profile. The jet exit inside diameter is 2.0 mm and the exit wall thickness is about 0.2 mm to eliminate recirculation effects. This fuel injector is similar to the one designed by Yuasa (1986). All flows investigated in this study have the following exit conditions: a velocity of 620 m/s, a Reynolds number of 12,200 based upon nozzle exit diameter, a total flow rate of 105 lpm at STP, a temperature of 283 K, and a Mach number of 0.48. For these conditions, three flow regimes are analyzed: an unswirled jet flame (axial flow = 105 lpm, $S = 0.0$), a weak swirl jet flame (axial flow = 52.5 lpm, tangential flow = 52.5 lpm, $S = 0.12$), and a strong swirl jet flame (no axial flow, tangential flow = 105 lpm, $S = 0.5$) where S is the geometric swirl number (Claypole and Syred, 1981). Based on visual observations, the lift-off height of all three flames is about 6 diameters. Mean and rms profiles of temperature and species concentrations in this flame have been reported by March et al. (1991).

4.2. Finite-rate chemistry effects

The instantaneous Raman/LIPF measurements of the reaction zone chemistry are compared to the chemistry predicted by laminar stretched flames. Strained laminar diffusion flame calculations by Chen (1992) for a Tsuji counterflow diffusion flame (Tsuji and Yamaoka, 1967) are shown in Fig. 7. Calculations of the variations

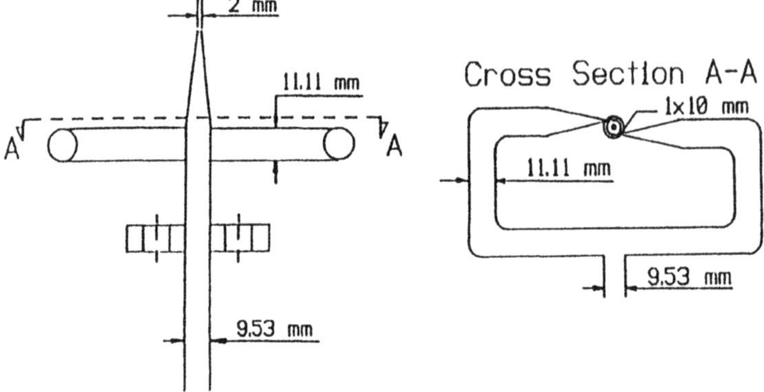

Fig. 6. Schematic of the subsonic swirl burner.

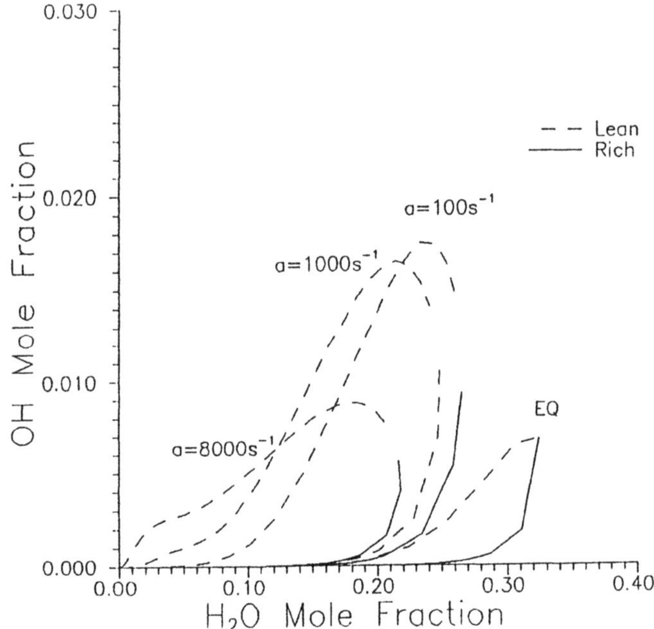

Fig. 7. Strained laminar flame calculations for the variation of OH and H_2O in a porous cylinder opposed flow diffusion flame at various values of the strain rate parameter: a = 100, 1000, and 8000 s^{-1}. The strain rate parameter is defined as $a \equiv 2U/R$ where U is the freestream velocity and R is the cylinder radius. The equilibrium curve is also shown. Lean and rich sides of the curves are denoted.

between H_2O and OH are shown for strain rate parameters of 100, 1000, and 8000 $s.^{-1}$ The strain rate parameter is defined as $a \equiv 2U/R$ where U is the freestream velocity and R is the cylinder radius. The equilibrium curve is also shown which is valid in the limit of fast chemistry and unity Lewis number (Bilger, 1980). Lean and rich sides of the curves are denoted separately. At strain rate parameters greater than $a = 11000\ s^{-1}$ the flame is extinguished. Thus in the OH vs. H_2O variation, there is a blank hole where no mixtures are expected for laminar diffusion flames.

All the scatter plots are measured in the flame zone where the average fuel/air ratio is stoichiometric. Scatter plots of the reaction zones in the lifted flame zone ($x/D = 6$) for the swirled and unswirled flames are shown in Fig. 8. In the unswirled flame (Fig. 8a), the variations have a blank hole characteristic of flamelets but the values lie between the flamelet curves and the equilibrium curves. The temperature is sub-equilibrium and the OH values are super-equilibrium due to slow three-body recombination.

The swirled flames in Fig. 8 show non-flamelet behavior. The OH values are super-flamelet and the H_2O values are sub-flamelet. The swirled flames for the OH vs. H_2O plots do not show a blank hole in the scattered region that is expected from a strained laminar diffusion flame (or from the equilibrium curve). This is probably due to premixing of the mixtures in the lifted flame zone. The premixing is seen more clearly in the mixture fraction scatter plots presented earlier (Pitz et al., 1991). This premixing has also been seen by Stårner, Bilger and Barlow (1992) in lifted hydrocarbon flames. They found evidence of premixing by comparing stoichiometric mixtures in the lifted zone to values expected from a stoichiometric premixed flame.

The results in the axial flame (Fig. 8a) do not show these intermediate values and exhibit a blank hole in the scattered plot. This is probably because these measurements were taken slightly downstream of the lifted flame zone in the axial flame. In similar measurements by Cheng, Wehrmeyer and Pitz (1992) in the lifted zone of an unswirled turbulent flame, intermediate values were found in the scatter plots that indicate premixing.

The effect of premixing will not account for the high values of OH seen in the lifted flame zone of the swirled flames. The maximum value of OH in a premixed stoichiometric opposed flow flame (Barlow, Dibble and Fourguette, 1989) is about 1.4%. Here values

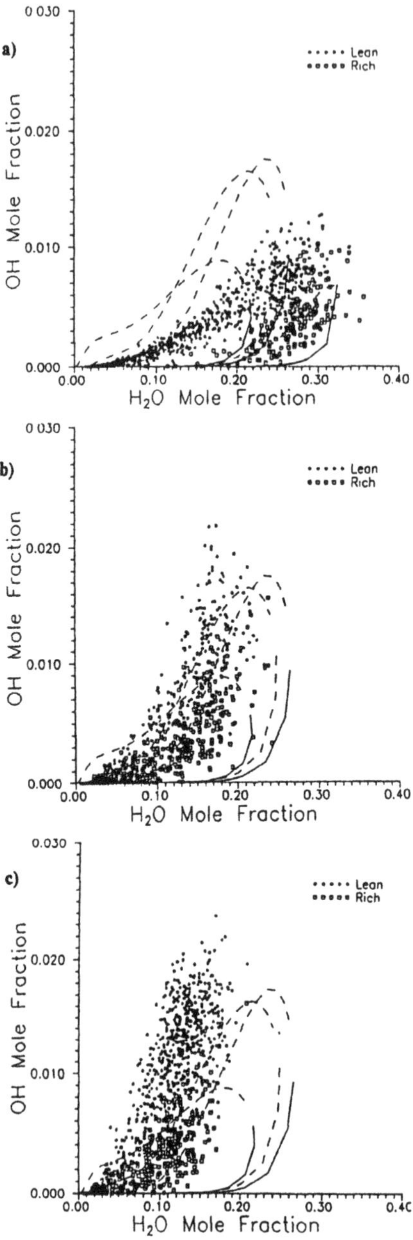

Fig. 8. Scatter plots of OH vs. H_2O in the lifted zone of a subsonic flame at $x/D = 6$. Strained laminar flame curves are shown from Fig. 7. a) Unswirled jet ($S = 0.0$), b) Weak swirl jet ($S = 0.12$), c) Strong swirl jet ($S = 0.5$).

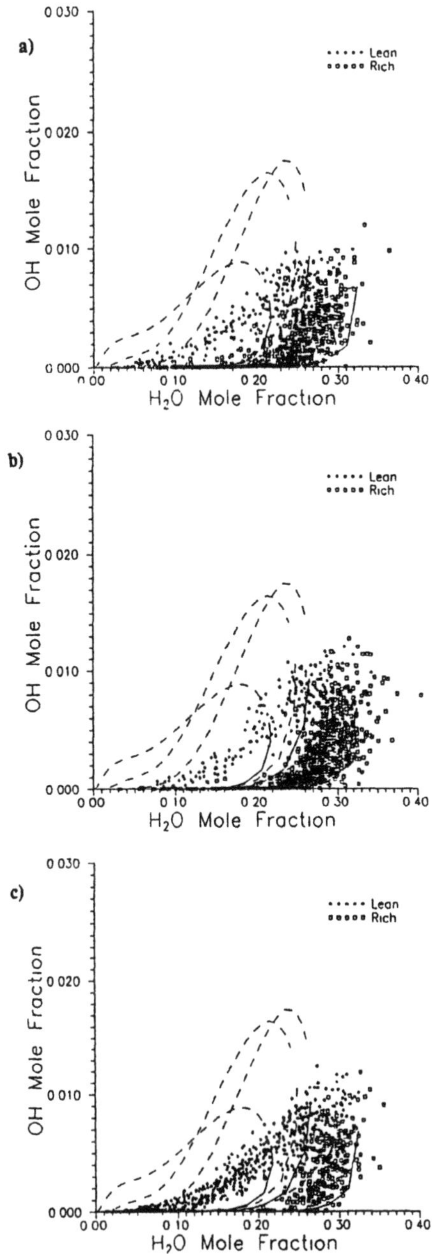

Fig. 9. Scatter plots of OH vs. H_2O in a subsonic flame at $x/D = 10$. Strained laminar flame curves are shown from Fig. 7. a) Unswirled jet ($S = 0.0$), b) Weak swirl jet ($S = 0.12$), c) Strong swirl jet ($S = 0.5$).

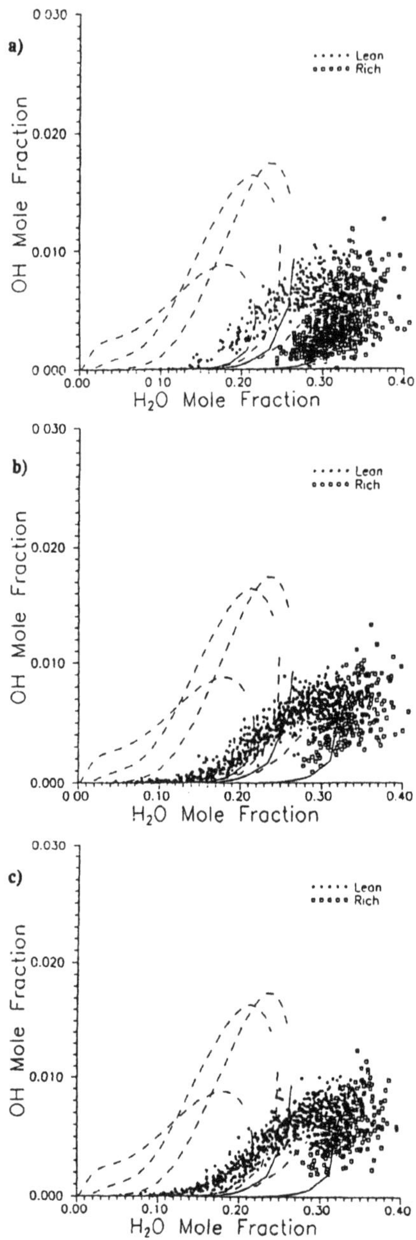

Fig. 10. Scatter plots of OH vs. H_2O in a subsonic flame at $x/D = 100$. Strained laminar flame curves are shown from Fig. 7. a) Unswirled jet ($S = 0.0$), b) Weak swirl jet ($S = 0.12$), c) Strong swirl jet ($S = 0.5$).

TABLE 1

Supersonic burner nominal operating conditions

Parameter	
Air mass flow rate (± 2%)	0.0735 kg/s
H_2 mass flow rate (± 2%)	0.00173 kg/s
O_2 mass flow rate (± 3%)	0.0211 kg/s
Fuel mass flow rate (± 3%)	0.000362 kg/s
Nozzle exit I.D.	17.78 mm
Fuel injector I.D.	2.36 mm
Fuel injector O.D.	3.81 mm
Stagnation conditions	
Pressure (± 4%)	778 kPa
Temperature	1752 K
Vitiated air exit conditions	
Pressure	107 kPa
Temperature	1250 K
Mach number	2.0
Velocity	1417 m/s
O_2 mole fraction	0.201
N_2 mole fraction	0.544
H_2O mole fraction	0.255
Fuel exit conditions	
Pressure	112 kPa
Temperature	545 K
Mach number	1.0
Velocity	1780 m/s
H_2 mole fraction	1.0

over 2% are found (see Figs. 8b and 8c) at low values of water vapor. Thus in the lifted flame zone of the swirled flames, the reaction zones do not correspond to 1D steady strained laminar flames (either premixed or diffusion flames). In this region, unsteady strain rate effects or flame curvature effects may account for these high values of OH.

Slightly downstream of the lifted flame zone at $x/D = 10$, all the scatter plots in the reaction zones look similar for both the unswirled and swirled flames (see Fig. 9). The OH vs. H_2O variations have the hole characteristic of flamelets but the variations lie between the flamelet and equilibrium curves.

Farther downstream at $x/D = 100$ where the flame is stoichiometric on the centerline (defined as the flame length), the scatter plots of the reaction zones in the unswirled and swirled flames are similar (Fig. 10). The variations of OH with H_2O are still non-flamelet and have shifted more toward equilibrium values. The scatter plots exhibit the blank hole characteristic of both the flamelet and equilibrium curves. These variations are consistent with a partial equilibrium model (e.g. Janicka and Kollmann, 1979).

Points far downstream of the flame length were not measured here but it is expected that the species variations will follow equilibrium curves at about $x/D = 200$. In the unswirled flame measured by Cheng, Wehrmeyer and Pitz (1992) (a slightly higher Reynolds number) equilibrium variations were found for major species, minor species, and temperature at $x/D = 175$.

5. Conclusions

Simultaneous measurements of mixture fraction, temperature, major species, and minor species concentration are obtained in subsonic and supersonic hydrogen jet diffusion flames by combining spontaneous Raman scattering with laser-induced predissociative fluorescence (LIPF). In the supersonic flame, the chemistry measurements are compared to the limits of frozen chemistry and equilibrium chemistry. The asymmetrical ignition of the supersonic flame in the lifted zone is clearly seen by the frozen mixtures on one side of flame and the reactive mixtures on the other side.

In subsonic flames, the reaction zone chemistry is compared to stretched laminar flamelets. Non-flamelet values of species concentrations are found in the reaction zones of both swirled and non-swirled lifted turbulent diffusion flames. In the lifted flame zone of the swirled flames, super-flamelet values of OH and sub-flamelet values of H_2O are found. These species concentrations do not correspond to values expected from 1D steady strained laminar flames. In the lifted zone, unsteady strain rate and flame curvature may affect the finite-rate chemistry. Downstream of the lifted zone, the concentration variations of OH and H_2O in the reaction zones are non-flamelet and lie between flamelet and the equilibrium curves. The variations correspond to a partial equilibrium model where the major species are in partial equilibrium and the minor species (OH) are in super-equilibrium due to slow three-body recombination. Finite-rate chemistry effects in these lifted hydrogen diffusion flames are not accurately described by an ensemble of 1D steady strained laminar flames.

Acknowledgement

The technical contribution of S. R. March in the design and construction of the swirl burner is gratefully acknowledged.

References

Barlow, R. S., Dibble, R. W. and Fourguette, D. C., 1989. "Departure from chemical equilibrium in a lifted hydrogen flame," AIAA/ASME/SAE/ASEE 25th Joint Propulsion Conference, AIAA 89-2523.

Barlow, R. S., Dibble, R. W., Stårner, S. H., Bilger, R. W., Fourguette, D. C., and Long, M. B., 1990. "Reaction zone structure in dilute methane jet flames near extinction," AIAA 28th Aerospace Sciences Meeting, AIAA-90-0732.

Beach, H. L., Jr., 1972. "Supersonic mixing and combustion of a hydrogen jet in a coaxial high-temperature test gas," AIAA/SAE 8th Joint Propulsion Specialist Conference, AIAA-72-1179.

Bilger, R. W., 1980. "Turbulent flows with nonpremixed reactants," in *Turbulent Reacting Flows*, P. A. Libby and F. A. Williams, editors, Springer-Verlag, New York, pp. 65-113.

Chen, J.-Y., 1992. Private communication.

Cheng, T. S., Wehrmeyer, J. A., Pitz, R. W., Jarrett, O., Jr., and Northam, G. B, 1991. "Finite-rate chemistry effects in a Mach 2 reacting flow," AIAA/SAE/ASME/ASEE 27th Joint Propulsion Conference, AIAA-91-2320. Accepted for publication in *Combust. Flame*, 1993. Also see NASA Report No. 189544, December, 1991.

Cheng, T. S., Wehrmeyer, J. A., and Pitz, R. W., 1992. "Simultaneous temperature and multispecies measurement in a lifted hydrogen diffusion flame," *Combust. Flame* **91**, pp. 323-345.

Claypole, T. C. and Syred, N., 1981. "The effect of swirl burner aerodynamics on NO_x formation," 18th Symposium (Int.) on Combustion, The Combustion Institute, Pittsburgh, pp. 81-89.

Eckbreth, A. C., 1988. *Laser Diagnostics for Combustion Temperature and Species*, Energy and Engineering Science Series, (A. K. Gupta and D. G. Lilley, ed.), Abacus Press, Cambridge, Vol. 7.

Ferri, A., Libby, P. A., and Zakkay, V., 1964. "Theoretical and experimental investigation of supersonic combustion," *Third Congress of the International Council of the Aeronautical Sciences*, Spartan Books, Washington, D.C., pp. 1089-1155.

Janicka, J. and Kollmann, W., 1979. "A two-variables formalism for the treatment of chemical reactions in turbulent H_2-air diffusion flames," 17th Symposium (Int.) on Combustion, The Combustion Institute, Pittsburgh, pp. 421-430.

Jarrett, O., Jr., Cutler, A. D., Antcliff, R. R., Chitsomboon, T., Dancey, C. L., and Wang, J. A., 1988. "Measurements of temperature, density, and velocity in supersonic reacting flow for CFD code validation," 25th JANNAF Combustion Meeting, CPIA Publication 498, Vol. 1, pp. 357-364.

Lezberg, E. A. and Franciscus, L. C., 1963. "Effects of exhaust nozzle recombination on hypersonic ramjet performance: I. Experimental measurements," *AIAA J.* **1**, pp. 2071-2076.

March, S. R., Cheng, T. S., Pitz, R. W., Wehrmeyer, J. A., and Debelak, K. A., 1991. "Temperature and species concentration measurements in a swirled hydrogen diffusion flame," 29th Aerospace Sciences Meeting, AIAA-91-0480.

Northam, G. B. and Anderson, G. Y., 1986. "Supersonic combustion ramjet research at Langley," AIAA 24th Aerospace Sciences Meeting, AIAA-86-0159.

Peters, N., 1984. "Laminar diffusion flamelet models in non-premixed turbulent combustion," *Prog. Energy Combust. Sci.* **10**, pp. 319-339.

Peters, N. and Williams, F. A. 1983. "Liftoff characteristics of turbulent jet diffusion flames," *AIAA J.* **21**, pp. 423-429.

Pitz, R. W., Cheng, T. S., March, S. R. and Wehrmeyer, J. A., 1991. "Effect of swirl on finite-rate chemistry in a lifted jet diffusion flames," AIAA/SAE/ASME/ASEE 27th Joint Propulsion Conference, AIAA-91-2319.

Pitz, R. W., Nandula, S., and Brown, T. M., 1992. "Comparison of reaction zones in turbulent lifted diffusion flames to stretched laminar flamelets," AIAA/SAE/ASME/ASEE 28th Joint Propulsion Conference, AIAA-92-3349.

Stårner, S. H., Bilger, R. W., and Barlow, R. S., 1992. "Raman/LIF measurements in a lifted hydrocarbon jet flame," in *Turbulent Shear Flows 8*, R. Friedrich, editor, Springer-Verlag, New York.

Tsuji, H. and Yamaoka, I., 1967. "The counterflow diffusion flame in the forward stagnation region of a porous cylinder," *11th Symposium (Int.) on Combustion*, The Combustion Institute, Pittsburgh, pp. 979-984.

Wehrmeyer, J. A., Cheng, T. S., and Pitz, R. W., 1992. "Raman scattering measurements in flames using a tunable KrF excimer laser," *Applied Optics* **31**, pp. 1495-1504.

Yuasa, S., 1986. "Effects of swirl on the stability of jet diffusion flames," *Combust. Flame* **66**, pp. 181-192.

DETONATIONS

LASER-INITIATED CONICAL DETONATION WAVE FOR SUPERSONIC COMBUSTION – A REVIEW[1]

George F. Carrier, Francis E. Fendell, and Mau-Song Chou

Center for Propulsion Technology and Fluid Mechanics
TRW Space & Electronics Group
Redondo Beach, California 90278

ABSTRACT

High-speed air-breathing combustion systems necessitate the release of chemical energy during the relatively brief residence time of reactants within a combustor of practical length. We explore an alternative to the mixing-controlled burning (supersonic diffusion flames) on which the scramjet is usually predicated.

Explicitly, we examine the concept of a supersonic combustor based on a nonintrusively stabilized oblique (conical) detonation wave. The conical wave is the resultant of the interaction of a train of spherical detonation waves. Each spherical wave is initiated (in a very-fast-flowing gaseous fuel/air mixture) by energy deposition, at a fixed site relative to the combustor, from a rapidly repeated, periodically pulsed laser. We note that trace-level additives may be required for efficiently sensitizing the gaseous reactive mixture to the incident laser radiation; in fact, judiciously selected sensitizers also may dissociate to form radicals that promote the chemical reaction. We address, expe.imentally and analytically, the current status and the remaining agenda for demonstrating proof-of-principle technology for such a supersonic combustor.

1. A Supersonic Combustor Based on a Nonintrusively Stabilized Conical Detonation Wave

1.1. Interaction of a train of spherical detonations in a supersonic stream

In addition to nonintrusive diagnostic instrumentation, the rapid evolution of laser technology affords novel opportunities for the non-

[1]This research is supported by the Air Force Office of Scientific Research, Air Force Systems Command, USAF, under Contract F49620-90-C-0070; the contract technical monitor is Julian Tishkoff.

J. Buckmaster et al. (eds.), Combustion in High-Speed Flows, 277–307.

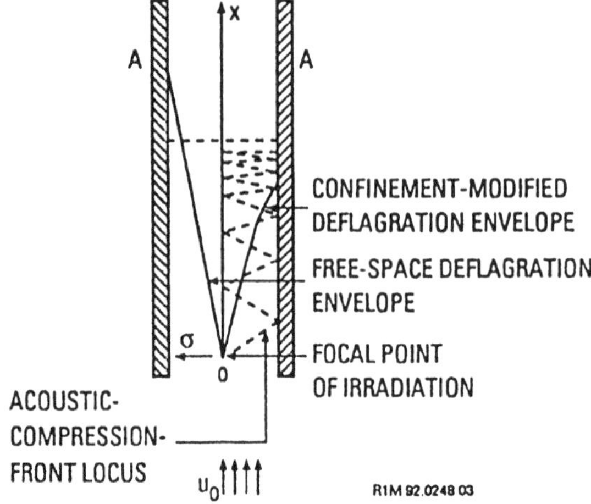

Fig. 1. The cone (and the modified cone, owing to confinement in a pipe of radius r_0) which subtend the burned-gas domain. Here, very-rapidly-repeated pulsed energy deposition at the origin, into a reactive mixture flowing at speed u_0, initiates an outwardly propagating deflagration.

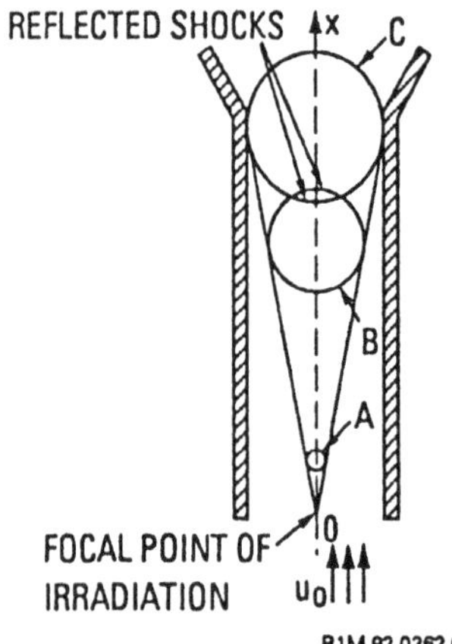

Fig. 2. Three intermittent pulses at the origin, in a reactive mixture flowing at speed u_0, have initiated outwardly propagating Chapman-Jouguet-type spherical detonations A, B, and C; the conceptual conical envelope for these detonations is indicated. Reflected shocks arise from the interactions of detonations B and C, but flaring of the pipe precludes reflected shocks from sphere/wall interaction. The speed u_0 exceeds the Chapman-Jouguet (CJ)-wave speed u_{CJ}. The cellular structure of observed detonations is not indicated.

intrusive, readily tightly focused deposition of monochromatic energy in very brief pulses into reactive mixtures. Most commonly, the brief-pulse energy deposition from a laser is used as an alternative to spark ignition of deflagration (Chou and Zukowski, 1991). However, in the context of a homogeneous supersonic stream of a reactive gaseous mixture, brief-pulse energy deposition for the initiation of a outwardly propagating, spherical deflagration results in a *very small* cone angle subtending the burned gas, because the flame speed $u_{flame} \ll u_0$, the convective speed of the cold mixture (Carrier et al., 1990; Fendell, Kung, and Sheffield, 1992). [See Fig. 1, in which a unit measure on the streamwise (x) axis signifies a distance far greater than a unit measure on the cylindrical-radial (σ) axis, and the origin of coordinates is the site of the brief-pulse energy deposition.] We take the admixture of fuel into the inlet airstream to be perfectly uniform, and to entail no flow deceleration, throughout most of this manuscript, because such a simplification serves the purposes of this preliminary investigation. We comment on the (in practice, critical) question of mixing only briefly near the conclusion of this manuscript, since other investigators can address mixing technology more knowledgeably. Meticulously, for spherical-deflagration initiation in a circular pipe by focused laser radiation, the conical envelope of a free-space deflagration is modified owing to the confinement, which gives rise to a multiply reflected acoustic-compression front (Fig. 1); however, the correction is negligible for present purposes. An attempt to use *many* deflagration-initiation sites in the origin-containing plane transverse to the streaming, in the hope that a myriad of small-angle cones may span the combustible stream with flame within a practical axial distance downwind of the origin, is logistically challenging. More significantly, such an attempt might result in a normal detonation, and hence large entropy rise – and the increase of entropy is a measure of the energy unavailable to do useful work.

Thus, the intermittent deposition of sufficiently large energy in a sufficiently large volume in a sufficiently brief interval of time (Abouseif and Toong, 1982), such that a sequence of spherical detonation waves is initiated (note spheres A, B, and C in Fig. 2), seems the effective means of spanning a supersonic stream with flame within a practically interesting distance downwind of the focal site. The direct initiation of a train of spherical detonations, the conceptual envelope of which is a cone with half angle related to the ratio of the Chapman-Jouguet (CJ)-wave speed u_{CJ} to the cold-mixture flow speed u_0, serves present purposes. The conceptual conical envelope would bulge at small distances downwind of the origin if the spherical waves were overdriven at initiation, and later decelerated to CJ-wave speed as they were convected downwind. The conceptual conical envelope would be depressed at small distances downwind of the origin if the spherical waves were underdriven, and later accelerated to CJ-wave speed as they were convected; while such acceleration might be possible, in practice, relying upon deflagration-to-detonation transition rather than direct (i.e., nearly instantaneous) initiation of a spherical detonation, probably results in an impractically long supersonic combustor. The conceptual conical envelope depicted in Fig. 2 pertains to a pulse with what we term the CJ energy for the direct initiation of a spherical detonation. Since the same element of combustible mixture can be converted but once by chemical reaction to form product, the interaction of neighboring spherical detonations (see spheres B and C in Fig. 2) results in reflected shocks that engender entropy rise in excess of the entropy rise associated with an isolated spherical CJ wave. We show below that this additional entropy rise, owing to reflected shocks, monotonically decreases as the interval between pulses is reduced, though deposition of two laser pulses into effectively the same "blob" of mixture is inefficient. Interaction of the spherical detonations with the pipe walls also engenders entropy-producing reflected shocks, so the walls of the container are flared downwind of the site of wave/wall interaction; a modest flaring (of a supersonic nozzle) hopefully suffices to permit discharge of the hot product gas at ambient pressure within a feasible length of combustor, with only modest further entropy rise. The circular locus of wave/wall interaction is a site of particular concern for heat transfer; judicious flaring also ought to (1) help suppress upwind transfer of information in the wall-contiguous, subsonic portion of the boundary layer, and (2) retain a streamlined geometry so the thrust exceeds the drag. It is premature to discuss incorporation of the combustor into a vehicle for flight, and, without such specification, discussion of thrust-to-drag ratio seems premature.

Thus, a train of periodic pulses, each of which directly initiates an outwardly propagating spherical detonation in a reactive stream flowing faster than the CJ-wave speed, results in the corrugated front depicted in Fig. 3. The corrugations decrease as the temporal interval between laser pulses decreases, and the idealized, nonintrusively

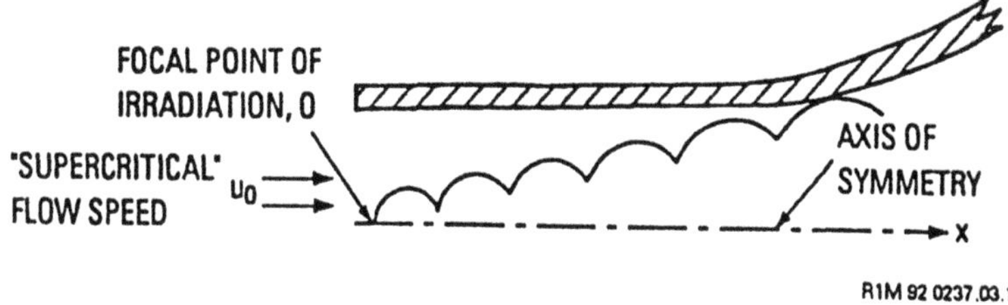

Fig. 3. Periodic pulsed energy deposition at the origin, in a reactive mixture flow at speed u_0, has resulted in a train of convected, outwardly propagating spherical detonations, the interaction of which leads to a "scalloped" wave.

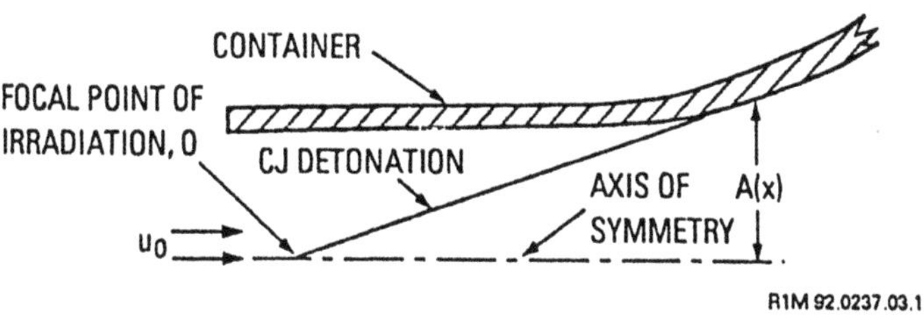

Fig. 4. For the scenario of Fig. 3, the "scalloped" wave resulting from the interaction of the spherical detonations has become a CJ conical detonation wave, in the limit of an indefinitely small time interval between laser pulses.

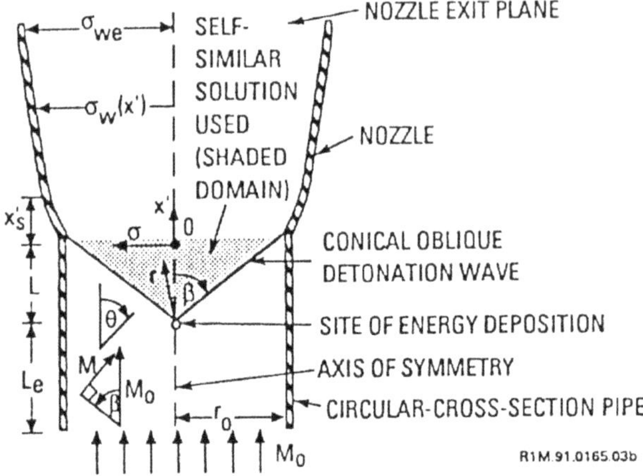

Fig. 5. A schematic of a supersonic combustor, based on the nonintrusive stabilization of a conical detonation wave by pulsed energy deposition at the origin of the spherical polar coordinates (r, θ), where the polar angle $\theta = \beta$ gives the inclination of the conical CJ wave. The freestream Mach number $M_0 = u_0/a_0$, where a_0 is the sound speed in the unreacted mixture; $M_{CJ} = u_{CJ}/a_0$. The detonated-gas flow is initially selfsimilar in terms of θ; cylindrical polar coordinates (σ, x') are convenient for the nozzle flow.

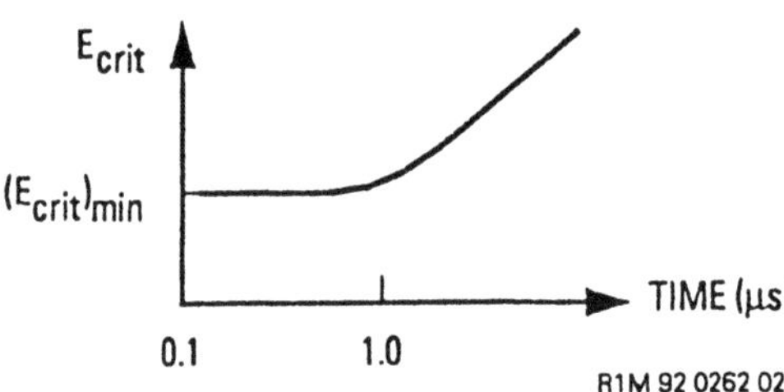

Fig. 6. A schematic of the observed variation (Lee, 1977) of the critical energy E_{crit} (for the direct initiation of detonation in a reactive mixture) as a function of the time interval for the energy deposition, for an electric-spark source.

stabilized conical detonation wave depicted in Fig. 4 results. Of course, any observed detonation has cellular structure, but if the dimensions of the cells are appreciably smaller than other physical dimensions of the combustor flow, then, for many engineering estimates, the cellular character may be ignored. The entropy rise across such an oblique (conical) wave may well be less than that for a strong detonation or even a weak detonation in a strictly one-dimensional flow, because only the component of the cold flow normal to the resultant, conical detonation is decelerated across the detonation – the tangential component is continuous. The flaring, downwind of the wave/wall-interaction site, of the cross-sectional area $A(x)$, with $A_x(x) \geq 0$, permits this reduced entropy rise. Of course, this comparison of entropy rise for alternative configurations pertains to oblique-detonation-wave engines (ODWE), and does not encompass mixing-controlled-combustion scramjets (Waltrup, 1987). In any case, even if the mixing-controlled scramjet were estimated to permit lower entropy rise, on the basis of a comparison of highly idealized modeling, the feasibility of any appreciably supersonic combustor is as yet undemonstrated, and the half-century-old concept of an ODWE remains of continued engineering interest.

Incidentally, the need for sensitization (via trace-level additives) of many fuel/air mixtures, otherwise virtually transparent to the incident radiation of available lasers, permits a way around the problem of significant laser-irradiation absorption all along the beam path. Confining the admixture of sensitizer to that portion of the cold combustible throughput near the axis of symmetry would preclude such beam degradation along its path to the intended site of focus. Nevertheless, for *nonintrusive* deposition in the *midst* of a supersonic stream of finite width, the use of very rapidly absorbed vacuum-ultraviolet radiation (e.g., for the efficient dissociation of oxygen) seems precluded.

1.2. Supersonic combustors based on a stabilized conical detonation wave

A rudimentary schematic (Fig. 5) of an axisymmetric supersonic combustor with a laser-initiated conical detonation wave (Carrier, Fendell, McGregor et al., 1992) displays a simple pipe-type inlet of cylindrical radius r_0 and axial length $(L_e + L)$, for a freestream flow of Mach number M_0. We reiterate that we account for no freestream deceleration for intake or fuel admixture to the captured air, to emphasize that minimization of flow deceleration and heating is desirable to avoid inadvertent preignition of the reactive mixture upwind of the site of laser-energy deposition. The above-discussed, nonintrusively stabilized, conical detonation wave is inclined at spherical polar angle $\theta = \beta$. The origin of the spherical coordinate system is the site of energy deposition; r is the spherical radial coordinate; $0 \geq \theta \geq \beta$ delimits the selfsimilar, just-detonated flow, where $\theta = 0$ is directly downwind (along the axis of symmetry) and $\beta = sin^{-1}(M/M_0)$, where: the Mach number M of the CJ wave is u_{CJ}/a_0; u_{CJ} is the CJ-wave speed; and a_0 is the speed of sound in the cold gas. Meticulously, the selfsimilar description of the detonated gas holds over a domain slightly greater than that shaded in Fig. 5 (because the characteristic "informing" the detonated gas of the finite lateral boundary extends downwind as it "runs" radially from pipe to axis). However, here the selfsimilar solution is adopted for just the shaded domain; the discharge of the detonated gas at ambient pressure is effected via a supersonic nozzle, the flow within which is conveniently described in terms of cylindrical polar coordinates. The cylindrical radial coordinate is denoted σ with $0 < \sigma < \sigma_w(x')$, where the wall contour $\sigma_w(x')$ is to be specified; the axial coordinate is x', with $0 < x' < x'_e$, where $x' = 0$ contains the plane of wave/pipe "intersection" and $x' = x'_e$ at the nozzle-exit plane, so $\sigma_w(x'_e) = \sigma_{we}$. The only family of nozzle contours to be examined here consists, in part, of a "materialization" of the detonated-flow streamsurface passing through the wave/pipe "intersection." This contour is adopted for $0 < x' < x'_s$, where, at $x' = x'_s$, a cone of prespecified half angle smoothly adjoins to the streamsurface contour. If the flow in the adjoined cone is not laterally uniformized (with the pressure equal to the ambient pressure p_0) within a "practical" axial distance x'_e, found in the course of solution, results are computed for another value for the cone angle for the nozzle (with a different corresponding value for x'_s) until satisfaction (if possible) of the condition on pressure at discharge is obtained with a nozzle of practical length.

The alternative ODWE scheme, explicitly, stabilizing a conical detonation wave by the introduction of an intrusive conical body (Cambier et al., 1989a, 1989b), incurs at least three disadvantages, relative to the use of a rapidly repeatedly pulsed laser for the direct initiation of interacting spherical detonations. First, additional frictional drag is incurred by the intrusive body. Second, even if we overlook any possible liabilities associated with the supports to maintain a symmetrically located conical body, the inevitable nose curvature and jitter of the intrusive body imply a locally normal detonation. The locally normal detonation might become a broadly normal detonation, owing to the upwind propagation of disturbances arising in the subsonic flow downwind of a strong detonation; a broadly normal detonation would incur a large rise of entropy. Third, and most importantly, the absence of a positive ignition device (such as a laser) permits a delay between the conical-body-induced shock wave and the initiation of reaction. The longer induction time for a low-pressure (high-altitude) cold flow into the combustor implies increased combustor length. In fact, firm experimental evidence of the stabilization of an oblique detonation wave by use of an intrusive body seems unreported. Thus, there is motivation to develop a efficient laser, with both adequate energy per (brief) pulse and adequate repetition rate to achieve nonintrusive stabilization of a conical detonation wave.

2. Direct Initiation of a Spherical Detonation by Nonintrusive Energy Deposition

2.1. Modeling related to a single pulse

The critical (minimum) energy E_{crit} for the direct initiation of a spherical detonation in a reactive mixture is proportional to the cube of the characteristic detonation-wave thickness (Zel'dovich et al., 1957). This spatial scale was formerly envisioned to be the reaction-induction-zone length ℓ (Bush and Fendell, 1971), but in light of observational results, now is taken to be the cellular dimension λ (Lee, 1984), on which chemical reaction is completed (where often $\lambda \approx 15\ell$, or more) (Lee, 1977; Strehlow, 1984). Furthermore, observationally, the critical energy must be deposited on a time scale less than the time scale on which pressure waves significantly disperse the deposited energy, or else the critical energy appreciably increases (Fig. 6); the dispersive time is here characterized by the ratio λ/u_{CJ}, typically a little more than a microsecond interval. It seems reasonable that the minimum initiation energy E_{crit} must be comparable to the energy chemically derivable from a volume of mixture characterized by the dimensional scale λ, for such exothermicity suffices to sustain any detonation-wave propagation which ensues a direct initiation.

The common means of direct initiation in a gaseous medium, whether by use of a brief laser spark with a wavelength to which the medium is mostly transparent or by setting off a solid explosive charge, is to generate an outwardly propagating spherical blast wave that decays in time to a CJ wave (if direct initiation is successful) (Daiber and Thompson, 1967; Bach et al., 1969, 1970; Lee and Knystautas, 1969; Klimkin et al., 1973; Lee and Matsui, 1977). The portion of the energy of the source pulse actually coupled efficiently into the detonable gaseous mixture may be appreciably less than the total energy of the source; e.g., much energy from the explosion of a solid charge may remain in the products of the charge, without transfer to the gaseous mixture on the pertinent temporal scale.

We seek a useful *sufficiency criterion* (i.e., reasonable upper bound) for the quantity E_{crit}. We consider a brief, localized pulse, which deposits energy E, in time t, in a blob of radius R_0, to be specified. The blob lies in an infinite expanse of stagnant detonable gas with initial density ρ_0. The first portion of the energy E is deposited in a mass $\rho_0 V_0 = \rho_0(4/3)\pi R_0^3$. The heated-region pressure rises, and the blob expands. We idealize the growth of the shock front, during the interval of deposition of energy, by adopting a constant rate of shock-front growth, c, to be identified; hence, $R(t) = R_0 + c\,t$, where the deposition begins at $t = 0$. So the energy E is shared with an augmented mass (of detonable medium) $\rho_0 V(t) = \rho_0(4/3)\pi R^3$, under the model that the gas outside of the sphere with radius $R(t)$ is unmoved. At the end of the interval t of energy deposition, we adopt the model that (1) there is no loss of any of the deposited energy E from the volume $V(t)$, and (2) there is no gain of energy within $V(t)$ via chemical reaction. If we ignore any role of kinetic energy, and *average*, over the domain $V(t)$, the specific enthalpy h of the heated mass, then

$$(h)_{ave} = \frac{E}{\rho_0 V} = \frac{E}{\rho_0 V_0}\left(\frac{V_0}{V}\right) = \frac{E}{\rho_0 V_0}\left(\frac{R_0}{R_0 + c\,t}\right)^3. \tag{1}$$

In accord with earlier remarks, we let: $R_0 = \lambda$; $c = u_{CJ}$; and $(h)_{ave} = q$, the exothermicity per mass of mixture. [The analogous choices for the minimum energy-deposition requirement for ignition of a deflagration would be $R_0 = D/u_{flame}$, $c = u_{flame}$, and $(h)_{ave} = q$, where D is the diffusion coefficient, for a gas with Lewis, Schmidt, and Prandtl numbers equal to unity, and u_{flame} is recalled to denote the laminar flame speed.] The choice $R_0 = \lambda$ is conservative, for consistency with the sufficiency criterion being sought, since deposition on a smaller scale might suffice for direct initiation (Klimkin et al., 1973). The choice $c = u_{CJ}$ is also conservative, since this is the fastest speed reasonably associated with the phenomenon under study, and adopting the model $c = a_0(< u_{CJ})$ might suffice. For such choices, from Eq. (1),

$$E \to E_{crit} = E_{CJ}[1 + (t/t_m)]^3, \tag{2}$$

where

$$t_m \equiv \lambda/u_{CJ}, \; E_{CJ} \equiv \rho_0[(4/3)\pi\lambda^3]q. \tag{3}$$

The significant increase in E_{crit} described by Eq. (2) if the deposition interval much exceeds time t_m, typically somewhat greater than a microsecond or so, is reminiscent of the observations summarized in Fig. 5. If $(t/t_m) \ll 1$, as would be the case for many pulsed lasers, then the critical energy is $E_{CJ} = O(10\ J)$ for a practically interesting, stoichiometric fuel/air mixture at atmospheric pressure and with a one-centimeter cell size. If $\rho_0 \sim p_0$, the cold-mixture pressure, and $\lambda \sim (p_0)^{-n}$, $n > 0$, then $E_{crit} \sim (p_0)^{1-3n}$, since q varies modestly with p_0. For $n = 1$ (Nettleton, 1987), $E_{crit} \sim p^{-2}$, a variation that has been suggested by other, less semi-empirical arguments (Westbrook and Urtiew, 1982). In any case, even conservative estimates for E_{crit} are about two orders of magnitude less than those associated with direct initiation by overdriving, i.e., by decay of a spherical blast wave initiated by an instantaneous point-source deposition of energy (Lee, 1984).

The factor E_{CJ} in Eq. (3) emphasizes that the pulse energy $\rho_0\, qV_0$, not the energy density $\rho_0 q$, is fundamental; the energy density of a laser pulse can be made to far exceed the chemical energy density of the reactive medium (Klimkin et al., 1973), but unless a sufficient volume is energized, a detonation wave may not be initiated.

2.2. Experiments related to a single pulse

Because numerical modeling is challenged to replicate the three-dimensional structure and detailed chemical kinetics of the direct-initiation process, we examine experimental results. A flat-flame burner, with a 5.8-cm-diameter stainless-steel porous disc and with a coflowing N_2 gas shroud (to avoid admixture of room air), is used to generate a steady, one-dimensional, highly subsonic stream of a

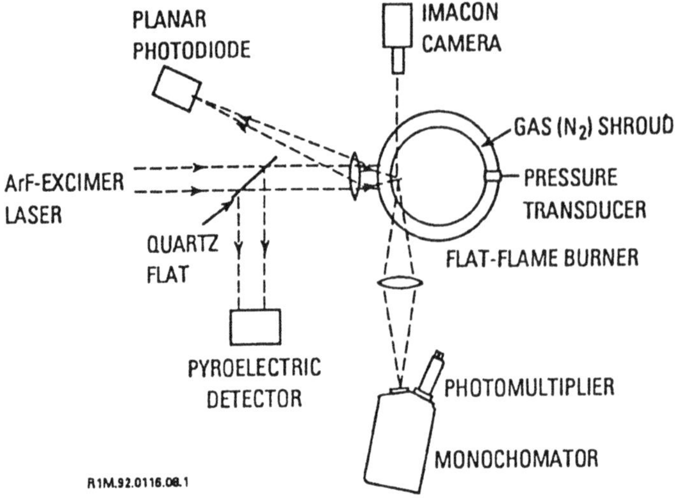

Fig. 7. A schematic of the experimental apparatus for investigating the direction initiation of detonation by energy deposition from a laser pulse into a homogeneous unconfined flowing gaseous mixture, by use of a flat-flame burner.

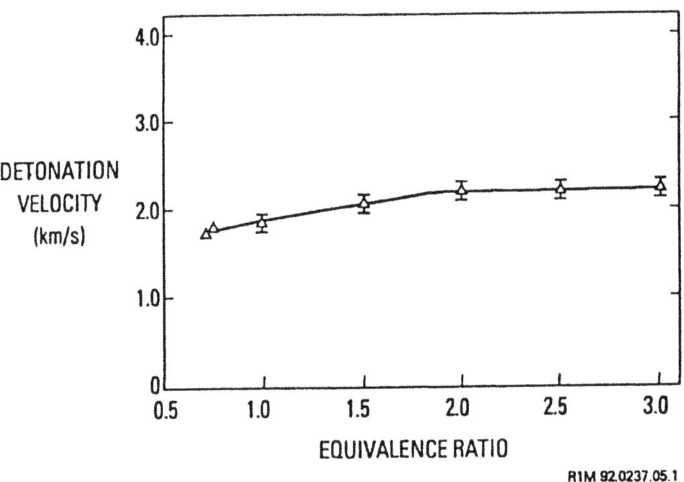

Fig. 8. The detonation-wave velocity in C_2H_2/O_2 mixtures, initially at atmospheric pressure, as a function of the equivalence ratio, from results obtained with the apparatus of Fig. 7.

uniform gaseous fuel/oxidizer mixture of known stoichiometry. The beam of an excimer laser (Lambda Physik Model ENG-150, with up to about 400 mJ in a 20-ns pulse), with an ArF dye (193-nm wavelength), is focused, by a 2.6-cm-focal-length lens, on to spot approximately 9 mm above the burner surface and approximately 5 mm from the nearest burner edge (Fig. 7). Focusing is constrained to a distance no more than about a centimeter above the burner, lest room air intrude on the mixture; focusing is constrained to a site about a half centimeter within the nearest portion of the conceptual cylindrical surface subtended by the burner face, because of the efficient absorption along the beam path by C_2H_2. Laser-beam energy is monitored via a quartz beam splitter and a pyrometer (Laser Precision Model 7200). The time of arrival of the laser pulse is monitored by a fast-response planar photodiode (Hamamatus Model R1193U-02) via light scattered from the focusing lens. The onset of detonation at the focal point is monitored, via OH emission at approximately 308 nm, by a monochromator and a photomultiplier tube. A high-speed image-converter (IMACON) camera (indicated in Fig. 7), with speed up to 0.1 μs per frame, monitors the early temporal evolution of the detonation wave. A pressure transducer (Kistler Model 211B2) is used to monitor the pressure at the site on the edge of the burner furthest from the laser focal spot. In the future, two identical lasers are to be used in oscillator-amplifier tandem to achieve an energy per pulse of about 1 J.

For a C_2H_2/O_2 mixture initially at one-atmosphere pressure, the detonation velocity, about 2 km/s, as derived from the onset of OH emission and the arrival time of the detonation wave at the pressure transducer, is nearly independent of equivalence ratio (Fig. 8). The wave velocity, the IMACON-camera pictures of frontal growth, and the pressures measured by the transducer distinguish when a detonation (as opposed to a deflagration) is initiated. The minimal deposition energy for direct initiation in initially-atmospheric-pressure C_2H_2/O_2 is minimal for equivalence ratio $\phi \approx 2.0$ (Fig. 9), and increases substantially with N_2 dilution (Fig. 10). The fuel species C_2H_2 not only is highly absorptive of the incident 193-nm radiation, but also photodissociates readily to form the active radicals C_2H and H, so that reaction is promoted (Lee et al., 1978). Hence, the requisite direct-initiation energy is significantly less than that reported in blast-overdriven tests. While H_2 is not absorptive at 193 nm, we speculate that use of even (say) 2% C_2H_2 in an H_2/air mixture may

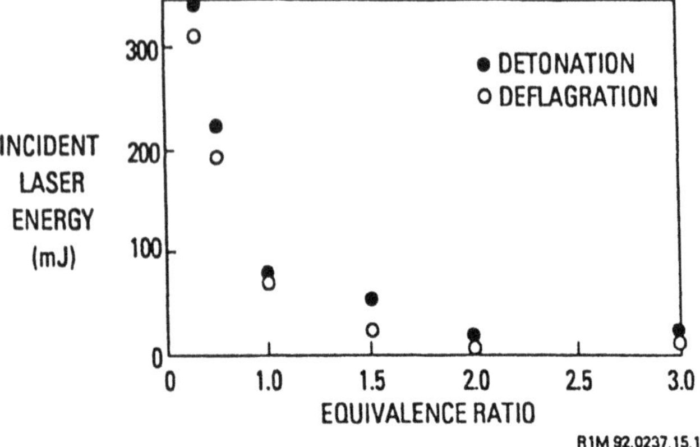

Fig. 9. The critical energy for the direct initiation of detonation in C_2H_2/O_2 mixtures, initially at one atmosphere, as a function of equivalence ratio, from results obtained with the apparatus of Fig. 7.

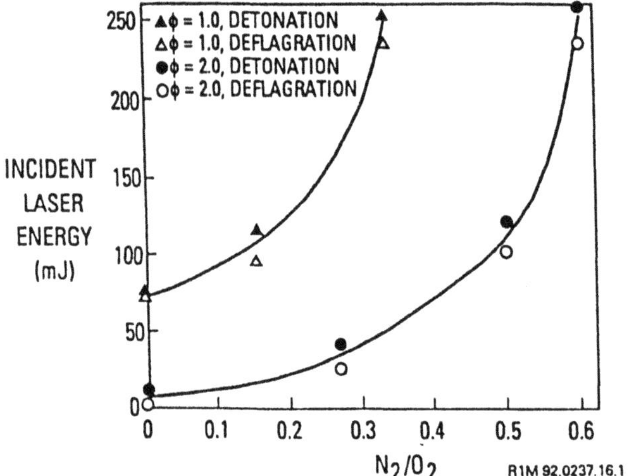

Fig. 10. The critical energy for the direct initiation of detonation in $C_2H_2/O_2/N_2$ mixtures, initially at atmospheric pressure, as a function of N_2/O_2, from results obtained with the apparatus of Fig. 7.

serve as a significant sensitization.

Experiments on direct initiation were also conducted in stoichiometric C_2H_2/O_2 mixtures over a range of cold-mixture pressures by ArF-excimer-laser deposition in a stainless-tube cylindrical tube of 1.6-cm-diameter cross-section, 1-m length, and about 0.16-cm wall thickness, with pressure-transducer instrumentation (Fig. 11). The laser beam is focused on a spot, on the axis of symmetry, at an axial distance of 5 mm from the end of the tube fitted with a quartz window for transmission. The critical energy for direct initiation varies inversely with pressure (Lee and Matsui, 1977; Lee et al., 1978); also, the delay time for detonation onset, inferred by extrapolating from detonation-wave arrival time at the pressure-transducer sites, decreases monotonically as pressure increases (Fig. 12). The critical energy for a stoichiometric C_2H_2/O_2 mixture initially at atmospheric pressure is about 60 mJ for the planar wave confined in the tube; this value is also approximately the critical energy measured for the initiation of a spherical wave in the open burner system.

3. Laser Repetition Rate and Entropy Rise from Reflected Shocks

We seek to identify the operating conditions for minimal entropy increase owing to the (previously discussed) reflected shocks which arise when neighboring spherical detonations in a train interact. The neighboring spherical detonation waves are of unequal radius (we concentrate on waves B and C in Fig. 13) because both propagate radially outward at the CJ-wave speed u_{CJ}, but the larger, downwind detonation was initiated at time interval t_1 earlier than the other, where t_1 is the interval between very brief, detonation-wave-initiating laser pulses. The reflected shocks (denoted by the dotted curves in Fig. 13) propagate at a speed other than the CJ-wave speed, so the spherical fronts become distorted. We adopt a frame of reference such that the cold mixture [moving at speed $u_0(> u_{CJ})$] is still. We concentrate on a particle p that has its first intersection with the propagating fronts B and C at the intersection of the two spherical detonations (Fig. 13). Of course, wave B was of smaller radius when particle p passed the depicted axial position of wave B, so indeed an unreacted particle at position p may exist.

If x denotes the axial coordinate, if ϖ denotes the cylindrical radial coordinate, if the origin of coordinates is the energy-deposition

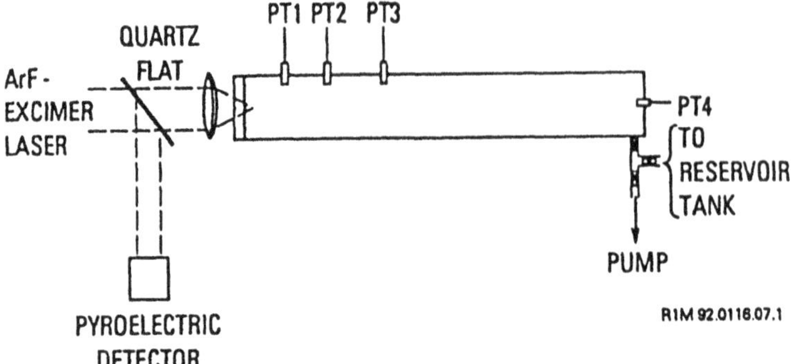

Fig. 11. Schematic of the apparatus for investigating the direct initiation of detonation by energy deposition from a laser pulse into a mixture confined in a tube of 1-m length and 1.6-cm diameter. The location of each of four pressure transducers (PT) is indicated.

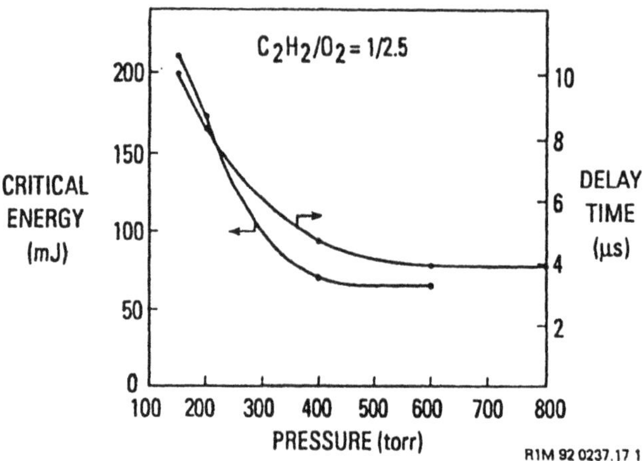

Fig. 12. The critical energy for the direct initiation of detonation in a stoichiometric C_2H_2/O_2 mixture, and the delay time for wave initiation as extrapolated from the arrival time at the pressure transducers, as a function of cold-mixture pressure (760 torr $\equiv$ 0.1 MPa). These data are from the apparatus sketched in Fig. 11.

site, and if $\overline{\sigma} = \sigma$ is the radius of the point p at time t, where $t = 0$ at the time of initiation of the smaller detonation (of radius R_2), then Fig. 14 presents geometry helpful for the analysis. The centers of the two spherical detonations are displaced axially by a distance $u_0 t_1$, and the angle ψ is defined by the two tangent planes to the radii extending from the center of each detonation to the point p.

We can interpret Fig. 13 as an instantaneous snapshot for an observer (moving with the flow speed u_0), who then sees the two spherical waves intersect at different values of the parameter σ as time t increases. With the increase of time, the spherical fronts advance (both at speed u_{CJ}) into unreacted mixture (that is stationary in the frame of reference). Various particles, situated behind one spherical detonation front, collide with various particles behind the other front, and the locus of the collision sites is the bisecting plane of the angle ψ. This plane in general is not perpendicular to the axis of symmetry. The particle speed just behind the detonation front is $(u_{CJ} - a_f)$, where a_f is the speed of sound for the equilibrium state in the just-reacted mixture, before any expansion occurs, and u_{CJ} (typically, appreciably) exceeds a_f.

When the two detonations are just interacting at the point p (Fig. 14), the state of the nearby gas {situated at a value of $y[\equiv \sigma/(u_{CJ} t_1)]$ just slightly smaller than the value holding for the point p} is that of the product gas of the CJ detonation. [It is convenient to introduce the parameter y in a dimensionless analysis of the phenomenon.] In Fig. 15, the bisector of the angle ψ is taken as a vertical line and the coordinate x is positive *toward the left.* Only the components (of the velocity of the just-detonated gas near the point p) that are perpendicular to the bisector of the angle ψ through the point p enter the analysis, because only these components are altered by the formation of the reflected shocks that propagate back through the detonated gas; the tangential components are continuous and unaltered. The subscripts 1 and 2 in Fig. 15 correspond to the subscripts 1 and 2 introduced in Fig. 14.

Just a small fraction of a second later, the close-up, one-dimensional picture of Fig. 15 evolves to the picture of Fig. 16, owing to the just-mentioned entropy-increasing reflected shocks, which propagate at speed s, to be found. These reflected shocks demarcate conditions in region 3 (detonated and shocked gas) from conditions in region 2 (detonated but not yet shocked gas). The above convention of denoting conditions in the detonated but unshocked gas with

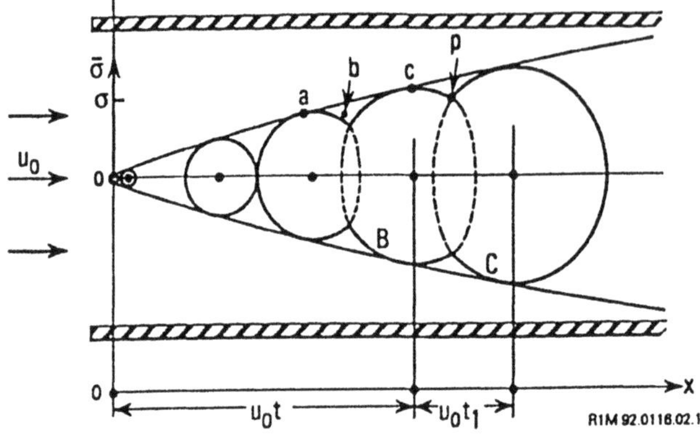

Fig. 13. Schematic of a train of CJ spherical detonation waves, initiated by periodic energy deposition at the origin and convected by a reactive mixture of speed $u_0 (> u_{CJ})$. The conceptual conical envelope is indicated, along with reflected shocks (dashed lines) formed from detonation-wave interactions. The time interval between brief pulses is t_1. The points a, b, c, d, and p indicate different possibilities for the first interaction of an element of unreacted mixture with an outwardly propagating detonation front.

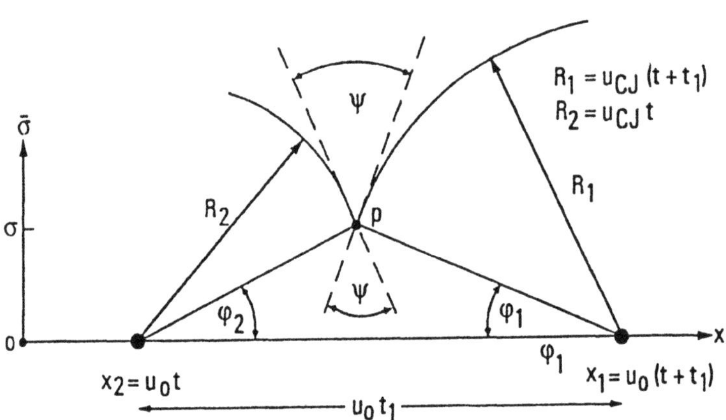

Fig. 14. Geometry for analyzing the interaction of two outwardly propagating CJ spherical detonations at time t, where one was initiated at $t = -t_1 (t_1 > 0)$, and one at $t = 0$, in a reactive mixture flowing at speed u_0. The point p here corresponds to that noted in Fig. 13. The angle ψ is defined by planes perpendicular to radii from the center of each detonation to the point p.

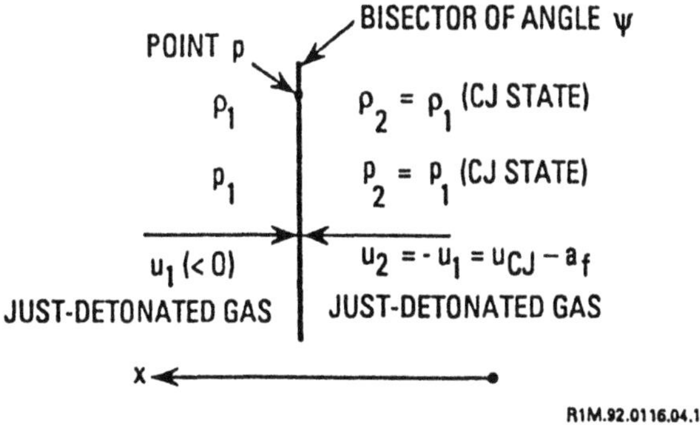

Fig. 15. A one-dimensional model of the just-detonated mixture very near the intersection of two CJ waves, in the frame of reference of the flowing cold mixture. The coordinate x, positive to the right in Fig. 14, is here positive to the left.

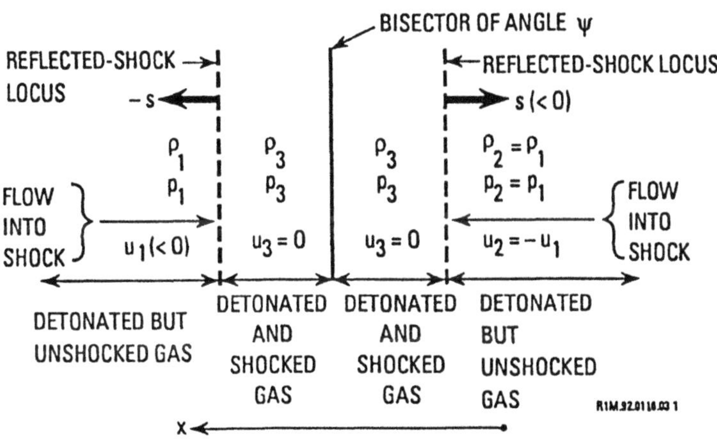

Fig. 16. Reflected shocks, propagating at speed $s(< 0)$, to be found, demarcate detonated unshocked gas from detonated shocked gas, in the frame of reference of the flowing cold mixture. The bisector of angle ψ is stationary in this frame. This scenario holds very shortly after that depicted in Fig. 15.

subscript 1 on one side of the angle-bisector plane (image plane), and with subscript 2 on the other side, because the sign of the velocity differs, is continued in Fig. 16. We anticipate that the shock speed $s < 0$, and we emphasize that, in the adopted frame of reference, the angle bisector remains fixed in position.

The adoption of constant values for u_2, p_2, and ρ_2, and for u_1, p_1, and ρ_1, limits, to very short time after the reflection, the rigorous application of the above one-dimensional model of the interaction of the two spherical detonations. The radial flow behind a (spherically) radially expanding detonation wave is nonuniform spatially: the flow monotonically decreases from a maximum value of $(u_{CJ} - a_2)$ just behind the spherical detonation to zero at a finite radial distance from the center. The pressure and density also monotonically decrease with decreasing radial distance from the origin of the radial coordinate, and a uniform thermodynamic state characterizes the just-mentioned stagnant region of finite radius about the center (Landau and Lifshitz, 1952, Sedov, 1959; Zeldovich and Kompaneets, 1960). However, we believe that the model is a qualitative indicator for a much longer time; in particular, we expect that the entropy rise for the adopted one-dimensional geometry exceeds the entropy rise for the actual spherical-wave interaction.

For fixed values of γ, $\alpha[\equiv q/(p_0/\rho_0)]$, and $\Lambda(> 1)$, execution (Carrier, Fendell, and Chou, 1992) of the analytic details reveals that the normalized-entropy increase (Liepmann and Rosko, 1957) across the reflected shock, $\Delta s/R$ (where R is the gas constant for the mixture), monotonically decreases as the parameter y increases, where it is recalled that $y \equiv \sigma/(u_{CJ}t_1)$ (Fig. 17). For a given cold mixture at a given thermodynamic state, reducing the time between pulses results in a smaller increase in entropy owing to the reflected shocks. Incidentally, practically interesting values of the dimensionless parameter α range from 20 to 30 or so, where $\alpha \equiv [\gamma/(\gamma - 1)]D_2$, the symbol D_2 denoting the second Damkohler number (the ratio of exothermicity per mass of mixture to the specific enthalpy of the unreacted mixture).

4. Nozzle Configuration for Exhaust of the Detonated Mixture

We briefly supplement earlier remarks (in Section 1.2) concerning the steady axisymmetric supersonic flow of the detonated gas through a nozzle (which we hope to design with "modest" divergence and length) to permit discharge at ambient pressure p_0, so that thrust may be developed.

Figure 19 presents the displaced origins of the axial coordinates x, x', and x''. Here, $x' = x - L$, where $L = r_0/\tan\beta$ and, in turn, $\sin\beta = M/M_0 = u_{CJ}/u_0$; it is recalled that $M \equiv u_{CJ}/a_0$, $M_0 \equiv u_0/a_0$, a_0 is the sound speed in the cold mixture, and r_0 is the cylindrical radius of the inlet pipe. Also, $x'' = x' - x'_s$, where x'_s is recalled to be the (relatively short) length of an upwind portion of the nozzle. The wall contour of this upwind portion is defined by the extension of the selfsimilar-flow streamsurface passing through the detonation-wave/inlet-pipe "intersection." Explicitly, if the velocity vector is denoted $\vec{v}$, then, in the selfsimilar flow, $\vec{v}(\theta) = u(\theta)\hat{r} + w(\theta)\hat{\theta}$, and the streamsurface of interest is described by (Figs. 18 and 19)

$$\frac{dr}{r\,d\theta} = \frac{u(\theta)}{w(\theta)}, \text{ with } r = r_0/\sin\beta,\ \theta = \beta; \tag{4}$$

the radial velocity component $u(\theta)$ and the spherical-polar-angle velocity component $w(\theta)$, for use in Eq. (4), are available from the solution (Carrier, Fendell, McGregor et al., 1992) for the just-detonated-gas flow, which (we reiterate) is selfsimilar in terms of the polar angle $\theta, 0 < \theta < \beta$. Figure 19 emphasizes that adoption of such a "streamsurface"-defined nozzle does achieve uniformization of the pressure, with a purely axial flow; these are *some* of the properties sought for the flow at the nozzle-exit plane (which we formally designate by the symbol x'_e). However, the value of the pressure [in the uniform-thermodynamic-state flow arising for the subdomain $0 \leq \theta \leq \theta_i$, where the value of the polar angle $\theta_i (< \beta)$ is to be found in the course of solution] typically far exceeds the ambient pressure p_0, whereas we seek to discharge at p_0. Thus, as indicated in Fig. 19, contouring a nozzle entirely by the use of Eq. (4) is an implausible design.

We have investigated, by means of a modified-method-of-characteristics calculation for the inviscid flow of an ideal gas, the flow field resulting from the smooth adjoining of a conical nozzle of

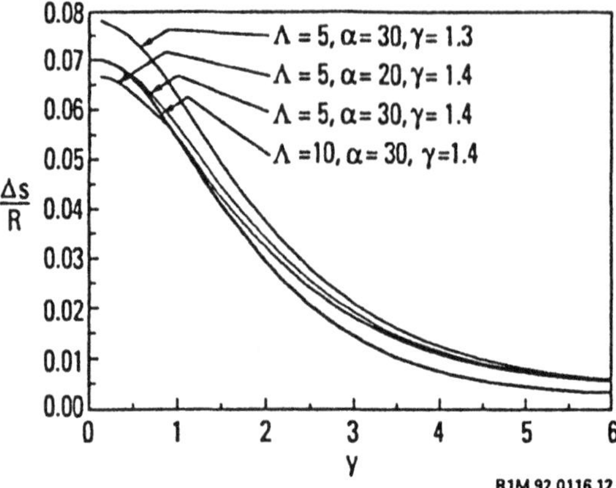

Fig. 17. The change in entropy Δs across one the reflected shocks depicted in Fig. 16, as a function of the parameter $y[\equiv \sigma/(u_{CJ}t_1)]$, where σ is the value cylindrical-radial coordinate of the point p. The parameter $\Lambda(\equiv u_0/u_{CJ}) > 1$; $\alpha = q/(p_0/\rho_0)$, where q denotes the chemical exothermicity per mass of mixture, and p_0 and ρ_0 denotes the cold-mixture pressure and density, respectively; and γ denotes the ratio of specific heats. The normalization factor R is the gas constant.

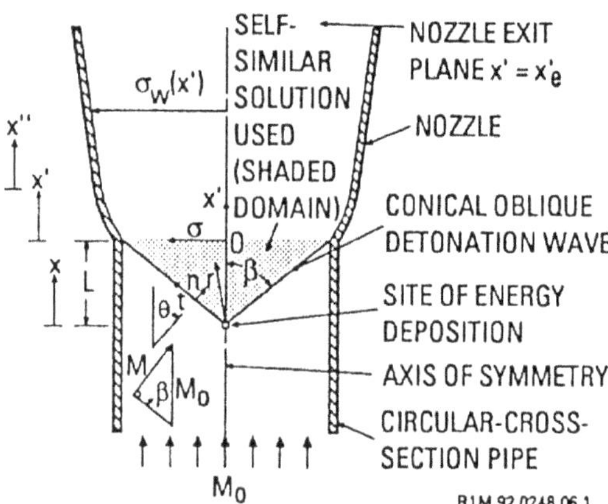

Fig. 18. A redrawing of Figure 5 to emphasize the definition of the origin of each of the three axial coordinates x, x', and x''. The cylindrical radial position of the nozzle wall at the nozzle exit $x' = x'_e$ is denoted $\sigma_w(x'_e) \equiv \sigma_{we}$.

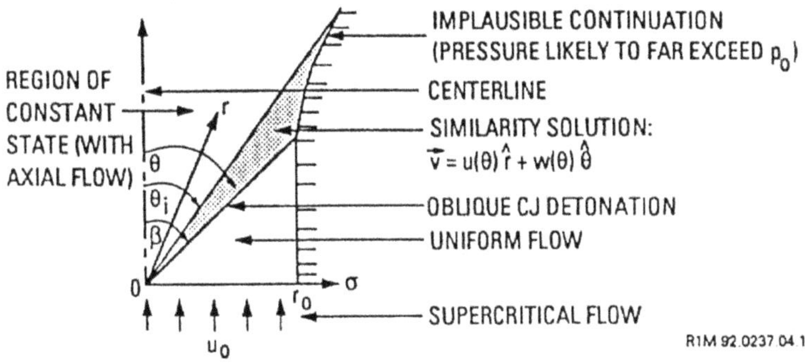

Fig. 19. A schematic of a possible choice (found to be implausible, unless modified) for the wall contour downwind of the oblique-CJ-detonation-wave/pipe-wall "intersection." The choice involves the "materialization" of that streamsurface (for the just-detonated flow, selfsimilar in the spherical polar angle θ) that passes through the intersection. The portion of the selfsimilar, detonated flow in $0 < \theta < \theta_i$ has uniform thermodynamic state and purely axial flow; the angle $\theta_i (< \beta)$ is to be found. In the shaded domain, the spatially nonuniform portion of the selfsimilar, detonated flow holds. Although this nozzle contour yields uniform axial flow and laterally uniform pressure at discharge, the exit pressure typically far exceeds ambient pressure p_0.

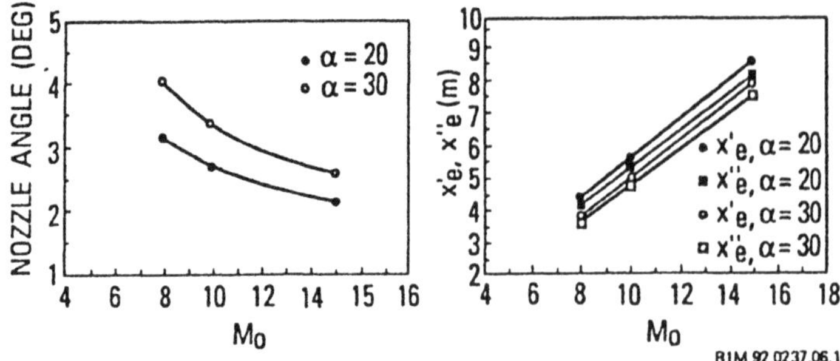

Fig. 20. The axial length x_e'' of a conical nozzle smoothly joined to a short segment of a streamsurface nozzle, the axial length x_e' of the total nozzle (streamsurface plus conical), and the half angle of the conical portion, as a function of ambient Mach number M_0, for $\gamma = 1.4$ and two values of α. The computations, for an ideal inviscid gas, adopt $p_0 = 100$ Pa, $T_0 = 272$ K, and $r_0 = 0.2$ m.

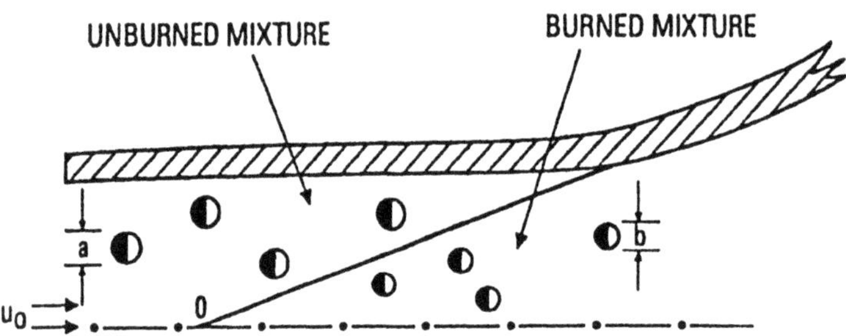

Fig. 21. An idealization for imperfectly mixed throughput, in which the well-mixed fraction is converted to product in the conical detonation wave. The inhomogeneous fraction is taken to consist of spheroids of fuel and air with the same stoichiometric balance as the well-mixed fraction. The spheroids, of equivalent radius a upwind of the wave, are compressed to equivalent radius b downwind, and are burned up on the diffusive time scale b^2/κ, where κ is the controlling transport coefficient. The radius of a spheroid is exaggerated relative to the radius of the container for ease of presentation.

prescribed modest angle, on to the just-described "streamsurface"-defined contour (Fendell, Mitchell et al., 1992). While other families of contours may be aerodynamically superior, trial-and-error calculation with this particular two-part family of nozzle contours yields results (displayed in Fig. 20) for the conical-portion half angle, the axial length of the conical portion x_e'' and the axial length of the entire (conical plus "streamsurface") nozzle x_e'. For these values, the exhaust is discharged uniformly at ambient pressure p_0. In these calculations, the pipe radius is 0.2 m, the ambient pressure p_0 is 100 Pa, and the ambient temperature $T_0 = 272$ K. Actually, the dependent and independent variables in the computations can be normalized in terms of p_0, T_0, and r_0, such that only three dimensionless parameters (α, M_0, γ) arise in a nondimensionalized mathematical formulation. Thus for any altitude in an isothermal atmosphere (in which $p_0 \sim \rho_0$), for fixed values of the parameters u_0, q, and γ, the results scale with r_0. The range of computed nozzle length and flaring seems feasible for supersonic combustors, although retention of real-gas properties and finite-rate kinetics would modify the results.

Since subscript e denotes conditions at the nozzle-exit plane, we adopt the symbol U_e to denote the axial flow component (virtually the total speed) at the nozzle-exit plane (where the nozzle radius is σ_{we}); then the thrust F is given by (Fendell, Mitchell et al., 1992)

$$F = \rho_0 u_0^2(\pi r_0^2)\left(\overline{\mu}\frac{U_e}{u_0} - 1\right) + (\pi\sigma_{we}^2)p_0\left(\frac{p_e}{p_0} - 1\right), \tag{5}$$

$$\overline{\mu} \equiv \frac{\rho_e U_e(\pi\sigma_{we}^2)}{\rho_0 u_0(\pi r_0^2)} \approx 1 \tag{6}$$

for most mixtures of interest here. Also, $p_e = p_0$ by design, so the finite thrust is well estimated by

$$F \approx \rho_0 u_0^2(\pi r_0^2)\left(\frac{U_e}{u_0} - 1\right), \tag{5a}$$

where $U_e > u_0$ for the conical-detonation-wave engine under study. Quantitative results for the thrust have been presented (Fendell, Mitchell et al. 1992).

5. Mixture Inhomogeneity

In practice, the admixture of fuel to the captured airstream will be imperfect, and the performance of the engine with an inhomogeneous mixture needs to be ascertained.

The source of entropy rise in a well-designed engine with a homogeneous mixture would arise predominantly in connection with the conical detonation wave. Poorly mixed fuel would be compressed but unreacted during traverse of the conical wave. Such fuel must be reacted with air downwind of the conical wave (but before chemical reaction is "frozen" in the supersonic nozzle – if such freezing occurs); otherwise, the wasteful situation in which unreacted fuel is exhausted arises. There is less entropy rise incurred across the conical wave if some of the mixture is not detonated, but the diffusive burning of this poorly mixed fraction entails entropy rise downwind of the conical wave. The tradeoff in entropy rise warrants quantification, because there even may be performance advantages in incurring some inhomogeneity.

For a conical-detonation-wave engine, one seeks to achieve rapid mixing of a fraction of the fuel sufficient to sustain the conical wave. Presumably the throughput near the axis of symmetry is especially high in well-mixed fraction to assist laser initiation. Also, the undesirable consequences of high temperature near the conical-wave/pipe-wall interaction suggests that container-material problems may be ameliorated if that vicinity is relatively reaction-free.

We consider one conceptual example of possibly advantageous "tailoring" of the scale and nature of the inhomogeneity in the imperfectly mixed fraction of the throughput. In this example (Fig. 21), the cold unmixed blobs consist of (say) spheres of radius a (where this use of the symbol a is not to be confused with its earlier use to designate the speed of sound). The spheres are taken to be composed of fuel and air in the same stoichiometric proportion as the well-mixed fraction of the throughput. The spheres of radius a are compressed after passage through the conical detonation wave, a phenomenon schematically accounted for by taking the unburned blobs in the burned gas to be spheres of radius $b (< a)$. These spheres are burned up on the diffusive time scale b^2/κ, where κ is the diffusive coefficient for the rate-controlling transport process, if a diffusive process is rate-controlling.

More generally, some insight concerning the effect of mixture inhomogeneity on the wave structure, and the entropy-rise and residual-fuel-combustion-before-expansional-freezing questions, may be achievable by modeling. However, achieving good mixing is so crucial for any oblique-detonation-wave-based supersonic combustor that no issue would seem to warrant higher priority on a research agenda than experiments on achieving good mixing without preignition-inducing static-temperature rise.

6. Concluding Remarks

The above-sketched supersonic combustor based on a laser-initiated conical detonation wave is conceived to be feasible at flight Mach numbers sufficiently large that the CJ-wave speed of a roughly stoichiometric mixture of air with the fuel of choice is less than the cold-mixture speed at the laser energy-deposition site (to preclude "detonative flashback"). On the other hand, the combustor is conceived to be feasible at flight Mach numbers sufficiently small that inevitable air deceleration associated with intake and fuel admixture does not result in preignition upwind of the laser-energy-deposition site. This range of feasible flight Mach numbers might lie in the range of (say) 6-12; though this range has been described as within the domain of relatively well understood technology (Bushnell, 1992), no air-breathing vehicle known to the authors currently is flying in this domain. Emphasis on developing an oblique-detonation-wave engine for the regime of flight Mach numbers greater than 14 or so encounters a myriad of fuel storage, mixing, preignition, materials, and other problems. In fact, for purposes of cruise, in view of sonic-boom and other environmental constraints, and in view of the size of this planet, flight in excess of Mach 12 seems of low priority. For purposes of launch to earth orbit, there may be economic and reliability advantages to a single-stage large unsophisticated pressure-fed liquid-rocket booster. But if a multistage vehicle for injection of a payload into orbit is considered, the use of a rocket engine above Mach number 12 or so (i.e., for the upper atmosphere, where the air is relatively low in density) is not unreasonable, so air-breathing earlier stage(s) for flight in the relatively dense, lower atmosphere might well be limited to flight Mach numbers below about 12. Thus, the plausible upper range for air-breathing high-speed propulsion perhaps may be for flight Mach numbers of 6-12 or so.

Even in this range of flight Mach numbers there are significant challenges. Choice of combustion-chamber pressure involves difficult tradeoffs, some of which are not unique to the conical-detonation-wave engine. Operation at lower combustor-inlet pressure results in: (1) higher critical energy (for direct initiation), which is inversely proportional to the pressure – so the energy-per-pulse requirements are more demanding; (2) larger detonation cell size, which tends to vary roughly inversely with the pressure, so the geometric scale of the engine is increased, because inclusion of several cell sizes is important for detonation-wave stability; and (3) more likely freezing of the flow in the nozzle. Alternatively, operation at higher combustion-inlet pressure results in: (1) increased possibility of preignition; (2) augmented heat-transfer problems in the vicinity of the detonation-wave/container-wall intersection; and (3) increased dissociation in the higher-temperature products of combustion, so the chemical energy derived from combustion is reduced.

The challenging requirements on the laser source involve not just the energy per pulse, but also the rate of repetition, which should be sufficiently rapid to minimize the reflected-shock-associated entropy rise from the interaction of neighboring spherical detonations. Further, we seek to have several spherical detonations coexistent in the conceptual conical envelope within the combustor; for flight Mach number 10, a repetition rate on the order of 10^5 Hz seems appropriate. Typically, Joule/pulse lasers currently do not exceed a repetition rate of about 10^2 Hz. Since, in current laser technology, the repetition rate tends to correlate inversely with the energy per pulse, there is strong motivation for experiments to identify the minimal energy-per-pulse requirement for direct initiation of detonation.

In Section 5 we have discussed some phenomena, associated with fuel admixture to the captured air stream, that seem addressable by modeling, as well as by experiment. We conclude by reiterating the high priority of experiments on the direct-initiation of detonations (without overdriving) in fuel/air mixtures of interest. For the H_2/air system frequently discussed for high-speed propulsion, a possibly noteworthy sensitizing additive that is absorptive at the 193-nm wavelength of the ArF-excimer laser, used in the above-reported experiments on direct initiation of detonation, is a dilute aqueous solution of H_2O_2. Adoption of such a sensitizing agent probably introduces no new product species. Also, H_2O_2 is not only absorptive of the incident radiation, but also photodissociates to form the highly reactive radical OH. For completeness, we reiterate that, if the introduction of carbon into the H_2/air system is acceptable, then C_2H_2 is absorptive at 193 nm and photodissociates to form the reactive radicals C_2H and H.

Acknowledgement

We are grateful for helpful suggestions and guidance concerning the literature by Prof. Joseph Shepherd of Rensselaer Polytechnic Institute.

References

Abouseif, G. E. and Toong, T. Y., 1982. "On direct initiation of gaseous detonations," *Combust. Flame* **45**, p. 39.

Bach, G. G., Knystautas, R., and Lee, J. H., 1969. "Direct initiation of spherical detonation in gaseous explosives," *Twelfth Symposium (International) on Combustion*, Combustion Institute, Pittsburgh, PA, p. 853.

Bach G., Knystautas, R., and Lee, J. H. S., 1970. "Initiation criteria for diverging spherical detonations," *Thirteenth Symposium (International) on Combustion*, Combustion Institute, Pittsburgh, PA, p. 1097.

Bush, W. B. and Fendell, F. E., 1971. "Asymptotic analysis of the structure of a planar steady detonation," *Combust. Sci. Tech.* **2**, p. 271.

Bushnell, D. M., 1992. "Mixing and combustion issues in hypersonic air-breathing propulsion," 2nd ICASE/NASA Langley Research Center Workshop on Combustion, Newport News, VA.

Cambier, J.-L., Adelman, H., and Menees, G. P., 1989a. "Numerical simulations of oblique detonation in supersonic combustors," *J. Propulsion Power* **5**, p. 482.

Cambier, J.-L., Adelman, H., and Menees, G. P., 1989b. "Numerical simulations of oblique detonation wave engine," *J. Propulsion Power* **6**, p. 315.

Carrier, G. F., Fendell, F. E., and Chou, M.-S., 1992. "Laser-initiated conical detonation wave for supersonic combustion. III," AIAA Paper 92-3247, American Institute of Aeronautics and Astronautics, Washington, DC.

Carrier, G., Fendell, F., McGregor, R., Cook, S., and Vazirani, M., 1992. "Laser-initiated conical detonation wave for supersonic combustion," *J. Propulsion Power* **8**, p. 472.

Carrier, G. F., Fendell, F. E., and Sheffield, M. W., 1990. "Stabilization of a premixed planar flame by localized energy addition," *AIAA J.* **28**, p. 625.

Chou, M.-S. and Zukowski, T. J., 1991. "Ignition of $H_2/O_2/NH_3$, H_2/air/NH_3 and $CH_4/O_2/NH_3$ mixtures by excimer-laser photolysis of NH_3," *Combust. Flame* **87**, p. 191.

Daiber, J. W. and Thompson, H. M., 1967. "Laser-driven detonation waves in gases," *Phys. Fluids* **10**, p. 1162.

Fendell, F. E., Kung, E. Y., and Sheffield M. W., 1992. "Flame configuration associated with localized energy addition to a flowing combustible mixture," *J. Propulsion Power* **8**, p. 464.

Fendell, F., Mitchell, J., McGregor, R., Magiawala, K., and Sheffield, M., 1992. "Laser-initiated conical detonation wave for supersonic combustion. II," AIAA Paper 92-0088, American Institute of Aeronautics and Astronautics, Washington, DC.

Klimkin, V. F., Soloukhin, R. I., and Wolansky, P., 1973. "Initial stages of a spherical detonation directly initiated by a laser spark," *Combust. Flame* **21**, p. 111.

Landau, L. D. and Lifshitz, E. M., 1952. *Fluid Mechanics*, Pergamon, New York, NY.

Lee, J. H. S., 1977. "Initiation of gaseous detonations," *Ann. Rev. Phys. Chem.* **28**, p. 75.

Lee, J. H. S., 1984. "Dynamic parameters of gaseous detonations," *Ann. Rev. Fluid Mech.* **16**, p. 311.

Lee, J. H. and Knystautas, R., 1969. "Laser spark ignition of chemically reactive gases," *AIAA J.* **7**, p. 312.

Lee, J. H., Knystautas, R., and Yoshikawa, N., 1978. "Photochemical initiation of gaseous detonation," *Acta Astronaut.* **5**, p. 971.

Lee, J. H. and Matsui, H., 1977. "A comparison of the critical energies for direct initiation of spherical detonations in acetylene-oxygen mixtures," *Combust. Flame* **28**, p. 61.

Liepmann, H. W. and Roshko, A., 1957. *Elements of Gasdynamics*, John Wiley, New York, NY.

Nettleton, M. A., 1987. *Gaseous Detonations – Their Nature, Effects, and Control*, Chapman and Hall, New York, NY.

Sedov, L. I., 1959. *Similarity and Dimensional Analysis in Mechanics*, Pergamon, New York, NY.

Strehlow, R. A., 1984. *Combustion Fundamentals*, 2nd ed., McGraw-Hill, New York, NY.

Waltrup, P. J., 1987. "Hypersonic air breathing propulsion: evolution and opportunities," *Conference Proceedings, Aerodynamics of Hypersonic Lifting Vehicles*, Advisory Group for Aerospace Research and Development, Neuilly sur Seine, France, p. 12-1.

Westbrook, C. K. and Urtiew, P. A., 1982. "Chemical kinetics prediction of critical parameters in gaseous detonations," *Nineteenth Symposium (International) on Combustion*, Combustion Institute, Pittsburgh, PA, p. 615.

Zel'dovich, I. B., Kogarko, S. M., and Simonov, N. N., 1957. "An experimental investigation of spherical detonation of gases," *Sov. Phys. – Tech. Phys.* **1**, p. 1689.

Zel'dovich, I. B. and Kompaneets, A. S., 1960. *Theory of Detonation*, Academic, New York, NY.

RECENT ADVANCES IN RAM ACCELERATOR TECHNOLOGY

A. Hertzberg, E. A. Burnham, and J. B. Hinkey

Aerospace and Energetics Research Program
University of Washington
Seattle, Washington 98195

ABSTRACT

Recent advances in ram accelerator technology are presented in this paper. The ram accelerator is an in-bore propulsion system in which a projectile shaped like the centerbody of a supersonic ramjet is accelerated down a stationary tube filled with a combustible gas mixture. Combustion on and behind the projectile body creates a pressure distribution which accelerates the projectile. Several different ram accelerator propulsive regimes have been experimentally demonstrated over a Mach number range of 3 to 8.5. In the subdetonative (thermally choked) regime projectiles have been accelerated to near the Chapman-Jouguet (C-J) detonation velocity with many different propellant mixtures. In the transdetonative velocity regime (90 to 110% of C-J speed), projectiles have smoothly transitioned from subdetonative to superdetonative velocities in a tube containing a single propellant mixture. In the superdetonative velocity range, the projectiles accelerate while always traveling faster than 110% of the C-J speed. Ram accelerator projectiles operating through these velocity regimes generate distinctive supersonic and hypersonic phenomena which can be studied very effectively in the laboratory. High spatial resolution tube wall pressure data in a single propellant mixture from the three regimes of propulsion are presented and reveal the three-dimensional character of the flow field induced by projectile fins and yawing of the projectile relative to the tube wall. Also presented for comparison to the experimental data are calculations made with an inviscid, three-dimensional CFD code. Experimental and computational examples provide useful benchmarks for confirmation of CFD codes which model three-dimensional, chemically reacting, steady or unsteady flows.

J. Buckmaster et al. (eds.), Combustion in High-Speed Flows, 309–344.

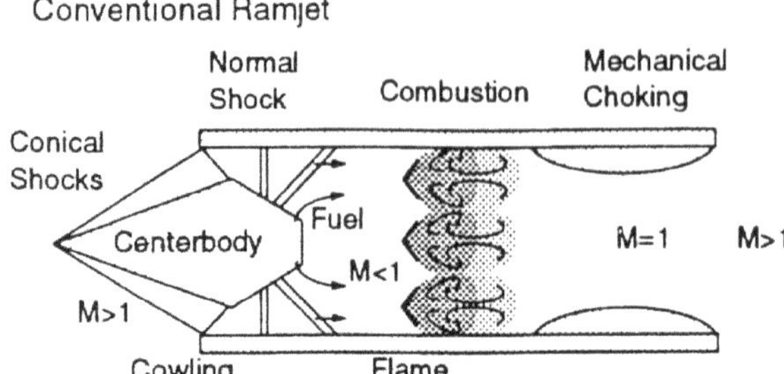

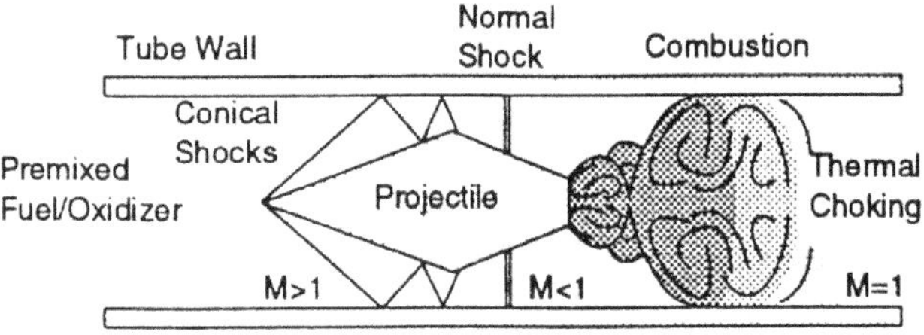

Fig. 1 Comparison of a conventional ramjet with the ram accelerator.

1. Introduction

At the University of Washington, experimental research on supersonic and hypersonic propulsion is being carried out using a ramjet-in-tube concept called the "ram accelerator" (Hertzberg et al., 1987; Hertzberg et al., 1988). The ram accelerator operates in a manner similar to a conventional supersonic airbreathing ramjet or scramjet. As shown in Fig. 1, a projectile which resembles the centerbody of a ramjet is accelerated through a stationary tube filled with a premixed gaseous fuel and oxidizer mixture. The projectile itself carries no propellant. The combustion process travels with the projectile, generating a pressure distribution which produces forward thrust. The pressure and composition of the gas mixture can be easily adjusted, controlling the Mach number and energy available for acceleration of the projectile. The ram accelerator is therefore a highly useful experimental tool with which to obtain data on supersonic and hypersonic propulsion over a wide range of Mach numbers, pressures, equivalence ratios, and accelerations. Furthermore, the ease with which the device can be scaled up in size gives it the potential to perform ground-based flight tests on relatively large, highly instrumented models and full-scale components at velocities and Reynolds numbers of interest to the NASP and other hypervelocity vehicles (Bruckner, Knowlen, and Hertzberg, 1992; Witcofski et al., 1990).

Three regimes of ram accelerator propulsion have been studied to date (Bruckner et al., 1987; Hertzberg et al., 1991). These are characterized by the projectile speed relative to the Chapman-Jouguet (C-J) detonation velocity of the combustible mixture. One of these regimes is the subdetonative or thermally choked ram accelerator (Fig. 2), which operates at velocities below 90% of the C-J detonation speed of the propellant mixture, with Mach numbers typically from 2.5 to 4.5. In this regime, thrust is provided by high projectile base pressure resulting from a normal shock system which is stabilized on the body by thermal choking of the flow at full tube area behind the projectile (Fig. 2). A theoretical model of the thermally choked mode predicts that the normal shock recedes along the body as the projectile Mach number increases. At some limiting Mach number the normal shock is predicted to fall off the base of the projectile and the acceleration should cease. In the hypothetical case where a projectile body tapers to a point and the flow is inviscid, the normal shock gradually falls back to the full tube area. A nor-

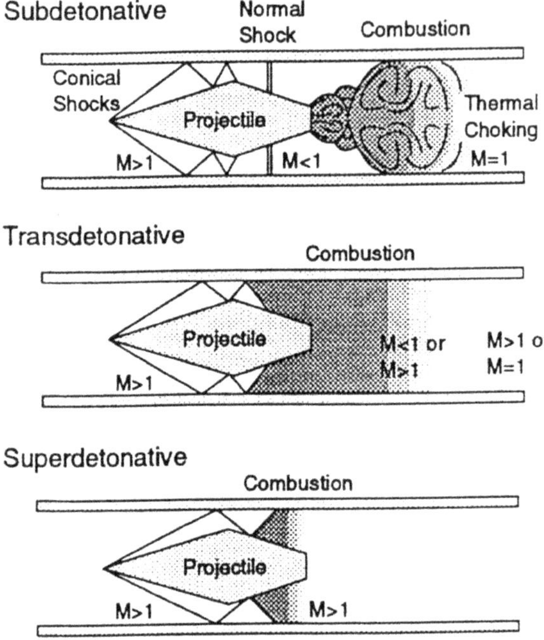

Fig. 2 Operational regimes of the ram accelerator.

mal shock in a constant area duct followed by heat addition and thermal choking in steady flow constitute a C-J detonation wave. Therefore, this theoretical model predicts that the thrust goes to zero as the projectile velocity approaches the C-J detonation speed of a particular propellant mixture.

It has been experimentally observed that while operating at subdetonative speeds (below 90% of the C-J detonation speed of the propellant gas) the thrust as a function of Mach number is predicted very well by the one-dimensional theoretical model of the thermally choked ram accelerator (Bruckner et al., 1991). As the projectile velocity exceeds 90% of the C-J detonation speed, however, the thrust on the projectile begins to exceed that predicted by theory. The projectile experiences a relative minimum in thrust when it has been allowed to accelerate up to the C-J detonation velocity of the combustible gas mixture, and it continues to gain velocity at acceleration levels that increase with increasing velocity. These experiments have shown that in this "transdetonative" velocity range (typically Mach 4.5 to 5.5), the projectile can accelerate smoothly from subdetonative to superdetonative speeds in a single gas mixture (Burnham et al., 1990). This transdetonative performance was an unexpected experimental discovery that has significant implications for many hypersonic propulsive concepts, not to mention single-stage high velocity ram accelerator applications.

At higher Mach numbers the superdetonative ram accelerator regime has been investigated. This regime occurs at velocities greater than 110% of the C-J detonation speed, with Mach numbers typically exceeding 5, by igniting the propellant mixture on the body of the projectile, as shown in Fig. 2 (Brackett and Bogdanoff, 1989; Yungster and Bruckner, 1992; and Yungster et al., 1991). The gasdynamic principles of the superdetonative ram accelerator are thought to be similar to those of scramjet and oblique detonation wave engines (Ostrander et al., 1987; Pratt, Humphrey, and Glenn, 1991).

Experiments have been conducted over the Mach number range of 2.5 to 8.5 using a variety of propellant mixtures at pressures of 3 to 44 atm. Over 1000 test firings have taken place to date. The thermally choked propulsive regime has been investigated at velocities of 700 to 2680 m/sec and Mach numbers of 2.5 to 4.5 (Hertzberg, Bruckner, and Bogdanoff, 1988; Bruckner et al., 1987; Bruckner et al., 1991). Combined thermally choked and transdetonative experiments have accelerated projectiles with entrance Mach numbers of $\sim$

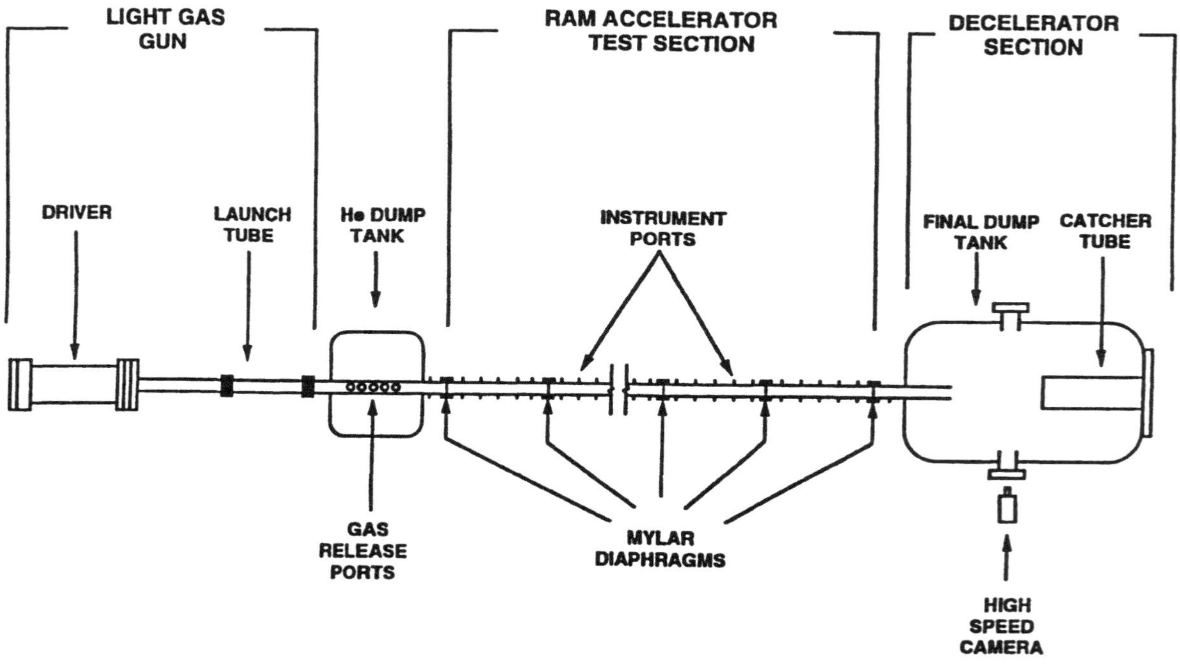

Fig. 3 Existing ram accelerator facility.

3 up to 6 within a single propellant mixture, achieving single-stage velocity gains of over 1000 m/sec (Burnham et al., 1990). Experiments conducted entirely in the superdetonative velocity range have successfully demonstrated positive thrust at Mach numbers up to 8.5 (velocities up to 150% C-J detonation speed) (Kull et al., 1989). In this paper we present the results of experimental and numerical investigations of the three-dimensional flow field induced by the projectile in the three propulsion modes discussed above, which were carried out in the Mach number range of 3.4 to 5.6 using methane based propellant mixtures.

2. Experimental Facility

The current ram accelerator facility (Fig. 3) consists of a light gas gun, ram accelerator section, final dump tank, and projectile decelerator. The 38 mm bore, 6 m long, single-stage light gas gun is capable of launching the obturator and projectile combination (typical combined mass approximately 60 to 100 g) to speeds up to 1300 m/s. The muzzle of the gas gun is connected to a perforated-wall tube that passes through an evacuated tank, which serves as a dump for the helium driver gas.

The 16 m long ram accelerator section consists of eight steel tubes, with a bore of 38 mm and an outer diameter of 100 mm. There are a total of 144 instrumentation ports at 40 axial locations, spaced at 40 cm intervals along the accelerator tube. At 24 axial stations there are four ports separated by 90° angles, and at 16 stations there are three ports separated by 120°. This permits the use of either three or four transducers at each station. Piezoelectric pressure transducers, electromagnetic transducers (copper wire coiled around a Lexan core), and fiber-optic light guides can be located in any of these observation stations. A 32 channel, 1 MHz digital data acquisition system is used to acquire the data. Multiplexing permits monitoring of up to 100 separate input signals.

The ram accelerator tube is designed to operate at propellant fill pressures up to 50 atm. Thin Mylar diaphragms are used to close off each end of the accelerator tube and to separate sections of the tube filled with different propellant mixtures. The fuel, oxidizer, and diluent gases are metered using sonic orifices and directed to the appropriate sections of the ram accelerator tube.

The end of the accelerator tube is connected by a 0.76 m long

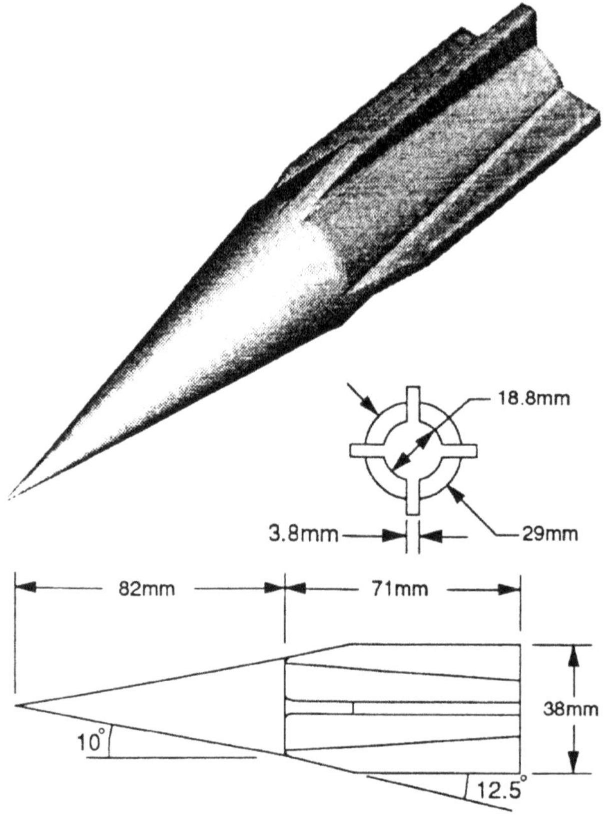

Fig. 4 Ram accelerator projectile.

drift tube to a 2.4 m long evacuated dump tank, where the projectile experiences free flight. The tank has a pair of 25 cm diameter viewing ports used for high-speed photography. The free-flying projectile impacts a metal witness plate and is brought to a stop in carpet remnants that are tightly packed in a 20 cm bore by 1 m long tube attached to the far end and inside of the final dump tank (Fig. 3).

The basic projectile geometry that has been used in the majority of the experimental work to date is illustrated in Fig. 4. It is fabricated of either magnesium alloy or aluminum alloy in two hollow pieces; the nose cone and the body, which are threaded together. Projectiles used to date range in mass between 45 and 90 g, depending on structural details. The fins serve to center the projectile in the tube. Thin magnetic disks are mounted in the nose-body joint and in the base of the body. When the projectile passes by the electromagnetic transducers in the accelerator tube, the magnets induce signals that are used to determine the time-distance ($t - x$) history of the projectile and, thus, its velocity and acceleration.

2.1. Highly instrumented tube section

A recent advance in ram accelerator technology is the development of a separate, high instrument density, short section of ram accelerator tube and a new tube coupling method (Hinkey et al., 1992a). The highly instrumented tube section (**HITS**) and tube couplers are shown in Fig. 5. The tube couplers (one with a right hand internal thread and one with a left hand internal thread) separately thread onto the existing tube segments and provide a sturdy flange to which the HITS is attached and allow the coupling of the entire assembly together. Unlike the original tube coupling system, the flanges allow the instrumentation stations in the HITS to be positioned extremely close to a diaphragm if desired. The length of the tube section was constrained by the maximum available spacing between any two adjacent 2 m long tubes in the system with the facility completely assembled. A 20 mm spacing between the instrumentation stations was chosen as a compromise between structural considerations and the need for high spatial resolution.

The 20 mm spacing between instrumentation stations and the available spacing between existing tubes allows eight instrumentation stations along the length of the HITS. The 20 mm spacing of these stations is approximately three times the 6.35 mm diameter

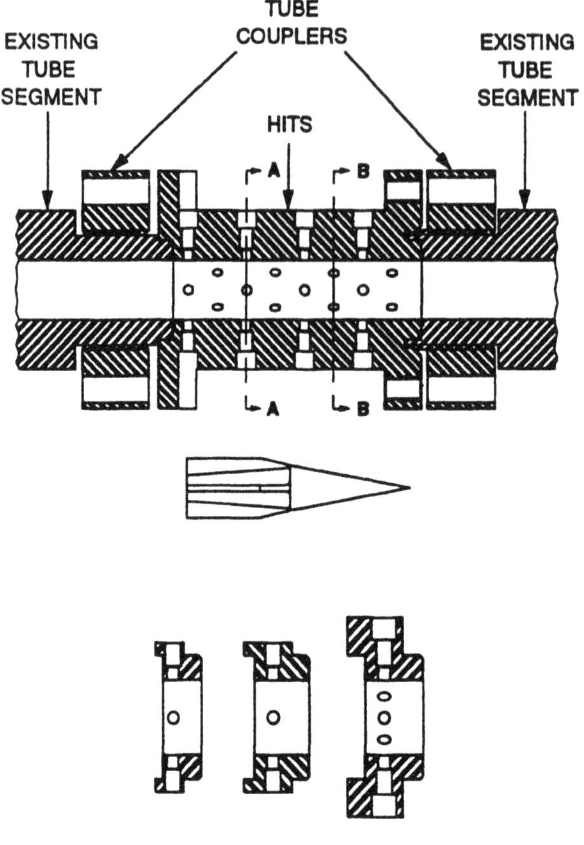

Fig. 5 Highly instrumented tube section and tube couplers. Projectile shown for scale.

of the instrumentation ports. This configuration allows for all eight instrumentation stations at some point during an experiment to be simultaneously located somewhere along the projectile body since the typical projectile length is 153 mm (see Fig. 4). At each instrumentation station there are four instrumentation ports separated by 90°. There are eight axial stations with each set of ports rotated by 45° relative to the previous station. Instrumenting two adjacent stations effectively provides a point measurement along the length of the tube with eight ports separated by 45° azimuthally and 20 mm axially. Under proper conditions and with appropriate instrumentation, these can be used to determine the projectile orientation, as well as infer the three-dimensional structure of the flow between the projectile and the tube wall.

Recently, a second type of high instrument density section was developed which allows twice the azimuthal resolution of the HITS. It consists of two very short sections of tube (Fig. 5) which each have 8 instrument ports separated by 45°. When the two short sections are combined or "stacked" together at different angles the effective angular resolution becomes 22.5° and with the 6.22 mm diameter of the pressure transducers, the entire tube circumference is effectively instrumented. These sections are referred to as "inserts" and are designed to obtain high resolution measurements of the flow field surrounding a projectile at one location or at multiple locations during a *single* experiment. When the inserts are used at several different locations along the tube, a high resolution run history of an entire experiment will result instead of piecing together data from several separate experiments.

2.2. Instrumentation

There are a total of 32 instrumentation ports in the HITS with four ports at each of the eight stations. Currently there are three types of instruments for data collection: piezoelectric pressure transducers, fiber optic luminosity probes, and electromagnetic sensors. The piezoelectric pressure transducers (from PCB Piezotronics) have a sensitive area of approximately 6.22 mm in diameter and a calibrated range from 0 to 80,000 psi. The fiber optic luminosity probes are sensitive to the visible part of the E-M spectrum and are logarithmically amplified so that the possibility of saturating the data acquisition system is reduced. The electromagnetic sensors detect

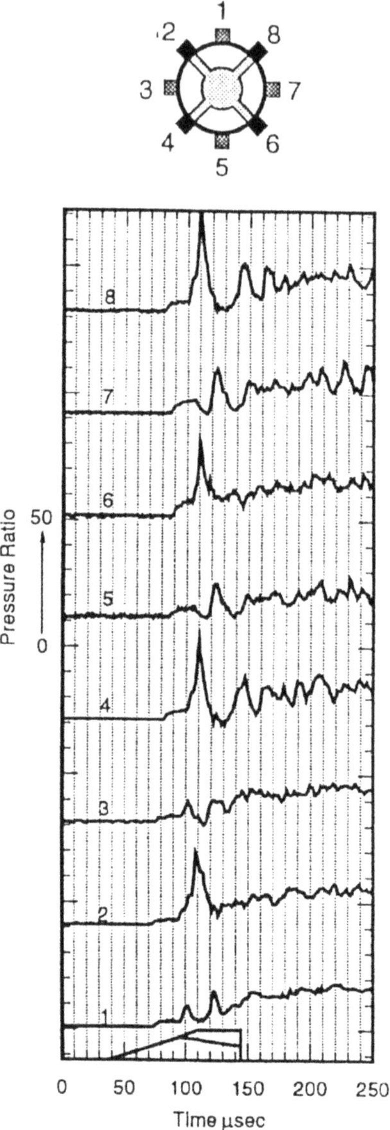

Fig. 6 Tube wall pressure traces from the subdetonative regime, 1440 m/s (Mach 4.0) or 82% of the C-J detonation speed.

the passage of a thin magnetic disc placed at the nose cone/body joint of the projectile and are used to locate the projectile relative to the instrumentation.

The resulting pressure traces from each angular position in the HITS are denoted by numbers 1 through 8 as denoted in Fig. 6. Figure 6 also shows the projectile fin orientation relative to the pressure traces with a small drawing above each set of traces. When possible, the remaining figures will contain one of these drawings with the projectile orientation for that experiment. Corresponding orientation drawings are also shown for the pressure traces generated by the inserts with numbers ranging from 1 to 16.

3. Experimental Results

Several series of experiments were performed in a single 16 m long, methane-based mixture consisting of $2.7CH_4 + 2O_2 + 5.8N_2$, which has an experimentally determined detonation speed of approximately 1750 m/s. Four-fin projectiles were manufactured from 7075-T651 aluminum alloy by computer numerical control (CNC) machines. These were used for this series of experiments because the CNC machines produced projectiles that were nearly geometrically identical. All projectile masses are approximately 74 gm with a maximum deviation of 0.3 gm. Initial tube fill pressures of 28 atm were used, except where noted. Using a light gas gun, the projectiles were pre-accelerated to a velocity of approximately 1150 m/s before entrance into the ram accelerator section of the facility. The HITS or pair of inserts were then installed at a position along the tubes where the projectile would have the desired velocity.

3.1. Subdetonative results

Several experiments were carried out in the subdetonative regime (thermally choked propulsive mode) at velocities of 1440 m/s to 1560 m/s (Mach 4.0 to 4.3) or 82% to 89% of the C-J detonation speed. The HITS was placed 4 m from the entrance to the ram accelerator. The first experiment was carried out with a fill pressure of 18 atm and a velocity of 1440 m/s. Eight pressure transducers were placed at 45° intervals in the second and third instrumentation stations (in the downstream or direction of projectile motion) of the HITS. The resulting tube wall pressure traces are shown in Fig. 6. A small out-

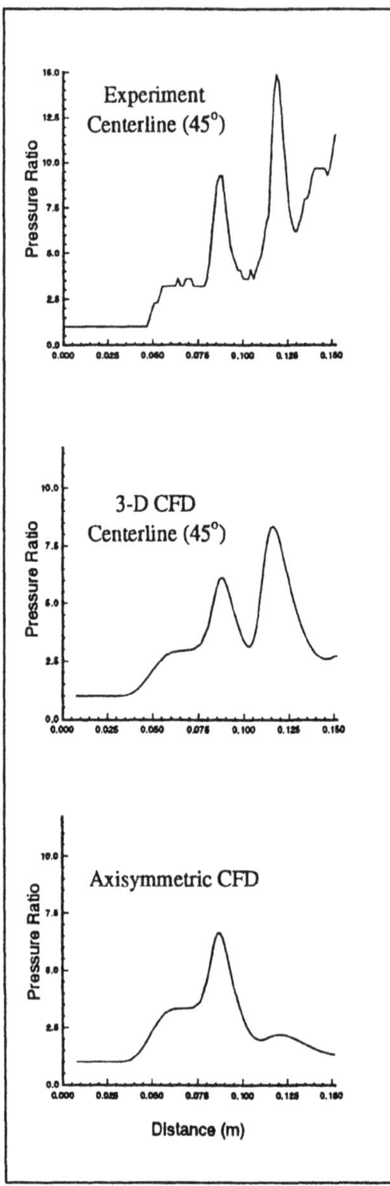

Fig. 7 Comparison of an experimental pressure trace taken along the centerline of a channel with those computed utilizing a three-dimensional CFD code and an axisymmetric CFD code.

line of one half of a projectile is placed at the bottom of the figure to indicate the position of the projectile body and fins relative to the pressure traces. The first rise in pressure, which appears similar in all traces, is the reflection from the tube wall of the conical shock generated by the projectile nose. Traces 2, 4, 6, and 8 are unlike those of 1, 3, 5, and 7 because of the large pressure spikes that appear slightly before the point where the projectile fins' leading surfaces[1] contact the tube wall. These pressure spikes coincide with the passage of the projectile fins directly over the pressure sensors. Traces 1, 3, 5, and 7 are then the tube wall pressures measured along the centers of the channels (the spaces between adjacent fins). The character of the flow appears to be radically different depending on the fin orientation relative to a sensor. The initial conical shocks are unaffected by the fin location; however, the fins have a tremendous influence on the remainder of the flow. A large pressure spike (pressure ratio $\sim$ 35) is formed near a fin while the pressure in the channel drops due to the expansion over the nosecone/body joint. Later in time a lower amplitude pressure spike is seen in the channel and the pressure over the fins is reduced. The high pressure spike near the fin in traces 2, 4, 6, and 8 of Fig. 6 is a result of the reflected shock system between the leading surface of the fin and the tube wall. The disturbances caused by each fin intersect and result in the high pressure spike seen later in the channel.

The measured tube wall pressure profile in the channel between the fins more closely matches those pressure traces produced by axisymmetric CFD simulations. The difference between the axisymmetric simulation and the experiment is shown by a large pressure rise after the initial set of conical shocks and is a result of the intersection of the disturbances caused by the high pressure regions generated by the flat leading surface of the fins. Figure 7 compares the tube wall pressure distributions from an experiment with two calculations using inviscid, nonreacting, axisymmetric and three-dimensional CFD codes. All three traces are from a projectile traveling at Mach 4.0. The strengths of the first two shock reflections from the tube wall are similar for all three plots, while the subsequent waves are dissimilar for the axisymmetric and three-dimensional traces. Con-

[1]The term "leading surface" is used for what is traditionally called a "leading edge" so as not to imply that the projectile fins in these experiments have a thin edge. The "leading edge" of a fin is in fact a flat surface having a width equal to the thickness of a fin.

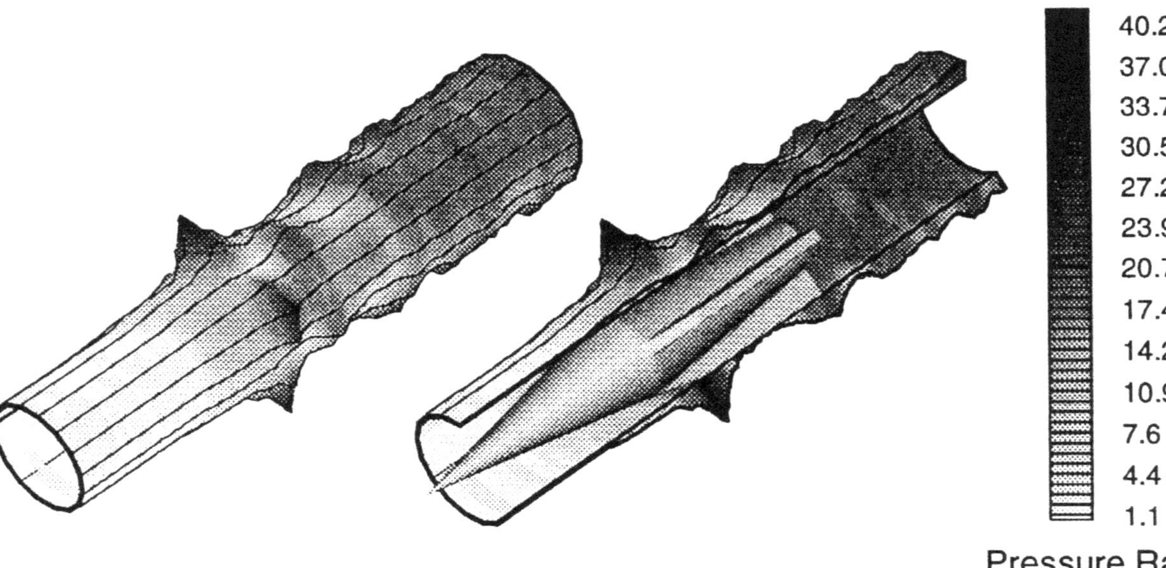

Fig. 8 Representation of measured tube wall pressure data as being proportional to radial distance from the surface of a cylinder representing the tube wall. Surface shading is also proportional to pressure. Subdetonative regime, 1440 m/s (Mach 4.0) or 82% of the C-J detonation speed. Projectile shown in right figure to correlate fin orientation with pressure distribution. Pressure normalized by the initial tube fill pressure.

versely, the experimental results are closely approximated by the three-dimensional simulation. Discrepancies between the experimental data and the three-dimensional simulation can be explained by a slight yawing of the projectile, differences in the fin gap[2] and/or localized combustion near the leading surface of the fin. These comparisons to experiment by nonreacting codes may be done because at this Mach number, in this mixture ($2.7CH_4 + 2O_2 + 5.8N_2$), the normal shock is at the very end of the projectile body, leaving most of the flow around the projectile supersonic and unreacted.

The pressure distribution on the tube wall, as the projectile passes through the tube, can be visualized by representing the measured (or calculated) tube wall pressure as the radial distance of a point away from the surface of a cylindrical tube representing the ram accelerator tube wall. The different angular stations in the HITS or the inserts correspond to the same angles on the hypothetical tube. The length of the tube represents time in the stationary (lab) reference frame or distance in the projectile (moving) reference frame (assuming steady flow). Another way to explain this visualization is as if the projectile were traveling through an infinitely thin, flexible tube that deforms radially outward proportional to the instantaneous pressure on the tube wall. Figure 8 represents such a visualization with the experimental pressure data of Fig. 6. Both the surface shading and radial distance from the tube wall are proportional to the pressure. Data between angular stations were linearly interpolated. A projectile with the correct fin orientation with respect to the pressure surface is also shown for reference. The surface is cut along the centerline of a channel and a fin in this and all similar figures. The fin and channel shocks are more apparent in this format as opposed to the line plots of Fig. 6.

Figure 9 shows the corresponding plot of the computed tube wall pressure distribution over the projectile utilizing the three-dimensional, inviscid, nonreacting CFD code. The computed distribution is much smoother than the experimental results, but they generally correlate very well. Both show the large pressure spikes associated with the leading surfaces of the fins, as well as the pressure reduction as the high pressure gas expands from the fin gap into the channel. The shocks from the leading surfaces of the fins intersect

[2]The fin gap is the space between the tube wall and the outer radius of a fin. The fin gap is approximately 0.025 mm for a projectile before an experiment.

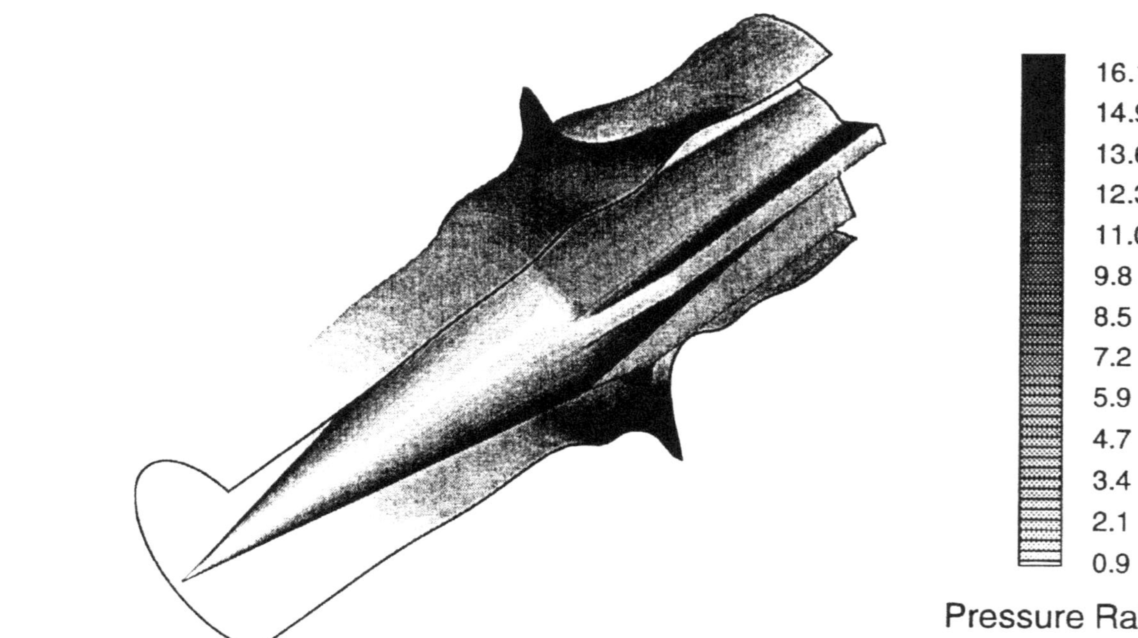

Fig. 9 Computed tube wall pressure distribution utilizing a three-dimensional CFD code. Pressure is proportional to radial distance from the surface of a cylinder representing the tube wall. Surface shading is also proportional to pressure. Subdetonative regime, 1440 m/s (Mach 4.0) or 82% of the C-J detonation speed. Pressure normalized by the initial tube fill pressure. Projectile shown to correlate fin orientation with pressure distribution.

along the center of the channel and form a pressure rise which then decays in both plots. These data illustrate that the flow around the projectile body is affected by the shocks generated at the leading surface of the fins; therefore, two-dimensional axisymmetric simulations are inadequate to model the details of the flow field around the projectile body.

Because the tube wall pressure distribution calculated with the three-dimensional CFD code correlates well with the experimental data, there is confidence in the calculated pressure distribution on the projectile body, shown in Fig. 10. The plot is of a three-dimensional projectile with the surface shading corresponding to the calculated pressure. Note the effect of the intersecting fin shocks on the body pressure distribution and the very high pressure on the faces of the leading surfaces of the fins. Also notice the high pressure regions on the conical part of the body near where the fin and body join. Experimentally, it has been observed that material is often eroded from this area of the projectile as well as from the leading surfaces of the fins. These observations lend support to the accuracy of the three-dimensional simulations.

3.2. Projectile yaw

The alignment of the projectile axis with respect to the tube axis can be inferred by comparing the shape and arrival time at the tube wall of the initial shock generated by the projectile nosecone. This is possible if one assumes no modification of the shock structure has occurred due to combustion in the region of the nosecone and that the nosecone has not suffered structural deformation. If the initial shock is very similar in magnitude and occurs at the same time at different angular locations, then the axis of the projectile is believed to be coincident with the axis of the tube, as depicted in Fig. 11(a). If the initial shock measured at the different angular locations has the same strength, but occurs at significantly different times, the projectile is believed to be simply translated to one side of the tube. If an axisymmetric projectile is traveling supersonically in an enclosed tube and is translated off the centerline without being rotated about an axis normal to the tube axis, the initial conical shock strength will not be affected. Conversely, the arrival times of the initial conical shock at different circumferential locations of the tube wall will be affected as shown in Fig. 11(b). The side of the tube nearest the projectile

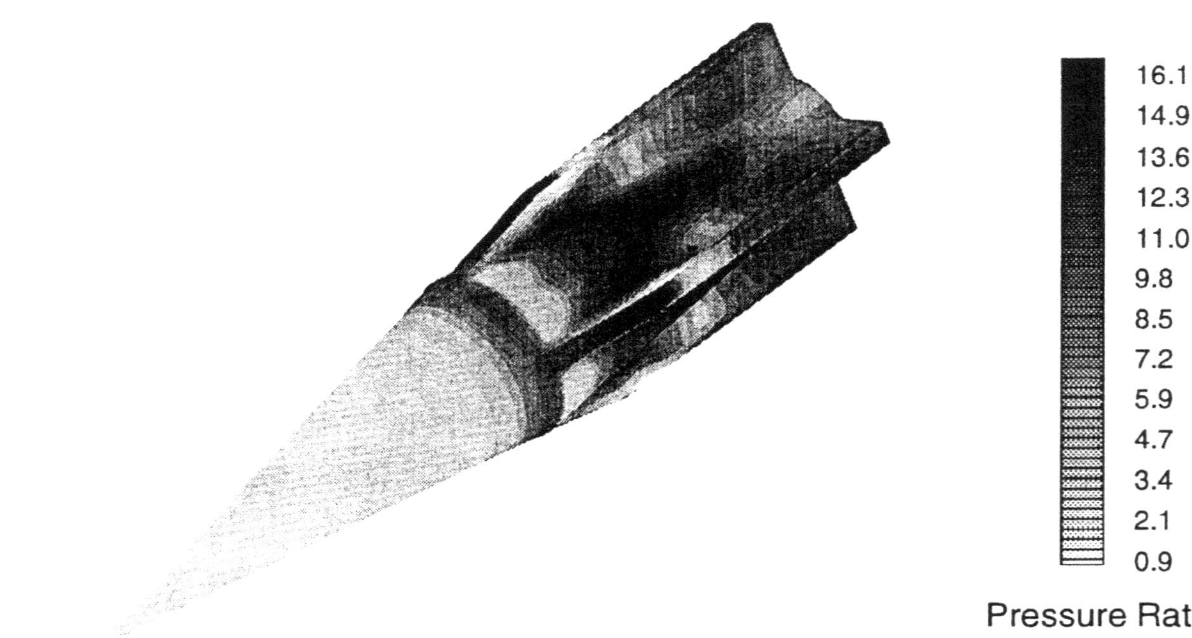

Fig. 10 Computed projectile body pressure distribution utilizing a three-dimensional CFD code. Pressure is proportional to surface shading. Subdetonative regime, 1440 m/s (Mach 4.0) or 82% of the C-J detonation speed. Pressure normalized by the initial tube fill pressure.

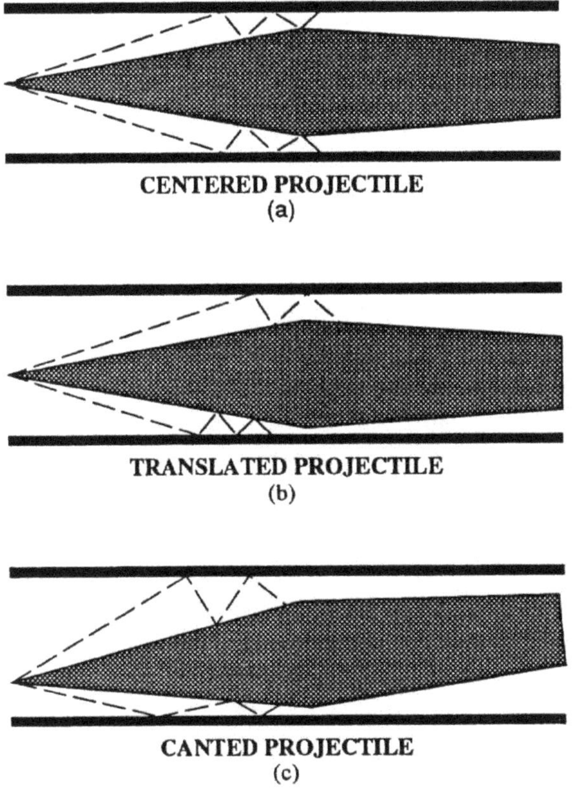

Fig. 11 Three possible projectile positions relative to the tube axis.

will have the initial shock arrive earlier than the opposite side. The case in which the initial shock from the nosecone reflects from the tube wall at different times and has different magnitudes is believed to be caused by the projectile being yawed at an angle relative to the tube axis. The yawing will cause the initial shock strength to vary over the cone since the flow effectively sees a different cone angle at different azimuthal points in the tube. The shock strength will then be larger on the higher angle "windward" side while becoming weaker on the opposite "leeward" side. This situation is shown in Fig. 11(c). The conical shock arrival times at the tube wall will be affected by the yawing and the translation. Figure 11(c) depicts the weaker initial shock from the leeward side arriving at the tube wall prior to the stronger initial shock from the windward side. In some instances experiments have shown that the yawing is severe enough such that on the leeward side of the cone a very weak shock or none at all is generated, while the windward side develops a very strong shock in which the pressure rises quite sharply.

A case of projectile yawing is illustrated by an experiment with a velocity of 1560 m/s (Mach 4.3) or 89% of the C-J detonation speed of the mixture. The measured tube wall pressures from eight transducers in the HITS are shown in Fig. 12. Note the significant variations in initial reflected shock strengths and arrival times. The leeward side is shown in traces 6 and 7 while the windward side is shown in traces 2 and 3. The arrival times differ by as much as 18μs which translates into approximately a 28 mm difference in the axial position where the initial shock reflects from the tube wall. The initial reflected shock strengths vary by a ratio of between 3 to 4 from the weakest to the strongest. It was roughly calculated that these conditions could correspond to an angular yawing of approximately 5°, based on the shock times and assuming a pivot point located where the fin leading surface meets the tube wall. The pressure traces in Fig. 12 do not show large pressure spikes near the leading surface of the projectile fins because the fins did not pass over any of the pressure transducers. Traces 2 and 8 do have pressure spikes that indicate that the fin passed close to the transducer. Upon inspection of the tube section after this experiment, fin streaks on the tube wall were discovered that indicated that the projectile was rotated approximately 6° (the half-width of a fin) from the midline of a sensor. Although this projectile was severely yawed, it attained a speed of 2060 m/s (Mach 5.7 and 118% C-J) in a single stage mixture. This

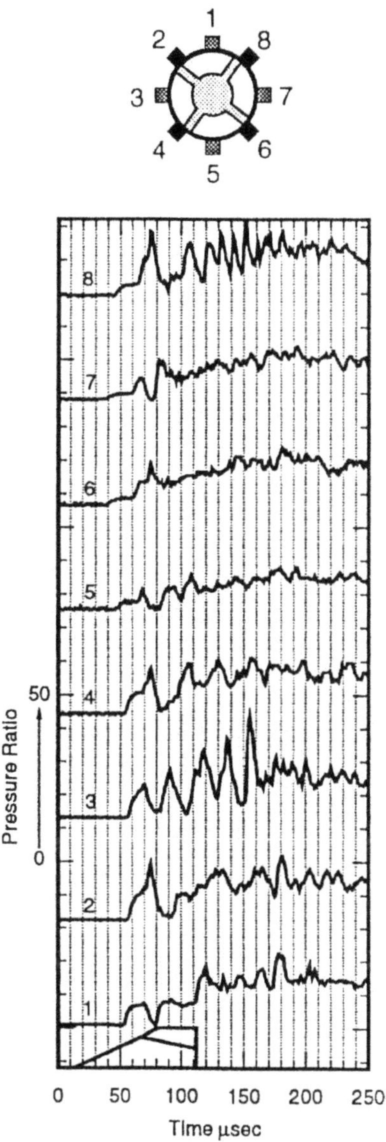

Fig. 12 Tube wall pressure traces using the HITS from the subdetonative regime, 1560 m/s (Mach 4.3) or 89% of the C-J detonation speed.

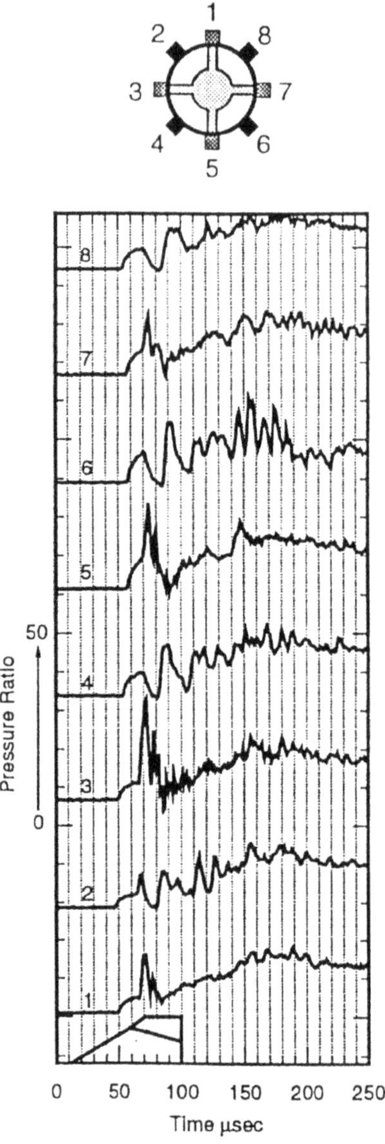

Fig. 13 Tube wall pressure traces using the HITS from the transdetonative regime, 1750 m/s (Mach 4.8) or 100% of the C-J detonation speed.

demonstrates the robustness of the gas dynamics which govern ram accelerator operation when such a highly yawed projectile can attain a significant velocity above the C-J detonation speed of the gas mixture.

The tube wall pressure traces shown in Fig. 12 have different characters with respect to variations in pressure as the projectile passes by the individual sensors. Particularly interesting are the large discrepancies in the character of the flow behind the projectile with some probes showing fairly smooth pressure profiles while others reveal large amplitude pressure variations. It has been demonstrated (Hinkey, Burnham, and Bruckner, 1992a) that the general character of the data from the pressure transducers in Fig. 12 is produced by real phenomena and is not a probe artifact. The cause of these pressure variations behind the projectile is not known at this time, but very similar pressure variations have been observed in axisymmetric, finite rate kinetics, Navier-Stokes calculations performed by Amtec Engineering, Inc. (Soetrisno, Imlay, and Roberts, 1992), using their INCA CFD code. Further experiments are planned to investigate this phenomenon.

3.3. Transdetonative results

An experiment was carried out in the transdetonative regime of operation, at a velocity of approximately 1750 m/s (Mach 4.8) or 100% of the C-J detonation speed. The HITS was placed 8 m from the entrance to the ram accelerator, corresponding to the middle of the 16 m facility. The gas mixture used was identical to that of the subdetonative experiments. Figure 13 shows the eight measured tube wall pressure traces from the pressure transducers in the second and third instrumentation stations in the HITS. The experimental data in Fig. 13 are shown in the same format as in Fig. 6. The projectile in this experiment was oriented such that a projectile fin passed over probes 1, 3, 5, and 7, which resulted in the characteristic large pressure spike near the area where each fin leading surface contacts the tube wall. The projectile appears to be slightly yawed in the tube since the arrival times of the initial shock differ by 10 μs (17 mm) and have a pressure ratio (strongest to weakest) of approximately 2. The high pressure spikes associated with the fins can clearly be identified on the odd numbered traces. Similarly, the channel shocks can be seen in the even numbered traces. It can be seen that the peak fin

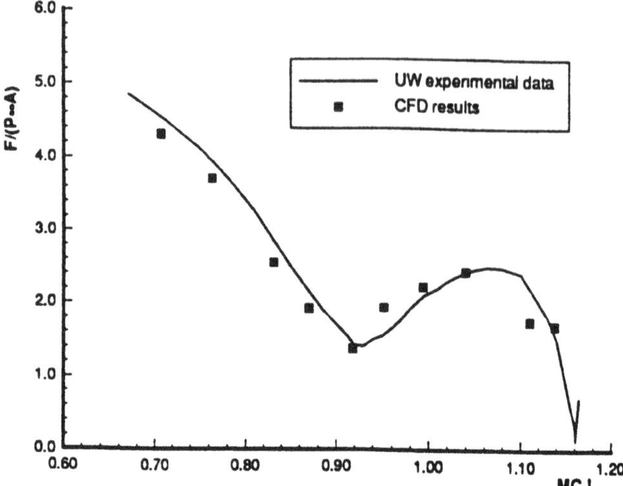

Fig. 14 Comparison of the experimentally determined thrust versus percentage of C-J Mach number with computations done by Amtec Engineering, Inc.

shock pressure ratio is lower than the subdetonative result shown in Fig. 6. For a projectile which retains its original fin diameter (no fin erosion) the higher speed experiment should exhibit higher relative pressures near the fins due to the increase in the total pressure of the flow. A possible explanation is that the projectile has suffered fin erosion, since numerical calculations indicate that as the fin gap is increased, the pressure rise due to the fin leading surface decreases.

A comparison of the experimentally determined thrust versus percentage of C-J Mach number with axisymmetric computations done by Amtec Engineering, Inc. with their INCA CFD code (Soetrisno, Imlay, and Roberts, 1992) is shown in Fig. 14. The thrust has been nondimensionalized by the initial tube fill pressure and tube area. The decrease in thrust, typical of the thermally choked regime, can be seen below 90% of the C-J Mach number. The increase in thrust seen experimentally in the transdetonative regime is well predicted by the Amtec code. The cause of the increase in thrust in the numerical simulations is from shock induced combustion releasing heat on the body of the projectile. This type of comparison demonstrates the value of using numerical and experimental results to complement and confirm predictions made about the flow field around the ram accelerator projectile.

3.4. Superdetonative results

Two experiments were also carried out in the superdetonative regime at velocities of 1960 m/s and 2020 m/s (Mach 5.4 and Mach 5.6) or 112% and 115% of the C-J detonation speed respectively. Two inserts were utilized in combination and were placed at a point 14 m from the entrance to the ram accelerator. Figure 15 shows the resulting 16 pressure traces from the experiment at 1960 m/s and 30 atm initial tube fill pressure. The orientation of the projectile is shown in the figure above the pressure traces, but no fin spike is evident in any of the traces. This may be a result of the fin gap being very large due to fin erosion. The initial conical shock reflection appears to be fairly uniform, but a subtle difference in arrival times of approximately 4 μs (8 mm) is found upon closer inspection. The second reflected shock appears to cycle from being noticeable in traces 2 through 15 and disappears in traces 1 and 16 lending further evidence that the projectile is not centered in the tube. Also readily noticeable are the series of large pressure variations behind the projectile. The position

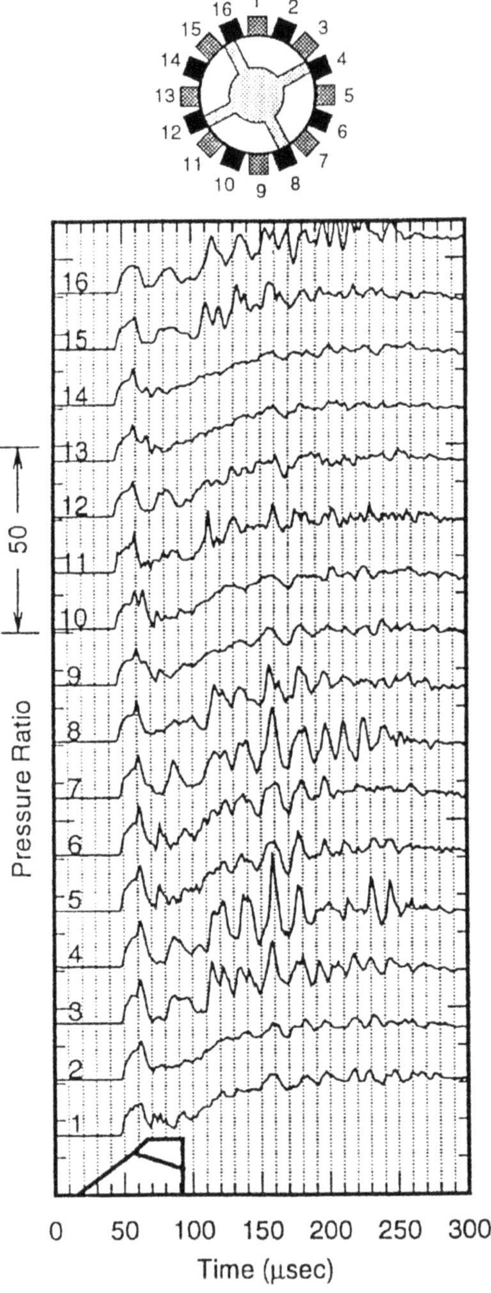

Fig. 15 Tube wall pressure traces from the pressure transducers in the inserts. Superdetonative regime, 1960 m/s (Mach 5.4) or 112% of the C-J detonation speed.

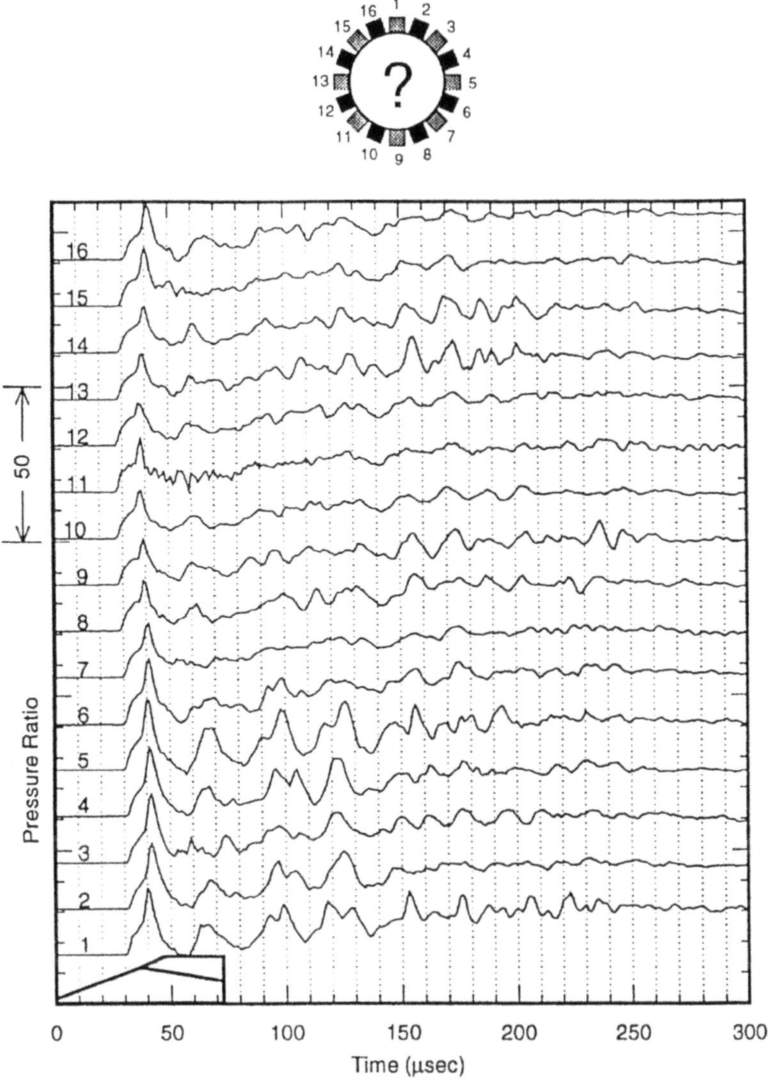

Fig. 16 Tube wall pressure traces from the pressure transducers in the inserts. Superdetonative regime, 2020 m/s (Mach 5.6) or 115% of the C-J detonation speed.

of these pressure variations seem to correlate from trace to trace and give clues to the structure of the flow behind the projectile. Not all of the transducers show these pressure variations, but they appear to be present only in certain portions of the tube behind the projectile. The reason for this variation is not known at this time, but it may be related to the yawing of the projectile.

The second experiment was carried out at a velocity of 2020 m/s and 32 atm initial tube fill pressure. Figure 16 shows the resulting 16 pressure traces. Again no pressure spikes caused by the fin leading surfaces are observed. The initial reflected conical shock wave appears to be fairly uniform with a 5 μs (10 mm) difference in arrival time. Again the second reflected shock is evident, but is of a much larger amplitude than in the previous experiment. The strength of the second shock also cycles in amplitude from a pressure ratio of 5 at trace 12 to a pressure ratio of 15 at trace 1, further suggesting that the projectile is yawed in the tube. To aid in the interpretation of the data in Fig. 16, the tube wall pressure distribution was reconstructed in the same format used in Fig. 8 and is shown in Fig. 17. A projectile is shown beside the reconstructed tube wall pressure distribution to provide scale only, since the projectile orientation could not be determined. Note the large pressure of the second reflected shock wave and the variation of its amplitude at different azimuthal locations. The pressure variations behind the projectile are also evident and seem to form coherent structures (Hinkey et al., 1992b).

It is also interesting to note that this projectile unstarted[3] approximately 0.4 to 0.8 m (200 to 400 μs) after passing through the pair of inserts. The large amplitude of the second reflected shock may in fact be the signature of a high Mach number unstart mechanism. The large pressure rise due to the second reflected shock in Fig. 16 may be the beginning of the unstart of the diffuser as has been similarly seen in starting experiments where the projectile has entered the ram accelerator section at a Mach number too low to start the diffuser (Burnham, Hinkey, and Bruckner, 1992). In this case a normal shock is disgorged from the projectile throat and travels in the direction of projectile motion, leaving the flow behind it subsonic and possibly combusting the propellant mixture. In the present experiment the projectile was intact after the unstart and

[3]The term "unstart" as used in this paper is a generic term for the cessation of ram acceleration due to a shock wave being disgorged from the projectile diffuser. The shock wave may occur for various reasons.

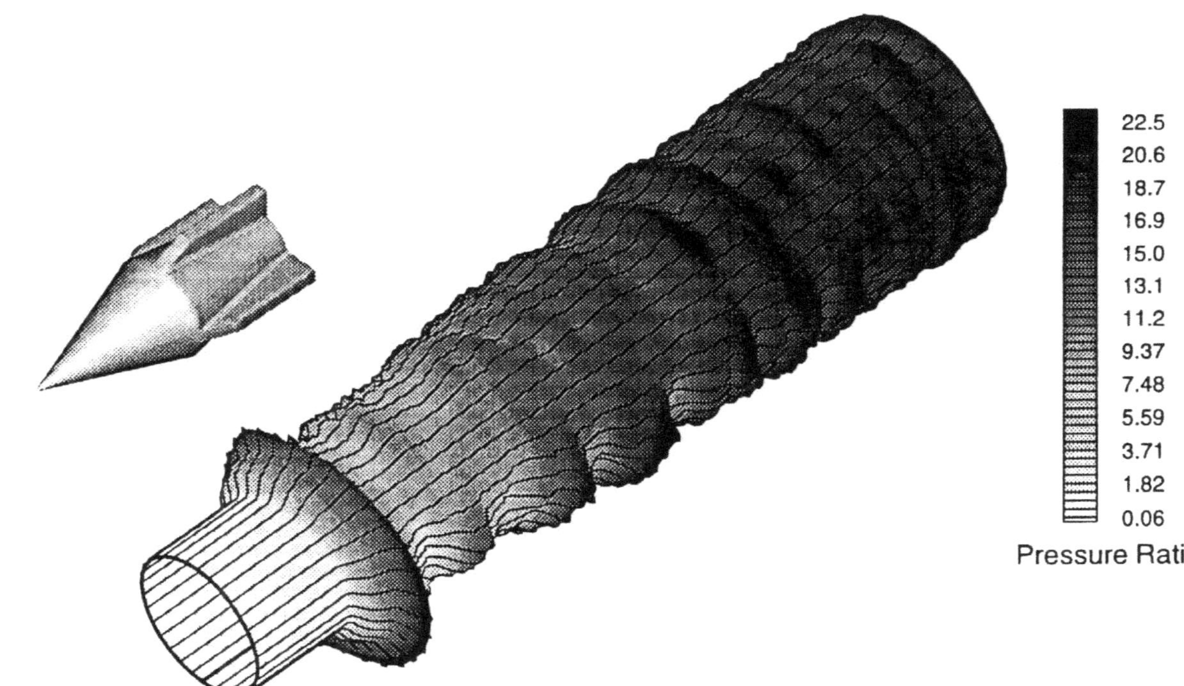

Fig. 17 Reconstructed tube wall pressure distribution from the superdetonative regime, 2020 m/s (Mach 5.6) or 115% of the C-J detonation speed. Pressure data are proportional to radial distance from the surface of a cylinder representing the tube wall. Pressure normalized by initial tube fill pressure.

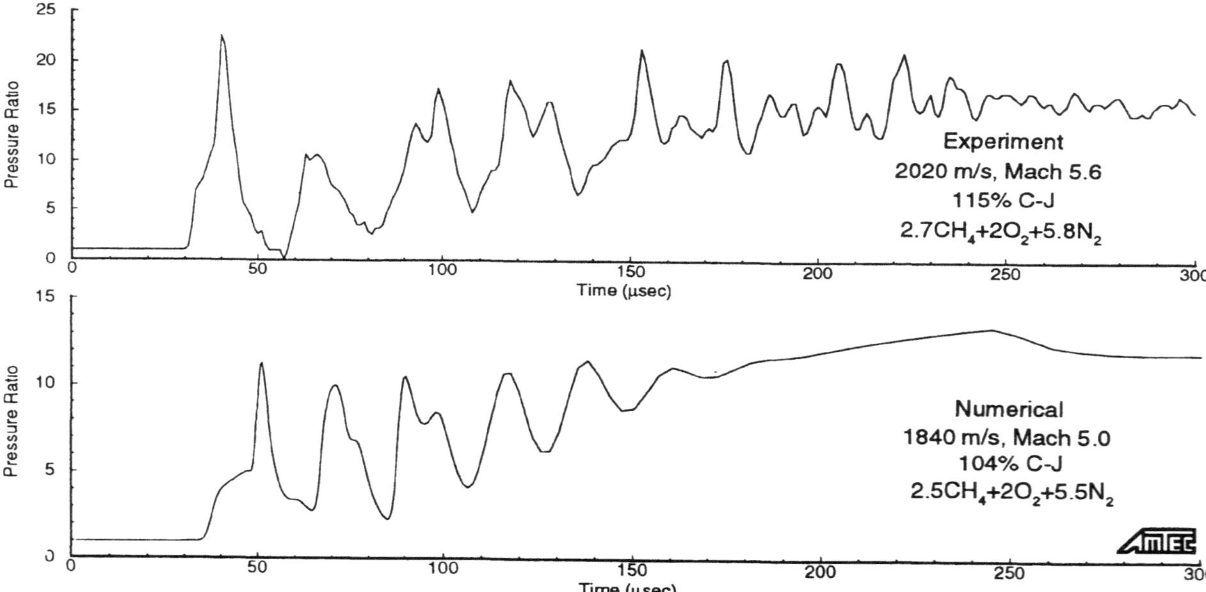

Fig. 18 Qualitative comparison of an experimental tube wall pressure trace with the axisymmetric numerical simulation of Soetrisno, Imlay, and Roberts, 1992 (used with permission). Experimental conditions are 2020 m/s (Mach 5.6) or 115% of C-J detonation speed with a mixture of $2.7CH_4+2O_2+5.8N_2$. Numerical conditions were 1840 m/s (Mach 5.0) or 104% of C-J detonation speed with a mixture of $2.5CH_4+2O_2+5.5N_2$.

therefore catastrophic structural failure (collapse of the hollow body and nose) did not cause the unstart. A similar pressure rise has also been seen in numerical simulations (Soetrisno, Imlay, and Roberts, 1992) at velocities significantly above the C-J speed of the mixture and is caused by combustion on the nosecone of the projectile steepening the initial and reflected conical shock waves. Figure 18 shows a comparison of the tube wall pressure distribution from a numerical simulation (Soetrisno, Imlay, and Roberts, 1992) with experimental data from Fig. 16. Although the chemistries and velocities are not identical, the general character of the flow seems to be captured by the simulation. The pressure spike in the region of the nosecone and the pressure variations behind the projectile correlate well. The pressure ratios are not directly comparable due to the different Mach numbers of the two cases.

The entropy rise from combustion and increased shock strengths *coupled with* projectile yawing may choke the diffuser, resulting in a gasdynamic unstart which quickly terminates projectile acceleration. Numerically, even after the flow has reacted on the nosecone, the projectile diffuser does not choke due to the high freestream Mach number (Soetrisno, Imlay, and Roberts, 1992) and therefore may require the mechanical choking of the flow provided by yawing of the projectile. A yawed projectile with pre-combustion on the nosecone may unstart a single channel (perhaps the channel on the side towards which the projectile is yawed) which then affects the adjacent channels causing them to unstart also. This may then complete the unstart of the entire projectile and lead to termination of ram acceleration. To prevent this possible type of unstart mechanism it is required to keep the projectile centered in the tube. This involves solving the projectile fin erosion issue by, for example, using advanced projectile designs and composite materials or high strength metals such as a titanium alloy. This unstart mechanism may also contribute to the upper velocity limits of the multiple stage experiments where projectiles fail at velocities below the C-J detonation speeds of the mixtures.

4. Conclusions

High spatial resolution measurements of the tube wall pressure around a projectile in a 38 mm bore ram accelerator were presented for the three regimes of ram accelerator operation in a single-stage mixture of $2.7CH_4 + 2O_2 + 5.8N_2$, at tube fill pressures of 18 to 32 atm and velocities of 1400 to 2040 m/s. It was shown that the fins which center the projectile in the tube must be taken into account when modeling the ram accelerator, since they have a large effect on the three-dimensional flow field. Also, the phenomenon of projectile yawing was revealed and shown to be common at all velocities. Good agreement between experimental and computational results were shown. Yawing of the projectile should also be modeled computationally to understand its effect on the full three-dimensional flow field. In addition, a possible high Mach number unstart mechanism was thought to be the choking of the diffuser due to pre-combustion on the nosecone in combination with projectile yawing.

Acknowledgements

The authors would like to acknowledge Andrew J. Higgins and Gilbert Chew for their assistance in performing the experiments. Thanks are due to S & J Tooling for manufacturing the HITS and inserts and to 4-D Manufacturing Inc. of Redmond, WA, for manufacturing the projectiles. Malcom Saynor deserves much appreciation for threading the tube couplers and machining some special instrumentation on short notice. We would also like to thank Moeljo Soetrisno and Scott Imlay, both of Amtec Engineering Inc., for providing data from their numerical simulations of the ram accelerator.

References

Brackett, D. C. and Bogdanoff, D. W., 1989. "Computational investigation of oblique detonation ramjet-in-tube concepts," *J. Propulsion and Power* **5**, pp. 276-281.

Bruckner, A. P., Bogdanoff, D. W., Knowlen, C., and Hertzberg, A., 1987. "Investigation of gasdynamic phenomena associated with the ram accelerator concept," AIAA Paper 87-1327.

Bruckner, A. P., Knowlen, C., and Hertzberg, A., 1992. "Application of the ram accelerator to hypersonic aerothermodynamic testing," AIAA Paper 92-3949.

Bruckner, A. P., Knowlen, C., Hertzberg, A., and Bogdanoff, D. W., 1991. "Operational characteristics of the thermally choked ram accelerator," *J. Propulsion and Power* **7**, pp. 828-836.

Burnham, E. A., Hinkey, J. B., and Bruckner, A. P., 1992. "Investigation of starting transients in the thermally choked ram accelerator," 29th JANNAF Combustion Subcommittee Meeting, Langley Research Center, Hampton, Virginia, October 19-23.

Burnham, E. A., Kull, A. E., Knowlen, C., Bruckner, A. P., and Hertzberg, A., 1990. "Operation of the ram accelerator in the transdetonative velocity regime," AIAA Paper 90-1985.

Hertzberg, A., Bruckner, A. P., and Bogdanoff, D. W., 1988. "Ram accelerator: A new chemical method for accelerating projectiles to ultrahigh velocities," *AIAA Journal* **26**, pp. 195-203.

Hertzberg, A., Bruckner, A. P., Bogdanoff, D. W., and Knowlen, C., 1987. "The ram accelerator and its applications," *Proceedings of the 16th International Symposium on Shock Tubes and Waves*, Aachen, West Germany, July 26-30, pp. 117-128.

Hertzberg, A., Bruckner, A. P., and Knowlen, C., 1991. "Experimental investigation of ram accelerator propulsion modes," *Shock Waves* **1**, pp. 17-25.

Hinkey, J. B., Burnham, E. A., and Bruckner A. P., 1992a. "High spatial resolution measurements of ram accelerator gas dynamic phenomena," AIAA Paper 92-3244, AIAA/SAE/ASME/ ASEE 28th Joint Propulsion Conference, Nashville, Tennessee, July 6-8.

Hinkey, J. B., Burnham, E. A., and Bruckner A. P., 1992b. "High spatial resolution measurements in a single stage ram accelerator," 29th JANNAF Combustion Subcommittee Meeting, Langley Research Center, Hampton, Virginia, October 19-23.

Kull, A. E., Burnham, E. A., Knowlen, C., Bruckner, A. P., and Hertzberg, A., 1989. "Experimental studies of superdetonative ram accelerator modes," AIAA Paper 89-2632.

Ostrander, M. J., Hyde, M. F., Young, R. D., and Kissinger, R. D., 1987. "Standing oblique detonation wave engine performance," AIAA Paper 87-2002.

Pratt, D. T., Humphrey, J. W., and Glenn, D. E., 1991. "Morphology of standing oblique detonation waves," *J. Propulsion and Power* **7**(5), pp. 837-845.

Soetrisno, M., Imlay, S. T., and Roberts, D. W., 1992. "Numerical simulations of the transdetonative ram accelerator combusting flow field on a parallel computer," AIAA Paper 92-3249, AIAA/SAE/ASME/ASEE 28th Joint Propulsion Conference, Nashville, Tennessee, July 6-8.

Witcofski, R. D., Scallion, W. I., Carter, D. J., and Courter, R. W., 1990. "A concept for a large hypervelocity aerophysics facility," Proceedings of the 41st Aeroballistic Range Association Meeting, San Diego, CA, October 22-25.

Yungster, S. and Bruckner, A. P., 1992. "Computational studies of a superdetonative ram accelerator mode," *J. Propulsion and Power* **8**, pp. 457-463.

Yungster, S., Eberhardt, S., and Bruckner, A. P., 1991. "Hypervelocity projectiles in detonable gases," *AIAA J.* **29**, pp. 187-199.

OBLIQUE DETONATIONS: THEORY AND PROPULSION APPLICATIONS

Joseph M. Powers[1]

University of Notre Dame
Notre Dame, Indiana 46556-5637

ABSTRACT

The oblique detonation, a combustion process initiated by an oblique shock, arises in most supersonic combustion applications including, most notably, the ram accelerator and the oblique detonation wave engine. Additionally, it is the generic two-dimensional compressible shocked reacting flow; consequently, its basic research value is inherent. The outstanding theoretical questions are also the fundamental practical questions: *e.g.* what conditions are necessary for steady solutions, what is the dependency of the steady propagation speed on the ambient condition, what is the susceptibility of the system to instability, and what is the behavior of the system in unsteady operation. A related topic which transcends all questions is the ability to describe these phenomena computationally. At this early stage, these issues are most clearly addressed with simple models. This paper will review the application of such models to oblique detonations and discuss their future relevance.

1. Introduction

This paper will attempt to demonstrate the utility of modeling simple systems in order to gain understanding of both oblique detonations and their application to propulsion devices. The plan of this paper is as follows. After a short review, some basic research issues are outlined followed by a summary of possible modeling approaches. Next a simple model is proposed which is both amenable to analysis and in a certain sense representative of real propulsion systems. A summary of the author's previous results in modeling aspects of this system is then presented. The paper concludes with

[1]This research has received support from the ASEE/NASA Summer Faculty Fellowship Program at the NASA Lewis Research Center, the Jesse H. Jones Faculty Research Fund, and the University of Notre Dame's Center for Applied Mathematics.

J. Buckmaster et al. (eds.), Combustion in High-Speed Flows, 345–371.

recommendations for future studies.

1.1. Review

The oblique detonation has received attention of late because of its role as the primary combustion mechanism in certain high-speed propulsion applications. One such device is the ram accelerator, (see Figure 1) first tested in recent years (Hertzberg, *et al.*, 1988, 1991). In this application, a high speed projectile is fired from a light gas

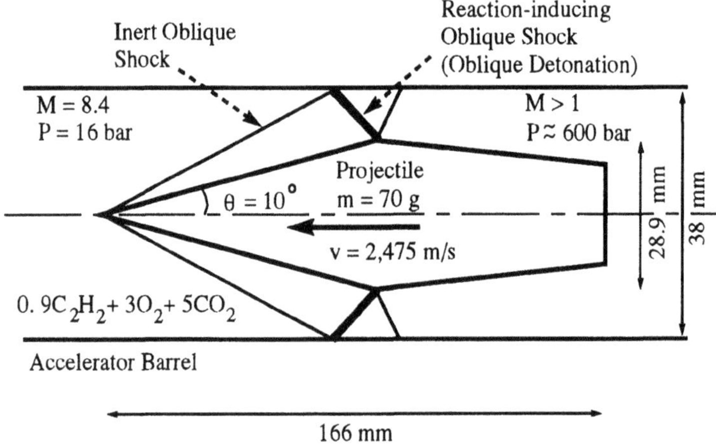

Figure 1: Schematic of ram accelerator, adopted from Hertzberg, *et al.* (1991).

gun into a tube filled with an unreacted mixture of combustible gases. Hertzberg, *et al.*, 1991, observed that upon entering a 16 m length, 38 mm bore tube filled in its first three stages with varying combinations of CH_4, O_2, N_2, and He at a pressure of 31 bar and in its final stage with $0.9C_2H_4 + 3O_2 + 5CO_2$ at a pressure of 16 bar, that a shock-induced combustion process accelerates a 70 g projectile from an initial velocity of near 1,200 m/s to a velocity of 2,475 m/s (corresponding to a Mach number, $M = 8.4$) at the end of the tube, at which location it was still accelerating. Downstream pressures in the neighborhood of 600 bar are measured. The diameter of the main body of the projectile was 28.9 mm. Its length was 166 mm and the leading edge conical half-angle $\theta = 10\,\text{deg}$. Four stabilizing fins (not shown) of diameter 38 mm were a part of the aft-body. A portion of the oblique shock train is sketched in Figure 1; the various expansion fans and wave interactions are not included. Figure 1 depicts the first reflected shock triggering significant chemical reaction; the temperature-sensitive reaction would be associated with the lead shock for faster projectile speeds, and with a downstream shock for slower speeds. For even slower speeds, the reaction would be downstream of the projectile. It was suggested that such a device can be scaled for direct launch to orbit, for hypervelocity impact studies, and for a hypersonic test facility.

These experiments have directly motivated further, primarily numerical studies: Brackett and Bogdanoff (1989), Bruckner, *et al.* (1991), Yungster, *et al.* (1991), Yungster and Bruckner (1992), Bogdanoff (1992), Yungster (1992), and Pepper and Brueckner (1993). In particular, the numerical calculations of Yungster and Bruckner (1992), predict that a ram accelerator can achieve a steady-state velocity such that a combustion-induced thrust force balances drag forces. This condition is achieved at a velocity near $9,600\ m/s$, corresponding to a Mach number near 12 for a mixture, initially at a pressure of $20.3\ bar$ and composed of $5H_2+O_2+4He$, flowing over an axisymmetric projectile of half angle $\theta = 14\,\mathrm{deg}$, diameter $29\ mm$, and overall length $190\ mm$ in a tube of diameter $38\ mm$. Direct comparisons cannot be made with the experiments of Hertzberg, *et al.*, as hydrogen rather than hydrocarbon combustion was modeled, because much higher flight velocities were modeled, and because the geometry, which contained an additional constant area mid-section, was slightly different.

Another relevant propulsion device is the proposed oblique detonation wave engine (ODWE). The idea of using an ODWE for supersonic combustion for a high-speed plane has existed for decades (*e.g.* Dunlap, *et al.*, 1958). The hypothesized operation is as follows (see Figure 2, adopted from Dunlap, *et al.*). Supersonic air enters the inlet. On-board fuel is injected downstream which mixes with the air without significant reaction. The mixture then encounters a downstream wedge. The oblique shock associated with the wedge compresses and ignites the mixture, generating a propulsive force. Relative to conventional air-breathing engines with subsonic combustion, Dunlap, *et al.* cite the ODWE's advantages as 1) simpler supersonic inlet diffuser design since the inherently supersonic oblique detonation does not require deceleration to a subsonic state, 2) reduced total pressure losses, 3) shorter combustion chamber length, 4) no ignition device other than the wedge, and 5) faster flight veloc-

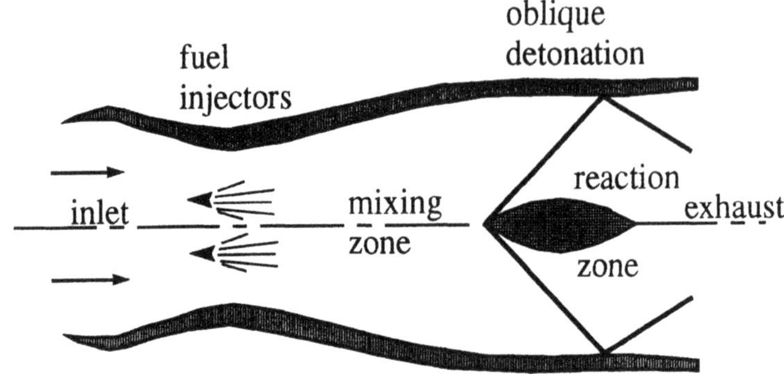

Figure 2: Envisioned oblique detonation wave engine, adopted from Dunlap, *et al.* (1958).

ities. Cited concerns are 1) the lack of static thrust, 2) uncertainty as to whether mixing lengths are practical, and 3) uncertainty with regards to the process's stability.

Renewed emphasis on high speed air-breathing propulsion alternatives led to modern studies of wedge-initiated ODWE's, (Cambier, *et al.*, 1989, 1990). Alternatively, laser-initiation has been studied (Carrier, *et al.*, 1992). Both wedge- and laser-initiated detonation engines contrast the more-studied Ferri engine in which two supersonic streams, one fuel and the other oxidizer, are brought together so that burning occurs in a convective-diffusive mixing layer.

Other more basic studies have relevance. Several give a Rankine-Hugoniot (RH) analysis of oblique detonations: *e.g.* Siestrunck, *et al.* (1953), Larisch (1959), Gross (1963), Oppenheim, *et al.* (1968), Chernyi (1969), Buckmaster and Lee (1990), and Pratt, *et al.* (1991). Other analyses are for either steady two-dimensional or unsteady one-dimensional flows with spatially resolved structure; many of these focus on the related topic of dissociation and vibrational relaxation: *e.g.* Clarke (1960), Moore and Gibson (1960), Sedney (1961), Spence (1961), Vincenti (1962), Capiaux and Washington (1963), Lee (1964), Spurk, *et al.* (1966), and Fickett (1984).

Recent analyses which this author and colleagues have performed have placed emphasis on oblique detonations with resolved reaction zone structure and the connections of these structures with the predictions of a RH analysis: Powers and Stewart (1992), Powers and Gonthier (1992a), and Grismer and Powers (1992). In addition, there is a new large body of general unsteady analyses of one- and two-dimensional detonations which though germane, have largely not been applied in propulsion studies, *e.g.* Bdzil and Stewart (1986), Stewart and Bdzil (1988), Buckmaster (1990), Clarke, *et al.* (1990), Lee and Stewart (1990), Jackson, *et al.* (1990), Lasseigne, *et al.* (1991), Bourlioux, *et al.* (1991), Bourlioux and Majda (1992), and Bdzil and Kapila (1992). These studies build largely on the Zeldovich, von Neumann, Doering (ZND) theory which has undergone continuous refinement since being introduced in the 1940's; extensive reviews exist, *e.g.* Fickett and Davis (1979).

Lastly, there exist fundamental experimental studies, *e.g.* Gross and Chinitz (1960), Nicholls (1963), Rubins and Rhodes (1963), Behrens, *et al.* (1965), Strehlow (1968), Strehlow and Crooker (1974), Lehr (1972), and Liu, *et al.* (1986). Of potential relevance, especially in light of Dunlap *et al.*'s concern, are dramatic observations of one- and three-dimensional detonation instabilities. One-dimensional instability can be observed when high speed projectiles are fired into reactive mixtures. An example is sketched in Figure 3, which is a representation of a photograph from Lehr. For this particular sketch, a 15 mm diameter projectile travels into a mixture with composition $2H_2 + O_2 + 3.76N_2$ at a pressure of 0.427 bar at an instantaneous velocity of $2,029\ m/s$ (corresponding to a Mach number of 5.04), slightly less than the Chapman-Jouguet (CJ) velocity of the mixture. The observed pulsations, which are at a frequency of 1.04 MHz, have been interpreted by Fickett and Davis as an essentially one-dimensional phenomena originating near the projectile tip that leaves traces which remain downstream. Evidence of three-dimensional detonation instability is given by Strehlow and Crooker. When the walls of a tube are coated with smoke, a detonation wave will sometimes leave a regular cell pattern on the walls. A sample pattern is sketched in Figure 4, which was traced from a photograph of Strehlow and Crooker. In this case the initial composition was $2H_2 + O_2 + 3Ar$ at a pressure of 0.077 bar. It is thought that the patterns are the result of a shock triple point leaving its trace on the coated wall. As with the unsteady analyses, the implications for high-speed propulsion of observed detonation instabilities have not been fully explored.

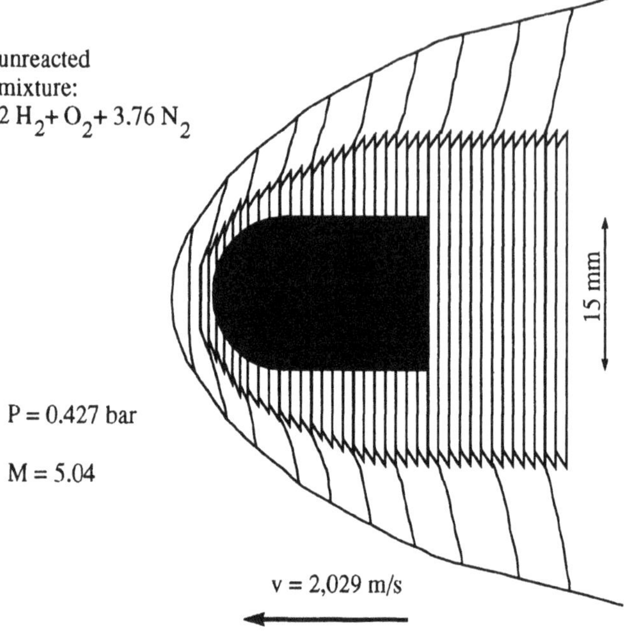

Figure 3: Sketch of observed combustion instability in projectile firing experiment, adopted from photographs of Lehr (1972).

1.2. Idealized oblique detonation definition

Before discussing detailed results, it is useful to have a working definition of an oblique detonation. As reviewed by Pratt, *et al.*, this has been a controversial topic. Here a definition is proposed which has been suitable for our studies. We define an oblique detonation as a combustion process which is initiated by an oblique shock in a flow field in which the fluid properties vary within length scales dictated primarily by the rate of chemical reaction. In such a process the oblique shock raises the temperature appreciably but is sufficiently thin to prevent significant combustion within the shock. Past the shock, the higher fluid temperature allows for significant reaction to occur in a spatially resolved reaction layer. The definition has the advantage of being device-independent as it does not require geometric length scales.

Though other scenarios are possible, one typically considers the oblique shock to be generated by the supersonic flow over a geometrical obstacle. It is illustrative to consider the flow over a straight wedge of half angle θ and semi-infinite length to frame some im-

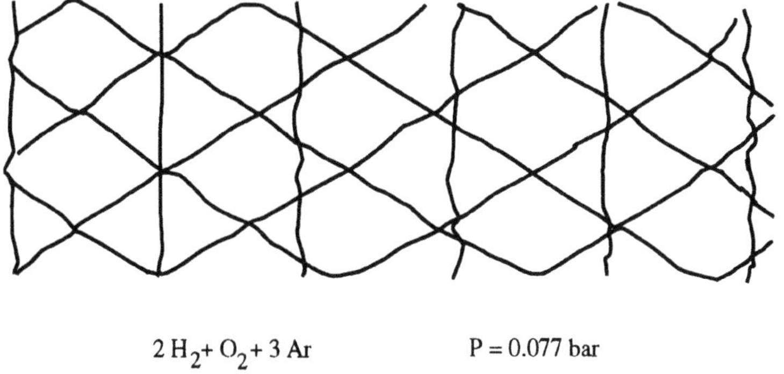

$2H_2 + O_2 + 3Ar$ P = 0.077 bar

Figure 4: Sketch of observed patterns on smoke-coated foils after passage of an unstable detonation wave, adopted from photographs of Strehlow and Crooker (1974).

portant issues. For such a geometry, as depicted in Figure 5, the shock angle near the wedge tip before significant reaction has occurred is that of an inert oblique shock. Far from the wedge tip the shock angle relaxes to a constant value, greater than the inert value. Powers and Stewart's linear analysis for small heat release shows this change can be attributed to the net effect of downstream local heat release disturbances. Such disturbances are propagated along characteristics which, in a complex reflection process between the wedge and the shock, strengthen the lead shock causing its inclination angle to increase with increasing distance from the wedge tip. Consequently, a region of shock curvature exists near the wedge tip which induces vorticity which is convected along streamlines in a layer near the wedge surface. Far from the shock and wedge surface, the flow relaxes to an irrotational, equilibrium, uniform state.

Such a definition allows the oblique detonation to be thought of as the two-dimensional analog of the ZND model for one-dimensional detonations (Fickett and Davis, 1979). The ZND model describes a reaction zone structure which links a shocked state to one of the three states identified by a RH analysis: a subsonic state (strong), a sonic state (CJ), or a supersonic state (weak). Energy release in the subsonic region behind the lead shock serves to drive the wave forward. Oblique detonation analogs for each case exist, which are repeated later in this paper. Also, the one-dimensional steady ZND detonation is often considered in the context of a piston problem; for strong solutions a portion of the energy to drive the wave comes

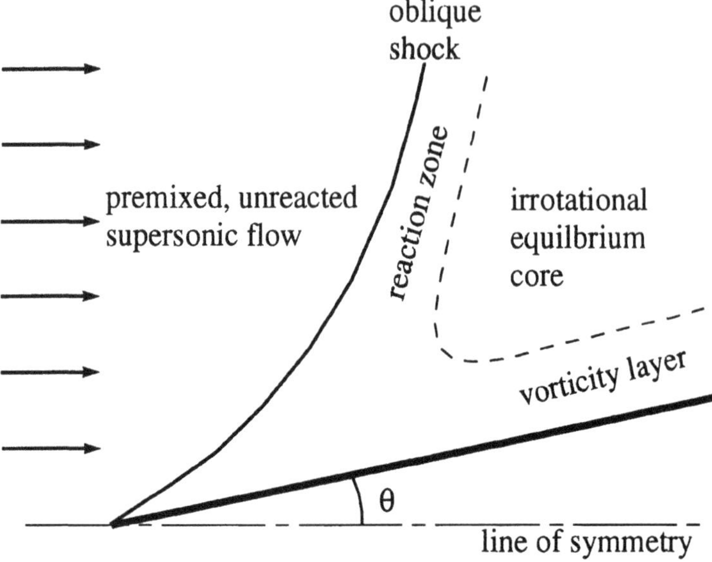

Figure 5: Straight wedge-curved shock oblique detonation configuration.

from the piston. For the oblique detonation of Figure 5, the wedge plays the role of the piston.

This definition is not universally accepted. In many oblique detonations the heat release only forms a small portion of the flow's total energy. In such case the wave is not primarily driven by the heat release; consequently, there is some reluctance to use the term "detonation." However, inasmuch as it is proper to describe a one-dimensional reactive wave driven by a supersonic piston as a strong (or overdriven) "detonation," it is proper to describe the corresponding two-dimensional waves as oblique "detonations." Also, as discussed by Pratt, *et al.*, distinctions are often drawn between an oblique detonation and "shock-induced combustion," in that the oblique detonation exists when the reaction occurs in a thin zone indistinguishable from an oblique shock, while the "shock-induced combustion" is characterized by an inert shock followed by a thick reaction zone. It is noted, however, that such a distinction requires the existence of an extraneous, independent, non-kinetic length scale, such as might be given by the diameter of a combustion chamber, in order to properly classify the phenomena. In addition, the distinction is inconsistent with the ZND characterization of a detonation as a shock followed by a resolved reaction zone.

1.3. Research issues

The configuration sketched in Figure 5 has both basic and applied value. Most importantly, it captures the essence of two-dimensional shocked reactive flows. As such, it seems necessary that this flow should form the basis for comprehension of more complex matters. Many of the basic research issues which remain for this flow are also issues of practical concern. Outstanding questions include what conditions are necessary for a steady state solution, what is the susceptibility of steady solutions to instability, and what is the fully transient behavior. Such questions bear directly on the operating characteristics of any propulsion device. These issues have been addressed in the detonation literature primarily for one-dimensional flows; for two-dimensional flows, there are relatively fewer studies.

1.4. Modeling approaches

A variety of modeling philosophies and techniques have been used to address these issues. One philosophy, the more Aristotelian, is to capture physical reality as much as possible. In ram accelerator or ODWE configurations, this typically involves modeling detailed geometries, detailed chemistry, diffusive transport, state dependent material properties, and turbulence. Such an approach, which necessitates a numerical solution, offers, significantly, the potential for predictions which closely mimic experiments to the extent that the computer becomes a substitute for the wind tunnel. As envisioned, all prototypes could be fully tested with the numerical model.

In the absence of verifying experiments or solutions from alternate techniques, caution must be used in this approach. First, many times the numerical results are as difficult to interpret as experiments because of the large number of simultaneously competing mechanisms. Moreover, it is often the case that the equations predict three-dimensional unsteady flows with phenomena occurring on widely varying scales. In combustion, scales are usually imposed by detailed kinetic models; time scales can range from 10^{-9} s to 10^{2} s and are typically far more severe than acoustic, diffusive, or turbulent scales (Maas and Pope, 1992). Capturing all of these scales can place severe demands on present computer resources. Additionally, the inherent non-linearity of the problem can give rise to a variety of coarse and fine scale structures. Striking evidence of these are given by the two-dimensional calculations of Bourlioux and Majda for a one-step reaction with Arrhenius kinetics. Furthermore, it can be shown (Yee, *et al.*, 1991 and Lafon and Yee, 1992) that discretization can actually mask the true solution features, and thus, insofar as the model represents physical reality, mask the actual flow physics. These papers, both of which are specifically addressed to modeling issues in hypersonic propulsion, apply typical discretization techniques to equations with known analytical solutions. Many dangers are discussed including the possibility of prediction of instability for known stable solutions, prediction of stability for known unstable solutions, and convergence to incorrect equilibria.

An alternative philosophy, the more Platonic, is to seek a complete understanding of a few selected phenomena. Typically, details are sacrificed at the discretion of the modeler so as to get to the essence of the problem. For propulsion applications, this may mean modeling simple geometries, simple chemistry, inviscid fluids, constant properties, and no turbulence modeling. Solution techniques are more varied and involve such methods as nonlinear analysis of dynamic systems, asymptotic analysis, and the method of characteristics. Advantages of this approach are that causality is easier to establish, quick and useful estimates are often provided, and best- and worst-case scenarios can sometimes be formed. Significantly, exact and asymptotic solutions are sometimes available, rendering determination of parametric dependencies and optimization easier. Such solutions also provide valuable test cases for numerical methods designed to solve more complex problems. An obvious disadvantage is that predictions are often far from physical reality. The remainder of this paper gives pertinent examples of this approach.

2. Idealized Propulsion Configuration

The configuration of Figure 5 is well-suited to study oblique detonations. However, because combustion on only the front side of the wedge is modeled, this represents a case where the force generated by combustion retards the body's motion. In order to achieve a propulsive force, one must consider the combustion over both sides of a finite projectile. It has been proposed (Powers and Gonthier, 1992b) to consider the geometry of Figure 6, similar to the geometry used by Yungster and Bruckner, but planar and with one fewer geometric length scale. Here a planar double wedge of half angle θ and length

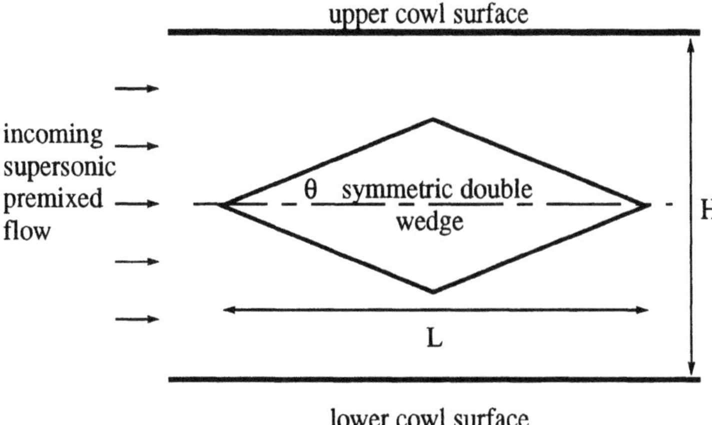

Figure 6: Schematic of idealized confined propulsion configuration, from Powers and Gonthier (1992b).

L is placed between confining walls separated by distance H. Such a geometry is representative of a ram accelerator if the confining walls are stationary and is representative of an ODWE if the confining walls move with the double wedge.

To further simplify, Powers and Gonthier (1992b), only considered the limit $H \rightarrow \infty$, Figure 7. While this geometry retains at most a rudimentary resemblance to actual devices, it is both potentially propulsive and amenable to analysis. The incoming flow is considered to be supersonic, premixed, and unreacted. An oblique or bow shock will exist at or near the leading edge. The shock should be of sufficient strength to initiate the induction phase of chemical reaction but not so strong that the reaction occurs immediately. The flow will expand in a rarefaction fan at the projectile apex. It is important that there be sufficient heat release to prevent the reaction from being quenched by the rarefaction. On the lee side, significant combustion should occur so that a force to counterbalance wave drag is generated. Finally to turn the flow, a shock at the trailing edge is required.

Of fundamental importance is the self-sustaining propagation velocity. Neglecting body forces, such a velocity is achieved when there is a balance of surface forces on the projectile, that is when

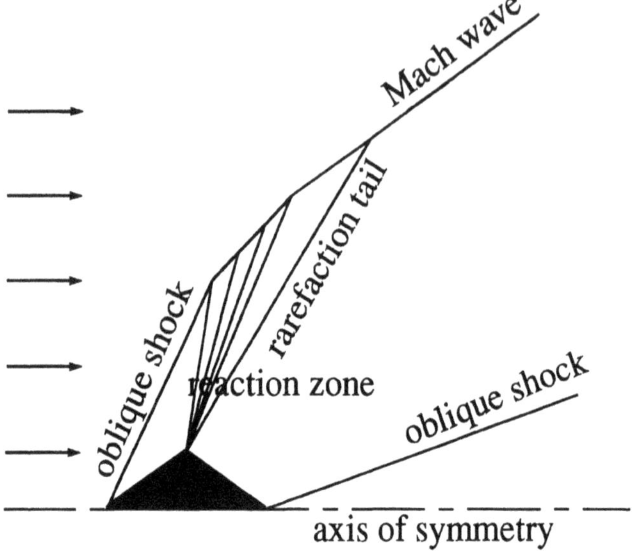

Figure 7: Schematic of idealized unconfined propulsion configuration, from Powers and Gonthier (1992b).

the thrust force induced by combustion equals the wave and viscous drag forces. As a solution technique for any particular fluid and combustion model, one can select the steady wave speed by a trial and error process. One can then determine the parametric dependency of the steady wave speed on geometry and material properties and also figures of merit such as propulsive efficiency. The steady solution also serves as a base state for stability and unsteady analyses.

Many models can be used to address such questions. Next, a simple model used by the author will be presented and, as an example of such a model's utility, the author's oblique detonation predictions from this model will be summarized.

3. Idealized Model

The model equations are taken to be the unsteady Euler equations and species evolution equation for a reactive calorically perfect ideal gas. These are expressed in dimensionless form as:

$$\frac{d\rho}{dt} + \rho \frac{\partial v_i}{\partial x_i} = 0, \tag{1}$$

$$\frac{dv_i}{dt} + \frac{1}{\rho}\frac{\partial P}{\partial x_i} = 0, \tag{2}$$

$$\frac{de}{dt} + \frac{P}{\rho}\frac{\partial v_i}{\partial x_i} = 0, \tag{3}$$

$$\frac{d\lambda}{dt} = \kappa(1-\lambda)\exp\left(\frac{-\Theta}{M_0^2 T}\right), \tag{4}$$

$$e = \frac{1}{\gamma - 1}\frac{P}{\rho} - \frac{\lambda q}{M_0^2}, \tag{5}$$

$$P = \rho T. \tag{6}$$

The variables contained in Equations (1–6) are the density ρ, the Cartesian velocity component v_i, the pressure P, the temperature T, the internal energy e, the reaction progress variable λ, and the Cartesian position coordinate x_i. Here the substantial derivative $\frac{d}{dt} = \frac{\partial}{\partial t} + v_i \frac{\partial}{\partial x_i}$ The freestream Mach number is M_0. Other dimensionless parameters include the ratio of specific heats γ, a kinetic parameter κ, the heat of reaction q, and the activation energy Θ. Equations (1–3) represent the conservation of mass, momenta, and energy, respectively. Equation (4) is a species evolution equation which incorporates an Arrhenius depletion model. Equations (5–6) are caloric and thermal equations of state. A single, first-order, irreversible, exothermic reaction is employed, $A \rightarrow B$. The reaction progress variable λ ranges from zero before reaction to unity at complete reaction. Species mass fractions, Y_i are related to the reaction progress variable by the formulae, $Y_A = 1 - \lambda$, $Y_B = \lambda$. Initial pre-shock conditions are specified as $\rho = 1$, $u = \sqrt{\gamma}$, $v = 0$, $P = 1/M_0^2$, and $\lambda = 0$.

Equations (1-6) have been scaled such that in the hypersonic limit ($M_0^2 \rightarrow \infty$) the pressure, density, and velocities are all $O(1)$ quantities behind the lead shock. The geometric length of the projectile (L) is chosen as the reference length scale. In terms of dimensional variables (indicated by the notation "$\tilde{}$") and dimensional pre-shock ambient conditions (indicated by the subscript "0"), the dimensionless variables are defined by

$$\rho = \frac{\tilde{\rho}}{\tilde{\rho}_0}, \qquad P = \frac{\tilde{P}}{M_0^2 \tilde{P}_0},$$

$$u = \frac{\tilde{u}}{M_0\sqrt{\tilde{P}_0/\tilde{\rho}_0}}, \qquad v = \frac{\tilde{v}}{M_0\sqrt{\tilde{P}_0/\tilde{\rho}_0}},$$

$$x = \frac{\tilde{x}}{L}, \qquad y = \frac{\tilde{y}}{L}. \tag{7}$$

Remaining dimensionless parameters are defined by the following relations:

$$q = \frac{\tilde{\rho}_0\tilde{q}}{\tilde{P}_0}, \qquad \Theta = \frac{\tilde{\rho}_0\tilde{E}}{\tilde{P}_0}, \qquad \kappa = \frac{\tilde{k}}{\frac{M_0}{L}\sqrt{\frac{P_0}{\rho_0}}}, \tag{8}$$

Here, $\tilde{E}$ is the dimensional activation energy, $\tilde{q}$ is the dimensional heat of reaction, and $\tilde{k}$ is the dimensional kinetic rate constant.

4. Summary of Results

To gain a basic understanding of two-dimensional reactive flows, and to gain insight into the possible propulsion situation in which the combustion is induced on the front side of the wedge, it is reasonable to study oblique detonations over semi-infinite wedges. In such a case, Equations (1-6) can be rescaled so that the kinetic rate defines the length scale. Figure 8 gives a diagram of a particularly simple type of such an oblique detonation. This was studied by Powers and Stewart for one-step kinetics [as in Equation (4)] and extended by Powers and Gonthier (1992a), to two-step kinetics [not written explicitly in Equations (1-6)]. The flow considered is an incoming unreacted gaseous mixture at supersonic Mach number, $M_0 > 1$, which encounters a straight shock, inclined at angle β to the horizontal, which is attached to a curved wedge. The mixture reacts downstream of the shock in the reaction zone. The special case in which the flow has variation in the direction normal to the shock, taken to be the x direction, but no variation in the direction parallel to the shock, taken to be the y direction, was considered. The origin was taken to be the wedge tip. The streamlines were taken to form an angle θ with the horizontal. At complete reaction, θ relaxes to a constant value. The flow has symmetry about the horizontal plane.

A RH analysis has been commonly used to restrict the potential equilibrium states which may be obtained in an oblique detonation. The RH analysis allows determination of both β, θ shock and detonation polars. If the dimensionless instantaneous heat release for

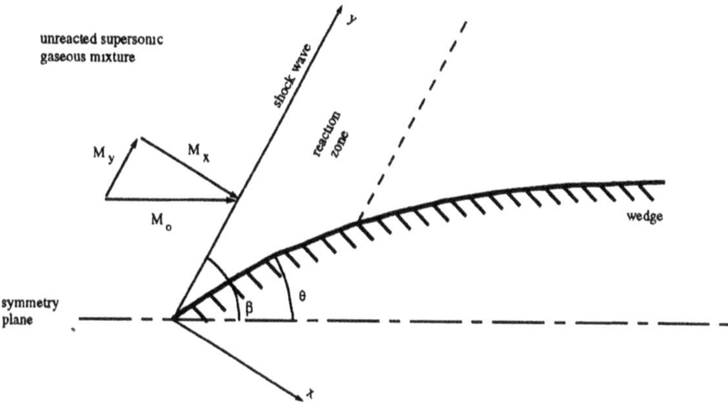

Figure 8: Curved wedge-straight shock oblique detonation configuration, from Powers and Gonthier (1992a).

one-step kinetics, Q, is taken to be

$$Q = \lambda q, \tag{9}$$

then for $M_0 = 10$, $\gamma = 7/5$, and $q = 25$, Figure 9 shows such polars for an inert oblique shock, $\lambda = 0, Q = 0$, and a complete reaction oblique detonation, $\lambda = 1, Q = 25$.

Following Pratt, *et al.*, the final value of the Mach number normal to the shock, M_x, and analogies with inert oblique shock nomenclature were used to classify oblique detonations. For shock angles below a critical value $\beta < \beta_{CJ}$, there is no real solution to the RH equations. For $\beta = \beta_{CJ}$, there is one solution which corresponds to the CJ solution of one-dimensional theory. For $\beta = \beta_{CJ}$, at complete reaction the normal Mach number is sonic, $M_x = 1$. For $\beta > \beta_{CJ}$, two solutions are obtained. The solution corresponding to the smaller wedge angle has a supersonic normal Mach number, $M_x > 1$, at complete reaction and is known as a weak underdriven solution. Its counterpart with the higher wedge angle is known as a weak overdriven solution if $\beta < \beta_{detach}$ and a strong solution if $\beta \geq \beta_{detach}$. For both weak overdriven and strong solutions, the final normal Mach number is subsonic, $M_x < 1$. Here β_{detach} is the shock angle corresponding to the wedge angle θ_{detach} beyond which there is no attached shock solution. The nomenclature "weak" and "strong" is suggested by oblique shock theory and is not consistent with the nomenclature of one-dimensional detonation theory.

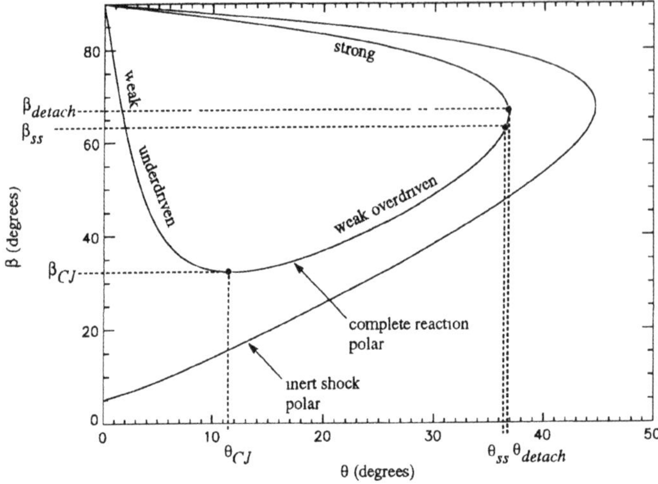

Figure 9: Inert ($Q = 0$) and complete reaction ($Q = 25$) shock polars, from Powers and Gonthier (1992a).

The two-dimensional steady flow can be further characterized by the hyperbolic or elliptic character of the governing partial differential equations. With the total Mach number M calculated from the velocity magnitude, the equations are elliptic if $M < 1$ and hyperbolic if $M > 1$. The subsonic to supersonic transition takes place at β_{SS} which is slightly less than β_{detach}. Strong solutions terminate at a subsonic point, $M < 1$. Weak overdriven solutions terminate at either subsonic or supersonic points: for $\beta_{CJ} < \beta < \beta_{SS}$, $M > 1$; for $\beta_{SS} < \beta < \beta_{detach}$, $M < 1$. Generally $\beta_{SS} \approx \beta_{detach}$; consequently the range of weak overdriven solutions with $M < 1$ is small. Weak underdriven solutions terminate at supersonic points, $M > 1$.

The conditions under which these solution classes, each of which satisfies the conservation principles and entropy inequality, could exist in nature is a question which has not been completely answered. A first step is to consider the resolved steady reaction zone structures and examine solution trajectories from an initial state to an equilibrium state in phase space. For a given kinetic scheme, this will disqualify certain classes of solutions. Those that remain should be subjected to the more rigorous test of hydrodynamic stability. What should result is a knowledge of the initial and boundary conditions which are necessary for a solution to exist. Based on analogies with inert theory which show that the existence of a strong or weak oblique shock depends on the downstream boundary conditions, it is hypothesized that there may be boundary conditions for each class of oblique detonation to exist. Given that in the course of its travels, both an ODWE and ram accelerator may encounter boundary conditions suitable for each class of oblique detonation, it stands to reason that each class should be subjected to systematic study.

With this philosophy in mind Powers and Stewart studied steady reaction zone structures. With the one-step kinetic model and for an oblique detonation characterized by a straight lead shock, it was shown that the reactive Euler equations admit strong, weak overdriven, and CJ solutions but do not admit weak underdriven solutions. The extension of Powers and Gonthier, 1992a, allowed for a two-step reaction with the first step exothermic and the second endothermic. For convenience, they define an equivalent Q for two-step kinetics,

$$Q = \lambda_1 q_1 + \lambda_2 q_2, \tag{10}$$

where $\lambda_1, \lambda_2, q_1$, and q_2 are the reaction progress $(0 \leq \lambda_1, \lambda_2 \leq 1)$ and heat release associated with the first and second reactions, respectively. It was shown that with such a model, steady solutions for all three classes are available and furthermore that the weak underdriven solution can be obtained for eigenvalues of shock angle.

Shock polars and reaction trajectories for all three classes are shown in Figure 10. Here $q_1 = 100$, $q_2 = -75$ so that at complete reaction $Q = 25$, as in Figure 9. However, due to the variable reaction rates, Q can and does take on larger values within the reaction zone. The results give the two-dimensional extension to the one-dimensional case described in detail by Fickett and Davis, pp. 168-173, which admits eigenvalue solutions. As such, straightforward analogies exist. It can be shown that lines of constant β correspond to Rayleigh lines and the shock polars correspond to partial reaction Hugoniot curves. For each class, strong (labelled **I**), weak overdriven (labelled **II**), and weak underdriven (labelled **III**), the reaction proceeds by shocking the fluid from the inert state **O** to the shocked state **N**. The reaction then proceeds along a line of constant β (on either **I**, **II**, or **III**), through the curve of maximum heat release (in this case $Q_{max} = 44.8$) until the reaction is complete at either the strong point **S**, the weak overdriven point **WO**, or the weak under-

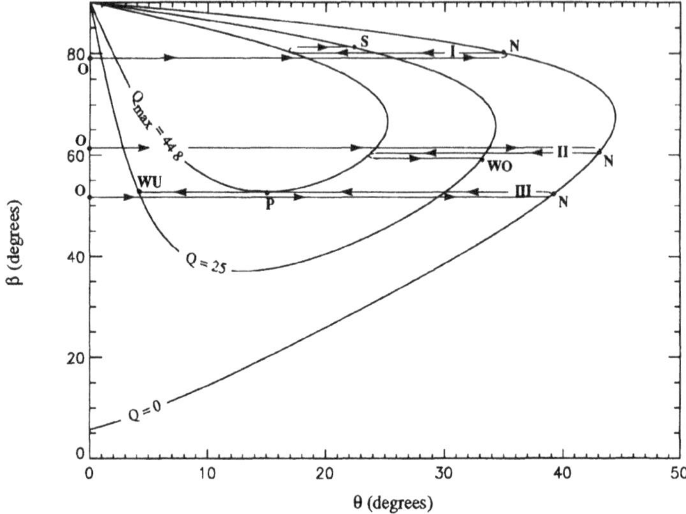

Figure 10: Inert ($Q = 0$), intermediate ($Q = 44.8$), and complete reaction ($Q = 25$) polars with reaction trajectories for strong (**I**), weak overdriven (**II**), and weak underdriven (**III**) cases, from Powers and Gonthier (1992a).

driven point **WU**. The state **WU** is accessible upon passage through the saddle point **P**. For this scenario it was shown that the eigenvalue wave angle is the minimum wave angle for a steady solution; thus, the CJ wave angle, which is lower, places an overly restrictive lower bound on oblique detonation wave angle. It was also inferred that more detailed kinetics could yield correspondingly more complex conditions for the existence of steady waves.

Powers and Stewart also considered rotational solutions in the asymptotic limit of high incoming Mach number, M_0. Here the depiction of Figure 5 was mathematically confirmed. The solution procedure was to linearize the equations in the limit of high Mach number, write them in characteristic form, and construct a solution which simultaneously satisfied the RH jump conditions and a kinematic downstream wall boundary condition. The rather detailed solution can be expressed as an infinite series.

Grismer and Powers then compared the rotational asymptotic solutions to full numerical solutions. The numerical solution was obtained with the RPLUS code (Shuen and Yoon, 1989), in development at the NASA Lewis Research Center, using standard available features to simulate the flow. A series of comparisons was performed in which only the incoming Mach number and the heat release were varied. For zero heat release, in which case the exact solution is available, it was deduced that at low supersonic Mach number the difference in the predictions of the asymptotic and numerical method was primarily attributable to the error in the asymptotic method, while at high Mach number the difference was primarily due to the numerical method. Similar results were inferred for flows with heat release in which there is no exact solution with which to compare. This is expected as the asymptotic solution should become more accurate as the ratio of heat release to flow kinetic energy becomes smaller while in the same limit, a point is reached when numerical errors overwhelm the effects of heat release. For very high Mach numbers, the numerical results become notably distorted while for very low (but still supersonic) Mach numbers, the same can be said for the asymptotic results.

A comparison of asymptotically and numerically predicted dimensionless pressure contours is shown in Figure 11. Here $M_0 = 20$ and $q = 10$. Assuming the ambient fluid is at temperature 300 K, this corresponds to a dimensional heat release of $\tilde{q} = 0.861\ MJ/kg$. The numerical values assigned to the three contours correspond to $\tilde{P} = 85.484\ bar$, $85.852\ bar$, and $86.168\ bar$ if the ambient pressure is assumed to be $\tilde{P}_0 = 1.000\ bar$. In this case it is seen that there is qualitative and quantitative agreement in the two methods' predictions. In order to achieve this agreement, it was necessary to study Mach numbers in a regime far from where the ideal gas, constant property model is valid. Such a step can be justified given that the purpose of this study was to develop a benchmarking procedure for reacting flow codes.

In an effort to better relate these models to propulsion applications, Powers and Gonthier (1992b), give a methodology to study of the configuration of Figure 7 along with a simple, non-rigorous analysis. The analysis divides the flow into six zones: 1) a pre-shocked region, 2) a post-shocked region, 3) a Prandtl-Meyer rarefaction region, 4) a post-rarefaction region, 5) a post-flame sheet region, and 6) a post-shock region. The transition from one zone to another is described by algebraic jump relations. The flame sheet is assumed *ad hoc* to be oriented normal to the lee wedge surface at such a location that a force balance exists. A thermal explosion theory is used to fix the flame location as a function of incoming Mach number.

Plausible results are obtained which are summarized in the bifurcation diagram of Figure 12. Here predictions of flight Mach number are plotted as a function of equivalence ratio $q/\tilde{Q}$ where $\tilde{Q}$ is the heat release associated with stoichiometric hydrogen-air combustion at atmospheric conditions. Below a critical heat release value, the heat release is insufficient to overcome wave drag, and there is no steady solution. Above this critical value, two solutions exist. The lower branch is unstable in a quasi-static sense in that a small perturbation of velocity gives rise to a force which moves the projectile away from equilibrium while on the upper branch a small perturbation in velocity gives rise to a restoring force. Thus one reaches the intuitively satisfying conclusion that an increase in energy released in combustion gives rise to an increase in flight speed. In making such stability conclusions, neither the inertia of the projectile or fluid has been taken into consideration. Finally, no correlation between the steady flight speed and CJ Mach number was found.

It is emphasized that these conclusions are based upon *ad hoc* modeling assumptions and that a more detailed study is required before ascribing any particular value of the predictions. Currently the author is studying numerical solutions to the flow over the double wedge which remove these difficulties (Powers *et al.*, 1993).

5. Recommendations

In conclusion, it is suggested that simple models continue to be used to address questions of relevance to the propulsion community. Though they cannot serve as a substitute for either comprehensive models or experiments (both of which have their difficulties), they can be useful guides for understanding. Examples of new configurations which could be considered are the reactive flow over a double-wedge at an angle of attack, flow including the effect of cowling, conical geometries, reactive flow through a Busemann biplane, and flow over a thin airfoil. Simple model extensions which deserve study include modeling of realistic chemistry with rationally reduced kinetic mechanisms, the inclusion of boundary layer effects, and the inclusion of inertial effects. Such studies should lead to a more fundamental understanding of propulsion systems.

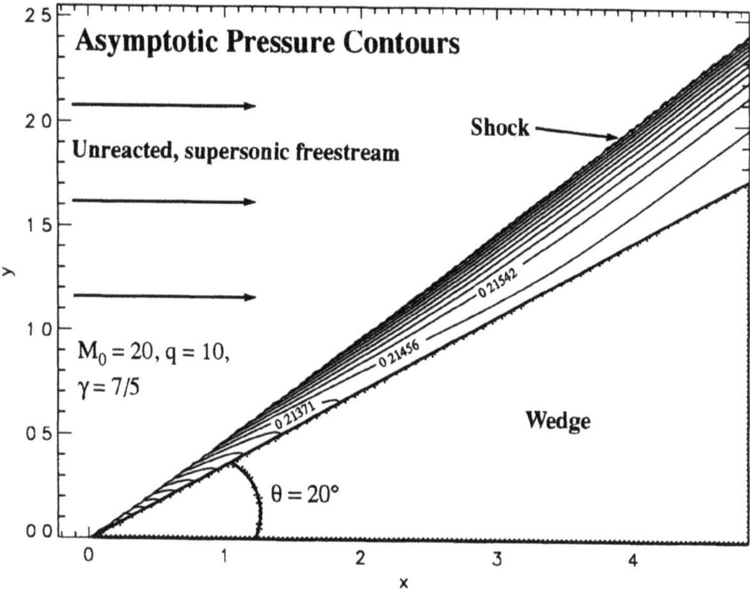

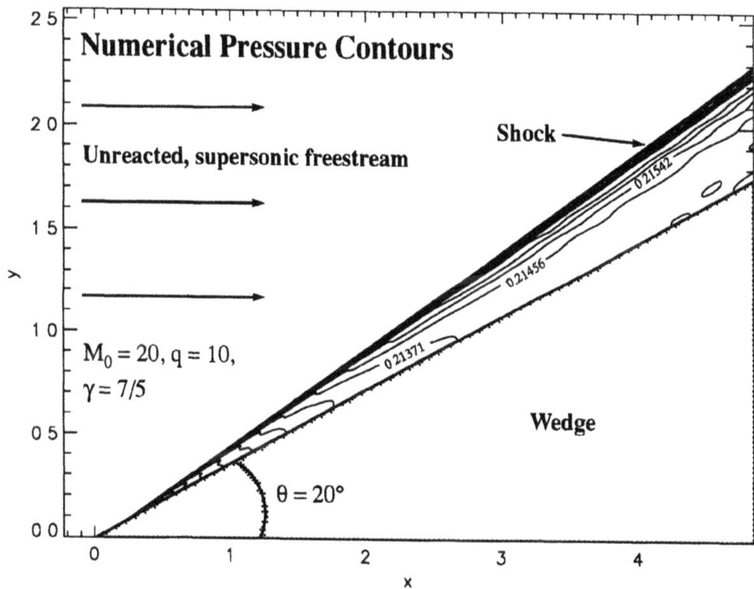

Figure 11: Dimensionless pressure contours predicted by asymptotic and numerical analysis, from Grismer and Powers (1992).

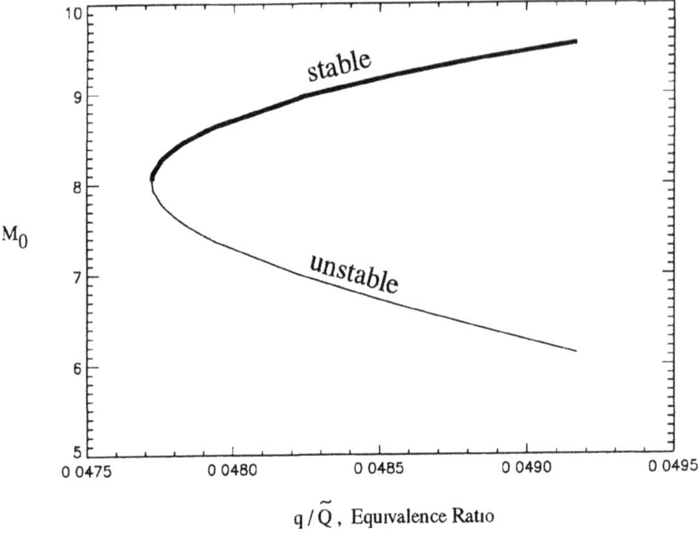

Figure 12: Bifurcation diagram for steady-state flight Mach number versus equivalence ratio, from Powers and Gonthier (1992b).

References

Bdzil, J. B. and Stewart, D. S., 1986. "Time-dependent two-dimensional detonation: The interaction of edge rarefactions with finite-length reaction zones," *J. Fluid Mech.* **171**, p. 1.

Bdzil, J. B. and Kapila, A. K., 1992. "Shock-to-detonation transition: A model problem," *Phys. Fluids A* **4**, p. 409.

Behrens, H., Struth, W., and Wecken, F., 1965. "Studies of hypervelocity firings into mixtures of hydrogen with air or with oxygen," *Proceedings of the Tenth Symposium (International) on Combustion*, The Combustion Institute: Pittsburgh, p. 245.

Bogdanoff, D. W., 1992. "Ram accelerator direct space launch system: new concepts," *J. Propulsion Power* **8**, p. 481.

Bourlioux, A., Majda, A. J., and Roytburd, V., 1991. "Theoretical and numerical structure for unstable one-dimensional detonations," *SIAM J. Appl. Math.* **51**, p. 303.

Bourlioux, A. and Majda, A. J., 1992. "Theoretical and numerical structure for unstable two-dimensional detonations," *Combust. Flame* **90**, p. 211.

Brackett, D. C. and Bogdanoff, D. W., 1989. "Computational investigation of oblique detonation ramjet-in-tube concepts," *J. Propulsion Power* **5**, p. 276.

Bruckner, A. P., Knowlen, C., Hertzberg, A., and Bogdanoff, D. W., 1991. "Operational characteristics of the thermally choked ram accelerator," *J. Propulsion Power* **7**, p. 828.

Buckmaster, J. and Lee, C. J., 1990. "Flow refraction by an uncoupled shock and reaction front," *AIAA J.* **28** p. 1310.

Buckmaster, J., 1990. "The structural stability of oblique detonation waves," *Combust. Sci. Tech.*, **72**, p. 283.

Cambier, J.-L., Adelman, H. G., and Menees, G. P., 1989. "Numerical simulations of oblique detonations in supersonic combustion chambers," *J. Propulsion Power* **5**, p. 482.

Cambier, J.-L., Adelman, H. G., and Menees, G. P., 1990. "Numerical simulations of an oblique detonation wave engine," *J. Propulsion Power* **6**, p. 315.

Capiaux, R. and Washington, M., 1963. "Nonequilibrium flow past a wedge," *AIAA J.* **1**, p. 650.

Carrier, G., Fendell, F., McGregor, R., Cook, S., and Vazirani, M., 1992. "Laser-initiated conical detonation wave for supersonic combustion," *J. Propulsion Power* **8**, p. 472.

Chernyi, G. G., 1969. "Supersonic flow past bodies with formation of detonation and combustion fronts," in *Problems of Hydrodynamics and Continuum Mechanics*, English Edition, SIAM: Philadelphia, p. 145.

Clarke, J. F., 1960. "The linearized flow of a dissociating gas," *J. Fluid Mech.* **7**, p. 577.

Clarke, J. F., Kassoy, D. R., Meharzi, N. E., Riley, N., and Vasantha, R., 1990. "On the evolution of plane detonations," *Proc. R. Soc. Lond. A* **429**, p. 259.

Dunlap, R., Brehm, R. L., and Nicholls, J. A., 1958. "A preliminary study of the application of steady-state detonative combustion to a reaction engine," *Jet Propulsion* **28**, p. 451.

Fickett, W., 1984. "Shock initiation of detonation in a dilute explosive," *Phys. Fluids* **27**, p. 94.

Fickett, W. and Davis, W. C., 1979. *Detonation*, Univ. California Press: Berkeley.

Grismer, M. J. and Powers, J. M., 1992. "Comparison of numerical oblique detonation solutions with an asymptotic benchmark," *AIAA J.* **30**, p. 2985.

Gross, R. A. and Chinitz, W., 1960. "A study of supersonic combustion," *J. Aero/Space Sci.* **27** p. 517.

Gross, R. A., 1963. "Oblique detonation waves," *AIAA J.* **1**, p. 1225.

Hertzberg, A., Bruckner, A. P., and Bogdanoff, D. W., 1988. "Ram accelerator: a new chemical method for accelerating projectiles to ultrahigh velocities," *AIAA J.* **26**, p. 195.

Hertzberg, A., Bruckner, A. P., and Knowlen, C., 1991. "Experimental investigation of ram accelerator propulsion modes," *Shock Waves* **1**, p. 17.

Jackson, T. L., Kapila, A. K., and Hussaini, M. Y., 1990. "Convection of a pattern of vorticity through a reacting shock wave," *Phys. Fluids A* **2**, p. 1260.

Lafon, A. and Yee, H. C., 1992. "On the numerical treatment of nonlinear source terms in reaction-convection equations," AIAA-92-0419, AIAA 30th Aerospace Sciences Meeting and Exhibit, Reno.

Larisch, E., 1959. "Interactions of detonation waves," *J. Fluid Mech.* **6**, p. 392.

Lasseigne, D. G., Jackson, T. L., and Hussaini, M. Y., 1991. "Nonlinear interaction of a detonation/vorticity wave," *Phys. Fluids A* **3**, p. 1972.

Lee, H. I. and Stewart, D. S., 1990. "Calculation of linear detonation instability: one-dimensional instability of plane detonation," *J. Fluid Mech.* **216**, p. 103.

Lee, R. S., 1964. "A unified analysis of supersonic nonequilibrium flow over a wedge: I. vibrational nonequilibrium," *AIAA J.* **2**, p. 637.

Lehr, H. F., 1972. "Experiments on shock-induced combustion," *Astro. Acta* **17**, p. 589.

Liu, J. C., Liou, J. J., Sichel, M., Kauffman, C. W., and Nicholls, J. A., 1986. "Diffraction and transmission of a detonation into a bounding explosive layer," *Proceedings of the Twenty-first Symposium (International) on Combustion*, The Combustion Institute: Pittsburgh, p. 1639.

Maas, U. and Pope, S. B., 1992. "Simplifying chemical kinetics: intrinsic low-dimensional manifolds in composition space," *Combust. Flame* **88**, p. 239.

Moore, F. K. and Gibson, W. E., 1960. "Propagation of weak disturbances in a gas subject to relaxation effects," *J. Aero/Space Sci.* **27**, p. 117.

Nicholls, J. A., 1963. "Standing detonation waves," *Proceedings of the Ninth Symposium (International) on Combustion*, Academic Press: New York, p. 488.

Oppenheim, A. K., Smolen, J. J., and Zajac, L. J., 1968. "Vector polar method for the analysis of wave intersections," *Combust. Flame* **12**, p. 63.

Pepper, D. W. and Brueckner, F. P., 1993. "Simulation of an oblique detonation wave scramaccelerator for hypervelocity launchers," in *Computers and Computing in Heat Transfer Science and Engineering*, W. Nakayama and K. T. Yang, eds., CRC Press: Boca Raton, Florida, p. 119.

Powers, J. M., Fulton, D. R., Gonthier, K. A., and Grismer, M. J., 1993. "Analysis for steady propagation of a generic ram accelerator/oblique detonation wave engine configuration," AIAA 93-0243, AIAA 31st Aerospace Sciences Meeting and Exhibit.

Powers, J. M. and Gonthier, K. A., 1992a. "Reaction zone structure for strong, weak overdriven, and weak underdriven oblique detonations," *Phys. Fluids A* **4**, p. 2082.

Powers, J. M. and Gonthier, K. A., 1992b. "Methodology and analysis for determination of propagation speed of high speed propulsion devices," Proceedings of the Central States Section Spring 1992 Technical Meeting of the Combustion Institute, Columbus, Ohio, p. 1.

Powers, J. M. and Stewart, D. S., 1992. "Approximate solutions for oblique detonations in the hypersonic limit," *AIAA J.* **30**, p. 726.

Pratt, D. T., Humphrey, J. W., and Glenn, D. E., 1991. "Morphology of a Standing Oblique Detonation Wave," *J. Propulsion Power* **7**, p. 837.

Rubins, P. M. and Rhodes, R. P., 1963. "Shock-induced combustion with oblique shocks: comparison of experiment and kinetic calculations," *AIAA J.* **1**, p. 2278.

Sedney, R., 1961. "Some aspects of nonequilibrium flows," *J. Aero/-Space Sci.* **28**, p. 189.

Shuen, J.-S. and Yoon, S., 1989. "Numerical study of chemically reacting flows using a lower-upper symmetric successive overrelaxation scheme," *AIAA J.* **27**, p. 1752.

Siestrunck, R., Fabri, J., and Le Grivès, E., 1953. "Some properties of stationary detonation waves," *Proceedings of the Fourth Symposium (International) on Combustion*, Williams and Wilkins: Baltimore, p. 498.

Spence, D. A., 1961. "Unsteady shock propagation in a relaxing gas," *Proc. R. Soc. Lond. A* **264**, p. 221.

Spurk, J. H., Gerber, N., and Sedney, R., 1966. "Characteristic calculation of flowfields with chemical reactions," *AIAA J.* **4**, p. 30.

Stewart, D. S. and Bdzil, J. B., 1988. "The shock dynamics of stable multidimensional detonation," *Combust. Flame* **72**, p. 311.

Strehlow, R. S., 1968. "Gas phase detonations: recent developments," *Combust. Flame* **12**, p. 81.

Strehlow, R. S. and Crooker, A. J., 1974. "The structure of marginal detonation waves," *Acta Astronaut.* **1**, p. 303.

Vincenti, W. G., 1962. "Linearized flow over a wedge in a nonequilibrium oncoming stream," *J. Méchanique* **1**, p. 193.

Yee, H. C., Sweby, P. K., and Griffiths, D. F., 1991. "Dynamical approach study of spurious steady-state numerical solutions of nonlinear differential equations. I. The dynamics of time discretization and its implications for algorithm development in computational fluid dynamics," *J. Comp. Phys.* **97**, p. 249.

Yungster, S., Eberhardt, S., and Bruckner, A. P., 1991. "Numerical simulation of hypervelocity projectiles in detonable gases," *AIAA J.* **29**, p. 187.

Yungster, S., 1992. "Numerical study of shock-wave/boundary-layer interactions in premixed combustible gases," *AIAA J.* **30**, p. 2379.

Yungster, S. and Bruckner, A. P., 1992. "Computational studies of a superdetonative ram accelerator mode," *J. Propulsion Power* **8**, p. 457.

DETONATION WAVES AND PROPULSION

Joseph E. Shepherd

Graduate Aeronautical Laboratories
California Institute of Technology
Pasadena, CA 91125

Abstract

The possibility of using a detonation wave as the key combustion system for supersonic propulsion is examined. A brief review of propagating detonations is provided first. This review emphasizes the unique and unstable nature of the coupling between reaction zone and shock waves that characterize detonations. The theory of idealized, steady, oblique detonation waves and their reaction zone structure are summarized. The evidence for the existence of stabilized or steady oblique detonations is discussed. Experiments with multiple layers of explosive and projectiles fired into explosive gases are examined. There are a variety of reasons that these previous studies have failed to produce stabilized detonations. A brief catalog of difficulties is provided and based on analogies with our knowledge of propagating detonations, a set of criteria are proposed for the existence and stability of stabilized detonations. The problems of initiation and instability are examined for the situation of a flow over a wedge.

1. Introduction

One of the most intriguing possibilities for supersonic propulsion is using a combustor based on a stabilized detonation wave. A number of studies of this concept were made in the 1950s and 1960s (a useful set of historical references are given in Pratt et al. 1991) and recently there has been a renewed interest in this subject. The concept is, in principle, quite simple as shown in Fig. 1. Fuel is mixed thoroughly into a supersonic flow of air within the engine duct and a detonation wave is stabilized by inserting a bend or obstruction into the flow. The thermal energy resulting from the combustion is converted to kinetic energy in the nozzle to produce thrust. Simple engine performance estimates (Dunlap et al. 1958, Sargent and Gross

J. Buckmaster et al. (eds.), Combustion in High-Speed Flows, 373–420.

1960, Ostrander et al. 1987 and Atamanchuk and Sislian 1990) and vehicle design studies (Atamanchuk and Sislian 1991) indicate that this concept may have some merit. However, it is speculative and the main focus of present-day high-speed propulsion research (Murthy and Curran 1991) is on the more traditional supersonic combustion ramjet or *scramjet* combustor approach.

Conceptually, the oblique detonation wave is analogous to an oblique shock wave (Thompson 1972) with chemical reaction. As shown in Fig. 2, the wave extends at an angle β into the flow and turns the flow by angle θ. Such a configuration is conceivable as long as the flow upstream is moving faster that the characteristic detonation wave speed, the Chapman-Jouguet (CJ) velocity. Since detonations are supersonic combustion waves that propagate with a relative Mach number between 4 and 10, only a modest amount of compression is required in the inlet diffuser in comparison with the traditional gas-turbine combustor. This is what makes the concept so attractive for propelling supersonic flight.

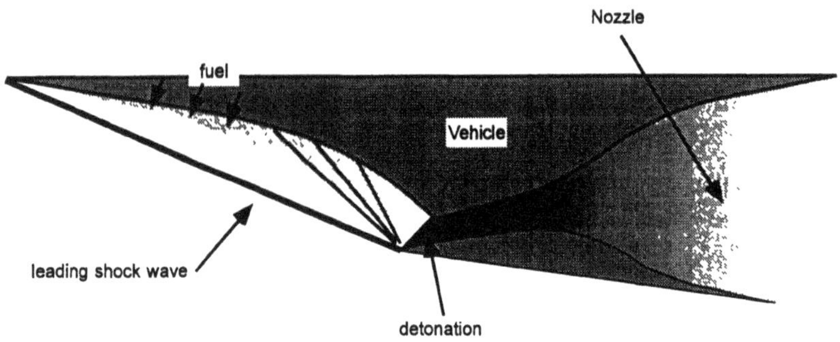

Figure 1: Schematic diagram of a propulsion system based on a standing detonation wave.

Despite the apparent simplicity of this approach, there are a number of practical issues that must be resolved before a detonation wave-based engine can be constructed. Foremost is the issue of detonation stability. Although propagating detonations have been extensively investigated, little is known about the initiation and stabilization of steady detonation waves. For propagating detonations, it is known that a minimum width duct and a minimum initiation energy are required. The precise corresponding criteria for steady detonation waves have yet to be defined. One of the aims of this paper is to suggest criteria based on our knowledge of propagating detonations.

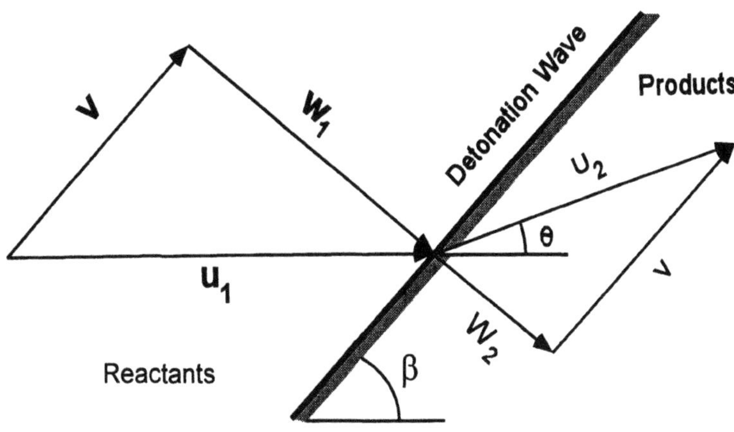

Figure 2: Flow deflection and velocity vectors associated with an idealized oblique detonation wave.

There are many other issues that are crucial to the performance of a detonation wave engine. Key among these are: the adjustment of the duct and inlet geometry as a function of flight Mach number; mixing of fuel with the air while controlling pre-ignition; conversion of gas chemical energy to kinetic energy within the exhaust nozzle (Harradine et al. 1990 and Sangiovanni et al. 1993). It is particularly important to have realistic analyses and reliable experimental data in order to construct performance models and predict the behavior of full-scale engines. This is due to the intrinsically marginal performance of airbreathing propulsion at hypersonic flight speeds. The potential energy addition due to combustion is a much smaller fraction of the stagnation enthalpy of the free-stream flow at hypersonic speeds as compared to traditional low-speed propulsion.

At the present time, it is not clear if the construction of an engine using detonation waves is feasible. A very substantial amount of engineering research and development is needed before this can be decided. Even the most basic issues such as the existence and stability of a steady detonation wave inside an engine duct are quite controversial. As we shall see, the experimental evidence is scant and contradictory and the theory is far from definitive. Our knowledge of detonations is almost exclusively based on propagating waves and almost all of the discussion about steady detonation wave engines is pure speculation. In this paper, the focus will be on steady waves although it should be pointed out that propagating detonations have been proposed as components in intrinsically unsteady wave engines (Eidelman et al. 1991, Voytsekhovskiy et al. 1964).

In the next section, an elementary review of propagating detonation phenomena will be given. The simple hydrodynamic discontinuity approach to oblique detonation and shock waves is discussed. The idealized structure of the reaction zone for steady oblique waves is then described. This structure is an analog of the standard Zel'dovich-von Neumann-Döring model. Results of numerical solutions to the reaction zone structure using detailed chemical reaction mechanisms are given for some representative cases that are of interest to propulsion. The experimental evidence for steady oblique detonations is presented.

A paradigm of the oblique detonation wave problem is a wave created by the flow deflection around a wedge. The uniqueness and stability of this solution are examined and a possible "test" for detonation-like behavior are proposed. Detonation stability is examined both from the viewpoint of microinstabilities and the potential for catastrophic or global instability of the entire flow. Initiation or failure of the detonation process is related to the presence of transients in the flow that are produced by unsteadiness or flow deflections.

2. Propagating Detonation Concepts

Propagating detonations in gases (Strehlow 1984, Fickett and Davis 1979, Zeldovich and Kompaneets 1960) are characterized by a self-sustaining configuration of shock waves and reaction zones, indicated schematically in Fig. 3 for an idealized steady configuration. The propagation velocity U is relatively constant for waves that are self-sustaining and is approximately equal to the *Chapman-Jouguet* (CJ) value U_{CJ}. The CJ velocity is obtained through purely thermodynamic considerations as the minimum velocity consistent with a steady wave separating reactants and equilibrium products.

If the combustion process results in the release of energy q per unit mass into the flow, then $U_{CJ} \sim \sqrt{2(\gamma^2-1)q}$. The flow following the detonation can then be treated using the standard techniques of nonreactive compressible flow. This flow often consists of a transient expansion wave which eventually brings the fluid back to rest.

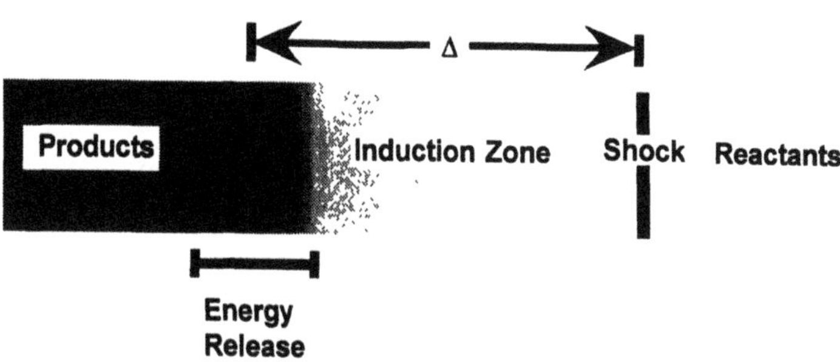

Figure 3: Idealized one-dimensional, steady-state detonation wave configuration (ZND model) consisting of shock wave followed by a reaction zone. The characteristic reaction zone length Δ is based on the location of the maximum in the chemical energy release rate.

The dynamic behavior of detonation waves is determined by the response of what is usually a relatively small portion of the flow, the shock wave-reaction zone complex located at the front of the wave. The intimate coupling between the shock wave and the chemical reactions occurring in the region immediately behind the shock plays a key role in this response. The shock wave produces the adiabatic compression which increases the gas temperature sufficiently to initiate the chemical reactions that result in the release of energy that drives the shock forward. This cycle of events is unstable since typical chemical reaction rates for hydrogen or hydrocarbon-air mixtures have an extreme sensitivity to the temperature. Small variations in the shock strength produce large variations in reaction rates in the flow directly behind the shock. After some time delay associated with the cumulative effects of chemical reaction and acoustic propagation of disturbances, the changes in reaction rates then result in variations in the shock strength since the flow is subsonic (relative to the shock) through most of the reaction zone. This creates an unstable feedback loop that results in the spontaneous and nonlinear instability of propagating detonations in gases.

This instability results in the breakdown of the idealized one-dimensional structure shown in Fig. 3, the production of transverse shock waves, an oscillatory motion of the main shock front and a turbulent flow field behind the detonation as shown in Fig. 4. A consequence of this almost universal instability of self-sustaining detonation waves in gases is the formation of quasi-periodic instability patterns associated with the motion of the intersections or triple-points between the transverse waves and the main front. The cellular appearance of these patterns motivated the term "cellular structure" for the instability and the characteristic transverse wavelength of the instability is referred to as the "cell width". The cell width λ is often used (Lee 1984) as length scale that defines an effective thickness of the detonation front. Another commonly used length scale is the calculated thickness Δ of an idealized one-dimensional, steady reaction zone structure.

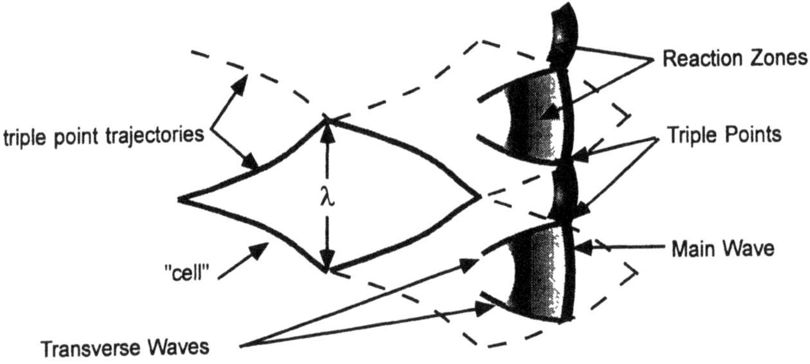

Figure 4: Instability pattern associated with propagating detonations, illustrating cellular pattern formation and characteristic cell width λ.

The existence of such a macroscopic intrinsic length scale λ or Δ distinguishes detonations from simple shock waves and other nonreactive fronts or interfaces. The processes of detonation initiation, stability of propagation, and behavior during transients such as diffraction can be correlated (Lee 1984) on the basis of relationships of these length scales to the characteristic physical dimensions of the confining boundaries. The correlations indicate that for each mixture composition and initial conditions there exist a set of critical length scales that define the limits within which propagating detonations can be produced. Representative configurations that have critical length scales are sketched in Fig. 5.

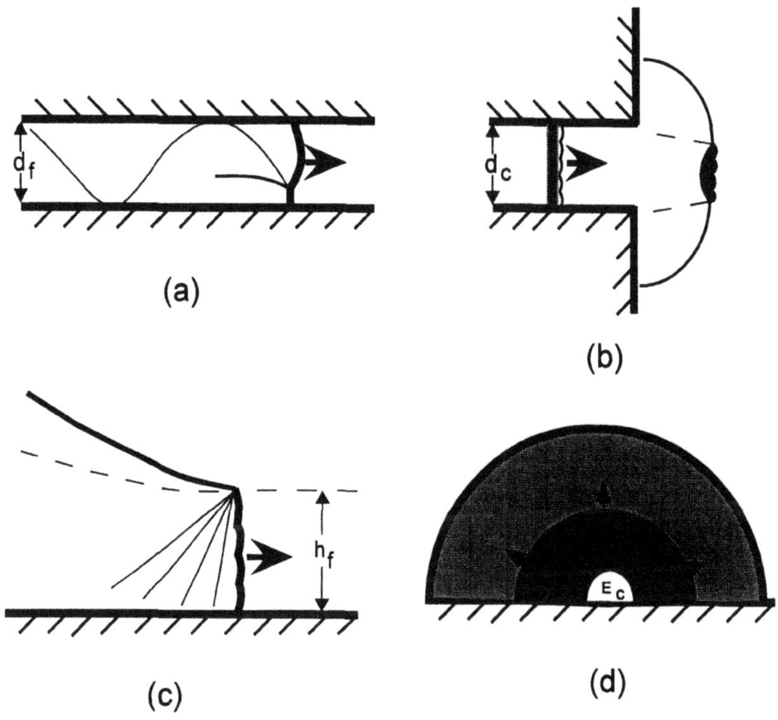

Figure 5: Configurations associated with critical length scales for detonation behavior. a) failure diameter d_f associated with a minimum tube size for confined steady propagation. b) critical tube or orifice diameter d_c associated with diffraction of a detonation. c) critical layer thickness h_f associated with steady unconfined propagation. d) Critical initiation energy E_c associated with the direct initiation of detonation by strong blast waves.

An alternative to correlating the behavior based on the experimentally measured cell width is to use the computed reaction zone length as based on the idealized ZND model of detonations. Extensive computations and comparisons with experiment (Westbrook and Urtiew 1982, Shepherd 1986) indicate that such a correlation is indeed feasible. Fig. 6 demonstrates the scaling relationships that have been determined in this fashion. Note that the scaling length for initiation is defined as $R_c = (E_c/P_o)^{1/3}$.

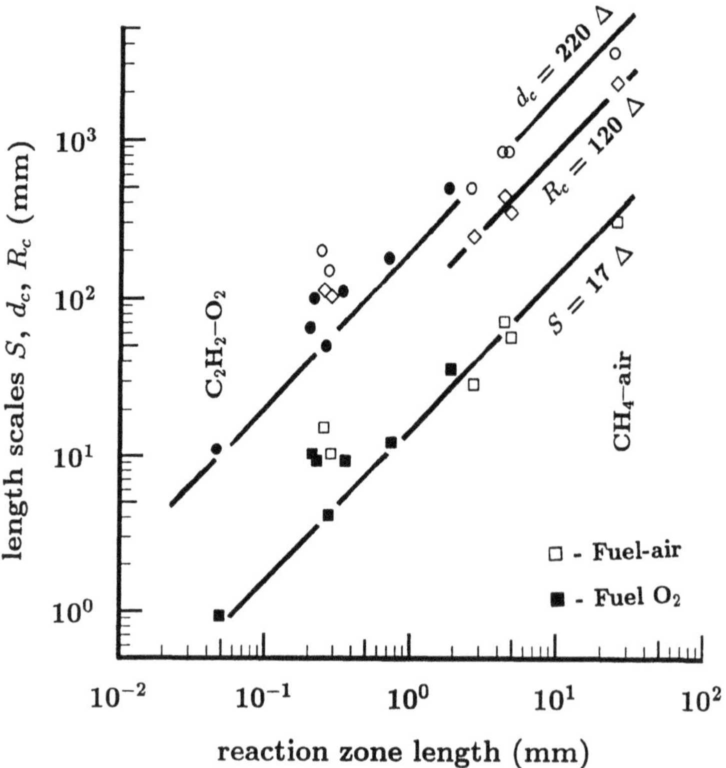

Figure 6: Scaling relationships between the computed ZND reaction zone length and the critical length scales for detonation propagation. The length scale S is the estimated cell width λ.

Such correlations are based on extensive experimentation and naive dimensional analysis. This simple idea is quite successful in correlating many overall trends but often fails to capture the nuances of behavior resulting from the interplay between fluid dynamics, chemical reactions and thermodynamics. On the other hand, direct numerical simulations are a powerful tool (Bourlioux and Madja 1992) but at present are only able to examine the simplest models of chemical reaction if the spatial resolution needed for resolving the instability is used.

What is the significance of these critical length scales to propulsion systems and oblique detonation waves in general? Extrapolating from propagating to stationary waves, we conclude that it will only be possible to observe detonation-like behavior under certain specific conditions. It is important to distinguish between conditions sufficient for existence of waves and those conditions required to initiate waves. First, the confining boundaries or the thickness of the combustible layer must exceed a certain minimum size in order that the detonation wave can exist. Second, a minimum amount of energy must be provided over some time period in order to initiate the detonation.

A certain difficulty immediately arises in attempting to extend notions concerning propagating detonations to stationary waves. How do we distinguish stationary detonations from mere shock-induced combustion? Is there a continuous spectrum of behavior or can distinct regimes be identified? I believe that the extreme sensitivity of most chemical reaction rates to temperature will in fact, result in a sharp distinction between the two phenomena: detonations vs. shock-induced combustion.

For propagating waves, the distinction between detonation and shock-induced combustion is clear when the reaction zone terminates in a sharp reaction front or explosion locus. This is the case for most gaseous reactions described by an Arrhenius rate law with at least a modest activation energy. In a detonation wave, the shock and reaction fronts must propagate at essentially the same velocity in order to be coupled. For self-sustaining waves, it is further observed that the wave speed is close to the CJ velocity and the front is unstable to transverse disturbances. In shock-induced combustion, the shock front and reaction front do not travel at the same speed and often become completely uncoupled, that is, the reaction front progressively lags behind the shock front as both waves decay. Shock-induced combustion is a transient process that either terminates in an explosive instability leading to transition to detonation or the decay of the wave system into a low-speed flame with a weak precursor shock wave.

Defining steady oblique detonations is apparently a more subtle problem than for propagating waves. For a steady flow, various wave configurations are possible depending on the starting and boundary conditions. In particular, the equivalent wave velocity does not have a unique minimum for steady flows. Many researchers have labeled any supersonic flow process that results in a shock wave followed by a reaction zone an "oblique detonation wave". Is there a simple way to distinguish oblique detonations? I suggest that there are two key tests. First, an oblique detonation must be a wave–like structure that can be enclosed by a control volume and satisfies the usual jump conditions relating upstream and downstream states. Second, the wave should be capable of self-sustained propagation, i.e., if the incoming flow was somehow stopped, the wave would propagate upstream. A clear symptom of the existence of a detonation is the presence of transverse waves. However, in the case of an oblique detonation, these waves may be suppressed if the detonation is sufficiently overdriven.

3. Oblique Detonation Concepts

Oblique detonations can be analyzed at several levels. The simplest type of analysis is to treat the detonation wave as a discontinuity analogous to the standard treatment of oblique shock waves (Thompson 1972). This is an essentially thermodynamic analysis that neglects the chemical reaction process and the structure of the detonation wave. A discontinuity analysis determines the locus of possible downstream equilibrium states for a given upstream state. The next more detailed level of analysis is to consider the reaction zone structure of a steady, oblique wave. This structure can be obtained as a transformation of the standard, one-dimensional ZND model of steadily propagating detonations.

The most complex situation for which some analytical considerations can be given is for a weakly-curved wave and the influence of curvature on the existence of detonation-like wave structures. Open questions about the stability and initiation of detonations in combustor geometries can probably only be addressed through experimentation or numerical simulations. The intrinsically multidimensional and transient nature of these processes indicates that direct numerical solutions of the equations of motion are necessary. It is possible to obtain some information by solving the simpler problem of computing the normal mode solutions to the linearized stability problem. Although quite well understood (Lee and Stewart 1990) for the case of one-dimensional disturbances, the stability of detonation waves to multi-dimensional disturbances is not well characterized. The linearized stability of oblique detonation waves is not at all understood.

3.1. Discontinuity Analysis

The discontinuity model is a purely hydrodynamic construct that considers the upstream state as a specified mixture of reactants and the downstream state as reaction products in a state of chemical equilibrium. An analysis of the conservation laws using a control volume across the wave front (Thompson 1972) yields a locus of solutions known as the *Hugoniot* relation or the *detonation adiabat.* Chemical reaction rate considerations, wave structure or instabilities are neglected in this type of analysis. The flow is considered to be steady and all front structure is considered to be contained within the control volume.

Detonation adiabats are essentially one-dimensional concepts that can be calculated without considering the oblique nature of the wave. Transforming to oblique wave coordinates then provides certain restrictions on the possible flow deflection angles as a function of the upstream flow state. This is usually represented in pressure-flow deflection coordinates as a *detonation polar.* There have been many studies of the detonation polars and the implications for oblique detonation waves, a review of this prior work is provided by Pratt et al. (1991). An outline of the analysis and its implications are given below.

Application of the integral balances for mass, momentum and energy within a thin control volume enclosing a section of the wave of Fig. 2 leads to a set of relationships between the upstream and downstream states across the oblique detonation wave. These relationships are conventionally referred to as *jump conditions* since a detonation or shock appears to be a sharp jump in properties according to the hydrodynamic discontinuity description. In terms of the mass density ρ, normal velocity w, tangential velocity v, pressure P and specific enthalpy H, the results are:

$$\rho_1 w_1 = \rho_2 w_2 \tag{1}$$

$$P_1 + \rho_1 w_1^2 = P_2 + \rho_2 w_2^2 \tag{2}$$

$$v_1 = v_2 \tag{3}$$

$$H_1 + \frac{w_1^2}{2} = H_2 + \frac{w_2^2}{2} \tag{4}$$

Given a set of upstream conditions (state 1) these equations can be solved once an equation of state is selected. The approximation of a mixture of ideal gases with realistic specific heats and heats of formation appears adequate for all air-breathing propulsion applications. The composition, given by amount of each species, is specified for the upstream state and must be computed for the downstream state. For a nonreactive shock wave, the composition is *frozen*, i.e., will not change across the wave. For the hydrodynamic model of detonation, the composition must be computed by solving the conditions for chemical equilibrium in the products.

Figure 7 shows a typical equilibrium detonation adiabat together with a shock adiabat. These curves show the relationship between pressure and specific volume obtained by eliminating velocity from the jump conditions, Eqns. 1-4, to obtain the *Hugoniot* relation

$$H_2 - H_1 = (P_2 - P_1)(V_2 + V_1)/2 \tag{5}$$

where $V = 1/\rho$ is the specific volume.

The frozen shock adiabat curve corresponds to the locus of possible states behind a non-reacting shock wave for upstream conditions. Each point corresponds to a given upstream normal velocity. The equilibrium shock adiabat curve is similar except that the points now correspond to the various equilibrium states attainable behind the detonation wave given the upstream thermodynamic condition, composition, and velocity. These results were obtained by numerical solution (Reynolds 1986) of the jump conditions with realistic thermodynamic properties and a full set of reaction products in the case of the equilibrium model. The CJ detonation velocity is about 1908 m/s for this mixture. Note that the detonation adiabat is substantially different than the usual rectangular hyperbola that results if the specific heat and heat of reaction are taken to be constant. The equilibrium and frozen solution curves cross when the normal velocity w_1 reaches about 3350 m/s. This is a consequence of the dissociation of the reaction products that occurs at high temperatures; the products actually have to cool down to reach equilibrium. This may play a role in limiting the operational regime for a detonation-based propulsion system.

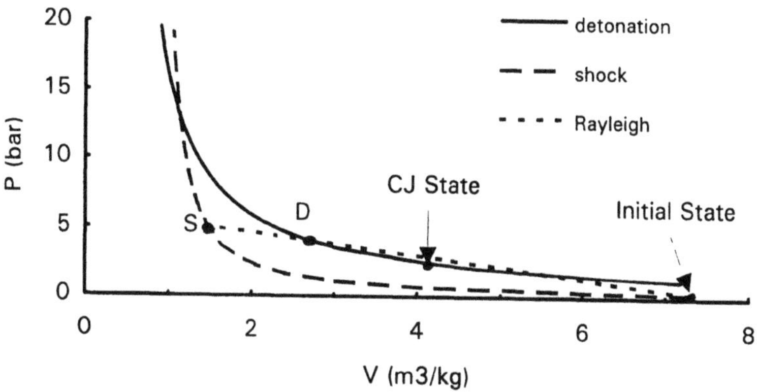

Figure 7: Shock and detonation adiabats for a stoichiometric hydrogen-air mixture with upstream conditions of 550 K and .3 bar, appropriate to the exit of the inlet diffuser for a detonation-based propulsion system.

3.2 ZND Model

The relevance of the shock adiabat is thus: An idealized model of the detonation structure can be formulated by supposing that a nonreactive shock precedes any chemical reactions. The basis for this idea is the notion that a shock occurs over only a few collision distances (mean free paths) of the reactant molecules while significant chemical reaction requires thousands or millions of collisions. This is the model first proposed by Zel'dovich, von Neumann and Döring (ZND model) to explain the role of shock waves and chemical reaction in detonation propagation.

During a steady reaction process, conservation of mass and momentum require the P–v locus to lie on a straight line:

$$P_2 - P_1 = -(\rho_1 w_1)^2 (V_2 - V_1) \tag{6}$$

These lines are drawn on Fig. 7 and are known as *Rayleigh lines.* The thermodynamic path from the initial state to the final equilibrium state can be conveniently traced out in Fig. 7. The fluid initial state is the low pressure and density condition at the lower right. The nonreactive shock preceding the reaction zone corresponds to moving along the Rayleigh line up to the frozen shock adiabat. As the fluid reacts, the state moves down along the Rayleigh line from the frozen shock adiabat to the point of intersection with the equilibrium shock adiabat. Note that the Rayleigh line will always have two intersections with the equilibrium shock adiabat except in the limiting case when the Rayleigh line is actually tangent to the equilibrium shock adiabat.

The point of tangency corresponds to the minimum upstream flow velocity that is required to reach the equilibrium shock adiabat. This minimum velocity is referred to as the Chapman-Jouguet velocity (w_{CJ}), and is only a function of the mixture composition and the conditions at the upstream state. Rayleigh lines corresponding to *overdriven* detonation waves have an upstream velocity larger than the CJ value, $w > w_{CJ}$. As long as the amount of overdrive is not excessive, the equilibrium shock adiabat is to the right of the frozen one and the temperature is higher in the equilibrium products than in the shocked reactants. If the detonation velocity is high enough, the postshock state will lie above the point where the adiabats cross and product temperature will be lower than the shocked fluid temperature.

If the upstream velocity is less than the CJ value, then apparently a steady detonation solution leading to complete combustion does not exist. There are two possibilities in this case. An unsteady flow leading to complete combustion consists of a shock wave followed by a slower reaction wave, termed *shock-induced combustion.* This configuration is intrinsically unsteady since the reaction wave progressively lags behind the shock in this situation (Zeldovich and Kompaneets 1960). The usual consequence of this unsteadiness is the decay of the shock wave and the eventual conversion of the reaction wave into a low-speed flame. Such a process may also be unstable since an explosion within the region behind the shock and the reaction wave will produce pressure waves that influence the leading shock. This can lead to a pulsating type of instability or transition to detonation in the case of a propagating detonation (Bach et al. 1969).

A steady but incomplete combustion process can also occur behind a steady shock wave that is curved (Bdzil and Stewart 1989).

The flow divergence behind the wave competes with the chemical reaction and results in a sonic point before the reaction is complete. This type of solution plays an important role in determining the combustion process produced behind the bow shock of a hypersonic projectile in a reacting gas. These solutions exist only if the wave curvature is not too large, which sets conditions on the existence of curved detonation waves.

The Rayleigh line shown in Fig. 7 corresponds to an overdriven wave with a normal velocity $w_1 = 2086$ m/s, about 10% higher than the CJ velocity. The figure shows two possible intersections between the Rayleigh line and the detonation adiabat. In principle, both intersection points are possible solutions: a "strong" one in which the flow behind the wave is subsonic, and a "weak" one in which the flow behind the wave is supersonic. In order to determine which of these solutions are actually realized, further considerations of the details of the chemical reaction mechanism are required. The supersonic solution is usually ruled out by these considerations since it is only possible in exceptional cases to smoothly pass through the sonic point that separates the two types of solutions (Fickett and Davis 1979). This is discussed in the subsequent section on reaction zone structure.

3.3 Application to Oblique Waves

An oblique detonation can be treated as a normal detonation by considering an orthogonal set of axes in the which the wave lies along one of the axes. In this case, only the normal component of velocity plays a role in the solution. According to the previous discussion of the jump conditions, the tangential velocity component remains unchanged. Note that for oblique detonation waves, the minimum upstream velocity requirement is now stated in terms of the normal component of the velocity, $w_1 \geq w_{CJ}$. From the geometry of Fig. 2, the equations for the upstream and downstream normal velocity components are given by:

$$w_1 = u_1 \sin \beta \tag{7}$$

$$w_2 = u_2 \sin(\beta - \theta) \tag{8}$$

The net velocities upstream and downstream of the wave are u_1 and u_2; β is the detonation wave angle, and θ is the flow deflection angle. The tangential component of velocity, v,

$$v = u_1 \cos\beta = u_2 cos(\beta - \theta) \tag{9}$$

is constant across the wave, which implies that:

$$w_2 = u_1 \cos\beta \tan(\beta - \theta) \tag{10}$$

Combining this last relation with the geometric transformations and the solutions to the normal detonation wave problem $w_2 = f(w_1)$ discussed earlier, the wave angle β and the flow deflection angle θ can be determined for a given upstream velocity u_1 and normal velocity w_1.

$$\beta = \sin^{-1}(w_1/u_1) \tag{11}$$

$$\theta = \beta - \tan^{-1}\left(\frac{w_2}{\sqrt{u_1^2 - w_1^2}}\right) \tag{12}$$

Realistic polar curves can be readily computed once the normal shock wave or detonation adiabats have been determined (Sabet 1990). The procedure is to first fit the computed relation between upstream and downstream normal velocity components to a polynomial curve. This relationship can then be used to evaluate β and θ from Eqns. 11 and 12. The other thermodynamic properties downstream of the wave can be determined from the jump conditions, Eqns. 1-4. Maximum flow deflection angle can be computed by first finding an analytical expression for the derivative of the flow deflection angle with respect to the upstream normal velocity. A numerical root-solver can then be used to solve for the value of the upstream normal velocity which makes this derivative zero.

This procedure enables the use of available computational tools for normal waves (Reynolds 1986, Gordon and McBride 1972) to be used to generate solutions for oblique waves without having to make any assumptions regarding the specific heats, energy release or equilibrium compositions. Once a shock adiabat is determined in the form $w_2 = f(w_1)$, solutions for any upstream velocity can be obtained readily by these simple transformations. Both the frozen and equilibrium states can be treated in this fashion in order to get the oblique analog of the ZND model. In order to complete this picture, the momentum flux conservation equation can be transformed to obtain the analog of the Rayleigh line which we term the *Rayleigh curve:*

$$P_2 = P_1 + \rho_1 u_1^2 \frac{\tan\theta}{\cot\beta + \tan\theta} \tag{13}$$

A set of shock and detonation polar curves and a connecting Rayleigh curve are shown in Fig. 8 for a freestream velocity of 2900 m/s and the upstream thermodynamic conditions used for the adiabats shown in Fig. 7. The Rayleigh curve shown corresponds to that of the slightly overdriven solution shown in Fig. 7 and results in a wave angle of about 46°. The locus of possible values β and θ for this example is shown in Fig. 9.

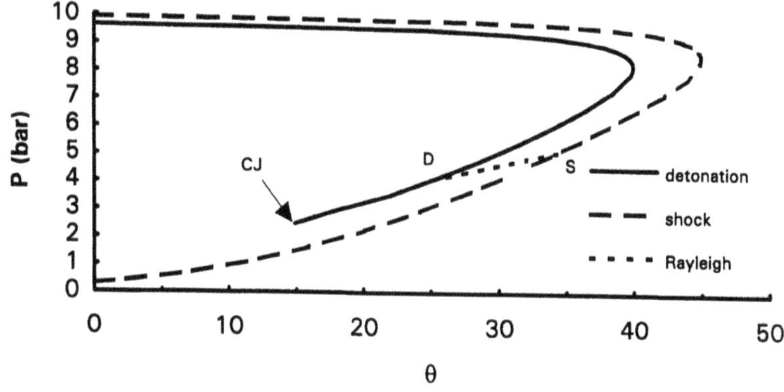

Figure 8: Shock and detonation polar curves for a stoichiometric hydrogen-air mixture with upstream conditions of 550 K, .3 bar and 2900 m/s. The Rayleigh curve for a idealized detonation structure is shown connecting the postshock and equilibrium states for a wave angle β of 46°.

Key points to note are the minimum θ_{CJ} and maximum θ_{max} flow deflection angles associated with transformation between uniform upstream and downstream states. These are only local limits on flow deflections. If the flow downstream must ultimately deflect outside these ranges, then there will be a nonuniform flow following the detonation and/or the detonation wave will be curved. The nature of the solutions in those cases will be discussed in a subsequent section.

A pair of representative streamlines and the flow deflection within an exothermic reaction zone is sketched in Fig. 10. This would be characteristic of the solution moving along the Rayleigh curve indicated in Fig. 8 for a solution with a slight amount of overdrive. The flow deflection history can be readily deduced from the structure equations given in the subsequent section. After passing through the shock, θ jumps to the value θ_S. As the chemical reaction proceeds, the flow deflects away from the wave for exothermic reactions and toward the wave for endothermic reactions. Ultimately, θ approaches the value θ_D as the reactions come to equilibrium.

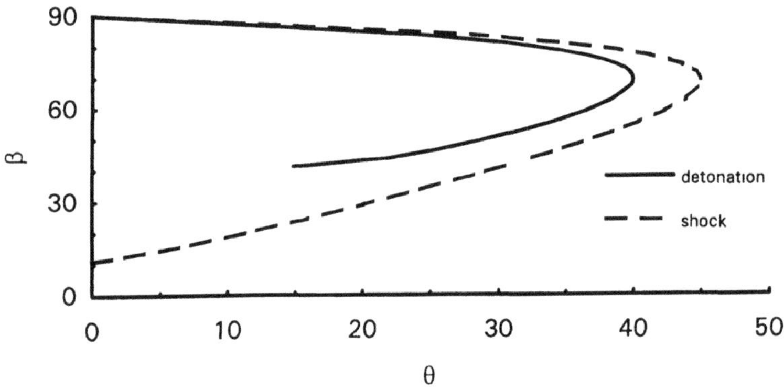

Figure 9: Possible range of wave angle β and flow deflection angle θ for an oblique detonation in a stoichiometric hydrogen-air mixture with upstream conditions of 550 K, 0.3 bar and 2900 m/s.

As the amount of overdrive becomes larger, $w_1 \gg w_{CJ}$, endothermic processes will become more significant and the flow deflection will be nonmonotonic within the reaction zone. If the overdrive is sufficiently large, the VN point will lie beyond the point where the two adiabats cross and the flow deflection throughout the reaction zone will be in the opposite sense to that shown in Fig. 10. In any case, the state of the flow will move along the Rayleigh curve, approaching the equilibrium adiabat as the reactions come to equilibrium in the flow behind the shock wave.

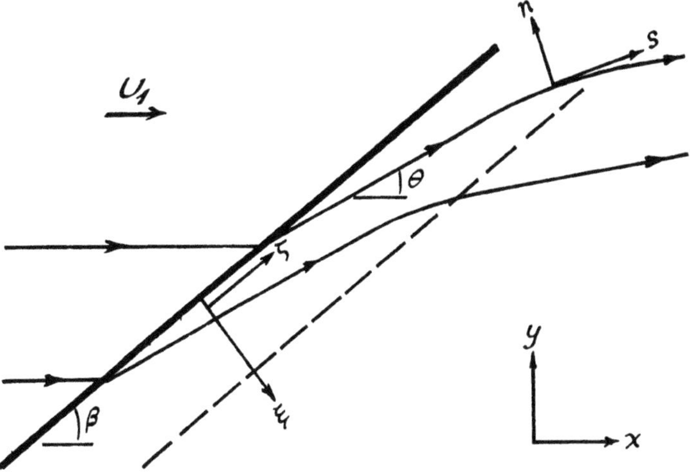

Figure 10: Flow deflection along a typical streamline with exothermic reactions within the reaction zone of an idealized oblique detonation. The natural (n, s) and wave-fixed (ζ, ξ) coordinates are also shown.

3.4 Reaction Zone Structure Equations

The flow within the reaction zone of an oblique detonation wave can be analyzed in more detail by considering the equations of motion for an inviscid, reacting, compressible flow. For two-dimensional flow, it is convenient to utilize a curvilinear system of coordinates as shown in Fig. 10. In such a system, one axis (s) is along the streamline, and the other axis (n) is orthogonal to it. Instead of considering two velocity components, we use the magnitude of the velocity u and the deflection angle θ relative to a reference axis. The steady conservation equations in this frame of reference are (Liepmann and Roshko 1957) for mass:

$$\frac{\partial(\rho u)}{\partial s} = -\rho u \frac{\partial \theta}{\partial n} ; \tag{14}$$

momentum components:

$$\rho u \frac{\partial u}{\partial s} = -\frac{\partial P}{\partial s}, \tag{15}$$

$$\rho u^2 \frac{\partial \theta}{\partial s} = -\frac{\partial P}{\partial n}; \tag{16}$$

energy:

$$H + \frac{u^2}{2} = H_o \tag{17}$$

and species:

$$\rho u \frac{\partial Y_k}{\partial s} = W_k \dot{\omega}_k \quad k = 1, \ldots, K \tag{18}$$

where Y_k is the mass fraction of species k, W_k is the molar mass of species k, and $\dot{\omega}_k$ is the net molar production rate of species k. A chemical reaction mechanism and a set of associated rate constants is required to compute the net molar production rate $\dot{\omega}_k$ of species k. The total enthalpy H_o has a constant value for a given streamline but may vary if the upstream state is nonuniform.

For a reacting flow, the fundamental property relation of thermodynamics is

$$dH = TdS + VdP + \sum_k \mu_k dN_k \tag{19}$$

where S is the specific entropy and N_k is the number of mols of species k per unit mass of material and μ_k is the chemical potential or partial Gibbs energy per mol of species k. This relationship and the momentum conservation equations can be combined to form the reacting flow extension to the Crocco-Vazsonyi equation (Thompson 1972) for steady, inviscid flows:

$$T\nabla S + \sum_k \mu_k \nabla N_k = (\nabla \times u) \times u + \nabla H_o \tag{20}$$

We conclude that the entropy changes will only be due to chemical reaction as long as the upstream fluid state is uniform $H_o = constant$ and the vorticity $\nabla \times u$ vanishes.

The conditions of Crocco's theorem will be satisfied by a straight, oblique detonation wave extending into a uniform flow. The direct relationship between entropy changes and species changes under these circumstances enables the substitution of the energy equation with the *adiabatic change equation* (Fickett and Davis 1979):

$$dP = a^2 d\rho + \rho a^2 \sum_k \sigma_k dY_k \tag{21}$$

where a is the local (frozen) speed of sound,

$$a^2 \equiv \left.\frac{\partial P}{\partial \rho}\right)_{S,Y_k} . \tag{22}$$

The thermicity components σ_k are the nondimensional energy release associated with production of species k. For an ideal gas mixture, these components are:

$$\sigma_k = \frac{W}{W_k} - \frac{H_k}{C_p T} \tag{23}$$

where W is the mixture average molar mass. The thermicity components are the coupling coefficients that determine the interaction between chemical reaction and gasdynamics. The first term represents the contribution of the changes in the number of mols and the second term represents the changes in the enthalpy of the mixture. Net exothermic and mol producing reactions result in positive values of σ_k, net endothermic and mol reducing reactions have negative values of σ_k. Note that the adiabatic change equation is the reacting flow extension of the isentropic relationship $dP = a^2 d\rho$ that is frequently used in nonreacting compressible flow analyses.

Vorticity can be expressed in natural coordinates as

$$\nabla \times u = u\frac{\partial \theta}{\partial s} - \frac{\partial u}{\partial n} \tag{24}$$

so that the irrotational flow condition $\nabla \times u = 0$ will be

$$u\frac{\partial \theta}{\partial s} = \frac{\partial u}{\partial n} \tag{25}$$

Further, if wave front is straight, then there is translational invariance along the wave and $\partial/\partial\zeta = 0$, which yields:

$$\frac{\partial}{\partial n} = \frac{-1}{\tan(\beta - \theta)}\frac{\partial}{\partial s} \tag{26}$$

Combining the conservation equations with the irrotational and translational invariance conditions results in the reaction zone structure equations in natural coordinates:

momentum:

$$\rho u \frac{\partial u}{\partial s} = -\frac{\partial P}{\partial s} \tag{27}$$

continuity:

$$\frac{\partial(\rho u)}{\partial s} = \frac{\rho u}{\tan(\beta - \theta)} \frac{\partial \theta}{\partial s} \tag{28}$$

irrotational:

$$u \frac{\partial \theta}{\partial s} = \frac{-1}{\tan(\beta - \theta)} \frac{\partial u}{\partial s} \tag{29}$$

adiabatic:

$$\frac{\partial P}{\partial s} = a^2 \frac{\partial \rho}{\partial s} + \rho a^2 \sum_k \sigma_k \frac{\partial Y_k}{\partial s} \tag{30}$$

species:

$$\rho u \frac{\partial Y_k}{\partial s} = W_k \omega_k \quad k = 1, \ldots, K \tag{31}$$

In general, a numerical solution of structure equations will have to be obtained if realistic reaction mechanisms and rates are considered. However, for a straight oblique wave, the natural coordinate formulation is unnecessarily complex. A simpler formulation can be obtained by incorporating the invariance of the transverse velocity v. Then we only have to consider variations in the flow velocity w normal to the wave. This is facilitated by defining the Mach number M_n of the flow normal to the wave front

$$M_n = M \sin(\beta - \theta) \tag{32}$$

where $M = u/a$ and the sonic parameter η is

$$\eta = 1 - M_n^2 \tag{33}$$

It is also conventional to carry out the numerical integration with Lagrangean time τ as the independent variable rather than the spatial coordinate. The conversion between Lagrangean time and distance along the streamlines is

$$\frac{ds}{d\tau} = u \tag{34}$$

The contribution of all the thermicity terms can be represented as a sum

$$\dot{\sigma} = \sum_k \sigma_k \frac{dY_k}{d\tau} \tag{35}$$

The Lagrangean-time version of the structure equations are:

$$\frac{dP}{d\tau} = -\rho w^2 \frac{\dot{\sigma}}{\eta} \tag{36}$$

$$\frac{d\rho}{d\tau} = -\rho \frac{\dot{\sigma}}{\eta} \tag{37}$$

$$\frac{dY_k}{d\tau} = W_k \frac{\dot{\omega}_k}{\rho} \quad k = 1, \ldots, K \tag{38}$$

The initial conditions for these equations are found by solving the jump conditions for the state on the frozen shock adiabat.

The question of flow deflection can be examined by combining these equations to obtain

$$\frac{\partial \theta}{\partial s} = -\sin 2(\beta - \theta) \frac{\dot{\sigma}}{\eta} \tag{39}$$

For the situation shown in Fig. 2, the argument of the sine function is positive and less than π. The rate of change of θ along the streamline is therefore opposite in sign to the thermicity $\dot{\sigma}$. This supports the previous assertion that the flow will turn away from the wave for an exothermic reaction and toward the wave for an endothermic reaction *when* the component of the flow normal to the wave is subsonic. This is always the case for the initial portion of the reaction zone behind a shock.

Note that these equations are identical to those (Fickett and Davis 1979) for steadily-propagating detonations with the idealized ZND structure. This establishes the correspondence between the reaction zone structure of idealized oblique waves and normally-propagating waves. The reaction zone structure of a planar, oblique wave is just that of the planar, normal wave rewritten in terms of coordinates relative to the wave front. Note that the introduction of upstream disturbances (Jackson, et al. 1990, Lasseigne and Hussaini 1993) or the curvature of the wave front (Bdzil and Stewart 1989) will invalidate this correspondence. In the case of a weakly-curved wave, an additional term can be introduced into the continuity equation to describe the effect of curvature on the flow within the reaction zone.

The known results (Fickett and Davis 1979) about the existence of solutions to the structure equations for the normally propagating wave can now be applied to the oblique wave problem. In particular, it is known that the minimum normal velocity w_1 for which solutions exist may be determined by the vanishing of the sonic parameter η rather than the CJ condition. Note that behind a shock wave, the normal component of the flow is always subsonic so that η will start out positive. If the reactions are exothermic then P and ρ will both decrease with increasing τ and vice versa for endothermic reactions. For exothermic reactions, the normal component Mach number will increase with increasing τ so that η will decrease and can approach zero as the flow comes to equilibrium. In order that the reaction zone structure equation solutions not be singular then $\dot{\sigma} = 0$ if $\eta = 0$ within the reaction zone. In general this will only occur for one particular value of the normal velocity, w_{min}.

If the reaction mechanism consists of an initially exothermic process followed by endothermic reactions, then the minimum normal velocity w_{min} will be larger than the CJ velocity. Effectively, only the energy released before the sonic point is added to the flow and plays a role in determining the minimum wave speed. For typical fuel-air mixtures, the computed minimum velocity is only slightly larger (3 to 7%) than the CJ velocity and the effects of instability and intrinsic unsteadiness mask this from being observed in propagating detonations. The actual observed minimum velocity of propagating detonations is usually 5 to 10% less than the computed CJ velocity. The is apparently due to the effects of the instability waves and losses to the boundaries. Most studies of detonations neglect the issue of reaction structure and assume that the minimum possible velocity corresponds to the CJ solution to the hydrodynamic model.

Nonsingular solutions with $w_1 < w_{min}$ are not possible for steady waves since $\dot{\sigma}$ will not vanish at the same time η does. Overdriven solutions $w > w_{min}$ will terminate with a subsonic normal velocity $1 > \eta > 0$. In the case of $w = w_{min}$, the flow can reach a supersonic normal velocity component at the end of the reaction zone by passing through the point $\eta = 0$. This corresponds to the lower portion of the detonation adiabat, below the CJ point. Reaching this is clearly an exceptional situation and ordinarily only the portion of the detonation adiabat above the CJ point is accessible. This is the reasoning behind terminating the detonation adiabats and polars at the CJ point in the previous discussion of the hydrodynamic model. Introducing wave curvature and upstream disturbances can reduce the minimum allowed velocity, since these influences effectively provide loss mechanisms that reduce the thermicity $\dot{\sigma}$.

3.5 Reaction Zone Structure Computations

In order to illustrate the nature of the solutions to these equations, numerical solutions have been computed (Shepherd 1986, Sabet 1990) using a detailed chemical mechanism for hydrogen-air combustion. The LSODE solver package (Hindmarsh 1983) for stiff ordinary differential equations was used together with the CHEMKIN subroutine package (Kee et al. 1990) for the chemical bookkeeping.

The ideal gas equation of state is used:

$$P = \rho R T \qquad R = \tilde{R}/W \qquad W = 1/\sum_{k=1}^{K} Y_k/W_k$$

where $\tilde{R}$ is the universal gas constant and W_k are the molar masses of each species k. The mixture specific enthalpy is given by:

$$H = \sum_{k=1}^{K} Y_k H_k(T)$$

where the individual species specific enthalpies H_k are computed from the NASA curve fits to the specific heat tabulations and the enthalpy of formation $\Delta H^0_{f,298}$ given in the JANAF compilation (Chase et al. 1985):

$$H(T) = \Delta H^0_{f,298} + \int_{298}^{T} C_p(T)\, dT$$

The reaction mechanism used is the standard hydrogen-air mechanism presented in Lutz et al. (1992). The reaction rate equations are of the modified Arrhenius form:

$$k = A[X_l]^{\nu_l}[X_m]^{\nu_m} T^n \exp\left(-E_a/\tilde{R}T\right)$$

where the pre-exponential term, A, the temperature exponent, n, and the activation energy, E_a, are given in Lutz et al. Reverse reaction rates are computed from forward rates by using detailed balancing and equilibrium constants. The molar concentrations of the species are denoted by $[X_i]$, and ν_l and ν_m are the stoichiometric coefficients of the species x_l and x_m in the elementary reaction formula:

$$\nu_l X_l + \nu_m X_m \longrightarrow \nu_i X_i + \nu_j X_j$$

Contributions of both forward and reverse reactions are included in the computation of the net molar reaction rate for each species.

Examples of the ZND structure for stoichiometric hydrogen-air reaction zones are shown in Fig. 11-13. The initial conditions of 550 K, 0.30 atm and a freestream velocity of 2900 m/s corresponds to combustor inlet conditions for a detonation-based propulsion system. The spatial variation of pressure P, temperature T and thermicity $\dot{\sigma}$ are shown for three degrees of overdrive $w/w_{CJ} = 1.09, 1.4$, and 1.8.

The reaction zone structure for the near-CJ case consists of a characteristic delay or induction period followed by a rapid release of energy (sharp maximum in $\dot{\sigma}$) corresponding to the formation of water molecules. The temperature and pressure are approximately constant in the induction region. Since the reactions are primarily exothermic, the temperature increases and the pressure decreases with increasing distance through the reaction zone.

The reaction zone structure is significantly different for the highly overdriven cases. The thermicity $\dot{\sigma}$ has several maxima and for large enough overdrive, the pressure will increase and the temperature decrease with increasing distance through the reaction zone. This is due to the increasing importance of dissociation processes with increasing shock normal velocity and consequently, postshock temperature. The first peak in $\dot{\sigma}$ corresponds to the formation of the water molecules. The formation of water in this case does not increase the temperature significantly since the thermal and kinetic components of the energy of the flow are so large that the chemical heat release becomes negligible by comparison. The subsequent decrease in $\dot{\sigma}$ corresponds the formation of intermediates such as H, O and OH which have large positive heats of formation. The second peak in $\dot{\sigma}$ corresponds the leveling off of the concentrations of the OH radical, a slight decrease in the concentration of the H and O atoms, and a slight rise in the concentrations of H_2 and O_2.

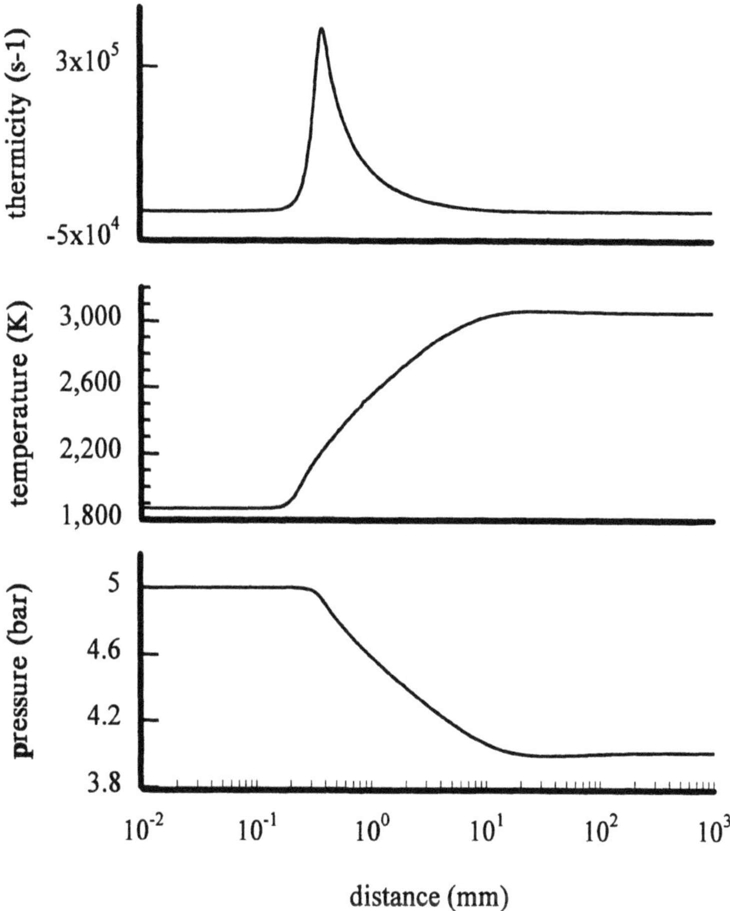

Figure 11: Calculated reaction zone structure for a near-CJ ($w = 1.09w_{CJ}$) stoichiometric hydrogen-air detonation with initial pressure of 0.3 bar and a temperature of 550 K.

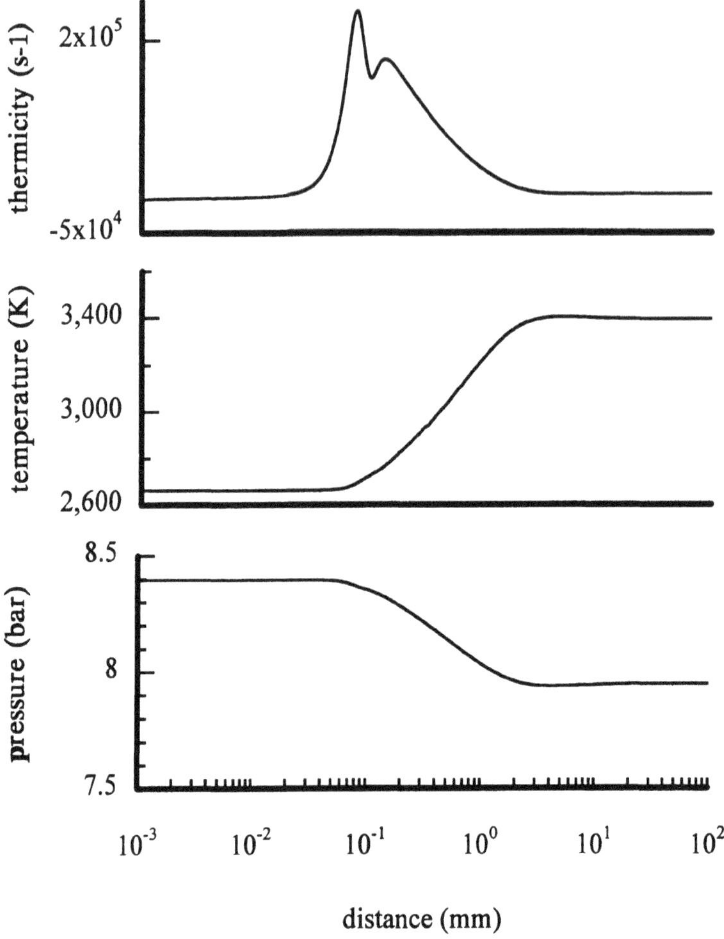

Figure 12: Calculated overdriven $(1.4w_{CJ})$ reaction zone structure for a stoichiometric hydrogen-air detonation with initial pressure of 0.3 bar and a temperature of 550 K.

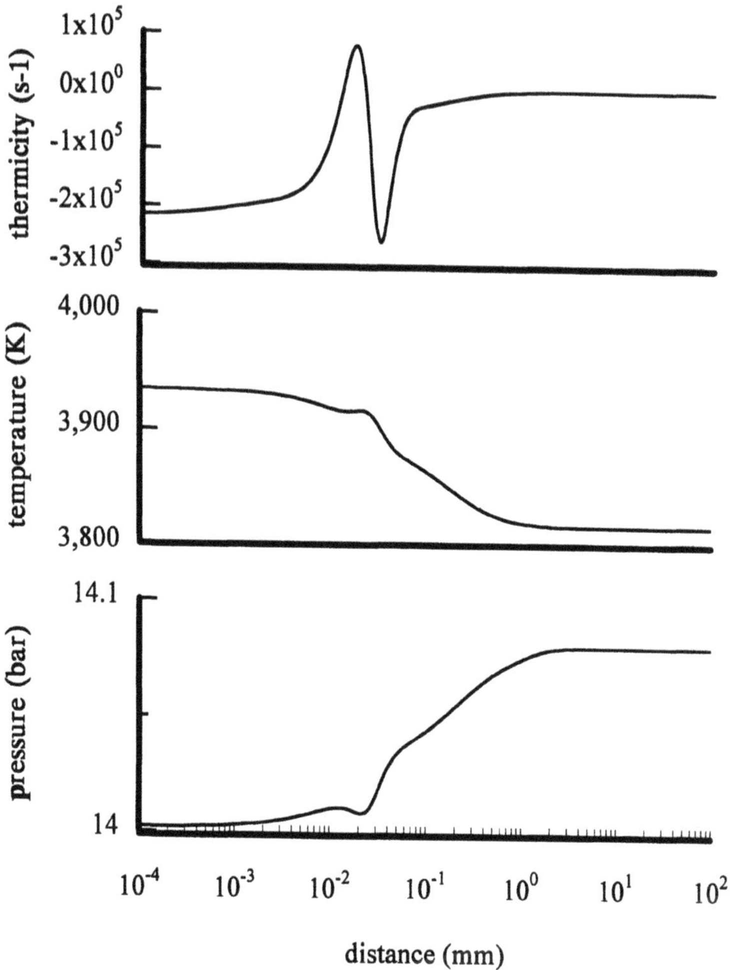

Figure 13: Calculated overdriven ($1.8w_{CJ}$) reaction zone structure for a stoichiometric hydrogen-air detonation with initial pressure of 0.3 bar and a temperature of 550 K.

4. Experiments on "Stationary" Detonation Waves

Four typical configurations that have been used to examine issues connected to stabilized or "stationary" detonations are shown in Fig. 14. Conceptually, flow over a wedge or a standing wave in a jet or nozzle is the simplest configuration. Early experiments were carried out by Gross and Chinitz (1960) using wedges and Nicholls et al. (1959) using jets. Although the terms "standing detonation wave" are often used in connection with these experiments, very little effect of combustion on the shock wave could be observed (Nicholls 1963) and the terms "shock-induced combustion" are more appropriate.

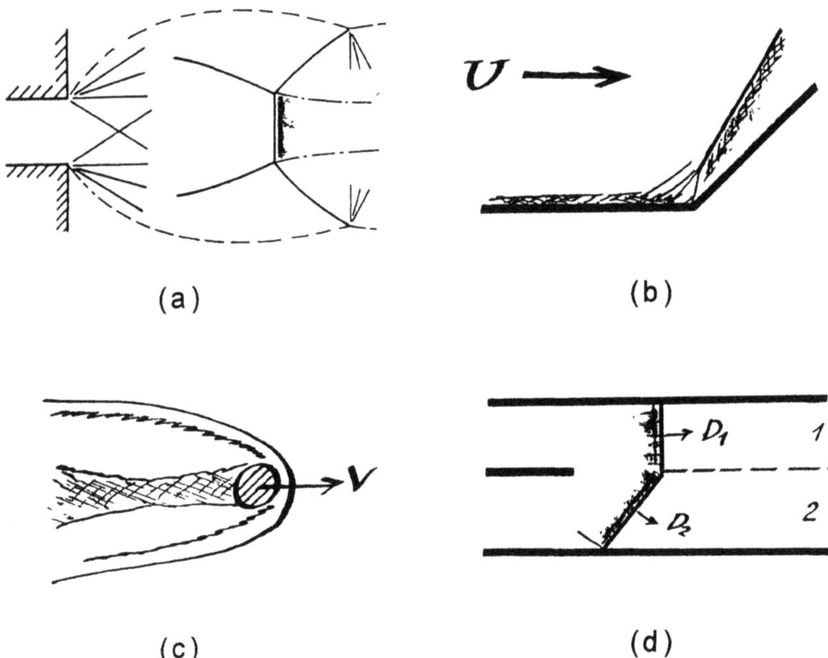

Figure 14: Four oblique detonation wave configurations. a) underexpanded supersonic jet with a Mach disk. b) wind or shock tunnel flow over a wedge. c) hypersonic projectiles d) two-layer propagating detonations.

The key problem is creating standing detonation waves is the difficulty of obtaining a fuel-air mixture with sufficient stagnation enthalpy H_o without getting preignition and burning upstream of the shock. A high stagnation enthalpy is needed so that the postshock temperature will be sufficient to initiate rapid chemical reactions and the energy release will couple with the shock front to create the detonation structure. The necessary stagnation enthalpy can be estimated by computing the stagnation states upstream of a normal CJ detonation. For a stoichiometric H_2-air mixture, $H_o \geq$ 2 MJ/kg and for stoichiometric H_2-O_2 mixture $H_o \geq 4.5$ MJ/kg. If the stagnation enthalpy is substantially less than these values, then the postshock temperatures will be very low, resulting in long reaction zone lengths and a decoupled reaction front and shock wave. In order to create a fuel-air mixture in a steady-flow supersonic wind tunnel with the required stagnation conditions, the air must be first heated and then mixed with the fuel after being cooled by expansion through the nozzle. Detonations can be established in wind tunnel test sections using mixtures with lower stagnation enthalpies, but the waves will be unsteady (Bellet and Deshayes 1970) since the Mach number in test section will be lower than the CJ value.

Early results that were interpreted as detonations (Gross and Chinitz 1960) were later shown (Rubins 1960) to be due to combustion upstream of the test section, near the fuel-injection point in the plenum. A similar combustion effect was found in the underexpanded jet studies (Nicholls et al. 1958). Later studies (Rubins and Rhodes 1964, Rhodes et al. 1964) eliminated this effect by relocating the fuel-injection point to a low–temperature portion of the flow. Oblique and normal shocks were produced in a nonuniform mixture of vitiated air (stagnation temperature of $\approx$ 2000 K) and hydrogen flowing with a Mach number of 3.1.

Combustion downstream of the shocks was observed in later tests, and this was properly interpreted as being shock-induced combustion rather than a detonation phenomena. This was due to the very narrow region of fuel-air mixture produced by the mixing of a fuel jet originating from an injector in the nozzle region of the tunnel. The reaction zone lengths in these experiments were comparable to the width of the jet. Under these conditions, little coupling would be expected between shock and the chemical reaction kinetics. Only when the flow behind the shock was confined by a tube or channel was significant coupling observed (Rubins and Cunningham 1965, Strehlow and Rubins 1969). However, no detonation-like structures were ever observed but only shock oscillations associated with choking within the duct.

Instead of attempting to create a high-speed, combustible flow, an alternative is to send a high-velocity projectile through a stationary flow. The first experiments of this type were carried out by Ruegg and Dorsey (1962) and later by Behrens et al. (1965), McVey and Toong (1971), Alpert and Toong (1972), and Lehr (1972). These experiments were carried out with different fuel-air and fuel-oxygen mixtures and small-diameter projectiles (10 to 20 mm diam.). In all cases, it appears that either the projectile was too small or the velocity was too low (less than the CJ value) to obtain detonations. In these situations, the reaction front decoupled from the shock front as shown in Fig. 15a.

Ruegg and Dorsey used stoichiometric hydrogen-air mixtures at 0.1, 0.25 and 0.5 atm. The spherical projectiles (20 mm diam.) were launched at various velocities between 1640 and 2665 m/s. Behrens et al. used 9 mm projectiles fired at velocities between 1500 to 3000 m/s into stoichiometric hydrogen-air mixtures at initial pressures of 0.25 and 0.55 atm. They demonstrated the correlation of the instability frequencies with the induction time in the stagnation region and also observed that most of the shocked gas is unburned due to the decoupling of the shock and reaction fronts that is shown in Fig. 15a. Although some velocities were in excess of the CJ velocity (1920 to 1960 m/s), the estimated cell widths (145 to 27 mm) were all substantially larger than the projectile so detonation would not be expected in either sets of experiments.

McVey and Toong fired 12.7 mm and 6.5 mm diam. projectiles into lean acetylene-oxygen mixtures at pressures from 50 to 200 Torr. Alpert and Toong examined 12.7 mm diam. spheres and cylinders fired into stoichiometric hydrogen-oxygen mixtures diluted with argon or nitrogen at pressures between 100 and 532 Torr. Similar conclusions about the cell width and projectile size apply to these experiments and explain the failure to observe a detonation mode of combustion.

Lehr's experiments came the closest to meeting the conditions for creating detonations but the projectile velocity was slightly too low even in the most favorable case. Among other tests, he fired 15 mm diam. sphere-cylinder projectiles into stoichiometric hydrogen-oxygen mixtures with an initial pressure of 186 Torr. Transient detonation waves were initiated for projectile velocities of 2160 and 2705 m/s but they were not stabilized on the projectiles since the highest velocity used was less than the CJ velocity of 2750 m/s. The cell width for this mixture is about 4 to 6 mm, less than half the width of the projectile. The characteristic cellular instability structure of a coupled reaction front-shock wave complex can be clearly observed in the schlieren photographs of these events (Lehr 1972). A hypothetical stabilized detonation configuration can be deduced from these tests and is shown in Fig. 15b.

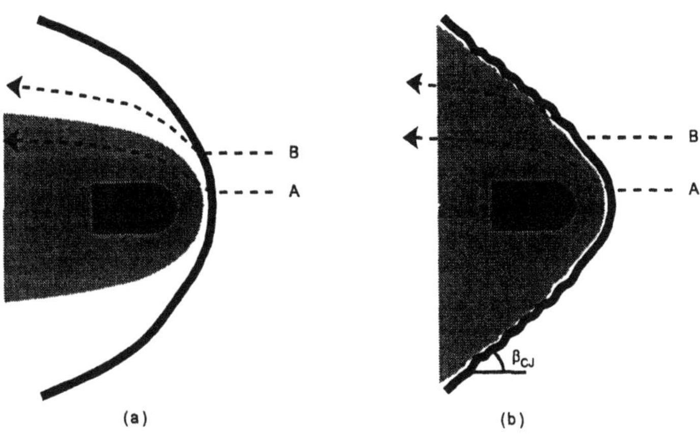

Figure 15: Combustion waves produced by hypersonic projectiles a) decoupled shock wave and reaction front configuration. b) hypothetical detonation mode configuration.

There have been a number of analyses of the flowfield produced by hypersonic projectiles in a detonable gas. The application of the hydrodynamic model of detonation and deflagration waves to this problem was extensively analyzed by Chernyi (Chernyi 1966, Chernyi and Gilinskii 1970, Gilinskii et al 1966) with a particular emphasis on self-similar flow fields about wedges and cones. Another aspect of this problem examined in these studies (Levin 1968, Chernyi et al. 1970) is the phenomenon of wave splitting and reaction quenching illustrated in Fig. 15a. This is particularly relevant to the problem of detonation initiation and stability for propulsion applications. As demonstrated by the experiments, the overdriven waves in the front of the projectile do not usually smoothly decay into the oblique Chapman-Jouguet waves far from the body but rather abruptly split into a shock and a trailing reaction zone that degenerates into a contact surface and eventually, can evolve into a flame. The expansion wave generated by the shoulder of the projectile produces the quenching action that is responsible for this effect. Oscillations of the shock and reaction front may occur during the decay processes. These oscillations are apparently precursors of the transverse instability observed for near-CJ detonation waves.

Sichel and Galloway (1967) and Galloway and Sichel (1969) demonstrated that a simple scaling parameter, $u_{\infty}\tau_{\sigma}/R_b$ can be used to classify the blunt body flows from a theoretical viewpoint. In this relation, R_b is the body radius and τ_{σ} is the characteristic chemical reaction time. For small values of this parameter, the hydrodynamic model of a detonation wave was useful and far from the projectile, the wave is predicted to asymptote to the CJ condition. For larger values, splitting and quenching of the reaction front is to be expected. This parameter is essentially a measure of the rate of quenching relative to the reaction rate. Models (Fickett and Davis 1979, Bdzil and Stewart 1989) of quasi-steady curved detonations also predict the existence of a critical wave curvature κ. Steady waves cannot exist if the curvature exceeds some critical level κ_f. Recognizing that the maximum value of $\kappa \propto 1/R_b$, we infer that wave splitting would be expected whenever R_b is significantly smaller than $1/\kappa_f$. We expect that these considerations could play an important role in determining the characteristics of stabilized curved detonation waves in combustors.

A number of researchers (McVey and Toong 1971, Matsuo and Fujiwara 1993, Wilson and Sussman 1993) have examined the periodic instabilities that occur in these flows. These instabilities are a consequence of the reflection of waves between the shock front, reaction front and the projectile. Convection of these disturbances into the wake of the body produce a striking periodic pattern. These patterns are not to be confused with the cellular instability of detonations. The leading shock waves in these experiments are smooth and free from any transverse instabilities since the reaction front lags far behind the shock in the region far from the projectile.

Oblique detonations have actually been observed in only one type of experiment, the two-layer detonations of Liu et al. 1987, Liu et al. 1988 and Dabora et al. 1991. A channel is divided longitudinally by a rigid partition terminating in a flexible film or diaphragm, Fig. 14d. The upper portion of the channel is filled with the "primary" explosive that has a high detonation velocity and the lower portion of the channel is filled with a "secondary" explosive with a lower detonation velocity. A detonation is initiated in the primary layer and propagates in the upper channel toward the film or diaphragm region. When the primary detonation propagates over the film or diaphragm, the high pressure gas behind primary detonation drives the film or diaphragm into the secondary layer with an oblique piston action. This produces a shock wave in the secondary explosive and initiates through a complex system of waves (Oran et al. 1992), an oblique propagating detonation in the secondary explosive. Reflections from the lower boundary of the secondary layer may play a significant role in the initiation process and a Mach stem may be created in the shock or detonation reflection process.

After the transient initiation processes have ceased, the oblique detonation wave angle can be predicted (Liu et al. 1987) by a simple steady flow analysis in the frame of the shock intersection located at the boundary between the primary and secondary layers. In general, a reflected wave (usually an expansion) is produced in the products of the primary explosive in order to match the flow deflection angle and pressure at the contact surface between the products of the primary and secondary explosive. This simple picture is complicated by the presence of an initiation region near the boundary and possibility of an irregular refraction process if the CJ detonation velocity is higher in the secondary explosive than in the primary.

Oblique detonations are not always produced in these experiments. In order to get coupling of the reaction front with the oblique shock, the reaction zone must be small enough compared to the lateral extent of the layer. The initiating shock must also be sufficiently strong to cause initiation without too much delay. The reaction zone length in the primary explosive must also be sufficiently small compared to the channel width so that the expansion wave created by the interaction does not quench the detonation in the primary layer. This is related to the critical tube diameter problem of propagating detonations that was mentioned previously. Failure to initiate the secondary detonation and in some cases, failure of the primary wave itself may result if these conditions are not satisfied.

Oblique detonations were produced using hydrogen-oxygen mixtures with a rich primary mixture and a lean secondary mixture by Liu et al. 1987 and Liu et al. 1988. The lateral extent of each channel was about 16 mm and the two mixtures were initially separated by a thin cellulose layer in the interaction region. A schlieren photograph of the interaction is shown in Fig. 16 from a more recent study by Tonello and Sichel (1993). Note the trailing transverse waves characteristic of cellular instability structure on both the primary and oblique waves. There is also a pronounced initiation shock near the boundary and the reflected waves generated by the interactions are clear. Interaction with a tertiary layer at the bottom of the channel is also visible in this photograph but not germane to the present discussion. The secondary wave angle abruptly increases due to the pressure waves generated by the explosion of the fluid elements in the secondary material close to the primary-secondary contact surface.

Figure 16: Schlieren photograph of the oblique detonation wave produced by a primary explosive of $2H_2 + 1/2O_2$ diffracting into a secondary explosive of $0.57H_2 + 1/2O_2$. Both explosives are at an initial pressure of 1 atm and a temperature of 300 K. From Tonello and Sichel 1993.

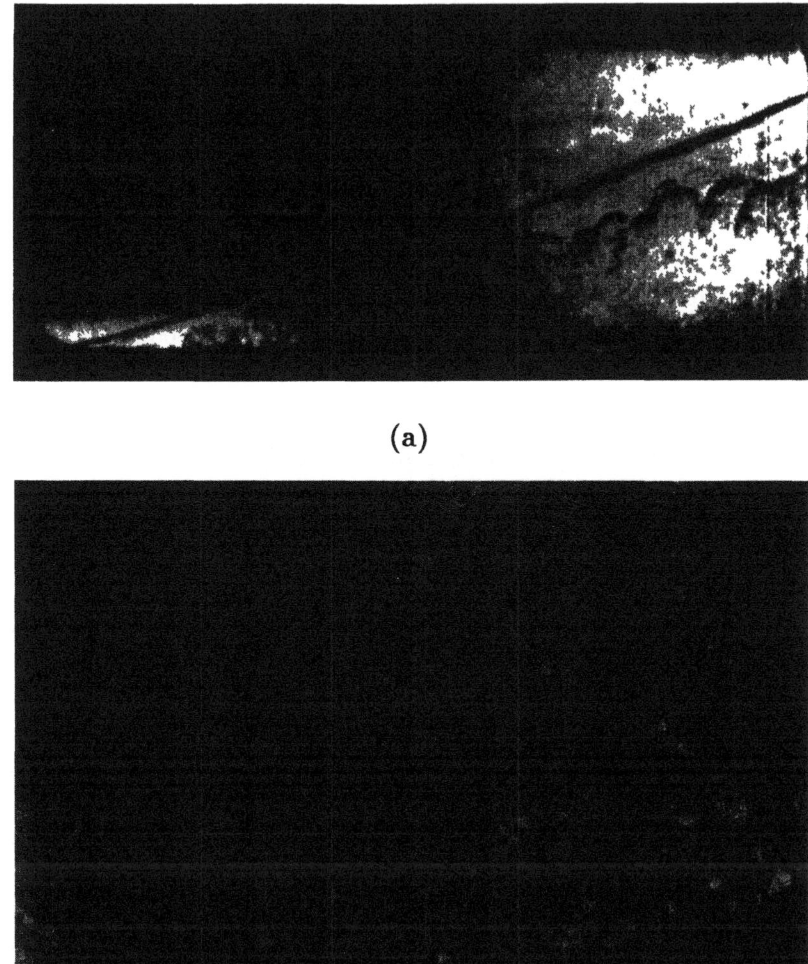

(a)

(b)

Figure 17: Schlieren photographs (Dabora et al. 1991) of the oblique waves produced by a primary explosive of $C_4H_2 + 3O_2 + 8He$ diffracting into a secondary explosive of $C_2H_2 + 2.52O_2 + 10.5Ar$. a) layers separated by a 38 μm aluminum film, no detonation in secondary layer. b) 2.5 μm aluminized film, oblique detonation in secondary layer.

An example in which a failure to obtain an oblique detonation is shown in Fig. 17a. Dabora et al. (1991) used a primary mixture of $C_2H_4 + 3O_2 + 8He$ ($U_{CJ} = 3050$ m/s) to drive an Al foil into a mixture of $C_2H_2 + 2.5O_2 + 10.5Ar$ ($U_{CJ} = 1750$ m/s). Each layer is approximately 56 mm high. Only an oblique shock is observed in the secondary layer with 38 μm thick foil. When the foil thickness is reduced to 2.5 μm aluminized mylar, an oblique detonation complete with the characteristic instability waves is observed (Fig. 17b). Broda and Dabora (1993) have demonstrated that thick foils slowly accelerate due to their large specific inertia. This produces a weak initial oblique shock that fails to initiate a detonation.

5. Waves on Wedges, Instabilities and Initiation

The simplest concept for creating oblique detonations is to just introduce a wedge or ramp into a uniform flow. Chernyi (1966) considered this flow and the axisymmetric analog, flow over a cone, in some detail. Pratt et al. (1991) considered the two-dimensional flow, particularly the situation with a uniform flow downstream. In addition, there have been a number of recent theoretical studies (Li et al. 1993, Powers and Stewart 1992, Cambier et al. 1989, Buckmaster and Lee 1990, Buckmaster 1990) of the two-dimensional flow over a wedge or a cone.

The possible steady oblique wave configurations can be determined by analyzing the shock polars (Figs. 8 and 9). The results are summarized in Fig. 18a on the velocity-flow deflection plot. First of all, the incoming flow velocity u_1 must be greater than the CJ velocity w_{CJ} in order for a stabilized detonation to exist. Otherwise, an unsteady detonation wave or shock-induced combustion will result. If the velocity is large enough, then there are several regime of stabilized waves, depending on the wedge angle θ_w.

A straight wave with uniform flow downstream (Fig. 18b) is possible only if the wedge angle is compatible with the flow deflection angle θ. Referring to the polar plot Fig. 8, this will occur only for $\theta_{CJ} < \theta_w < \theta_{max}$, which is a function of the incoming flow velocity as shown in Fig. 18. This is the case considered in most studies. Note this region is rather narrow near the CJ velocity but is quite substantial at higher velocities. However, for large u_1 and θ_w, solutions are obtained (region to the right of dashed line in Fig. 18a) for which the reaction zone is net endothermic, an undesirable situation for propulsion purposes.

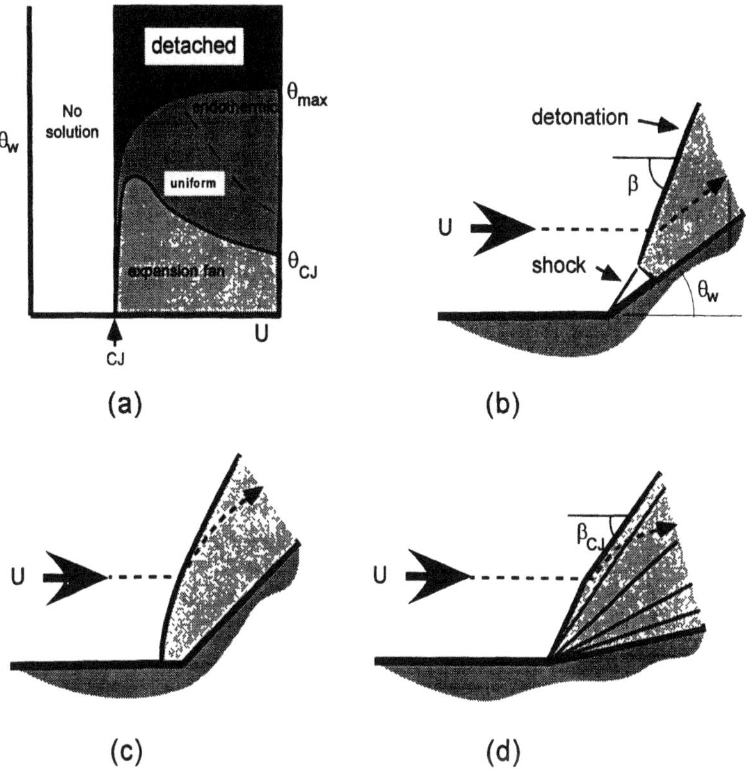

Figure 18: Regimes of oblique detonation stabilization on wedges. a) possible configurations as determined from a polar analysis. b) transition to uniform downstream flow after a shock initiation transient. c) detached wave over a steep wedge. d) CJ wave followed by an expansion fan.

Solutions with curved waves and nonuniform flow downstream are possible for wedge angles outside this range. For steep wedges, $\theta_w > \theta_{max}$, the wave will be detached (Fig. 18c) just as in ordinary shock waves over blunt bodies. For shallow wedges, $\theta_w < \theta_{CJ}$, the detonation wave will be followed by an expansion fan (Fig. 18d) and the wave angle will be given by the CJ value. This is the steady analog of the self-sustained detonation. The conical version of this flow was examined by Chernyi (1966) and more recently in the context of propulsion by Carrier et al. (1992).

The polar discussion is based on the notion of a thin wave with negligible reaction zone structure effects. Reaction zone effects will be of two sorts. First, the detonation wave will possibly be unstable. Second, there will be some sort of initiation transient.

The usual transverse wave instability of propagating detonations is certainly possible and has been observed on oblique detonations in the two-layer experiments discussed previously. Based on experience with propagating waves, we speculate that unless the cell width becomes comparable to the combustor transverse dimension, this instability will not cause the detonation wave to fail. The extent of the instability depends crucially on the amount of overdrive, w/w_{CJ}. Experiments with overdriven detonations (Meltzer et al. 1993) indicate that the cell width decreases by a factor of 10 with an overdrive ratio of 1.35 and the instability is completely suppressed at an overdrive ratio of 1.4. Recent experiments with the two-layer configuration (Broda and Dabora 1993) indicate a similar decrease in cell width with overdrive. Although a slight overdrive might be beneficial in a propulsion system, a significant loss in performance for a propulsion system may be incurred by operating at too high an overdrive. The total pressure loss across the wave increases rapidly with overdrive and therefore the net thrust will decrease.

The initiation transient is observed as a precursor shock (see Figs. 16 and 17) in the two-layer experiments discussed previously. A similar precursor would be expected in front of the wedge, as indicated in Fig. 18b. Structures of this sort are visible in the computations of Li et al. (1993). Inspection of the polar curves (Figs. 8 and 9) reveals that a nonreactive shock with a given flow deflection angle is much weaker than the leading shock portion of a detonation with a given flow deflection angle in the products. In effect, the exothermic reaction increases the effective wedge angle. This has rather severe implications for starting and maintaining oblique detonations. Since the key reaction rates depend strongly on temperature, a weak precursor shock will result in a much longer reaction zone (up to 10^5 times larger than for near-CJ waves!) than behind the leading shock of a steady oblique detonation wave on the same wedge.

Consequences of this include postshock reaction zones that are so long in comparison to the combustor width that no detonation is ever established. Another possibility is that the detonation wave is catastrophically unstable to disturbances that decouple the leading shock from the energy-releasing portion of the reaction zone. Such decoupling occurs in the projectile flows (Fig. 15a) and is another manifestation of the sensitivity of the reaction zone length to post-shock temperature. This type of instability effect for wedge flows has been analyzed by Buckmaster (1990) and a polar analysis of the decoupled configuration was given by Buckmaster and Lee (1990). Clearly, some special measures will be needed to initiate the detonation wave in this type of flow. External initiation or programmed excursions in the ramp angle may be required in order to successfully start an oblique detonation. These considerations may also apply to certain types of numerical simulations.

6. Conclusions

There are many unresolved issues related to using detonation waves as the basis combustion process in a hypersonic propulsion system. A few of these have been examined in this paper. The theoretical ideas that have been presented are rather rudimentary and focus on idealized models of detonation wave structure. It is apparent that resolution of many issues requires investigation into unsteady and multidimensional flows. Sophisticated experimentation and numerical simulation will certainly be required to make further progress. However, the simple considerations of this paper indicate that there are some key parameters and physical phenomena that should be carefully considered in future studies.

The role of the reaction zone structure and the spontaneous instability of detonation waves appear to be crucial factors that complicate any study, experimental or numerical. The reaction zone length or detonation cell width provide key scaling parameters that should be considered in the choice of experimental setup or computational domain. The characteristic dimensions of the combustor should be large enough to accommodate at least several cell widths λ or several hundred times the ZND reaction zone length Δ. In addition, the upstream flow velocity and stagnation enthalpy must be sufficiently high in order to initiate and maintain a stabilized detonation. Flow disturbances must be minimized in order not to cause decoupling and catastrophic failure. Finally, even with carefully designed experiments, there may be more than one steady state depending on the nature of the initiation process.

Acknowledgements

E. K. Dabora and M. Sichel generously provided photographs and shared the results of their oblique detonation wave experiments with me. I would like to thank P. M. Rubins for providing his unique perspective and papers about the early research on shock-induced combustion. J. H. S. Lee pointed out many connections and provided references to related work. B. McBride graciously provided an early version of the NASA high-temperature thermodynamic property fits.

Several of my students, J. Meltzer, R. Akbar, and A. I. Sabet, have contributed through their studies on detonation-related problems. In particular, A. I. Sabet did many of the computations on oblique wave polars and reaction zone structures. Most of this work and writing was carried out at Rensselaer Polytechnic Institute in Troy, NY. A. Kapila, M. Y. Hussaini, J. Quirk, and T. Jackson provided encouragement and a forum for these ideas at ICASE.

References

Alpert, R. L. and Toong, T. Y., 1972. "Periodicity in Exothermic Hypersonic Flows about Blunt Projectiles," *Astronautica Acta* **17**, pp. 539–560.

Atamanchuk, T., and Sislian, J., 1990. "On-and-Off Design Performance Analysis of Hypersonic Detonation Wave Ramjets," 26th Joint Propulsion Conference, AIAA-90-2473.

Atamanchuk, T., and Sislian, J., 1991. "Hypersonic Detonation Wave Powered Lifting-Propulsive Bodies," 3rd International Aerospace Planes Conference, AIAA-91-5010.

Bach, G. G., Knystautas, R., and Lee J. H., 1969. " Direct initiation of spherical detonation in gaseous explosives," *12th Symp. Comb.*, The Combustion Institute, Pittsburgh, PA, pp. 853-864.

Bdzil, J. B., and Stewart, D. S., 1989. "Modeling Two-Dimensional detonations with detonation shock dynamics," *Phys. Fluids* **A1**(7), pp. 1261-1267.

Behrens, H., Struth, W., and Wecken, F., 1965. "Studies of hypervelocity firings into mixtures of hydrogen with air or oxygen,"

10th Symp. (Intl.) Comb., The Combustion Institute, Pittsburgh, PA, pp. 245-252.

Bellet, J. C., and Deshayes, G., 1970. "Structure and Propagation of Detonations in Gaseous Mixtures in Supersonic Flow," *Astronautica Acta* **15**, pp. 465-469.

Bourlioux, A. and Madja, A. J., 1992. " Theoretical and Numerical Structure for Unstable Two-Dimensional Detonation," *Combust. Flame* **90**, pp. 211-229.

Broda, J.–C., and Dabora, E. K., 1993. "Recent Experimental Results on Oblique Detonation Waves," Presented at the 14th ICDERS Symposium, August 1993, Coimbra, Portugal.

Buckmaster, J. D. 1990. "The Structural Stability of an Oblique Detonation Wave," *Combust. Sci. Tech.* **72**, 283-296.

Buckmaster, J. D., and Lee, C. J., 1990. "Flow Refraction by an Uncoupled Shock and Reaction Front," *AIAA J.* **28**, pp. 1310-1312.

Cambier, J.-L., Adelman, H. G., Menees, G. P., 1989. "Numerical Simulations of Oblique Detonations in Supersonic Combustion Chambers", *J. Propulsion Power*, **5**(4), pp. 482-491.

Carrier, G., Fendell, F., McGregor, R., Cook, S., and Vazirani, M., 1992. "Laser–Initiated Conical Detonation Wave for Supersonic Combustion," *J. Propulsion Power* **8**, pp. 472-480.

Chase, M. W., Jr., Davies, C. A., Downey, J. R., Jr., Frurip, D. J., McDonald, R. A., and Syverud, A. N., 1985. **JANAF Thermochemical Tables**, Third Edition. *J. Phys. Chem. Ref. Data* **14**, Supplement No. 1.

Chernyi, G. G., 1966. "Self-similar Problems of Combustible Gas Mixture Flow Past Bodies," *Mekhanika Zhidkosti i Gaza* (translated in english as Fluid Dynamics) **1**(6), pp. 10-24.

Chernyi, G. G., and Gilinskii, S. M., 1970. "High-Velocity Motion of Solid Bodies in Combustible Gas Mixtures," *Astronautica Acta* **15**, pp. 539-545.

Chernyi, G. G., Korobeinikov, V. P., Levin, V. A., and Medvedev, S. A., 1970. "One-Dimensional Unsteady Motion of Combustible Gas Mixtures Associated with Detonation Waves," *Astronautica Acta* **15**, pp. 259-256.

Dabora, E. K., Desbordes, D., Guerraud, C., Wagner, H. Gg., 1991. "Oblique Detonations at Hypersonic Velocities" *Prog. Aero. Astro.* **133**, 187-204.

Dunlap, R., Brehm, R. L., and Nicholls, J. A., 1958. "A Preliminary Study of the Application of Steady State Detonative Combustion to a Reaction Engine," *Jet Propulsion* **28**, pp. 451-456.

Eidelman, S., Grossman, W., and Lottati, I. 1991 "Review of Propulsion Applications and Numerical Simulation of the Pulsed Detonation Engine Concept," *J. Propulsion Power* **7**, pp. 857–865.

Fickett, W. and Davis, W. C., 1979. **Detonation**, University of California Press, Berkeley, California.

Galloway, A. J., and Sichel, M., 1969. "Hypersonic Blunt Body Flow of H_2-O_2 Mixtures," *Astronautica Acta* **15**, pp. 89-105.

Gilinskii, S. M., Zapryanov, Z. D., and Chernyi, G. G., 1966. "Supersonic flow of a Combustible Gas Mixture Past a Sphere," *Mekhanika Zhidkosti i Gaza* (translated in english as Fluid Dynamics) **1**(5), pp. 8-13.

Gordon, S. and McBride, B. J., 1976. "Computer Program for the Calculation of Complex Chemical Equilibrium Compositions, Rocket Performance, Incident and Reflected Shocks and Chapman-Jouguet Detonations." NASA SP-273.

Gross, R. A. and Chinitz, W., 1960. "A Study of Supersonic Combustion," *J. Aero. Science* **27**(7), pp. 517-524.

Harradine, D. M., Lyman, J. L., Oldenborg, R. C., Schott, G. L., Watanabe, H. H., 1990. "Hydrogen/Air Combustion Calculations: The Chemical Basis of Efficiency in Hypersonic Flows" *AIAA J.* **28**, pp. 1740-1744.

Hindmarsh, A. C., 1983. "ODEPACK, A Systemized Collection of ODE Solvers," *IMACS Trans. on Scientific Computation* **1**, pp. 55-64.

Jackson, T. L., Kapila, A. K., and Hussaini, M. Y., 1990. "Convection of a pattern of vorticity through a reacting shock wave," *Phys. Fluids* **A2** 1260-1268.

Kee, R. J., Rupley, F. M., and Miller, J. A., 1989. "CHEMKIN II: A FORTRAN Chemical Kinetics Package for the Analysis of Gas-Phase Chemical Kinetics," Sandia National Laboratories Report SAND90-8009.

Lasseigne, D. G., and Hussaini, M. Y., 1993. "Interaction of disturbances with an oblique detonation wave attached to a wedge," *Phys. Fluids* **A5**, 1047-1058.

Lee, J. H., 1984. "Dynamic Parameters of Gaseous Detonations," *Ann. Rev. of Fluid Mech.* **16**, pp. 311-336.

Lee, H., and Stewart, D. S., 1990. "Calculations of Linear Detonation Instability," *J. Fluid Mech.* **216**, pp. 103-132.

Lehr, H. F., 1972. "Experiments on Shock-Induced Combustion," *Astronautica Acta* **17**, pp. 589-597.

Levin, V. A., 1968. "Transition of a Plane Overdriven Detonation Wave to the Chapman Jouguet Regime" *Mekhanika Zhidkosti i Gaza* (translated in english as Fluid Dynamics) **3**(2), pp. 50-55.

Liepmann, H. W. and Roshko, A., 1957. **Elements of Gasdynamics**, John Wiley, New York.

Li, C., Kailasanath, K., and Oran, E. S., 1993. "Structure of Reaction Waves Behind Oblique Shocks," *Prog. Astro. Aero.* **153**, pp. 231-240.

Liu, J. C., Liou, J. J., Sichel, M., Kaufmann, C. W., and Nicholls, J. A., 1987. "Diffraction and Transmission of a Detonation into a Bounding Explosive Layer," *21st Symposium (International) on Combustion*, The Combustion Institute, Pittsburgh, PA, pp. 1639-1647.

Liu, J. C., Sichel, M., and Kaufmann, C. W., 1988. "The lateral interaction of detonating and detonable gaseous mixtures," *Prog. Astro. Aero.* **114**, AIAA, pp. 264-283.

Lutz, A. E., Kee, R. J., Miller, J. A., Dwyer, H. A., and Oppenheim, A. K., 1988. "Dynamic Effects of Autoignition Centers for Hydrogen and $C_{1,2}$-Hydrocarbon Fuels," *22nd Symp. (Int.) on Combust.*, pp. 1683-1693, The Combustion Institute, Pittsburgh, PA.

Matsuo, A. and Fujiwara, T., 1993. "Numerical Prediction of Mechanism on Oscillatory Instabilities in Shock-Induced Combustion," *Progress in Astonautics and Aeronautics* **154**, AIAA, Washington, DC, pp. 516-531.

McVey, J. B. and Toong, T. Y ., 1971. "Mechanism of Instabilities of Exothermic Hypersonic Blunt-Body Flows," *Combust. Sci. Tech.* **3**, pp. 63-76.

Meltzer, J., Shepherd, J. E., Akbar, R., Sabet, A., 1993. "Mach Reflection of Detonation Waves," *Progress in Astronautics and Aeronautics* **153**, AIAA, Washington, DC, pp. 78-94.

Murthy, S. N. B., and Curran, E. T., Editors. 1991 **High-Speed Flight Propulsion Systems** in *Progress in Astronautics and Aeronautics* **137**, AIAA, Washington, DC.

Nicholls, J. A., Dabora, E. K., and Gealer, R. L., 1959. "Studies in Connection with Stabilized Gaseous Detonation Waves," *7th Symp. (Intl.) Comb.*, Butterworths, London, pp. 766-772.

Nicholls, J. A., 1963. "Standing Detonation Waves," *9th Symp. (Intl.) Comb.*, Academic Press, pp. 488-498.

Oran, E. S., Jones, D. A., and Sichel, M., 1992. "Numerical simulations of detonation transmission," *Proc. Roy. Soc. Lond. A* **436**, pp. 267-297.

Ostrander, M. J., Hyde, J. C., Young, M. F., Kissinger, R. D., and Pratt, D. T., 1987. "Standing Oblique Detonation Wave Engine Performance," AIAA paper 87-2002.

Powers, J. M., and Stewart, D. S., 1992. "Approximate Solutions for Oblique Detonations in the Hypersonic Limit," *AIAA J.* **30**, pp. 726-736.

Pratt, D. T., Humphrey, J. W., Glenn, D. E, 1991. "Morphology of Standing Oblique Detonation Waves," *J. Propulsion Power* **7**(5), pp. 837–845.

Reynolds, W. C., 1986. "The Element Potential Method for Chemical Equilibrium Analysis: Implementation in the Interactive Program STANJAN; Version 3", Department of Mechanical Engineering, Stanford University.

Rhodes, Jr., R. P., Rubins, P. M., and Chriss, D. E., 1964. "Effect of Heat Release on Flow Parameters in Shock Induced Combustion," *SAE Transactions* **72**, pp. 87-95.

Rubins, P. M., 1960. "Installation and Calibration of a Supersonic Combustion Tunnel," AEDC-TN-60-162.

Rubins, P. M., and Rhodes, Jr., R. P., 1964. "Shock-Induced Combustion with Oblique Shocks: Comparison of Experiment and Kinetic Calculations," *AIAA J.* **1**(12), pp. 2778-2784.

Rubins, P. M., and Cunningham, T. H. M., 1965. "Shock-Induced Supersonic Combustion in a Constant-Area Duct," *J. Spacecraft Rockets* **2**(2), pp. 199-205.

Ruegg, F. W. and Dorsey, W. W., 1962. "A Missile Technique for the Study of Detonation Waves," *J. Research of the NBS* **66C**(1), pp. 51-58.

Sabet, A. I., 1990."Investigation of Equilibrium and Chemical Kinetic Behaviour of Detonation Waves in Hydrogen-Air Mixtures and an Evaluation of the Oblique Detonation Wave as the Combustor for a Scramjet", Masters Thesis, Rensselaer Polytechnic Institute, Troy, New York.

Sangiovanni, J. J., Barber, T. J., and Syed, S. A., 1993. "Role of Hydrogen/Air Chemistry in Nozzle Performance for a Hypersonic Propulsion System," *J.Propulsion Power* **9**, 134-138.

Sargent, W. H., and Gross, R. A., 1960 "Detonation Wave Hypersonic Ramjet," *ARS J.* **30**, pp. 543-549.

Shepherd, J. E., 1986. "Chemical Kinetics of Hydrogen-Air-Diluent Detonations," *Prog. Astro. Aero.* **106**, pp. 263–293.

Sichel, M., and Galloway, A. J., 1967. "Regimes of Exothermic Blunt Body Flow," *Astronautica Acta* **13**, pp. 137-145.

Strehlow, R. A., 1984. **Combustion Fundamentals**, McGraw-Hill, NY.

Strehlow, R. A. and Rubins, P. M., 1969. "Experimental and Analytical Study of H_2-Air Reaction Kinetics Using a Standing Normal Shock Wave," *AIAA J.* **7**, pp. 1335-1344.

Thompson, P. A., 1972. **Compressible Fluid Dynamics**, McGraw-Hill, NY.

Tonello, N. A. and Sichel, M., 1993. Personal communication. See also Tonello, N. A., Sichel, M., and Kaufmann, C. W., "Mechanisms of Detonation Transmission in Layered H_2-O_2 Mixtures" Presented at the 14th ICDERS Symposium, August 1993, Coimbra, Portugal.

Voytsekhovskiy, B. V., Mitrofanov, V. V., and Topchiyan, M. Ye., 1964. "Struktura Fronta Detonatskii v Gazakh," Novosibirsk. Available in english as "The structure of a Detonation Front in Gases," Air Force Translation FTD-MT-64-527.

Westbrook, C. K. and Urtiew, P. A., 1982. "Chemical Kinetic Prediction of Critical Parameters in Gaseous Detonations," *19th Symposium (Intl.) on Combustion*, pp. 615–623, The Combustion Institute, Pittsburgh, PA.

Wilson, G. J., and Sussman, M. A., 1993. "Computation of Unsteady Shock-Induced Combustion Using Logarithmic Species Conservation Equations," *AIAA J.* **31**, pp. 294-301.

Zeldovich, Ia. B., and Kompaneets, A. S., 1960. **Theory of Detonation**, Academic Press, NY.

STUDIES ON DETONATION DRIVEN HOLLOW PROJECTILES

P. A. Thibault, J. D. Penrose, and A. Sulmistras

Combustion Dynamics Ltd.
203, 132 4th Avenue, S.E.
Medicine Hat, Alberta T1A 8B5, Canada

S. B. Murray and J. L. D. S. Labbé

Defence Research Establishment Suffield
Box 4000
Medicine Hat, Alberta T1A 8K6, Canada

ABSTRACT

This study explores the concept of a detonation driven projectile by performing CFD computations for a simple hollow projectile propelled in a pre-mixed detonable mixture. The particular projectile selected for the calculations has a 203 mm outer diameter and is propelled in a stoichiometric acetylene-air mixture at atmospheric pressure. The calculations model the experiments that will be performed at DRES using a 203-mm diameter, 55-caliber gun to fire the hollow projectile, and a long 2-m diameter thin plastic bag to contain the pre-mixed combustible mixture.

Fully transient inviscid 2-D computations, performed with a simple one-step combustion model, indicate that a detonation front can be stabilized in the projectile nozzle for an appropriate choice of projectile geometry and inlet Mach number. The detonation structure inside the projectile is relatively complex and involves three connected detonation branches, two of which interact with the conical bow shock generated by the leading edge of the projectile. The central branch is overdriven due to the flow contraction produced by the other two oblique detonation branches. Fully coupled computations that include projectile acceleration effects indicate that the diameter of the central branch approaches a diameter of 44 mm when the projectile reaches a terminal Mach number of 6.43. Based on detonation cell size measurements for a CJ detonation, and ZND calculations for an overdriven detonation, this corresponds to a diameter to cell size ratio $D/S = 18$ which is large enough to suggest good stability properties for this branch.

J. Buckmaster et al. (eds.), Combustion in High-Speed Flows, 421–443.

1. Introduction

The development of efficient supersonic air-breathing engines will require modern techniques to mix and burn the fuel-air mixture. Due to the high flow velocities involved, these processes must be cleverly accelerated in order to minimize the overall size of the engine. Conventional SCRAM engine designs that rely on turbulence to promote mixing and combustion may not be effective in achieving complete combustion at high velocities and altitudes. These problems may be resolved by triggering flow instabilities upstream of the combustor to promote fuel-air mixing, and by burning the mixture as a detonation rather than as a turbulent flame.

The concept of using detonation combustion for supersonic engines has been discussed by numerous authors including Sargent and Gross (1960), Morrison (1978), Ostrander et al. (1987) and Atamanchuk and Sislian (1991). Many of the proposed concepts involve Oblique Detonation Wave Engines (ODWE) that rely on an oblique detonation wave to achieve positive thrust. The onset of detonation in such engines requires sufficient shock heating to ignite the combustible gas near the throat of the engine nozzle. This can be achieved using repeated shock reflection in a constant angle inlet section (Fig. 1a) or by the merging of multiple shock waves in a double angle inlet geometry (Fig. 1b). The presence of an oblique detonation has also been proposed as a possible mechanism for the "superdetonative" mode of the ram accelerator developed at the University of Washington (Hertzberg et al., 1988, and Hinkey et al., 1992). In this case, the detonation rides on the leeward side of the projectile and accelerates the latter to velocities higher than the Chapman-Jouguet detonation velocity for the mixture (Fig. 1c).

Studies related to detonation driven engines have thus far focused on oblique detonations and ram accelerators. Lehr (1972) has performed free-flight experiments of solid projectiles fired into pre-mixed mixtures. These studies have revealed a fascinating spectrum of combustion modes and instabilities. The solid projectiles used however were not conducive to producing positive net thrust and acceleration.

The purpose of the present study is to investigate a hollow projectile geometry that is propelled in an unconfined pre-mixed gas and accelerated by a stabilized detonation front inside the projectile. As illustrated in Figure 1d, the simple hollow projectile geometry consists of a short round tube with conical inlet and nozzle sections.

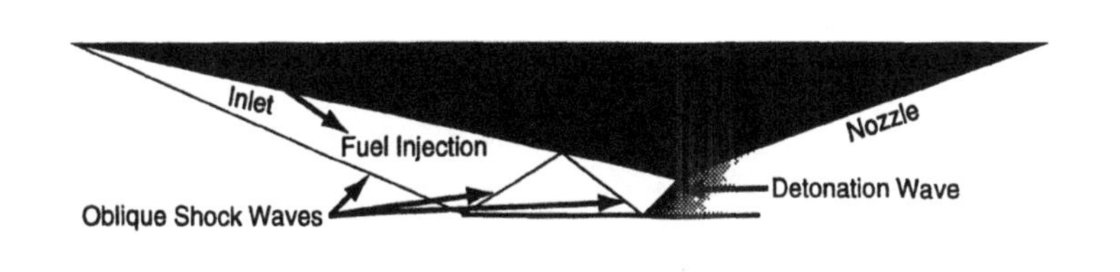

Figure 1a Oblique Detonation Wave Engine (ODWE)
Ostrander et al., 1987

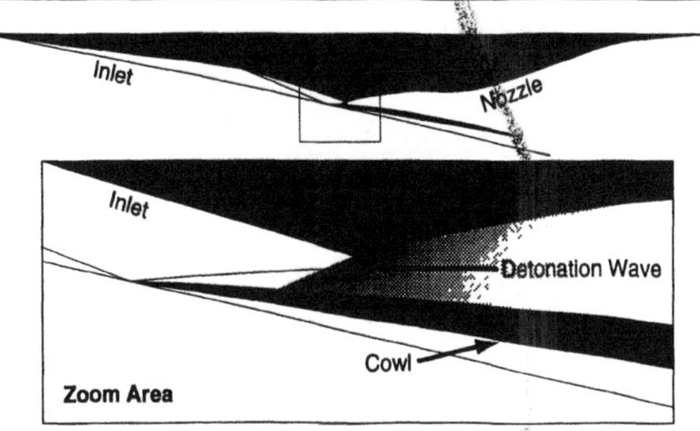

Figure 1b Hypersonic Detonation Powered Lifting-Propulsive Body
Atamanchuk and Sislian, 1991

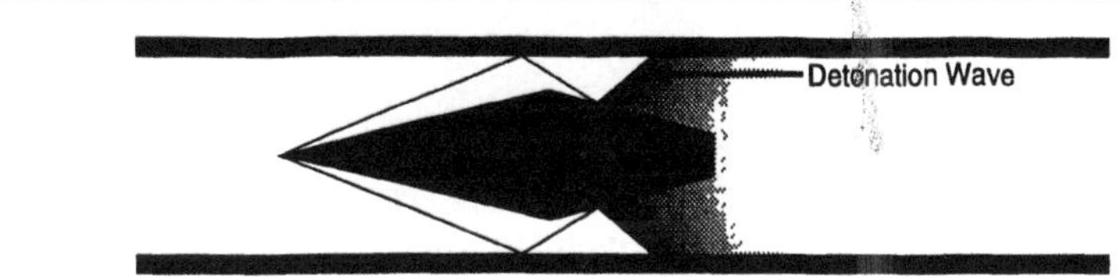

Figure 1c RAM Accelerator - Superdetonative Regime
Hinkey et al., 1992

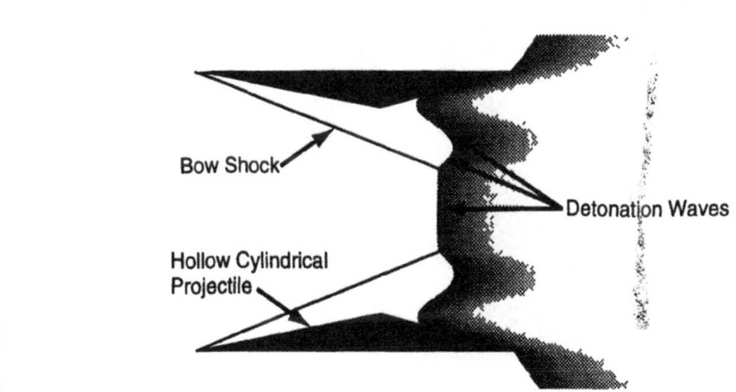

Figure 1d Detonation Driven Hollow Projectile

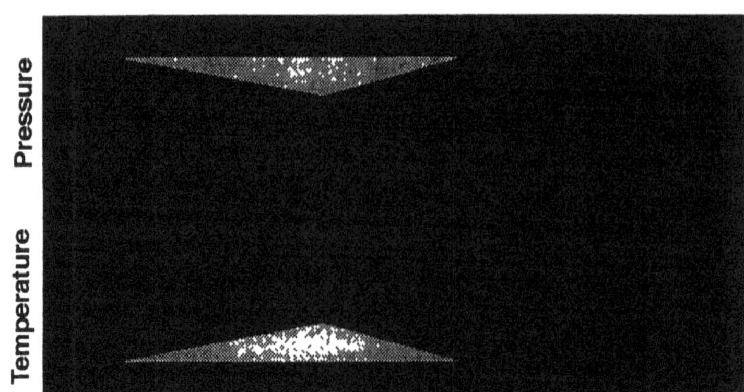

Figure 2a Detonation Driven Hollow Projectile
Bow Shock Implosion

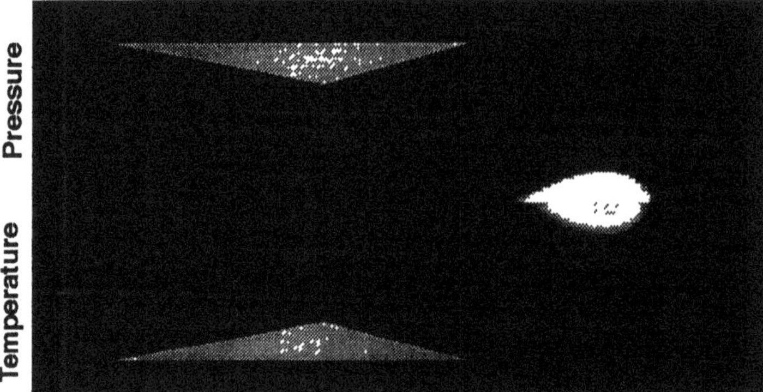

Figure 2b Detonation Driven Hollow Projectile
Detonation Initiation

Figure 2c Detonation Driven Hollow Projectile
Detonation Stabilization

This geometry is essentially the ram accelerator geometry of Figure 1c turned inside out and is also similar to tubular projectiles used for artillery training (Evans and Wardlaw, 1989). This paper presents the results of a series of calculations that were performed as an initial feasibility study for large-scale experiments that will be performed at Defence Research Establishment Suffield, using a 203-mm diameter hollow projectile fired from a 203 mm - 55-caliber gun.

2. Detonation Driven Hollow Projectile Concept

The hollow projectile geometry provides many advantages over solid projectiles for self-sustained propulsion. Firstly, the flow field inside the projectile consists of a conical bow shock that implodes with increasing strength toward the axis of symmetry (Fig. 2a). As illustrated in Figure 2b, this shock focusing serves as a natural detonation initiation source when the projectile enters a combustible gas. Projectile acceleration then becomes possible when the detonation wave expands and stabilizes in the nozzle section (Fig. 2c).

2.1. Parameters related to projectile design

As with all detonation driven engine designs, the feasibility of achieving positive thrust in a hollow projectile depends on a variety of parameters associated with the mixture, the geometry, and the initial velocity of the projectile. Direct detonation initiation depends on the energy deposited by the shock implosion process, and the reactivity of the mixture. Energy focusing is controlled by the inlet angle, the inflow Mach number, and the size of the projectile. Mixture reactivity is determined by the detailed chemical kinetics, and can be quantified in terms of the detonation cell size of the mixture which decreases with increasing mixture sensitivity. Previous studies on detonation propagation and transmission suggest that the onset of detonation by an imploding bow shock becomes more probable when the projectile diameter, D, is large compared to the detonation cell size, S (Edwards et al., 1979, and Murray and Lee, 1986). The stability of the detonation front inside the nozzle may also be expected to depend on the ratio D/S and the regularity of the mixture's detonation cellular structure (Moen et al., 1986).

Even in the limit where $D \gg S$, sustained propulsion may not be possible due to an inappropriate combination of projectile geometry

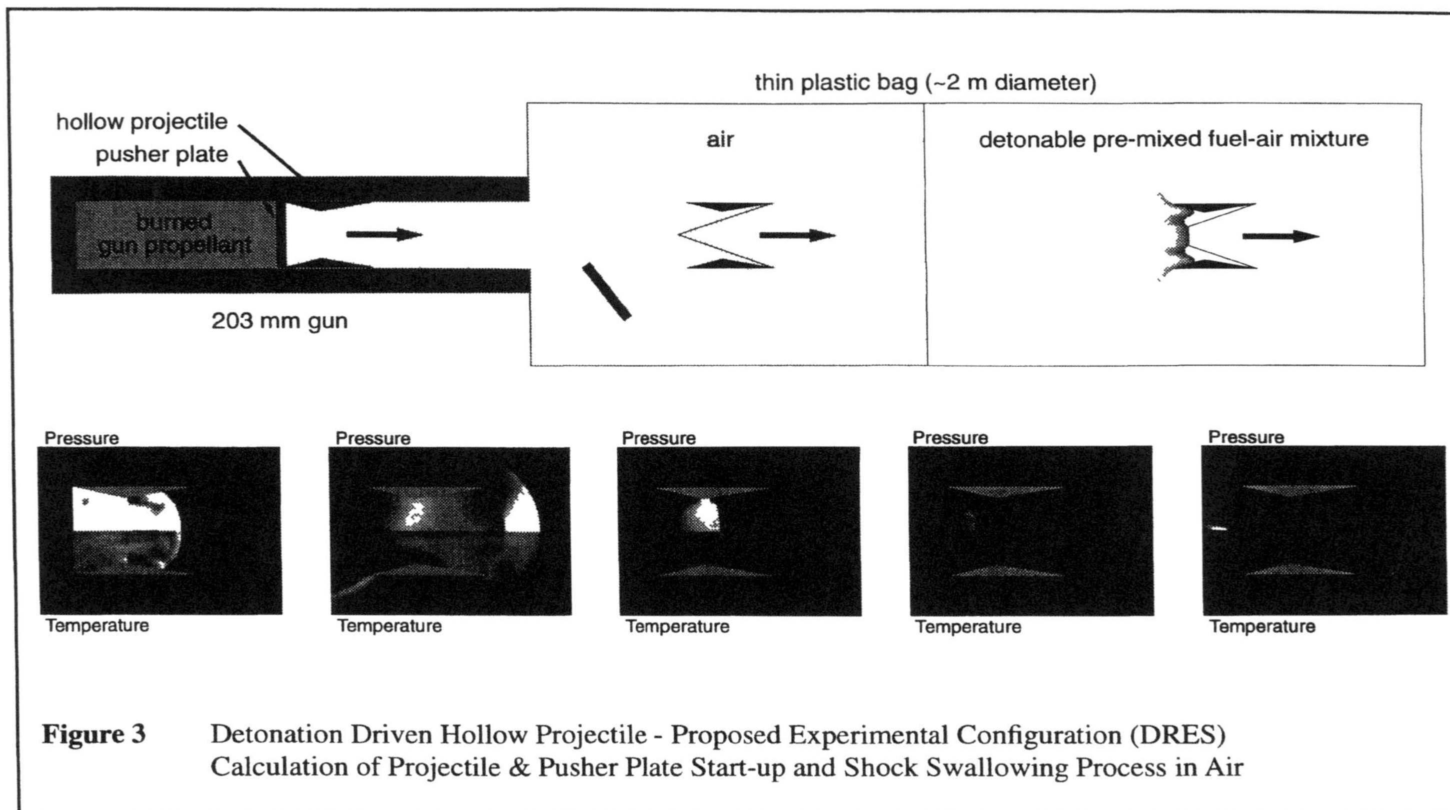

Figure 3 Detonation Driven Hollow Projectile - Proposed Experimental Configuration (DRES) Calculation of Projectile & Pusher Plate Start-up and Shock Swallowing Process in Air

or initial velocity that can cause the detonation to either trail behind the projectile or to initiate prematurely in the inlet section. The key geometric parameters for the hollow projectile are the throat to outer diameter ratio, d/D, and the relative position of the throat, l/L. The initial inflow velocity may be characterized in terms of the ratio V_p/V_{CJ} where V_p is the initial projectile velocity and V_{CJ} denotes the Chapman-Jouguet detonation velocity for the mixture.

2.2. Parameters related to experiment design

As illustrated in Figure 3, the proposed experiments are to be performed using a 203 mm projectile that will be fired from a 203 mm - 55-caliber gun into a cylindrical bag of pre-mixed fuel-air mixture. The experiments introduce additional parameters such as the diameter of the bag which must be sufficiently large compared to the projectile diameter to eliminate any possible confinement effects of the plastic enclosure. The bag should also be sufficiently wide to accommodate the vertical displacement of the projectile due to gravity, and long enough to observe its acceleration and approach to terminal velocity. The distance required to reach this velocity is controlled by the total mass of the projectile.

The projectile will be ejected from the muzzle using a pusher plate to separate it from the burned gun propellant. The resulting flow blockage inside the projectile will produce a bow shock ahead of it as it leaves the muzzle. A sufficient free-flight distance in air must therefore be allotted for removal of the pusher plate and to allow the bow shock to be "swallowed" before the projectile enters the combustible mixture. Finally, the gun barrel will be rifled to spin the projectile and improve its aerodynamic stability.

3. Computational Method and Validation

Initial proof of concept computations were performed using a fully transient and inviscid CFD code based on the Lax-Wendroff finite difference scheme in conjunction with the 6th order (phase error) version (Book and Fry, 1984) of the Flux Corrected Transport (FCT) algorithm (Boris, 1976). The code uses a structured rectangular grid with a special slanted boundary algorithm that imposes appropriate free-slip boundary conditions on surfaces that are not aligned with the grid.

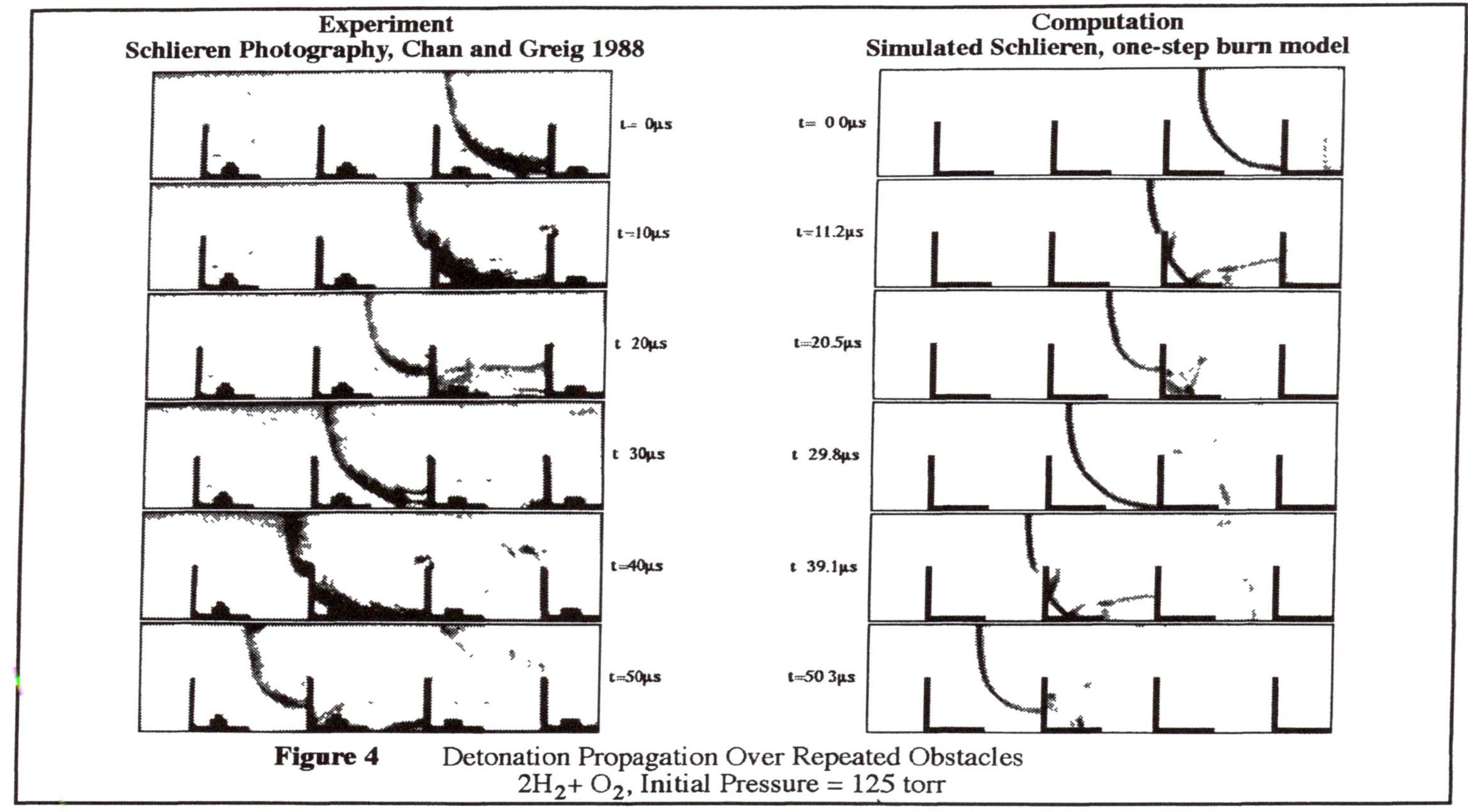

Figure 4 Detonation Propagation Over Repeated Obstacles
$2H_2 + O_2$, Initial Pressure = 125 torr

The purpose of the calculations was to follow the propagation and stabilization of the detonation wave without resolving the transverse wave structure of the detonation front or attempting to model the detailed onset of detonation initiation (or failure). As previously mentioned, the latter phenomena are intimately related to the ratio of the projectile diameter, D, over the characteristic detonation cell size, S, of the mixture. In the limit where D/S is assumed to be very large, detonation initiation and propagation are essentially guaranteed when the temperature is raised above the ignition temperature of the mixture. In this limit, the computations can be performed with a very simple reactant depletion model with a constant reaction time and a temperature threshold for ignition. During the reaction process, the chemical energy released is in proportion to the reactant depleted, and the specific heat ratio γ varies continuously from the unburned to the burned gas values. Due to the absence of a temperature dependent term, this model inhibits the onset of gasdynamic-chemical instabilities that would require a prohibitively expensive fine grid to be accurately resolved. All computations were performed using a chemical energy per unit mass, and a specific heat ratio that are consistent with chemical equilibrium computations for a Chapman-Jouguet detonation in a stoichiometric acetylene-air mixture.

The simple one-step burn model described above has been used successfully in computing the propagation of hydrogen-oxygen detonations over repeated obstacles. Figure 4 compares computational results with experiments performed by Chan and Greig (1988). The computed and experimental Schlieren images reveal excellent agreement at the detonation front. Both the detonation diffraction over the obstacles and the focusing in the bottom corners are correctly represented. The calculations clearly capture complex shock reflection patterns that are also evident, though more diffused, in the experiments. The most significant difference between the experiments and computations appears in the fourth frame where the experimental Schlieren image suggests a slight decoupling between the diffracted shock wave and the reaction front. This local detonation failure is due to the weakening of the wave as it diffracts around the obstacle. Its occurrence depends on the sensitivity of the mixture and can result in a very distinct cycle of detonation failure and re-initiation in less sensitive mixtures. Such "quasi-detonation" cycles cannot be quantitatively computed without resolving the detonation

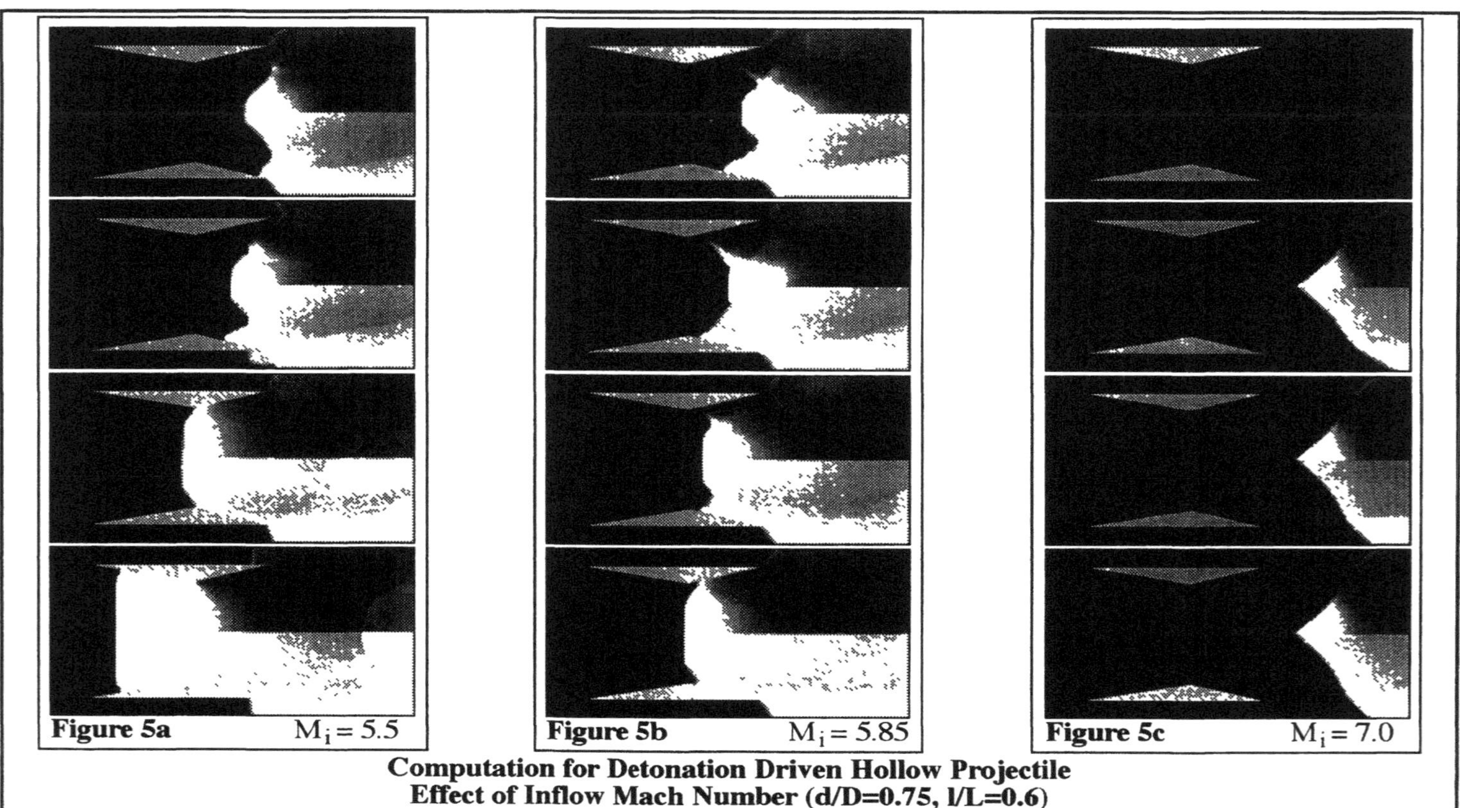

Computation for Detonation Driven Hollow Projectile
Effect of Inflow Mach Number (d/D=0.75, l/L=0.6)

cellular structure and the turbulent flow downstream of the obstacle.

4. Computational Results

In order to investigate the effect of geometry and initial Mach number on the stabilization of the detonation front in the projectile nozzle, initial computations were performed using a resolution of 40 computation cells per projectile radius and approximately 3 cells per reaction-zone length. For a 203-mm diameter projectile, this corresponds to a resolution of 2.5 mm per computational cell. These calculations were followed by higher resolution computations with 80 cells per radius, and as many as 30 cells per reaction zone. This provided a more detailed view of the relatively complex detonation front structure. The main computational grid was surrounded by a geometrically expanding grid that contained 25 additional computational cells in both directions. This extended the grid by an additional 76 and 38 projectile radii for low and high resolution computations respectively. "Flow-through" boundary conditions were imposed at the end of the expanded grid. The entire grid was initialized to an ambient pressure, $P = 1$ atm., temperature, $T = 298.15$ K, and inflow Mach number, M_i.

4.1. Effect of Mach number

Figure 2 shows the results of the calculations with an inflow Mach number, $M_i = 5.85$ for a stoichiometric acetylene-air mixture with a CJ detonation Mach number of 5.4. In this case, the detonation is initiated along the axis of symmetry behind the projectile and propagates into the nozzle to finally stabilize near the throat. The detonation propagation process is shown in greater detail in Figure 5 which compares pressure and temperature contours for $M_i = 5.5, 5.85$ and 7. The successful engine start-up configuration for $M_i = 5.85$ (Fig. 5b) reveals a detonation front that is constructed of three connected branches. The central branch represents a normal overdriven detonation that propagates upstream along the axis of symmetry in the undisturbed region of the flow field. This branch connects to an oblique detonation that propagates in a mixture that has been precompressed by the bow shock. The third branch propagates along the nozzle wall into a low-pressure region behind the throat. The three branches move relative to each other during the overall engine start-

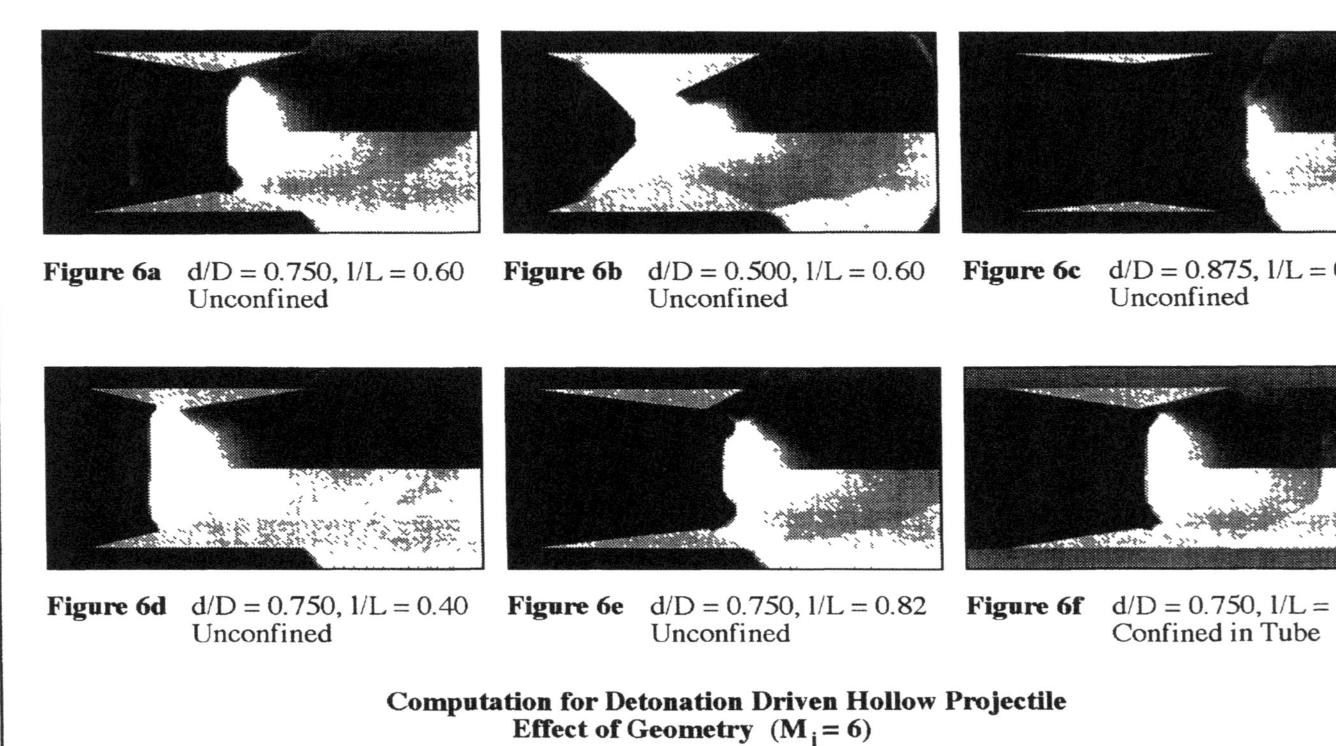

Figure 6a d/D = 0.750, l/L = 0.60 Unconfined

Figure 6b d/D = 0.500, l/L = 0.60 Unconfined

Figure 6c d/D = 0.875, l/L = 0.60 Unconfined

Figure 6d d/D = 0.750, l/L = 0.40 Unconfined

Figure 6e d/D = 0.750, l/L = 0.82 Unconfined

Figure 6f d/D = 0.750, l/L = 0.60 Confined in Tube

Computation for Detonation Driven Hollow Projectile Effect of Geometry ($M_i = 6$)

up process. The central branch initially moves ahead of the other branches until the outer branch comes in contact with the projectile wall. The latter branch then rapidly accelerates along the nozzle surface to overtake the central branch and eventually stabilizes just downstream of the throat (Frame 2). At this point, it is overtaken by the central branch (Frame 3) which stabilizes just upstream of the throat (Frame 4).

The same "seesawing" behavior between the central and outer detonation branches is observed for the case of a lower inlet Mach number $M_i = 5.5$ (Fig. 5a). In this case, however, the central detonation penetrates through the throat causing the engine to unstart. The three branch structure is then reduced to two branches with the central branch remaining overdriven while the second branch propagates in the pre-compressed region behind the bow shock. It is interesting to note that the hollow projectile engine unstarts for an inlet Mach number that is slightly larger than the Chapman-Jouguet Mach number for the mixture. This observation is consistent with the analysis of Pratt et al. (1987) which indicates that the CJ state represents "a lower bound for possibly stable oblique detonations."

The detonation propagation sequence for a large inlet Mach number, $M_i = 7$, is displayed in Figure 5c. In this case, the detonation spreads radially from the axis of symmetry but cannot propagate upstream to provide thrust to the projectile. In a free-flight situation, the projectile would decelerate due to drag and allow the detonation to catch up.

4.2. Effect of geometry

The detonation propagation phenomena described above depend not only on the inlet Mach number but also on the internal geometry of the projectile. Even in the case of non-reacting flow, the shock structure inside the projectile depends greatly on the ratio, d/D, of projectile's throat diameter over its outer diameter. For a given inlet Mach number, there exists a minimum value of d/D below which the flow restriction causes the flow to become subsonic at the throat. This creates a detached shock upstream of the projectile and causes a sudden deceleration of a projectile in free flight (Shapiro, 1953 and Evans, 1989).

The effect of area blockage on the start up of a hollow projectile engine is displayed in Figures 6a, b, and c for an inlet Mach number

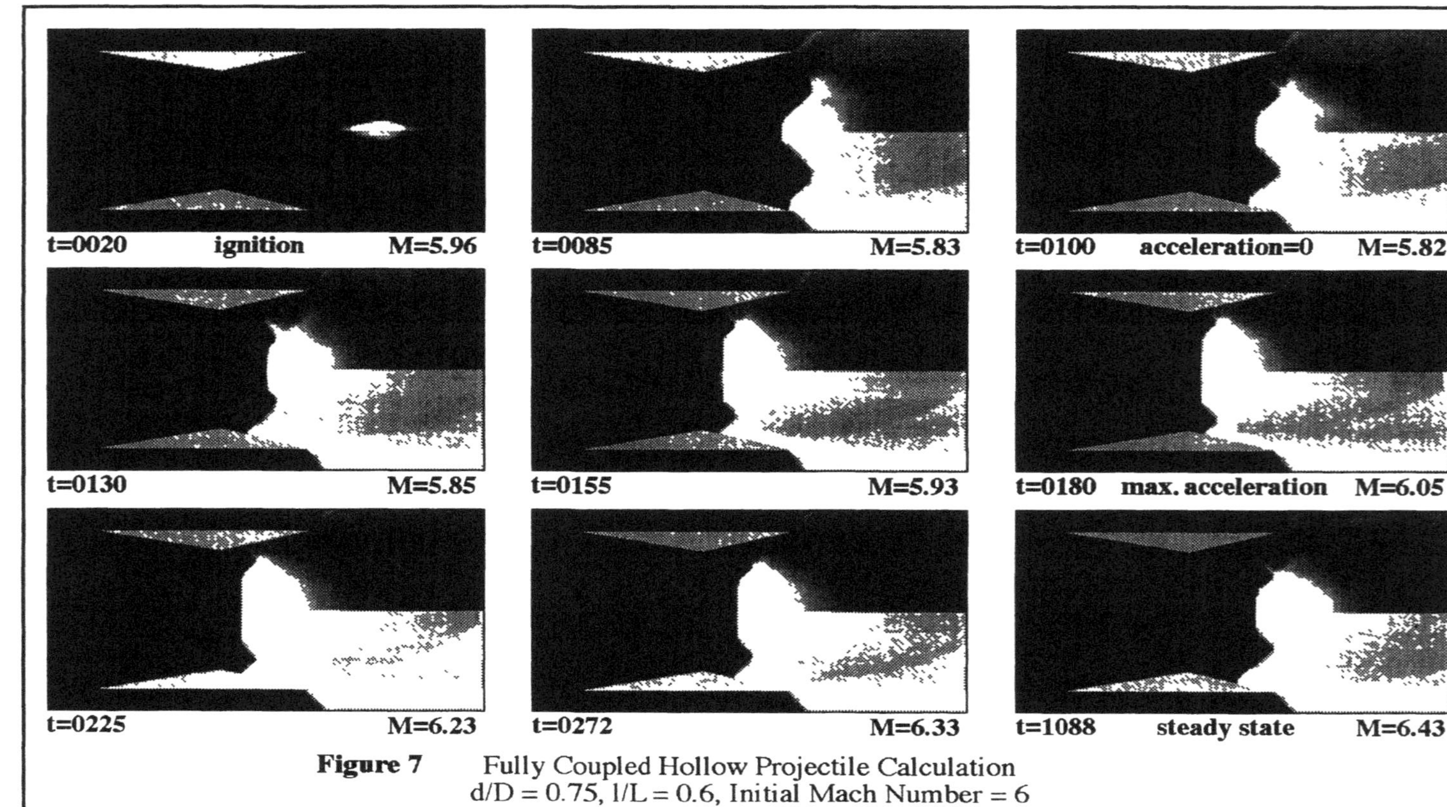

Figure 7 Fully Coupled Hollow Projectile Calculation
d/D = 0.75, l/L = 0.6, Initial Mach Number = 6

$M_i = 6$ and diameter ratios $d/D = .75$, $.5$ and $.875$. As in the previous comparison for different inlet Mach numbers, this comparison reveals a successful start up for $d/D = .75$, an unstart for $d/D = .5$, and detonation standoff for $d/D = .875$. The unstart condition for $d/D = .5$ is due to an increase in flow restriction, and to a larger inlet angle which results in a stronger bow shock and ignition upstream of the throat section. A similar unstart condition also occurs for $d/D = .75$ if the throat position is moved upstream (Fig. 6d). Figure 6e, on the other hand, shows that the throat position may be moved a great distance downstream without causing a standoff between the detonation and the nozzle surface. It is interesting to note that the geometries causing unstart in Figures 6b and 6d do not result in shock detachment in a "cold flow" calculation with the same initial and boundary conditions. Once again, this observation is consistent with the studies of Pratt et al. (1987) which show that the maximum turning angle is smaller for a standing oblique detonation than for an oblique shock.

The final frame in Figure 6 shows a successful start-up condition for the same projectile and inlet flow velocity as in Figure 6a with the exception that the projectile is confined to propagate in a solid tube. The confined and unconfined geometries display a very similar detonation structure. The main difference occurs downstream of the detonation front where the confined case exhibits a strong Mach disc structure.

4.3. Fully coupled computations

The above calculations have investigated various projectile configurations assuming that the projectile velocity remains constant throughout the transient flow process. An actual projectile may be expected to decelerate due to drag until the detonation front impinges on the nozzle surface. At this point, the positive net thrust will accelerate the projectile to a terminal velocity. This process is illustrated in Figure 7 which displays computational results for a fully coupled calculation that takes into account the acceleration of the projectile. Due to the long computational times involved, the approach to steady state was accelerated by using an artificially light projectile with a density 10 times less than aluminum. After an initial deceleration phase, the projectile suddenly accelerates when the detonation reaches the nozzle wall and propagates upstream to

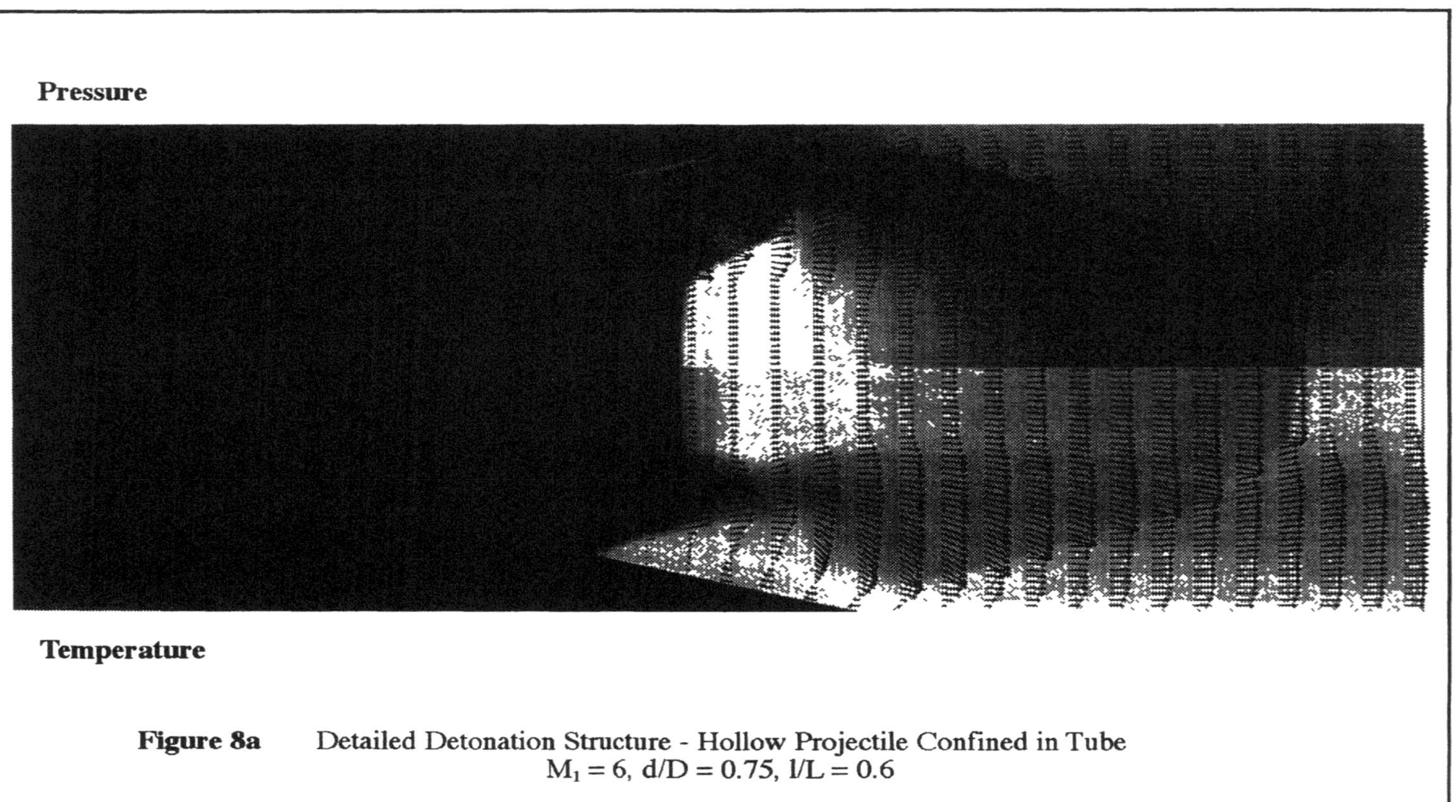

Figure 8a Detailed Detonation Structure - Hollow Projectile Confined in Tube
$M_1 = 6$, $d/D = 0.75$, $l/L = 0.6$

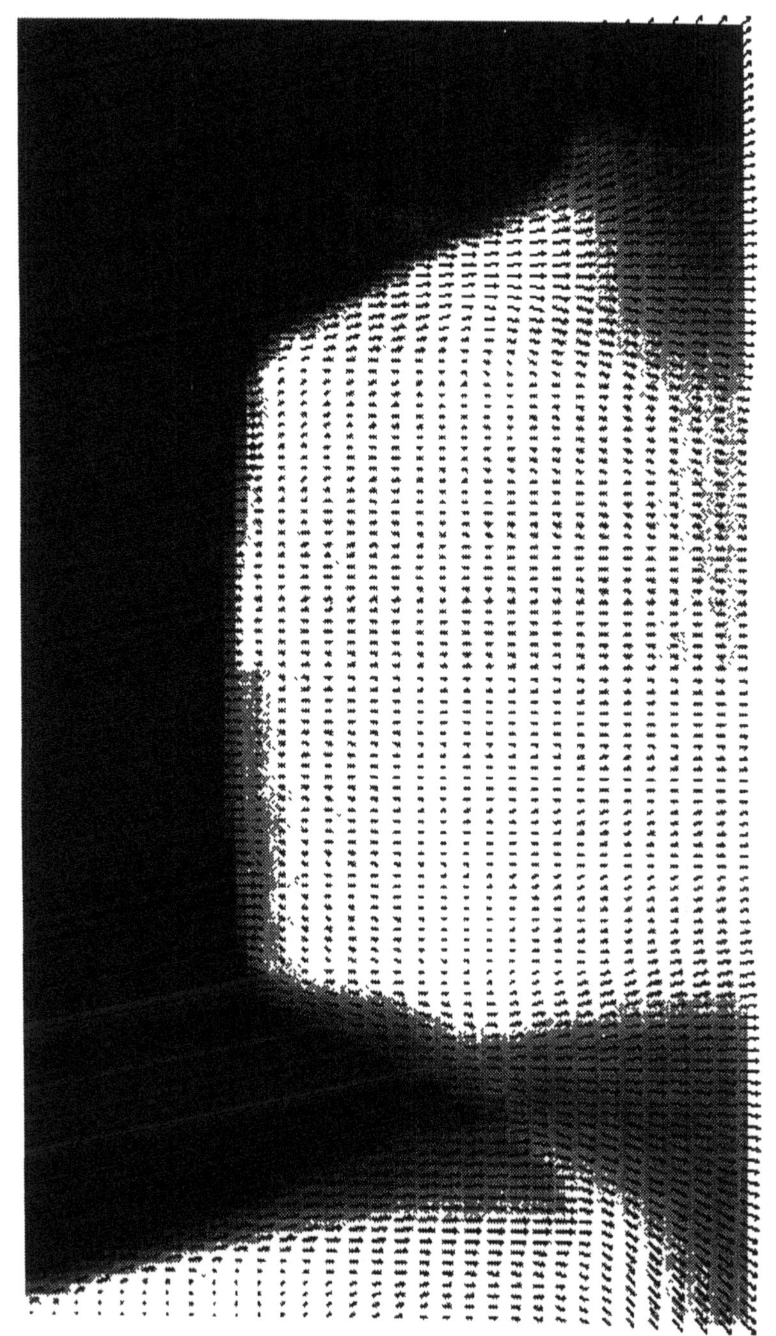

Figure 8b Detailed Detonation Structure (Zoomed Area)
Hollow Projectile Confined in Tube
$M_i = 6$, $d/D = 0.75$, $l/L = 0.6$

a position near the throat where maximum acceleration is achieved. Subsequent acceleration of the projectile then causes the detonation front to recede downstream in the nozzle. This gradually reduces the thrust surface and the acceleration until the projectile reaches a terminal Mach number of 6.43, which corresponds to an 18% overdrive of the central detonation branch.

4.4. Analysis of the detonation structure

In order to shed further light on the detonation structure in the hollow projectile, higher resolution calculations were performed for the case of a fully confined projectile with $M_i = 6.0$ and $d/D = .75$. The exothermic reaction time was also increased to further increase the number of computational cells in the reaction zone. Figure 8a displays the steady-state pressure and temperature contours overlaid on a velocity vector plot. The pressure distribution is also displayed as a 3-D carpet plot in Figure 9a. The overall flow field is characterized by a three-branch detonation front with a central and oblique detonation interacting with the imploding conical bow shock. The third detonation branch is anchored on the projectile just downstream of the throat and interacts with the oblique detonation to produce a large pressure peak at the intersection point. An underexpanded jet structure with a characteristic recompression shock and Mach disc is established downstream of the central and oblique detonation branches. A burned gas recirculating flow pattern is also produced near the nozzle surface behind the third detonation branch.

A closer view of the detonation front structure (Figs. 8b and 9b) highlights the interaction between the detonation branches and the conical bow shock. The flow behind the central detonation branch temporarily expands radially within the reaction zone and re-compresses due to the high pressure generated by the interaction of the two remaining branches. The latter "pinching" effect is responsible for the 11% velocity overdrive of the central detonation branch. The exact nature of the other two branches is more difficult to categorize due to the complex structure of the detonation front and the relatively thick reaction zone.

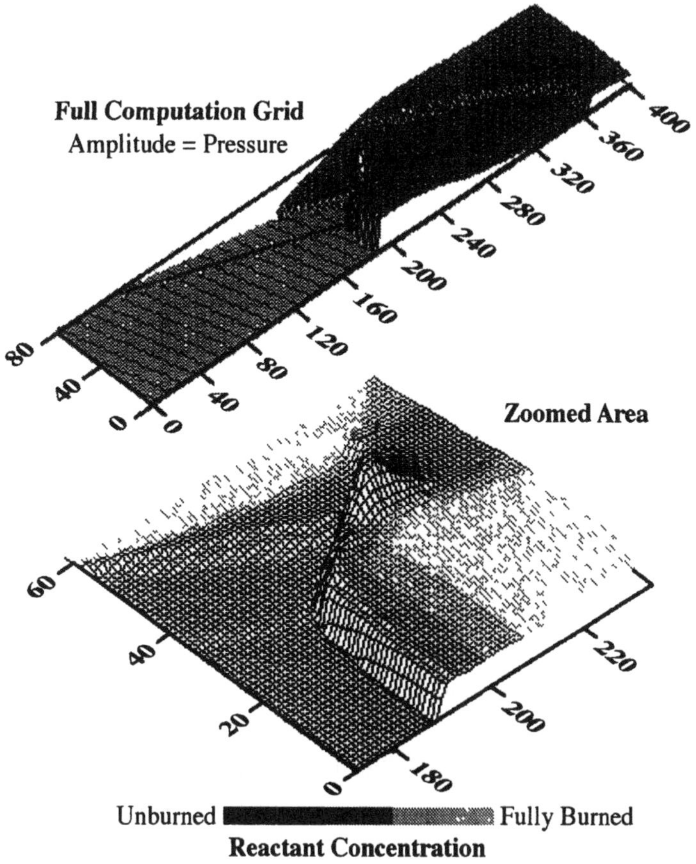

Full Computation Grid
Amplitude = Pressure

Zoomed Area

Unburned — Fully Burned
Reactant Concentration

Figure 9 Detailed Detonation Structure

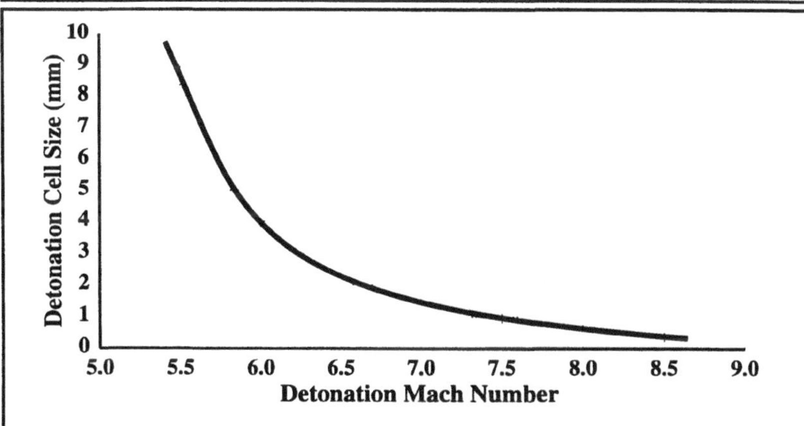

Figure 10 Detonation Cell Size Estimates - Overdriven Acetylene-Air (Stoichiometric, Pressure = 1 atm., Temperature = 298.15 K)

4.5. Detonation cell size considerations

The above calculations have not resolved the important transverse wave structure of the three branches of the detonation front. The detonation cell size for a CJ detonation in stoichiometric acetylene-air with initial conditions $P_o = 1$ atm. and $T_o = 298.15$ K is approximately $S = 10$ mm (Moen et al., 1985). Due to the increase in shock pressure and temperature, the detonation cell size may be expected to decrease as the detonation becomes overdriven. The influence of overdrive on the cellular structure of fuel-oxygen detonations has been investigated by Meltzer et al. (1991), Edwards et al. (1984), and Gravilenko and Prokhorov (1982). Unfortunately, similar measurements have yet to be performed for overdriven fuel-air mixtures. In the absence of suitable experimental data, detonation cell sizes for overdriven acetylene-air detonations may be estimated on the basis of reaction-zone lengths computed using detailed kinetics and a one-dimensional ZND model. Using, as a reference point, the experimental cell size for a CJ detonation, the overdriven detonation cell size may then be computed as a function of the overdrive ratio M_{OD}/M_{CJ}. This approach assumes a constant ratio between the detonation cell size and the ZND reaction zone. Figure 10 shows the results of such an analysis based on computations performed by Shepherd (1993). These calculations indicate that the detonation cell size decreases from 10 mm to 2.4 mm for a 20% velocity overdrive. For a projectile outer diameter of 203 mm, this corresponds to a diameter to cell size ratio $D/S = 85$. In the fully coupled calculations displayed in Figure 7, the diameter of the central detonation branch, D_c, decreases by a factor of 2 from 88 mm to 44 mm during its acceleration to the steady-state velocity, whereas the ratio D_c/S decreases by only 22% from 23 to 18. These large ratios are certainly encouraging since they suggest that the central detonation branch will remain stable during the projectile acceleration process. More detailed computations would be required to assess the fate of the other two branches.

5. Conclusions

Numerical studies have investigated the feasibility of achieving detonation driven propulsion of a hollow tubular projectile in a premixed fuel-air mixture. The inviscid CFD computations based on a simple one-step combustion model indicate that such a concept may indeed be feasible with the appropriate selection of a projectile geometry and an initial velocity which must be larger than the CJ detonation velocity of the mixture. The detonation structure inside the projectile is relatively complex and involves three connected detonation branches, two of which interact with the bow shock. The central branch is overdriven due to the flow contraction produced by the other two detonation branches. The flow field includes a recirculation zone near the nozzle wall which also contributes to the overall flow contraction downstream of the throat. A more accurate analysis taking into account viscous effects and turbulence will undoubtedly be required to accurately model the recirculation zone, and to investigate possible ignition in the inlet boundary layer. The combined use of adaptive gridding and detonation trackers would also be useful to explore the complex interactions between the detonation fronts, the bow shock, the recirculation zone, and the boundary layer.

Acknowledgement

The authors would like to thank Dr. J. E. Shepherd for his very helpful comments and for providing ZND reaction-zone computations that were used for detonation cell size estimates.

References

Atamanchuk, T. and Sislian, J., 1991. "Hypersonic detonation wave powered lifting-propulsive bodies," *AIAA Third International Aerospace Planes Conference*, Orlando, Florida.

Book, D. L. and Fry, M. A., 1984. "Airblast simulations using flux-corrected transport codes," Naval Research Laboratory Memorandum 5334.

Boris, J. P., 1976. "Flux-corrected transport modules for solving generalized continuity equations," Naval Research Laboratory Memorandum 3237.

Chan, C. K. and Greig, D. R., 1988. "Structure of fast deflagrations and quasi-detonations," *22nd Symposium (International) on Combustion*, p. 1733.

Edwards, D. H., Thomas, G. O., and Nettleton, M. A., 1979. "The diffraction of a planar detonation wave at an abrupt area change," *J. Fluid. Mech.* **95**, pp. 79-96.

Edwards, D. H., Walker, J. R., and Nettleton, M. A., 1984. "On the propagation of detonations along a wedge," *Archivum Combustion* **4**(3), pp. 197-209.

Evans, J. R. and Wardlaw, A. B., Jr., 1989. "Prediction of tubular projectile aerodynamics using the zeus euler code," *Journal of Spacecraft and Rockets* **26**(5), pp. 314-321.

Gravilenko, T. P. and Prokhorov, E. S., 1982. "Compressed detonation wave in a real gas," *Combustion, Explosion, and Shock Waves* **17**(6), pp. 689-692.

Hertzberg, A., Bruckner, A. P., and Bogdanoff, D. W., 1988. "Ram accelerator: A new chemical method for accelerating projectiles to ultrahigh velocities," AIAA Journal **26**(2), pp. 195-203.

Hinkey, J. B., Burnham, E. A., and Bruckner, A. P., 1992. "High spatial resolution measurements of ram accelerator gas dynamic phenomena," *AIAA/SAE/ASME/ASEE 28th Joint Propulsion Conference and Exhibit*, Nashville, Tennessee, AIAA 92-3244.

Lehr, H. F., 1972. "Experiments on shock-induced combustion," *Astronautica Acta* **17**, pp. 589-597.

Meltzer, J., Shepherd, J. E., Akbar, R., and Sabet, A., 1991. "Mach reflection of detonation waves," *13th ICDERS*, Nagoya, Japan.

Moen, I. O, Funk, J. W., Ward, S. A., Rude, G. M., and Thibault, P. A., 1985. "Detonation length scales for fuel-air explosives," *Progress in Astronautics and Aeronautics* **94**.

Moen, I. O., Sulmistras, A., Thomas, G. O., Bjerketvedt, D., and Thibault, P. A., 1986. "The influence of cellular regularity on the behavior of gaseous detonations," *Progress in Astronautics and Aeronautics* **106**, pp. 220-243.

Morrison, R. B., 1978. "Evaluation of the oblique detonation wave ramjet," Universal Systems Inc., NASA CR NAS1-14771.

Murray, S. B. and Lee, J. H., 1986. "The influence of physical boundaries on gaseous detonation waves," *Progress in Astronautics and Aeronautics* **106**, p. 325.

Ostrander, M. J., Hyde, J. C., Young, M. F., Kissinger, R. D., and Pratt, D. T., 1987. "Standing oblique detonation wave engine performance," Aerojet TechSystems Co., PRA-SA-AFAL, published by the American Institute of Aeronautics and Astronautics Inc.

Pratt, D. T., Humphrey, J. W., and Glenn, D. E., 1987. "Morphology of a standing oblique detonation wave," *Colloquium on Supersonic Combustion, AIAA/ASME/SAE/ASEE 23rd Joint Propulsion Conference*, Paper No. AIAA-87-1785.

Sargent, W. H. and Gross, R. A., 1960. "Detonation wave hypersonic ramjet," *ARS Journal.*

Shapiro, A. H., 1953. "The dynamics and thermodynamics of compressible fluid flow, Volume 1," John Wiley and Sons.

Shepherd, J. E., 1993. Private communication.

IGNITION AND STRUCTURE

THE ROLE OF MATHEMATICAL MODELING IN COMBUSTION

J. Buckmaster

University of Illinois
Urbana, Illinois 61801

Preface

My original intent had been to spend 15-20 minutes making some general remarks about modeling before proceeding with the session on Ignition and Structure that I was chairing. But the withdrawal at the last minute of one of our speakers, and my perception from remarks made during some of the talks that modeling is often not understood, prompted me to expand the presentation. It was prepared purely for oral presentation and the editors have given me permission to reproduce it here in its original colloquial style.

Many years ago, when I was an assistant professor at New York University, in the mathematics department - this was at the old Heights campus, and we were, amongst other things, a service department for the engineering college - a distinguished professor of engineering came over to tell us about a problem that he was interested in. I don't remember the details, it was a long time ago, but it was a mess of detached boundary layers and eddies. He showed us pictures, drew the configuration, and after a long presentation stepped back from the blackboard and asked for our help.

The exercise was futile. We mathematicians sat there bewildered that anyone would think that we could solve such a complicated problem. And no doubt our engineering colleague returned to his department convinced that mathematicians weren't worth a brass farthing when it comes down to *real* problems.

What we had was a problem in communication. He gave no indication that he understood the limitations of what we could do (which were and are substantial); and we made no effort to force him to break down the problem into the kind of bite size pieces that we could think about applying our skills to.

All too often, in the intervening years, this scenario has been repeated for me. I find myself sitting in a small group, which it is hoped can help in some way, whilst an experimentalist, investigating a hard

J. Buckmaster et al. (eds.), Combustion in High-Speed Flows, 447–459.

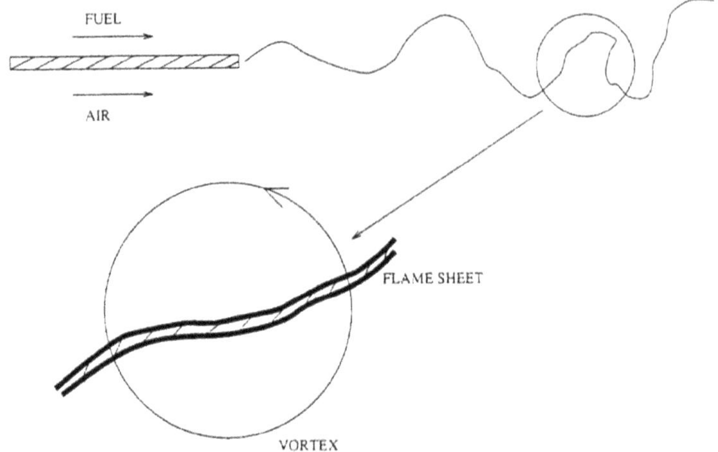

Figure 1. The Marble Problem - Wrap-up of a flame-sheet by a vortex.

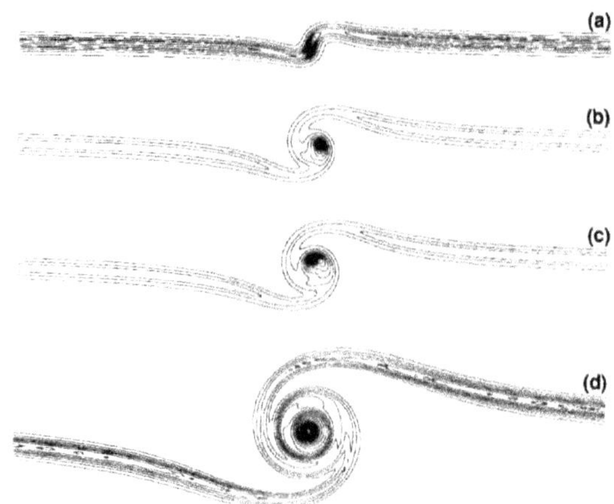

Figure 2. Vortex wrap-up with ignition - isotherms: the sequence (a) - (d) corresponds to increasing vortex Reynolds number.

problem, shows the complex details that he sees in his laboratory as movies, schlieren pictures and the like, and then says there, that's my problem, what do you think? The answer to that is, inevitably, not much. The process of reduction is essential to the modeling game. The best we can do, as modelers, is examine a very limited number of physical ingredients interacting in an idealized fashion, and so the key is to identify problems of this kind which are relevant to the big picture. And this is something where the experimentalist, with the physical insights that he has generated, can be enormously helpful.

To give an example of what I mean, consider the problem of non premixed combustion in which mixing is enhanced by turbulence. Suppose we have two parallel streams that form a mixing layer, and this layer is turbulent, Fig. 1. This is a hard problem, and people like me can't be of direct help with it. But if we look locally, in parts of this combustion field we have a thin flame that is wrapped up by an eddy, and perhaps if we understood that problem it would help us to understand the larger problem. And here, perhaps, I can help. I am probably going to have to do a planar problem; and I am probably going to have to assume that the velocity field is assigned since calculating the flow field generated by the wrap up in this variable density flow looks difficult, but perhaps I can keep enough ingredients so that the answer will be of value. This is, of course, the well known Marble problem (Marble, 1985), and it has triggered a great deal of theoretical work. A recent example is ignition in a vortex by Macaraeg, Jackson, and Hussaini (1992). Fig. 2 shows the isotherms generated during the wrap-up process, and Fig. 3 shows ignition locations for different values of the vortex Reynolds number.

This process of extracting from the real problem, which is hard, a much simpler problem that is relevant to the hard problem, is the key ingredient in modeling, and is really what we should be focusing on when experimentalists and modelers get together. Its the natural interface between the two groups. The experimentalist has some idea of the physical ingredients which necessarily must be retained and he should explain what these are and not simply exhort us to "solve the real problem." And the mathematician is very much aware of his limitations and knows how simple the thing must become before there can be any hope of progress.

Extraction of the simple from the complex is the natural meeting ground of the experimentalist and the modeler.

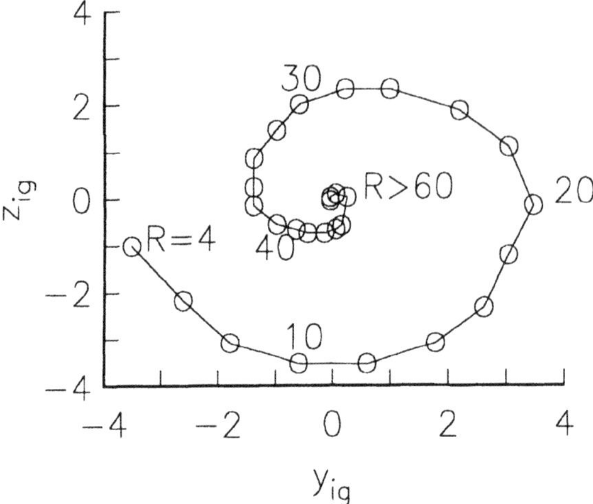

Figure 3. Vortex wrap-up with ignition - ignition locations for various values of the vortex Reynolds number.

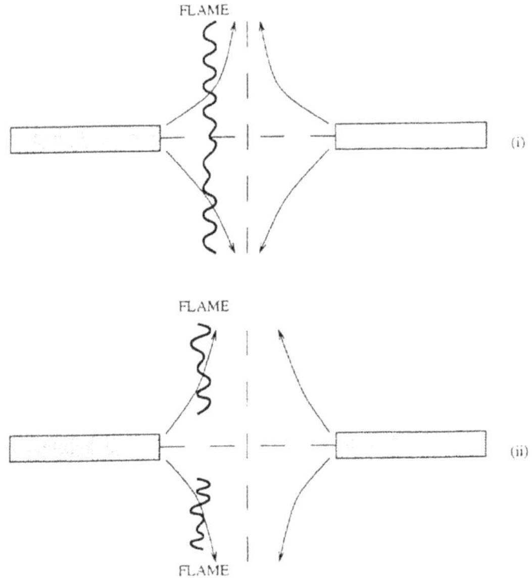

Figure 4. Diffusion flame in a hydrogen/air counter flow: (i) no extinction, (ii) extinction in the neighborhood of the axis and formation of an annular flame.

I mentioned a flame in an eddy as a reduced problem. Let me briefly mention another example. I was at a meeting recently and saw a fine poster display on supersonic mixing and combustion. Beautiful pictures of the flow field, but it suggested nothing in the way of a modeling strategy. Then when I was here at ICASE this year I looked at some experiments carried out by Gerry Pellett on hydrogen combustion which he is doing because he believes they are relevant to the supersonic combustion problem (Pellett, Northam, and Wilson, 1991). He looks at hydrogen/air counterflows. Normally what he sees is an unbroken thin flame. But if the counterflows are strong enough a hole appears in the flame to form an annulus, Fig. 4. Clearly the leading edge of this annular flame is a very interesting creature. The overall body of the flame is a diffusion flame, but at the leading edge some premixing must occur, and this premixed/diffusion flame structure must have a well defined propagation speed so that it maintains its position in the radial flow. This is a sharply focused problem, with limited ingredients, that modeling can say something about. And it might be relevant to the big problem, that of supersonic combustion. There are, presumably, leading edge flame structures at the ignition point. And perhaps, under some circumstances, holes can be torn in the established flame and knowing something about the behavior of these holes will tell us whether the flame will be blown out or not.[1] These kinds of questions are thrust upon us when we look at a simple basic experiment like Pellett's; they are less obvious if we look at real supersonic mixing. Simple experiments are the natural stimulus for modeling. With complicated experiments modelers need much more help in the formulative stage.

Simple basic experiments are much more likely to provide input for modelers than are complicated ones.

Let me turn to something else, the tools that modelers use. I once got a review in which the reviewer took a general swipe at mathematical modelers by pointing out that he had a colleague of this persuasion who - would you believe it? - said that he only solved problems that he knew how to solve. Well I suspect that experimentalists only do experiments that they know how to do, but I think that the complaint he was trying to make, despite his difficulties with the language, is that there are all these really important phys-

[1] W. Dold is one of the leading authorities on structures of this kind, see, for example, Dold, Hartley, and Green (1991).

ical problems that we would like to understand, and all too many times we modelers look at them and say we can't help.

It is, I think, important to bear in mind that the tools of the modeler are very limited, and there is little point in examining a problem where these tools do not suffice. There is a danger of looking only at hard complicated issues, where modeling has limited impact, and coming to the conclusion that modelers are a waste of time and money. The fact is that where our tools work we can be of enormous value, where they don't we are useless, and we should be judged by our contribution to the former, and not our lack of contribution to the latter.

Tools are limited; where they work modeling can be irreplaceable; where they don't, modeling is useless.

What are these tools? The list is quite short:

- Simplified geometry.
- Limited ingredients.
  (a) eliminate small effects (rational approximation)
  (b) eliminate ingredients that don't control the problem (extra rational approximation)
  (c) substitute the simple for the complex (extra rational approximation)
- Coping with non linearities.
  (a) elimination (linearization)
  (b) weak non linearities (e.g. bifurcation analysis)
  (c) asymptotic methods
  (d) special methods (e.g. inverse scattering transform)
- Computing.

Let us consider this list in detail.

Geometry. This needs little discussion. In recent years I have been engaged in the modeling of microgravity experiments, and one reason microgravity experiments are a good resource for the kind of thing that modelers do, is that without buoyancy it is easier to have simple geometries. I might add that another reason is that microgravity experimentalists are forced, by the competitive review process and the practical limitations of the arena, to define simple clean experiments - Gerry Pellett style experiments rather than ramjets - and so they have already done an important part of the modeling.

Rational simplifications. Here also, little enlargement is necessary. We all did billiard ball problems in our dynamics classes assuming that a billiard ball is a perfect sphere.

Extra rational approximations. These are not always understood. I once heard it said at a meeting that, in modeling, a wrong theory can give the right answer. I think I know what was meant by that, but not everyone does. Later a speaker was challenged on a questionable assumption, and invoked this statement. This was fine humor, but it has the danger of reinforcing the belief of those inclined to think that mathematicians will do anything, no matter how bizarre, to get equations that they can solve.

There is an old joke that applied mathematicians tell about themselves. It concerns a syndicate formed to apply scientific principles to horse race betting. A large number of experts are hired whose job it is to provide ingredients for a strategy, or model, to be assembled by an applied mathematician. After due passage of time a symposium is held in which the experts and the mathematician present the results of their efforts to the syndicate. A genealogist gives a dazzling multi-media presentation describing the blood-line data base she has assembled for all registered horses; a physiologist explains how, from knowledge of the muscle tissue, lung capacity and the like one can say something about an animal's potential performance; a statistician describes his performance data base and how this can be used to weight the odds; a game-theorist.....but you get the picture. Finally, at the end of all of these precise scientific presentations incorporating computer graphics and other high-tech tools, the applied mathematician stands up to explain how he has incorporated all of this information into the model. He goes to the blackboard; draws a circle; turns to the audience and says: "Let us model the horse by a sphere... ."

This is a fine joke when told well by an applied mathematician, but those engaged in my kind of work are less amused when it is told by others.

To get back to 'wrong theories.' A statement more to the point and less likely to be misunderstood is the following quote from the American poet Robinson Jeffers which appears as an epigram in the book on numerical combustion by Elaine Oran and Jay Boris: 'The mathematicians and the physics men have their mythology; they work alongside the truth never touching it; their equations are false but the things work.'

Modelers write down false equations and extract useful, physically relevant information from them.

A splendid example is the constant density approximation.

I was at a meeting some years ago at which several talks were given in which constant density was assumed in studies that told us a great deal about the physics of premixed flame stabilities, an active subject in the mathematical combustion community at that time. And then a non mathematician got up to describe some numerical calculations of a flame propagating in a tube. He got to a slide showing the density variations and, pointing to it, turned to us accusingly and said: "See! the density is not constant... ." Apparently he believed we thought the density could indeed be close to a constant value in a real flame. Well perhaps some of us did. It can not be denied that the mathematics community does have members who have never opened Lewis and von Elbe. Yet most of us were well aware that the constant density equations are false. And we were also aware that the famous paper by Burke and Schumann on diffusion flames was no less important because it assumed constant density. And we were also well aware of the important physics that was revealed by the work of the day using those false equations. The things worked, and the reason they worked is that a lot of important physics is not controlled by density variations, and so these do not affect the qualitative picture. One of the cleverest things we do as modelers is write down false equations that work.

Asymptotics

Of the various tools that we have for dealing with non linearities, asymptotics is arguably the most important in the modeling of difficult continuum systems. There has been an enormous amount of progress in the theoretical treatment of low Mach number combustion in the past 20 years or so, and the reason for this can fairly be attributed to the use of activation energy asymptotics. This is a strategy that is now familiar to many. It starts with modeling the chemistry by 1-step Arrhenius kinetics:

e.g. for premixed flames

$$Y \rightarrow \text{products}: \quad \text{rate} \sim Ye^{-E/RT};$$

for diffusion flames

$$X + Y \rightarrow \text{products}: \quad \text{rate} \sim XYe^{-E/RT}.$$

And then continues by carrying out rational asymptotics valid in the limit when the non-dimensional activation energy goes to infinity. In dealing with premixed flames, a suitable non dimensional parameter is E/RT_* where T_* is the flame temperature.

The asymptotic description is characterized by a thin flame sheet (reaction zone) and gauge function expansions both within and without the flame sheet.

In real physical systems the activation energy defined in this way is in the range 10-20, which is not very large, and it is intriguing that the method works as well as it does. This is often explained with the observation that the exponential provides a magnifier - 1/10 is not small but exp(-10) is. This is probably a reasonable explanation for why the reaction zone can be supposed thin, but does not account for the success of series truncation. And indeed, the method sometimes fails us.

It would obviously be useful if we could characterize those problems for which the method works. Experience suggests that a necessary condition for premixed flames is near similarity between T and Y in the sense

$$T + c_1 Y \simeq c_2 \quad (c_1, c_2 \text{ constants}).$$

This is satisfied in the case of adiabatic deflagration when the Lewis number is close to 1, for then

$$T + \frac{(T_b - T_f)}{Y_f} Y \simeq T_b$$

(T_f = supply temperature, Y_f = supply mass fraction, T_b = burnt gas temperature).

Meaningful stability results can be obtained in this case for example.

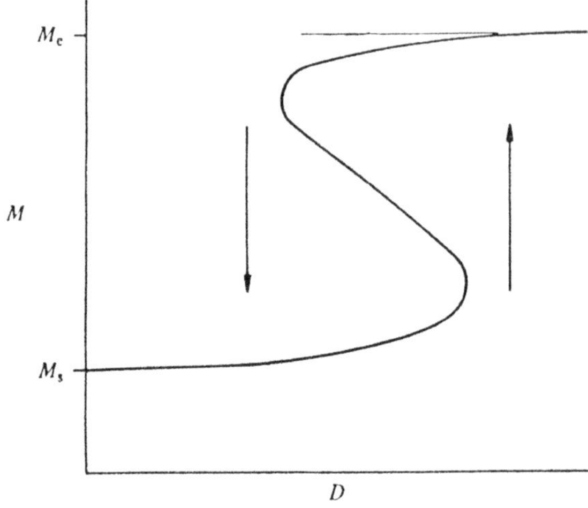

Figure 5. S-shaped response for a diffusion flame. When the Lewis number is less than 1 the upper and lower branches "attract" and are stable.

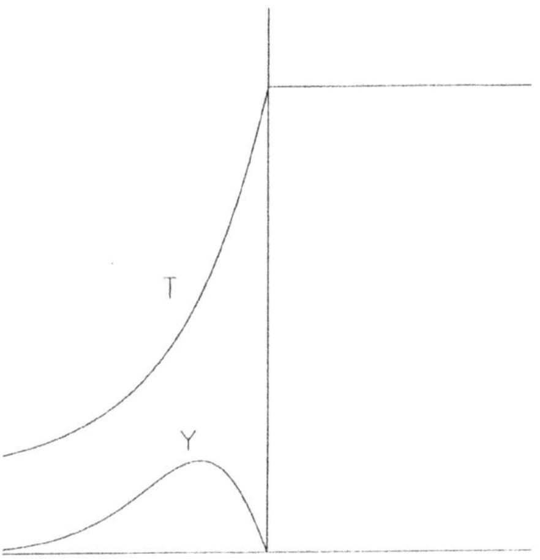

Figure 6. Distribution of temperature and fuel concentration typical of particle-cloud flames.

An example of a problem which does not satisfy the near-similarity condition and leads to non-physical results is that of a fuel-drop burning in a hot atmosphere. If the ambient temperature is close to the flame temperature there is significant leakage of oxygen through the flame sheet and the sheet structure is that of a premixed flame. The solution can be expressed as a graph of evaporation rate vs. Damkohler number, and this curve is, for some parameter values, S-shaped. When the Lewis number is less than 1, the asymptotic treatment predicts that the middle branch is unstable, the top and bottom branches stable, as one would (correctly) expect, Fig. 5. But if the Lewis number is greater than one this situation is reversed, a non-physical result (Buckmaster and Ludford, 1982). This is not to say that the asymptotics is wrong. There is no reason to believe that it fails to provide a mathematically proper description if the activation energy is large enough. But here 'large enough' puts us well beyond the range of physical values.

Another example, good and bad, is afforded by particle- (or drop-) cloud flames. These are premixed flames in which the fuel is supplied in the form of particles that gasify in the preheat zone, mixes with the air in which the particles are transported, and burns in a premixed reaction zone. Sensible results can be obtained if near-similarity is enforced. This requires (i) gasification complete near the cold- boundary, (ii) small particles so that there is insignificant lag between the velocities and temperatures of the two phases, (iii) a Lewis number of the stoichiometrically deficient component of the mixture close to 1.

These conditions are not satisfied in many applications of interest (Buckmaster and Clavin, 1992), Fig. 6, and then the results of a formal asymptotic treatment are not good. Fortunately there is an alternative strategy which works well. The thin flame-sheet approximation is retained; but the series truncation is omitted and the activation energy is treated as a finite parameter. This is known as the delta-function model (Buckmaster and Ludford, 1983).

One of the problems we have with detonations is that activation energy asymptotics does not work well. It tends to generate 'caricature behavior' rather than good physical behavior. As a consequence we are forced to rely on strategies which can have analytical components - linear stability, bifurcation for example - but in the end rely heavily on numerical ingredients. One of the few problems of detonation physics which comes readily to mind in which asymptotics is the central tool is the work of Stewart and Bdzil (1988) on slowly varying detonations in which the small parameter is the detonation thickness divided by a length scale on which the wave is evolving. The point is that if there is not a built in small parameter then typically we are reduced to special cases - very short wave disturbances (e.g. Choi and Majda, 1989), long wave disturbances, local analyses near bifurcation points, etc. This leaves a large range of important problems for which the only strategy is a numerical one.

I am not prepared to say too much about numerical approaches but, roughly speaking, these can be classified as (i) simulation, or (ii) modeling. By simulation I mean an effort to retain as much of the physics and quantitative detail as will lead to a numerical facsimile of the experiment. This is the kind of work that might, someday, put experimentalists out of business. In modeling one adopts strategies similar to those used in analytical modeling which were listed earlier. Physical insights come from performing numerical experiments - removing this physics, adding that. The greatest physical insight comes when analysis is intertwined with the numerical work as much as is possible. A good example is provided by Buckmaster, Smooke, and Giovangigli (in press), a numerical study of flame-balls. Detailed numerical simulations of steady flame-balls provide the necessary quantitative confirmation of their existence. Analysis reveals the importance of one- and three-dimensional stability analyses in explaining experimental observations.

Acknowledgments

I am grateful to M. Y. Hussaini and T. Jackson for helpful advice during the preparation of this talk. My combustion research is supported by AFOSR and the NASA Lewis Research Center.

References

Buckmaster, J. and Clavin, P., 1992. "An acoustic-instability theory for particle-cloud flames," Proceedings of the 24th International Symposium on Combustion.

Buckmaster, J. D. and Ludford, G. S. S., 1982. *Theory of Laminar Flames*, Cambridge University Press.

Buckmaster, J. and Ludford, G. S. S., 1983. *Lectures on Mathematical Combustion*, SIAM Press.

Buckmaster, J., Smooke, M., and Giovangigli, V., in press. "Analytical and numerical modeling of flame-balls in hydrogen-air mixtures," *Combustion & Flame*.

Choi, Y. S. and Majda, A., 1989. "Amplification of small-amplitude high-frequency waves in a reactive mixture," *SIAM Review* **31**, pp. 401-427.

Dold, J. W., Hartley, L. J., and Green, D., 1991. "Dynamics of laminar triple-flamelet structures in non-premixed turbulent combustion," in *Dynamical Issues in Combustion Theory*, eds. P. C. Fife, A. Linan, and F. Williams, *IMA* **35**, Springer-Verlag, pp. 83-105.

Macaraeg, M. G., Jackson, T. L., and Hussaini, M. Y., 1992. "Ignition and structure of a laminar diffusion flame in the field of a vortex," *Combustion Science and Technology* **87**, pp. 363-387.

Marble, F. E., 1985. "Growth of a diffusion flame in the field of a vortex," in *Recent Advances in Aerospace Sciences*, ed. by C. Cassei, pp. 395-413.

Pellett, G. L., Northam, G. B., and Wilson, L. G., 1991. "Counterflow diffusion flames of hydrogen, and hydrogen plus methane, ethylene, propane, and silane vs. air: Strain rates at extinction," AIAA Paper 91-0370.

Stewart, D. S. and Bdzil, J. B., 1988. "The shock dynamics of stable multidimensional detonation," *Combustion & Flame* **72**, pp. 311-323.

IGNITION AND FLAME SPREAD IN LAMINAR MIXING LAYERS

Amable Liñán

Escuela Técnica Superior de Ingenieros Aeronáuticos
Universidad Politécnica de Madrid

ABSTRACT

In order to identify some of the mathematical problems encountered in analyzing the ignition and flame spread in mixing layers, we shall describe the structure of the laminar mixing layer between two parallel streams of a fuel and air, initially separated by a splitter plate, undergoing an Arrhenius reaction.

If the activation energy of the reaction is lower than a critical value, there is only one steady solution of the problem, showing a transition from nearly frozen mixing to diffusion controlled combustion downstream of the plate.

For higher, typical, values of the activation energy we may find a multiplicity of solutions, depending on the value of the Damköhler number D, characterized by the temperature of the hotter of the two streams. For values of the Damköhler number lower than a critical value D_c, there is a solution where a thermal runaway is found to occur, after an induction length, at a point that serves as the origin of premixed flames that do not propagate upstream. For values of D larger than a critical lift-off value $D_l(< D_c)$ we find a solution with diffusion controlled combustion in a diffusion flame. This flame is anchored, with a triple-flame structure, in the near wake of the splitter plate, where upstream heat conduction to the plate plays a dominant role.

In the interval $D_l < D < D_c$ there is a third, unstable, solution. This solution determines where an external ignition source should be placed so that, by means of upstream triple-flame propagation to the splitter plate, transition to diffusion controlled combustion can take place.

1. Introduction

The chemical reaction in the mixing layer between two streams of fuel and oxidizer will take place in a thin diffusion flame, without

J. Buckmaster et al. (eds.), Combustion in High-Speed Flows, 461–476.

significant effects of the finite rate of the chemical reaction, only after spontaneous or artificially triggered ignition and only if the mixing layer is not subject to an excessive strain that may extinguish the flame.

Spontaneous ignition occurs, for one-step irreversible Arrhenius reactions, in the form of a thermal runaway due to the enhancement of the reaction rate by the heat release in the nearly frozen region of the mixing layer downstream of the splitter plate. A hot spot created by this process serves as the steady origin of premixed flames, if they are not fast enough as to propagate upstream to the nearly frozen region. The premixed flames separate the upstream nearly frozen region from two downstream regions of near-equilibrium, one without fuel and the other without oxidizer that are separated by a diffusion flame. An analysis of this spontaneous ignition process, and of the triple flame system which is set up in this way, was carried out by Liñán and Crespo (1976) for the unsteady mixing layer and extended later by Jackson and Hussaini (1988) and Grosch and Jackson (1991) to the analysis of compressible supersonic mixing layers. See also the review paper by Jackson (1992).

Although the spontaneous ignition may be expected to be the mechanism characterizing the transition from nearly frozen mixing to diffusion controlled combustion in the mixing layers encountered in supersonic combustion, this is not always the case in combustion in unpremixed systems. If we ignite locally the mixing layer, by means of a spark or a hot body, we create a premixed flame front that moves in the mixing layer, relative to the fluid, both upstream and downstream, leaving behind a diffusion flame. The premixed flame front, which together with the diffusion flame forms a triple flame, may propagate all the way up to the splitter plate, leaving the diffusion flame attached to the plate. If propagation up to the plate is not possible, then, the flame remains lifted-off or is blown away. See Liñán (1988) and the early and the more recent of work of Dold (1989) and coworkers on triple-flame propagation. See also the numerical work of Prasad and Price (1992) and the review of Takahashi and Schmoll (1990).

One could expect that the lift-off distance should be determined as the distance from the injector where the triple flame front velocity equals the local flow velocity. However this configuration is stable only if the flow velocity is decreasing downstream from the lift-off distance. In addition, the flame front velocity in the partially mixed, typically turbulent, mixing layers is different from the laminar premixed flame velocity for a stoichiometric mixture. The triple flame system must move along very thin mixing layers strongly distorted and strained by turbulence. Propagation along these layers is only possible if there is a large enough fraction of thin mixing layer regions not excessively strained to extinguish the diffusion flames. Peters and Williams (1983) used ideas of percolation theory to estimate the lift-off height in turbulent jet diffusion flames, where the evaluation of flame front propagation is difficult.

The lift-off height may, however, be determined in some cases by a spontaneous ignition process, as indicated before. The main purpose of this work is to show, analyzing as an example the low Mach number laminar mixing layer downstream of a thin splitter plate, how to identify the mechanism that determines the flame lift-off height and when the flame is attached to the splitter plate.

2. Characteristic Scales in the Near Region of the Mixing Layer

We shall analyze the structure of the steady laminar mixing layer between two parallel streams of fuel and oxidizer, separated by a thin plate. The Reynolds number R_B of the flow, based on the higher free stream flow velocity U_o, kinematic viscosity ν_o, and the thickness δ_B of the corresponding boundary layer just upstream of the end of the splitter plate, is considered to be large. Therefore, the boundary layer approximation can be used for the description of the mixing process between the two streams, thus neglecting the effects of upstream heat conduction and diffusion.

Jackson (1992) has called our attention to the role that the wake of the splitter plate and the initial boundary layer thickness may have in determining the ignition length and in the stability of the diffusion flame. The triple deck theory, introduced by Stewartson (1969) and Messiter (1970), for the analysis of the trailing edge region in boundary layer flows should be used, as done by Daniels (1977), to describe the viscous mixing in the trailing edge.

According to this theory, the effects of upstream heat conduction and diffusion always play a role in a small Navier-Stokes region embedded in the larger lower deck of the much larger triple deck region. Transverse heat conduction, diffusion and viscous effects are confined to the lower deck; but the pressure gradients, produced by

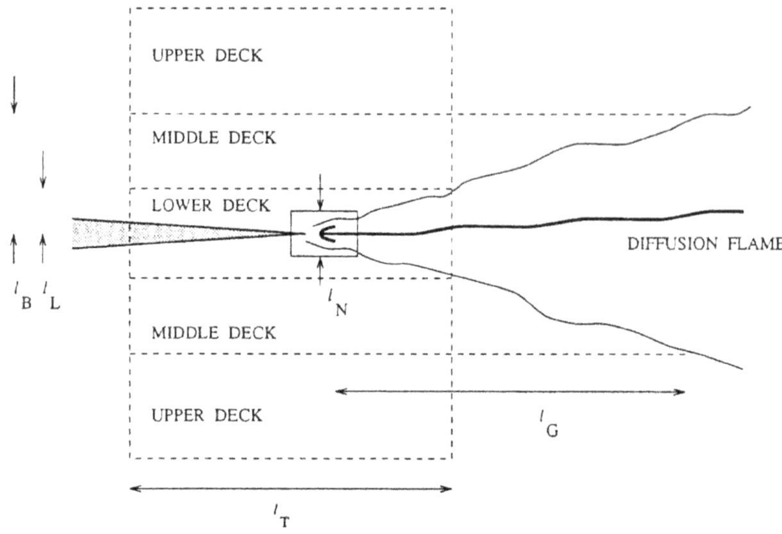

Figure 1. Skematic showing the Triple-deck, Navier-Stokes, and the Goldstein regions.

the displacement induced in the outer decks by the lower deck, also determine the flow velocity in the lower deck.

The analysis of the viscous lower deck determines the air and fuel stream velocity gradients A_o and A_F at the wall, at the edge of the splitter plate. These gradients determine the boundary conditions and, therefore, the flow structure in the Navier-Stokes region.

The characteristic sizes of the triple decks, sketched in Fig. 1, are given in terms of the Reynolds number R_B based on velocity U_o, boundary layer thickness l_B, and free stream kinematic viscosity ν_o by the relations

$$l_T/l_B = R_B^{1/4}, \qquad l_L/l_B = R_B^{-1/4} \tag{1}$$

determining the characteristic values of the size and streamwise extent l_T of the upper deck and of the thickness l_L of the lower deck. Let the wall velocity gradient on the air side just upstream of the Navier-Stokes region be A_o, of order l_B/U_o. Then, the size l_N of the Navier-Stokes region and the characteristic velocity U_N in this region are given by the relations

$$U_N/l_N = A_o, \qquad U_N l_N/\nu_0 = 1 \tag{2}$$

so that

$$l_N/l_B = U_N/U_o = R_B^{-1/2}. \tag{3}$$

Notice that $1/A_o$ is the residence time of the fluid particles in the Navier-Stokes region.

Mixing between the fuel and the oxygen of the air begins in the Navier-Stokes region and continues downstream, first in the lower of the decks and then in the viscous sublayer of the Goldstein region.

In the Goldstein region the velocity u_G along the dividing streamline and the thickness l_M of the mixing layer vary initially with the distance x from the splitter plate as

$$u_G/U_o = l_M/l_B = (x/l_G)^{1/3} \tag{4}$$

where l_G is the extent of the Goldstein region, given by the relation $l_B = \sqrt{\nu_o l_G/U_o}$ or

$$l_G/l_B = R_B. \tag{5}$$

3. Analysis of the Navier-Stokes Region

If the diffusion flame is attached to the splitter plate, the flame attachment region is in the Navier-Stokes region. There, upstream heat conduction losses from the flame to the colder splitter plate play an important role.

We shall, for simplicity in the presentation, consider that the thickness of the plate l_P is small compared with the size l_N of the Navier-Stokes region. Otherwise, the ratio l_P/l_N will affect the flame attachment process. We shall also consider that the plate conductivity is high enough so that its temperature will be equal to T_o, the temperature of the ambient air and fuel streams.

We shall write the conservation equations in non-dimensional form, using as scales: l_N for the spatial coordinates, U_N for the velocities, the free stream air values ρ_o, T_0 and α_o for the density, temperature, and thermal diffusivity. We shall measure with $\rho_o U_N^2$ the spatial pressure variations p' from the trailing edge value given by the outer triple-deck structure. Here we redefine l_N and U_N in terms of α_o and the wall velocity gradient A_o by the relations

$$l_N = U_N/A_o = \sqrt{\alpha_o/A_o}. \tag{6}$$

We shall consider that the reaction between the oxygen and the fuel is an Arrhenius reaction consuming a mass s of oxygen and releasing an energy q per unit mass of fuel. The mass rate of consumption of the fuel per unit volume is

$$w_F = \rho B Y_F Y_o \, exp(-E/RT) \tag{7}$$

in terms of the frequency factor B and activation energy E.

We shall measure Y_F and Y_o in terms of their free-stream values Y_{F0} and Y_{O0} to obtain Y_1 and Y_2.

Then the conservation equations take the form

$$\nabla \cdot (\rho \vec{v}) = 0 \tag{8}$$

$$\rho \vec{v} \cdot \nabla \vec{v} = -\nabla p' + P_r \nabla \cdot \tau \tag{9}$$

$$\rho \vec{v} \cdot \nabla Y_1 - \frac{1}{L_1} \nabla \cdot (\rho \alpha \nabla Y_1) = -\rho D Y_1 Y_2 e^{-\beta/T} \tag{10}$$

$$\rho \vec{v} \cdot \nabla Y_2 - \frac{1}{L_2} \nabla \cdot (\rho \alpha \nabla Y_2) = -S \rho D Y_1 Y_2 e^{-\beta/T} \tag{11}$$

$$\rho\vec{v}\cdot\nabla T-\nabla\cdot(\rho\alpha\nabla T)=\gamma(S+1)\rho DY_1Y_2e^{-\beta/T} \tag{12}$$

and the equation of state

$$\rho T=1 \tag{13}$$

if the mean molecular mass is constant.

In these equations, τ is the non-dimensional viscous stress tensor, and the main parameters are the Lewis Numbers, L_1 and L_2, of the fuel and of the oxygen, the Prandtl number P_r, together with the Damköhler number, or non-dimensional value of the frequency factor, $D=Y_{oo}B/A_o$, the activation energy $\beta=E/RT_o$, the heat release $\gamma(S+1)=Y_{FO}q/c_pT_o$, and the stoichiometric ratio $S=sY_{FO}/Y_{oo}$.

These equation are to be solved with the boundary conditions

$$Y_1=Y_2-1=T-1=u-y=v=p'=0 \tag{14}$$

for $y\to\infty$ or $y>0$, $\quad x\to-\infty$

$$Y_2=Y_1-1=T-1=u=v=p'=0 \tag{15}$$

for $y\to-\infty$ or $y<0$, $\quad x\to-\infty$.

At the splitter plate $y=0^+$ or $y=0^-$, with $x<0$

$$T=1,\qquad \partial Y_1/\partial y=\partial Y_2/\partial y=0. \tag{16}$$

We have written here the boundary conditions for the case where the velocity of one of the streams is negligible compared with the other. In this case, in first approximation, there is no upstream perturbation in the triple-deck region, and A_o is the trailing edge value of the unperturbed, boundary layer velocity gradient at the wall. If the two stream velocity gradients are comparable, we should replace the condition $u=0$ in Eq. (15) by $u+ay=0$, and then an additional parameter $a=A_F/A_o$ appears in the equations.

Downstream, the solution must evolve toward the boundary layer similarity solution associated with the near plate region of the lower deck, with the reaction taking place in a diffusion controlled way if D is large enough.

It should be noticed that the structure of the Navier-Stokes diffusion flame attachment region is locally two-dimensional, due to its small size, and given by the previous equations, even if the global flow is unsteady or turbulent as long as the lower of the characteristic times $1/A_o$ or $1/A_F$ associated with the wall velocity gradients are small compared with the time characterizing their time variations.

Notice that the two wall velocity gradients A_F and A_o, just upstream of the edge of the splitter plate, are determined by the outer triple-deck structure. If the diffusion flame is anchored at the Navier-Stokes region, the displacement velocities induced by the thermal expansion in the lower deck will produce overpressures in the trailing edge region. These overpressures decrease the values of the wall velocity gradients from their unperturbed values, if a is not negligible or large compared with 1. In fact, the displacement effects may even lead to boundary layer separation upstream of the splitter plate, and then, the character of the flow structure will change significantly.

The solution of the problem of Eqs. (8)-(16) should give the transition to the diffusion controlled combustion downstream, although this transition will take place for values of x of order unity only if D is large enough.

If the Lewis numbers can be considered as equal to 1, we can derive, from the conservation Eqs. (10)-(12), two conservation equations,

$$\rho\vec{v} \cdot \nabla Z - \nabla \cdot (\rho\alpha\nabla Z) = 0 \tag{17}$$

$$\rho\vec{v} \cdot \nabla H - \nabla \cdot (\rho\alpha\nabla H) = 0 \tag{18}$$

for the mixture fraction,

$$Z = (SY_1 + 1 - Y_2) / (S + 1), \tag{19}$$

and total enthalpy H,

$$H = Y_1 + Y_2 + (T - T_e)/(T_e - 1), \tag{20}$$

where T_e is the non-dimensional adiabatic flame temperature, $T_e = 1 + \gamma$. H and Z satisfy the boundary conditions $H = 0$, $Z = 1$ on the fuel free stream and $H = Z = 0$ in the air free stream.

When the reaction can be modelled by an Arrhenius reaction, the non-dimensional activation energy β is typically large compared with 1, but the value of D will also be very large. We can take advantage of the large values of β and D to carry out an asymptotic description of the problem.

Let us define a non-dimensional characteristic temperature T_c, by the relation

$$(T_c/\beta)^3 D\exp(-\beta/\mathrm{T_c}) = 1. \tag{21}$$

T_c is such that at this temperature the reaction term in Eqs. (10)-(12) is of the same order as the diffusion terms. The factor $(T_c/\beta)^3$ has been introduced to account, partially, for the small transverse size of the reaction zone and of the values of the reactant concentrations there.

We can now re-write, for the large activation energy analysis, the reaction term in Eqs. (10)-(12) in the form

$$DY_1Y_2\exp(-\beta/T_c) = (\beta/T_c)^3 Y_1Y_2\exp\{\beta(T-T_c)/TT_c\}$$

and then use the fact that $\beta/T_c \gg 1$ to describe the solution of the problem in terms of the infinite activation energy solution $\beta/T_c \to \infty$, with T_c fixed. In this limit, $\beta \to \infty$, T_c fixed, the chemical reaction is frozen for $T < T_c$ and infinitely fast for $T > T_c$. In the region of low temperature, $T < T_c$, the chemical reaction is frozen and the reactants can mix. In the region where $T > T_c$ the reactants can not coexist; one of the reactants has been completely consumed.

Therefore, we find three regions in the flow field where the chemical reaction term is zero. Namely,

i) A frozen region, Ω_f, where $T < T_c$ and $Y_1Y_2 \neq 0$.

ii) An equilibrium region, Ω_1, without oxygen, where $T > T_c$ and $Y_2 = 0$.

iii) An equilibrium region, Ω_2, without fuel, where $T > T_c$ and $Y_1 = 0$.

The regions Ω_f and Ω_1 are separated by an infinitely thin, rich, premixed flame, where all the oxygen reaching the flame by diffusion is burned with part of the fuel reaching the flame by diffusion. The regions Ω_f and Ω_2 are similarly separated by a lean premixed flame. The regions Ω_1 and Ω_2 are separated by an infinitely thin diffusion flame.

The reaction terms appearing in Eqs. (10)-(12) take the form of Dirac delta functions along these flame sheets with strengths determined by the conditions of continuity of the temperature and concentrations there, together with the condition $Y_1 = 0$ in Ω_2 and $Y_2 = 0$ in Ω_1. Notice that, in this limit of infinite activation energies, all the kinetic information is retained only in the parameter T_c, the premixed flame temperature that is very precisely determined by Eq. (17).

If the flame temperature T_c, defined by Eq. (21), is lower than the plate temperature, namely for values of $D > D_d$ defined in order of magnitude by the relation

$$D_d \beta^{-3} e^{-\beta} = 1 \tag{22}$$

then, the quench region disappears, and the diffusion flame reaches the plate. If the resulting value of T_c is larger than the non-dimensional adiabatic flame temperature, $T_e = 1 + \gamma$, the premixed flame does not lie in the Navier-Stokes region. Or, in other words, for values of the Damköhler number D lower than a lift-off Damköhler number D_l, defined in order of magnitude by the relation

$$D_l \left\{\beta/(1+\gamma)\right\}^{-3} \exp\left\{-\beta/(1+\gamma)\right\} = 1 \tag{23}$$

the diffusion flame is lifted off, away from, the Navier-Stokes flame attachment region.

The condition (23) for the flame lift off is roughly equal to the condition that the residence time through the stoichiometric, fuel/air, premixed flame is equal to the residence time, $1/A_o$, in the Navier-Stokes region.

For values of the Damköhler number in the interval (D_l, D_d) we therefore encounter a combustion regime, where the diffusion flame is attached to the Navier-Stokes region and ends in a V shaped premixed flame system. The premixed flame temperature is T_c, larger than the plate temperature and lower than the adiabatic flame temperature T_e. When the Lewis numbers of the reactants are equal to 1, the values of the concentrations and of the temperature are related to the local value of H. Due to the heat losses by heat conduction to the splitter plate, H is negative in the Navier-Stokes region and rises to zero when we move away from the region. Thus, the temperature T_f along the diffusion flame, that lies on the stoichiometric surface $Z = Z_s = 1/(S+1)$, rises from the value T_c at the tip of the flame to the adiabatic flame value T_e at distances from the plate large compared with l_N.

For the precise evaluation of the values D_d and D_l, we should carry out a more refined asymptotic analysis, which is difficult, or a numerical integration of the equations with the Damköhler number D replaced by δD_d in one case and δD_l in the other to lessen the influence of β in the resulting value of δ.

For values of $D < D_l$, there is another weakly burning solution of the problem, for which if the activation energy is large the chemical reaction can be neglected in the Navier-Stokes region. For values of D in the interval (D_l, D_c), with D_c to be defined later, we may encounter three solutions of the problem for the typically large values of the activation energy larger than a critical value β_c.

4. Triple Flame Downstream of the Navier-Stokes Region

For large values of the activation energy, there is a second, unstable, solution of the problem of Eqs. (8)-(16) for values of the Damköhler number D larger than D_l. In this solution a triple-flame front (also formed by a premixed flame, with rich and lean branches, and a trailing diffusion flame) is established at a distance l_F from the splitter plate that, when measured with l_N, is large compared with unity if D is moderately large compared with D_l.

When D begins to rise above D_l, the leading edge l_F of the unstable flame front begins to move downstream of the Navier-Stokes to the lower deck region in the wake of the plate and, for larger values of D, to the Goldstein region.

When the, unstable, triple-flame front lies, for values of D close to D_l, in the Navier-Stokes region, the flame feels the effects of heat losses to the splitter plate. These effects determine, also for the main stable solution, the flame front position and shape.

The heat lost to the splitter plate from the unstable flame becomes negligible when, for larger values of D, the flame front moves downstream to distances large compared with l_N. In this case, in the triple-flame front structure, we must retain upstream heat conduction and diffusion effects, but only in a preheat transport zone upstream of the reaction sheet of thickness l_p small compared with the local thickness l_M of the mixing layer. The local velocity of the premixed flame relative to the flow, determined by the upstream concentration and the temperature T_o, takes its maximum value where the upstream mixture is close to stoichiometric and decreases rapidly away from this surface. For this reason the flame front is curved with a radius of curvature l_M/β.

The boundary layer approximation can be used to describe the concentration field upstream of the triple-flame front region.

The flow in the nose region of the flame front, at distances from it of the order l_M/β, is rotational downstream of the thin premixed flame with overpressures that deflect the incoming streamlines and slow the flow in the stoichiometric streamline. For this reason the flame front propagation velocity U_p, relative to the unperturbed flow, is larger than, although of the order of, the planar stoichiometric flame velocity. U_p is given, in order of magnitude, in dimensional variables by the relation

$$U_p^2 = \alpha_o (RT_e/E)^3 BY_{FO} e^{-E/RT_e}. \tag{24}$$

This velocity is equal to the characteristic flow velocity in the viscous sublayer of the Goldstein region, given by Eq. (4) or, equivalently, by

$$u_G/U_N = (x/l_N)^{1/3}. \tag{25}$$

Therefore, the unstable flame front distance l_F, when it lies in the Goldstein region, is given, using Eqs. (24) and (25), by the relation

$$(l_F/l_N)^{2/3} = (\alpha_o/U_N^2)(RT_e/E)^3 BY_{FO} e^{-E/RT_e}. \tag{26}$$

or, equivalently, by the relation

$$l_F/l_N = (D/D_l)^{3/2}. \tag{27}$$

The solution with the triple flame in the Goldstein region is unstable but determines where we should place (at $x < l_F$) an ignition source to insure that upstream propagation of the generated flame front to the splitter plate is possible.

The Goldstein region ends when the thickness of the mixing layer becomes equal to the initial boundary layer thickness l_B, that is, at a distance $l_G = R_B l_B$. The characteristic fluid velocity along the stoichiometric surface, which should be used to calculate the flame front lift-off distance, becomes constant for values of the distance larger than l_G. Thus, for a value D_b/D_l of D/D_l, of order R_B, the unstable triple-flame leaves the Goldstein region, and the lift-off distance position is not well determined until, due to the merging of mixing layers in jet flows, the velocity along the stoichiometric surface begins to decrease with the downstream distance. Thus, at $D = D_b$, we encounter a bending bifurcation to a new, triple-flame, stable, solution branch with higher values of l_F for $D < D_b$. See Chung and Lee (1991).

5. Spontaneous Ignition Regime

For values of the Damköhler number below a critical value, D_c, that will be determined below, there is a third type of solution of the problem of Eqs. (8)-(16).

In this regime, analyzed by Jackson and Hussaini (1988) for laminar supersonic mixing layers without accounting for the wake effect associated with the original boundary layer thickness, the effects of the chemical reaction can be neglected in the Navier-Stokes region where upstream diffusion and conduction are important. The boundary layer approximations can be used downstream to describe how the heat release, due to the reaction, slowly increases the temperature to finally produce a thermal runaway. If this occurs within the Goldstein region, the distance l_I from the splitter plate at which the thermal runaway occurs is defined, in order of magnitude, by equating the residence time in the ignition region to the adiabatic explosion time. Namely, using the relation

$$1/t_I = u_G/l_I = \gamma(S+1)\beta Y_{oo} B e^{-E/RT_o} \tag{28}$$

together with Eq. (25). That is, l_I is given by

$$(l_I/l_N) = (A_o t_I)^{3/2} = \{\gamma(S+1)\beta D e^{-\beta}\}^{-3/2} \tag{29}$$

or

$$l_I/l_N = \left\{A_o \frac{(RT_o)^2}{qE} \frac{\exp(\mathrm{E}/\mathrm{RT_o})}{\mathrm{Y_{FO}Y_{oo}B}}\right\}^{3/2}. \tag{30}$$

This ignition length is to be compared with the unstable flame front lift-off distance l_F. If $l_I < l_F$ for values of the Damköhler number larger than a critical value D_c,

$$\frac{D_c}{D_l} = \frac{(1+\gamma)^{3/2}}{\beta^2\sqrt{\gamma(S+1)}} \exp\{\frac{\gamma\beta}{2(1+\gamma)}\}, \tag{31}$$

determined, in order of magnitude, by the condition $l_I = l_F$. The calculated spontaneous ignition length (30) does not have physical meaning because upstream flame propagation from the point of thermal runaway is possible.

6. Concluding Remark

We summarize the results in Fig. 2 where the flame stand-off distance for the V flame, the unstable triple-flame, and the spontaneous ignition lengths are represented schematically in terms of the Damköhler number D/D_l.

We find the results that we anticipated in the introduction.

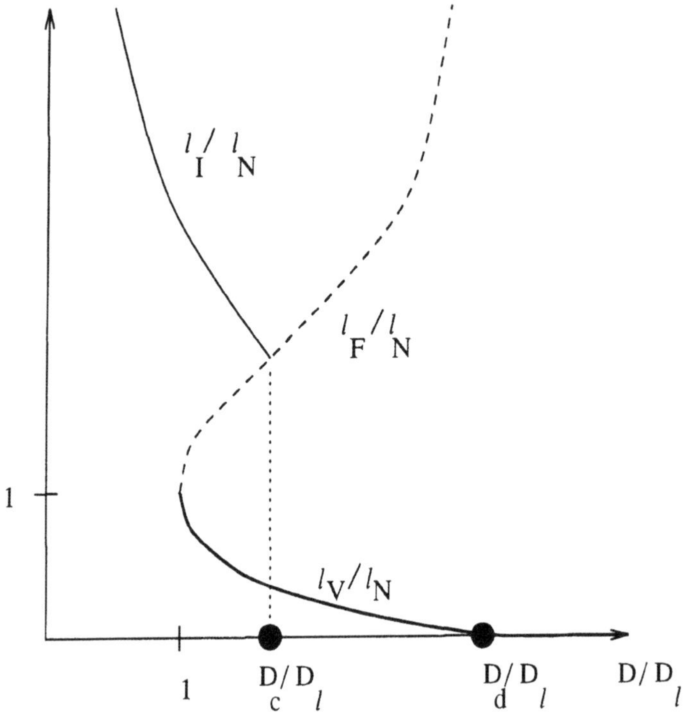

Figure 2. The flame lift-off distance in terms of the Damköhler number.

Acknowledgment

This research has been partially supported by the Spanish CICYT under Project No. ESP-90-013001.

References

Chung, S. H. and Lee, B. J., 1991. "On the characteristics of laminar lifted flames in a non-premixed jet," *Comb. and Flame* **86**, pp. 62-72.

Daniels, P. G., 1977. "Viscous mixing at a trailing edge," *Quart J. Mech. & Appl. Math.* **30**, pp. 319-42.

Dold, J. W., 1989. "Flame propagation in a non-uniform mixture: Analysis of a slowly varying triple flame," *Comb. and Flame* **76**, pp. 71-88.

Dold, J. W., Hartley, L. J., and Green, D., 1991. "Dynamics of laminar triple-flamelet structures in non-premixed turbulent combustion," in *Dynamical Issues in Combustion Theory*, Fife, P. C., Liñán, A., and Williams, F. A. (Eds.), IMA Volumes in Mathematics and its Applications, 35, Springer Verlag, pp. 107-126.

Grosch, C. E. and Jackson, T. L., 1991. "Ignition and structure of a laminar diffusion flame in a compressible layer with a finite rate chemistry," *Phys. Fluids* **A3**, pp. 3087-97.

Jackson, T. L., 1992. "Stability of laminar diffusion flames in compressible mixing layers," Published in *Major Research Topics in Combustion*, Springer-Verlag, Hussaini, Kumar and Voigt, eds., pp. 131-161.

Jackson, T. L. and Hussaini, M. Y., 1988. "An asymptotic analysis of supersonic reacting mixing layers," *Comb. Sci. Tech.* **57**, pp. 129-140.

Liñán, A., 1988. "Diffusion flame attachment and flame front propagation along mixing layers," Edited by C. M-Brauner and C. Schmidt-Lain, NATO ASI Series E, Vol. 140, pp. 151-154, Martinus Nijhoff.

Liñán, A. and Crespo, A., 1976. "An asymptotic analysis of unsteady diffusion flames for large activation energies," *Comb. Sci. Tech.* **14**, pp. 95-117.

Messiter, A. F., 1970. "Boundary layer flow near the trailing edge of a flat plate," *SIAM J. Appl. Math.* **18**, pp. 241-257.

Peters, N. and Williams, F. A., 1983. "The lift-off characteristics of turbulent jet diffusion flames," *AIAA J.* **21**, pp. 423-429.

Prasad, K. and Price, E. W., 1992. "A numerical study of the leading edge of laminar diffusion flames," *Comb. and Flame* **90**, pp. 155-173.

Stewartson, K., 1969. "On the flow near the trailing edge of a flat plate," *II Mathematika* **16**, pp. 106-21.

Takahashi, F. and Schmoll, W. J., 1990. "Lifting criteria of jet diffusion flames," 23rd Symposium (Int.) on Combustion, The Combustion Institute, pp. 677-683.

NUMERICAL AND ASYMPTOTIC ANALYSIS OF IGNITION PROCESSES

C. Treviño[1] *and A. Liñán*

E.T.S.I. Aeronáuticos, UPM
Plaza del Cardenal Cisneros 3, 28040 Madrid, Spain

ABSTRACT

In this paper the transient process leading to ignition of a combustible mixture of lean hydrogen/air is studied using both Computational Singular Perturbation (CSP) techniques as well as asymptotic methods that allow us to deduce the most important regimes in the process as well as the characteristic times and the appropriate reduced kinetic mechanism involved. The CSP technique developed by Lam and co-workers has been used to follow closely the temperature and species concentration evolution with time up to the ignition event. For lean mixtures, the most important intermediate species is the atomic oxygen; the other radicals follow the quasi-steady state behavior. A closed form solution is then obtained for the ignition delay time, which compares very well with the numerical solution with full kinetics.

1. Introduction

To obtain of reduced kinetic schemes for processes with complex chemical kinetics is very attractive for several reasons. It reduces computer time and can be used to study more complex fluid mechanical problems as turbulent combustion with realistic kinetics. The simplest case is the ignition process of hydrogen/air mixtures where the detailed reaction mechanism is very well known (Baulch et al., 1972; Warnatz, 1984; and Dixon-Lewis et al., 1977). Several analyses have been published regarding the ignition of H_2/O_2 mixtures (Brokaw, 1965; Gardiner et al., 1981; Kailasaneth et al., 1982; Wiriyawit and Dabora, 1984; and Maas and Warnatz, 1989). The Zeroth dimensional analysis had been found to be enough to describe the ignition phenomena. Recently, efforts have been conducted in order to deduce a reduced kinetic mechanism that globally contains

[1]Currently, Facultad de Ciencias, UNAM, Mexico

J. Buckmaster et al. (eds.), Combustion in High-Speed Flows, 477–490.

the essential features of the studied process (Peters, 1985). The procedure to obtain the reduced mechanism has been improved by the CSP (Computational Singular Perturbation) technique developed by Lam and Goussis (1989). Here, the chemical reactions are reordered in reaction groups associated to the same order of magnitude of the characteristic times. These reaction groups can be active, dead or dormant depending on the relation of their characteristic times to the actual relevant characteristic time of the studied process. The dead groups are related to fast chemical reactions from which the appropriate quasi-steady and partial equilibrium assumptions can be derived. The dormant groups are related to slow chemical reactions playing no role at that moment. From the active and dead groups the global mechanism can be obtained. Recently, Treviño (1990) studied the zeroth-dimensional ignition of H_2/O_2 mixtures in an isochoric adiabatic reactor. The reduced kinetic mechanism has been obtained after solving the evolution equations. Three different regimes were identified for the ignition process, depending on the initial temperature and pressure. For initial temperatures larger than the critical temperature (temperature that makes reaction rates of the chain branching reaction $H + O_2 \rightarrow OH + O$ and the chain terminating reaction $H + O_2 + M \rightarrow HO_2 + M$ equal in magnitude), the ignition can be described as a typical chain branching explosion with no heat release in a first approximation. As the temperature decreases to values close to the critical temperature, the heat release has to be retained in order to improve the ignition time. As the temperature decreases further, the chain branching reaction loses in importance and the intermediate species evolution control the process. A thermal runaway is produced in this regime characterizing the ignition process at low initial temperatures. In all three regimes, an analytical expression has been obtained for the ignition delay time. In the reduced kinetic mechanism for the high temperature regime, where the ignition process consumes 50% more fuel as given by stoichiometry, it is assumed that the rate limiting chemical reaction is $H + O_2 \rightarrow OH + O$. This is true for stoichiometric and rich mixtures where abundant H_2 exists. In this case all H_2 consuming shuffle reactions are very fast. However, this reduced kinetic scheme is not valid for lean mixtures as found for example in the ignition of hydrogen in mixing layers in supersonic combustion engines, where the air temperature is much larger than that of the fuel.

The main objective of the present study is to deduce the relevant reduced kinetic mechanism valid for the high temperature ignition of lean H_2/air mixtures, using the CSP technique in order to obtain analytical expressions for the ignition delay time, using asymptotic methods.

2. Governing Equations

The zero-dimensional reacting governing equations of an homogeneous hydrogen/oxygen mixture in an adiabatic reactor are given by

$$\rho_0 c_p dT/dt - dp/dt + \sum_{i=1}^{N} \dot{C}_i h_i M_i = 0 \tag{1}$$

$$dC_i/dt = \dot{C}_i, \; i = 1, N, \tag{2}$$

where ρ_0 corresponds to the mixture density assumed to be constant; c_p is the specific heat at constant pressure; T is the mixture temperature; t corresponds to the time; p is the pressure; C_i corresponds to the molar concentration (mol/volume) of the specie i; h_i is the specific enthalpy of species i; M_i is the molecular mass of species i. N corresponds to the total number of species taking part in the ignition process. The mixture density is related to the species molar concentrations by

$$\rho_0 = \sum_{i=1}^{N} C_i M_i = \text{ constant.} \tag{3}$$

Assuming an ideal gas mixture, the state equation is given as

$$p = RT \sum_{i=1}^{N} C_i, \tag{4}$$

where R is the universal gas constant, $R = 8.314$ KJ/(mol K). From Eqs. (1) and (4), the energy equation (1) transforms to

$$\rho_0 c_v dT/dt + \sum_{i=1}^{N} \dot{C}_i (h_i M_i - RT) = 0, \tag{5}$$

where c_v corresponds to the specific heat at constant volume. The species production rate, $\dot{C}_i$ is given by

$$\dot{C}_i = \sum_{j=1}^{M} K_j(T) \nu'_{ij} \Pi_{k=1}^{N} C_k^{\nu_{kj}}, \text{ with } K_j(T) = A_j T^{n_j} \exp(-E_j/RT). \tag{6}$$

Here, M corresponds to the total number of elementary chemical reactions; v'_{ij} and v_{ij} corresponds to the stoichiometric coefficients of the species i in the chemical reaction j, as product and as reactant, respectively. The constants A_j, n_j and E_j are given in Table 1 for all chemical reactions considered.

The solution of the governing equations (2) and (5), with the corresponding initial conditions,

$$T = T_0 \text{ and } C_i = C_{i0} \text{ at } t = 0, \tag{7}$$

is obtained using conventional codes for stiff differential equations (DGEAR). This numerical code is driven by a CSP routine, built parallel to the main program. The procedure employed in order to obtain the reduced kinetic mechanism is by dropping out that reactions with a participation index (Lam and Goussis, 1989) lower than a critical value in the whole event leading to the ignition of the mixture. This critical value is the maximum value obtained when no appreciable change in the ignition time results after dropping these chemical reactions. The resulting set of chemical reactions can be reduced after the introduction of steady-state behavior of several species. In the present study 38 elementary chemical reactions are employed as the detailed kinetic mechanism as in Maas and Warnatz (1989).

3. Asymptotic Analysis

The ignition process of hydrogen with oxygen changes dramatically depending on the initial gas temperature. A critical temperature is obtained from the competition between the chain branching reaction (1) and the chain braking reaction (9). This crossover temperature is denoted by T_C. From the CSP data, the following elementary reactions are important for high initial temperatures and lean mixtures. At the beginning (no radicals present), reaction 14

$$(14) \qquad H_2 + O_2 \rightarrow HO_2 + H$$

is the important initiation step. Once trace amounts of radicals are present, this reaction has no more influence on the process in this high temperature limit. From this point up to the end, the three forward shuffle reactions are very important and have to be retained in the ignition process. These reactions are:

$$(1) \qquad H + O_2 \rightarrow OH + O$$

$$(3) \qquad H_2 + O \rightarrow OH + H$$

$$(5) \qquad H_2 + OH \rightarrow H_2O + H.$$

Due to the fact that H_2 is consumed rapidly in the ignition process of lean mixtures, reaction (7) becomes the most important OH consumer reaction

$$(7) \qquad OH + OH \rightarrow H_2O + O$$

HO_2 is produced mainly through reaction (9), whose rate is very slow compared with the other reactions for high initial temperatures

$$(9) \qquad H + O_2 + M \rightarrow HO_2 + M.$$

The HO_2 is being consumed mainly by reactions (15) and (19)

$$(15) \qquad HO_2 + OH \rightarrow H_2O + O_2$$

$$(19) \qquad HO_2 + O \rightarrow OH + O_2.$$

For very lean mixtures, reactions (3) and (5) are not very fast due to the low concentration of H_2 and become the rate limiting reactions in the ignition process. Reaction (1) now is very fast and makes the hydrogen atom to behave in quasi steady-state. A second run of the computer program with only these reactions (14, 1, 3, 5, 8, 9, 15, 19), reproduces the same ignition delay time as with all the reactions included. The backward reactions (mainly radical-radical) have no influence on the ignition process.

From the CSP data it is observed that the radicals OH, H and HO_2 are produced and consumed at rates much larger than the radical O. It means that the production-consumption of these radicals are contained in the dead reaction groups for all times up to ignition. Therefore, the reaction rate of reactions (3) and (5) are limited by the radical pool concentration and the species H, OH, and HO_2 can be assumed to be in steady-state. The kinetic mechanism reduces to

$$(I) \qquad H_2 + O_2 \stackrel{3+5+14}{\rightarrow} H_2O + O$$

$$(II) \qquad O + O \stackrel{9}{\rightarrow} O_2,$$

where the number above the arrow represents the rate of the elementary reaction associated with the global step. From the steady-state assumptions for H and OH, the concentration of these radicals are related to that of atomic oxygen as

$$\frac{C_H}{C_{H2}} = \frac{K_{14}}{(K_1 + \overline{K}_9)} + \frac{K_3 C_O}{(K_1 + \overline{K}_9) C_{O2}} + \frac{K_5 K_3^{1/2} C_{H2}^{1/2} C_O^{1/2}}{(K_1 + \overline{K}_9) K_7^{1/2} C_{O2}} \tag{8}$$

$$C_{OH} = \frac{K_3^{1/2} C_{H2}^{1/2} C_O^{1/2}}{K_7^{1/2}}, \tag{9}$$

where $\overline{K}_9$ corresponds to the equivalent two-body reaction rate of the three body reaction (9). The reaction rate of the two global steps are then given by

$$w_I = K_{14} C_{O2} C_{H2} + K_3 C_O C_{H2} + K_5 K_3^{1/2} K_7^{-1/2} C_{H2}^{3/2} C_O^{1/2} \tag{10}$$

$$w_{II} = \frac{\overline{K}_9}{(K_1 + \overline{K}_9)} w_I. \tag{11}$$

For high initial temperatures, the reaction rate of reaction (9) is very small compared reaction (1), that is $K_1 \gg \overline{K}_9$ and then the second global step is very slow and can be neglected in the ignition process in a first approximation. Global step I corresponds to a chain branching step with radical O as the chain branching species. We introduce the following nondimensional variables

$$x = \frac{C_O}{(C_{H2})_0}; \; y = \frac{C_H}{(C_{O2})_0}; \; u = \frac{C_{H2}}{(C_{H2})_0}; \; s = (K_3)_0 (C_{H2})_0 t \tag{12}$$

where $(\;)_0$ is the value of the variable at $t = 0$ and $T = T_O$. From coupling relationships we obtain

$$u = 1 - x; \; y = 1 - 2\phi x.$$

The evolution equation for the nondimensional concentration of the atomic oxygen can be written as

$$\begin{aligned} \frac{dx}{ds} = {} & \frac{K_3}{(K_3)_0} x(1-x) + \frac{K_{14}}{(K_3)_0} \frac{1}{2\phi} (1-x)(1-2\phi x) \\ & + \frac{K_5 K_3^{1/2}}{(K_3)_0 K_7^{1/2}} (1-x)^{3/2} x^{1/2}, \end{aligned} \tag{13}$$

where ϕ corresponds to the equivalence ratio and then

$$(C_{H2})_0 = \frac{p}{RT}\frac{2\phi}{(2\phi+4.76)}; \; (C_{O2})_0 = \frac{p}{RT}\frac{1}{(2\phi+4.76)}. \qquad (14)$$

For high initial temperatures and very lean mixtures of hydrogen, the energy release can be neglected in a first approximation. The ignition is then a chain branching process. The ignition criterion used in this analysis is related with the condition for partial equilibrium of one of the shuffle reactions. The election of one or the other brings only very small errors in the ignition delay time. Assuming reactions (7) and (8) reach partial equilibrium then the final nondimensional atomic oxygen concentration is then

$$x_f = \frac{K_3}{K_3 + K_8}. \qquad (15)$$

Introducing a normalized nondimensional atomic oxygen concentration, $z = x/x_f$, Eq. (13) takes the form

$$\frac{dz}{ds} = \gamma_3 z(1-x_f z)+\gamma_{14}(1-x_f z)(1-x_f\phi z)+\gamma_5(1-x_f z)^{3/2}z^{1/2} \qquad (16)$$

where

$$\gamma_3 = \frac{K_3}{(K_3)_0}; \; \gamma_{14} = \frac{K_{14}}{(K_3)_0}\frac{1}{2\phi x_f} \text{ and } \gamma_5 = \frac{K_5 K_3^{1/2}}{(K_3)_0 K_7^{1/2} x_f^{1/2}}.$$

In Eq. (5) γ_{14} is very small and γ_5 is small compared with unity. Neglecting the effect of reaction (5), due to the fact that both C_{OH} and C_{H2} are very small for very lean mixtures ($\phi \to 0$), Eq. (16) can be integrated in this limit, resulting

$$s_{ig} = \frac{1}{1+2\gamma_{14}x_f} Ln\left\{\frac{1+\gamma_{14}}{\gamma_{14}}\frac{1+\gamma_{14}x_f}{1-x_f(1-\gamma_{14})}\right\}. \qquad (17)$$

In physical units, the ignition delay time is given by

$$t_{ig} = \frac{RT_0(2\phi+4.76)}{2\phi p K_3} s_{ig} \approx \frac{RT_0(2\phi+4.76)}{2\phi p K_3} Ln\left\{\frac{2K_3^2\phi}{K_{14}K_8}\right\}. \qquad (18)$$

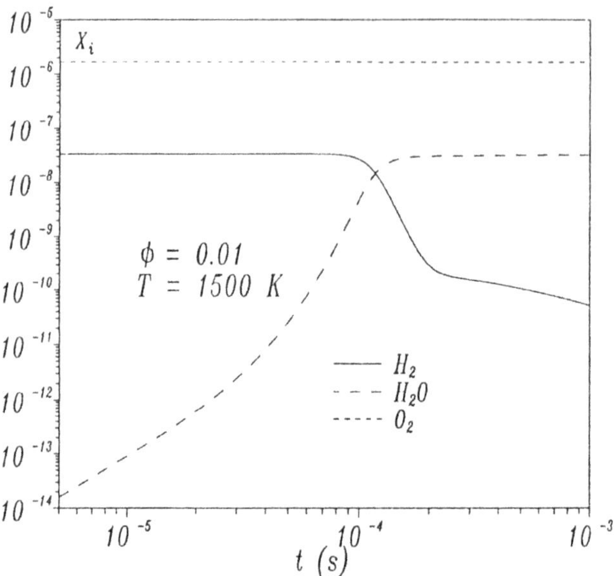

Fig. 1. Main species concentration evolution for an hydrogen/air mixture with $\phi = 0.01$ and $T_0 = 1500\ K$.

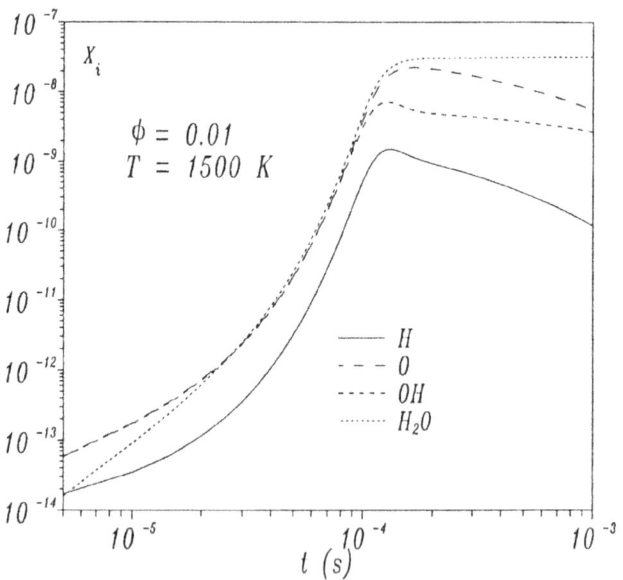

Fig. 2. Products and intermediate species concentration evolution for an hydrogen/air mixture with $\phi = 0.01$ and $T_0 = 1500\ K$.

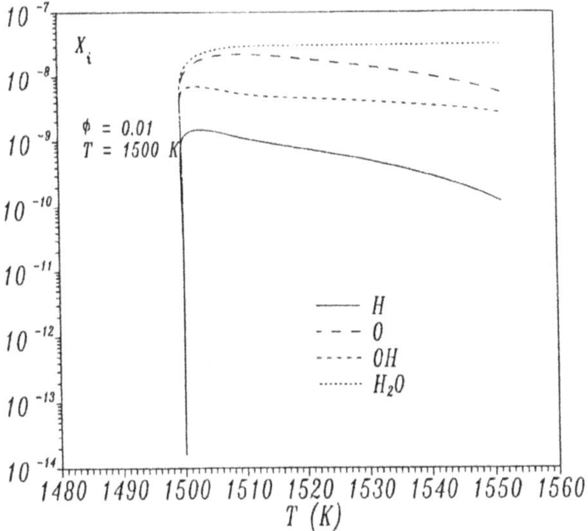

Fig. 3. Phase-plot of product and intermediate species concentration as a function of temperature for an hydrogen/air mixture with $\phi =$ 0.01 and $T_0 = 1500\ K$.

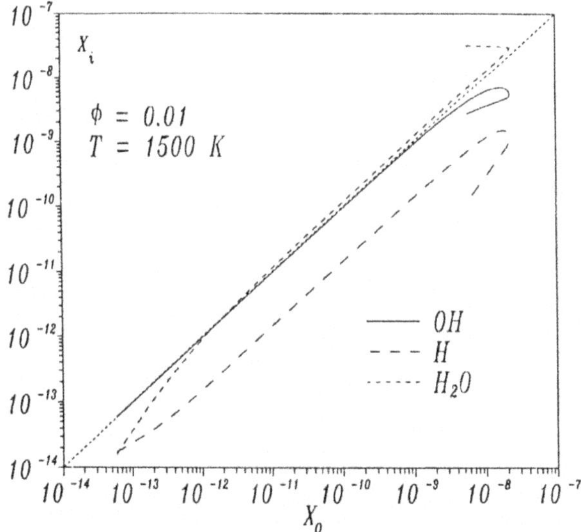

Fig. 4. Phase-plot of product and intermediate species concentration as a function of the **O** radical concentration for an hydrogen/air mixture with $\phi = 0.01$ and $T_0 = 1500\ K$.

				A	n	E
(1,2)	$H + O_2$	=	$OH + O$	$2.00\ 10^{14}$	0.0	70.30
(3,4)	$H_2 + O$	=	$OH + H$	$5.06\ 10^{04}$	2.7	26.30
(5,6)	$H_2 + OH$	=	$H_2O + H$	$1.00\ 10^{08}$	1.6	13.80
(7,8)	$OH + OH$	=	$H_2O + O$	$1.50\ 10^{09}$	1.1	0.40
(9,10)	$H + O_2 + M$	=	$HO_2 + M$	$2.30\ 10^{18}$	-0.8	0.00
(11,12)	$HO_2 + H$	=	$OH + OH$	$1.50\ 10^{14}$	0.0	4.20
(13,14)	$HO_2 + H$	=	$H_2 + O_2$	$2.50\ 10^{13}$	0.0	2.90
(15,16)	$HO_2 + OH$	=	$H_2O + O_2$	$6.00\ 10^{13}$	0.0	0.00
(17,18)	$HO_2 + H$	=	$H_2O + O$	$3.00\ 10^{11}$	0.0	7.20
(19,20)	$HO_2 + O$	=	$OH + O_2$	$1.80\ 10^{11}$	0.0	-1.70
(21,22)	$HO_2 + HO_2$	=	$H_2O_2 + O_2$	$2.50\ 10^{11}$	0.0	-5.20
(23,24)	$OH + OH + M$	=	$H_2O_2 + M$	$3.25\ 10^{22}$	-2.0	0.00
(25,26)	$H_2O_2 + H$	=	$H_2 + HO_2$	$1.70\ 10^{10}$	0.0	15.70
(27,28)	$H_2O_2 + H$	=	$H_2O + OH$	$1.00\ 10^{11}$	0.0	15.00
(29,30)	$H_2O_2 + O$	=	$OH + HO_2$	$2.80\ 10^{11}$	0.0	26.80
(31,32)	$H_2O_2 + OH$	=	$H_2O + HO_2$	$5.40\ 10^{10}$	0.0	4.20
(33,34)	$H_2 + M$	=	$H + H + M$	$6.76\ 10^{16}$	-1.0	436.26
(35,36)	$H + OH + M$	=	$H_2O + M$	$2.20\ 10^{20}$	-2.0	0.00
(37,38)	$O + O + M$	=	$O_2 + M$	$2.90\ 10^{15}$	-1.0	0.00

Table 1. Mechanism of the H_2/O_2 mixtures.

Units: A (cm,mole,s); E (KJ/mole); $K = A\ T^n \exp(-E/RT)$

4. Results and Conclusions

Fig. 1 shows the concentration of the main species as a function of time for a very lean mixture with an equivalence ratio of $\phi = 0.01$ and an initial temperature of $T_0 = 1500K$. The molecular oxygen concentration changes very little during the ignition process. Molecular hydrogen disappears almost completely, allowing water vapor concentration to increase. The concentration of active radicals and water vapor as a function of time are shown in Fig. 2 for the same initial conditions. The concentration of atomic hydrogen is always very low, due to the fast reaction (1). The concentration of radicals OH and O increase almost at the same rate, together with that of the water vapor, but OH reaches a maximum (steady-state behavior) before ignition. Fig. 3 shows a phase-plane plot of the concentration of radicals and water vapor as a function of the temperature, for the same initial conditions. We see here that the concentration of all products and intermediate species increase with no heat release up to the ignition condition. The temperature then increases slowly because radical recombination was not considered in the reduced kinetic scheme developed here (no radical- radical reactions). The ignition event is then a typical isothermal process for very lean mixture. Fig. 4 shows the product and intermediate species concentration as a function of the concentration of the main radical O. Here again, it is shown how the concentrations of O and H_2O increase almost at the same rate up to the ignition point. The steady-state behavior of OH and H is also shown clearly in the figure. Fig. 5 shows a comparison of the radical concentration evolution with time for a stoichiometric and a very lean mixture. In the stoichiometric mixture, the most important radical is the atomic hydrogen with the atomic oxygen following the steady-state behavior. For very lean mixtures the reverse is obtained, with the atomic oxygen as the main radical. Ignition occurs for lower radical concentration. Finally, Fig. 6 shows the ignition delay time for stoichiometric and lean hydrogen/air mixtures computed numerically with full kinetics (solid line) as a function of the equivalence ratio. The asymptotic solution given by Eq. (18) is also shown in this figure, giving very good agreement with the numerical results and following the same trend with the equivalence ratio. The asymptotic relationship obtained using the overall global step

$$3H_2 + O_2 \rightarrow 2H_2O + 2H$$

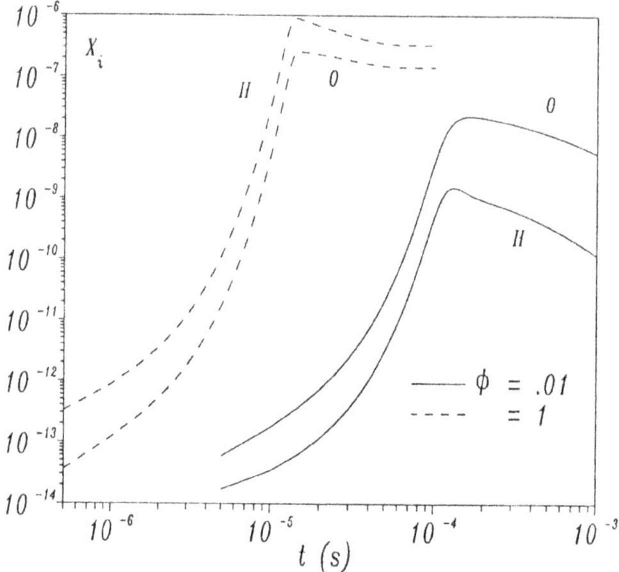

Fig. 5. Radical concentration evolution for two different equivalence ratios and $T_0 = 1500\ K$.

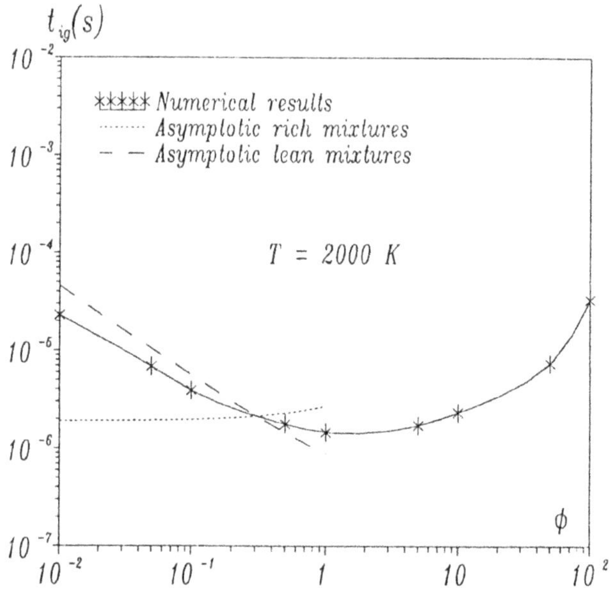

Fig. 6. Numerical and asymptotic solution for the ignition delay time as a function of the equivalence ratio for an initial temperature of 2000 K.

obtained for stoichiometric and rich mixtures is also plotted. This relationship shows the inverse behavior of the ignition delay time with the equivalence ratio.

The transient analysis leading to the ignition of lean mixtures of hydrogen and oxygen is studied in this paper using both numerical analysis based in Computational Singular Perturbation (CSP) and asymptotic analytical methods. The reduced kinetic mechanism is obtained after solving the evolution equations and evaluating the CSP data. For lean mixtures with high initial temperatures, fuel consumption is very important and the limiting reaction switches from reaction (1) for stoichiometric and rich mixtures to reaction (3) for lean mixtures due to the low concentration of molecular hydrogen. A simple formula for the ignition delay time is obtained in this case showing a very good agreement with numerical results with full kinetics.

Acknowledgements

César Treviño thanks the DGICYT of Spain for supporting a sabbatical leave in Madrid.

References

Baulch, D. L., Drysdale, D. D., Horne, D. G., and Lloyd, A. C., 1972. "Evaluated kinetic data for high temperature reactions, Vol. 1: Homogeneous gas phase reactions of the $H_2 - O_2$ system," *Butherworths*, London.

Brokaw, R. S., 1965. "Analytic solutions to the ignition kinetics of the hydrogen-oxygen reaction," 10th (Symposium) International on Combustion, The Combustion Institute, pp. 269-278.

Dixon-Lewis, G. and Williams, D. J., 1977. *Comprehensive Chemical Kinetics, Vol. 17: The Oxidation of Hydrogen and Carbon Monoxide*, Elsevier, Oxford, C. H. Bamford and C. F. H. Tipper (eds.).

Gardiner, W. C., Wakefield, C. B., and Walker, B. F., 1981. *Shock Waves in Chemistry and Chemical Technology*, Marcel Dekker, New York, A. Lifshitz (ed.).

Kailasaneth, K., Oran, E. S., Boris, J. P., and Young, T. R., 1982. "Numerical methods in laminar flame propagation," *Vieweg*, Braunschweig, p. 152.

Lam, S. H. and Goussis, D. A., 1989. "Understanding complex chemical kinetics with computational singular perturbation," XXII Int. Symposium on Combustion, The Combustion Institute, in press.

Maas, U. and Warnatz, J., 1989. "Ignition processes in hydrogen-oxygen mixtures," *Combustion and Flame*, in press.

Peters, N., 1985. "Numerical and asymptotic analysis of systematically reduced reaction schemes for hydrocarbon flames," *Lecture Notes in Physics* **241**, Glowinski et al. (eds.), Springer, New York, pp. 90-109.

Treviño, C., 1990. "Ignition phenomena in H_2/O_2 mixtures," *Progress in Astronautics and Aeronautics* **131**, p. 19.

Warnatz, J., 1984. "Rate coefficients in the $C/H/O$ system," *Combustion Chemistry*, W. C. Gardiner (ed.), Springer, New York, pp. 197-360.

Wiriyawit, S. and Dabora, E. K., 1984. "Modeling the chemical effects of plasma ignition in one-dimensional chamber," 20th Symposium (International) on Combustion, The Combustion Institute, pp. 179-186.

UNSTEADY BEHAVIOR

STEADY AND UNSTEADY ASPECTS OF DETONATION INITIATION

J. W. Dold

School of Mathematics, University of Bristol
Bristol BS8 1TW, Britain

J. F. Clarke

College of Aeronautics, Cranfield Institute of Technology
Cranfield, Bedfordshire MK43 0AL, Britain

M. Short

Department of Astronomy, The University of Edinburgh
Blackford Hill, Edinburgh EH9 3HJ, Britain

1. Introduction

This article outlines a number of basic notions concerning the peculiarities of detonation initiation in mixtures described by one-dimensional reactive Euler equations having a thermally-sensitive rate of chemical reaction. It also mentions some strange properties of such equations and touches on a few matters remaining unresolved.

An initiation of a detonation requires significant interaction between the release of energy by some chemical means and the compressible dynamics of a system. Such a situation may be brought about, for example, by the passage of a shock wave into a reactive mixture or by the self-ignition of some already reacting medium over a large enough scale for major pressure changes to be induced.

Unsteady evolutions are almost an inevitable consequence of conditions of this kind, at least for general initial and boundary conditions, and a number of such evolutions are discussed. These may involve chemical and gasdynamic interactions, with or without significant wave steepening and shock formation. While the chemistry remains thermally (or more generally, thermodynamically) sensitive, surprisingly large-amplitude interactions involving repeated ignitions and shock collisions can be predicted on theoretical grounds.

Nevertheless, unsteady initiations of detonation can involve many steady, or quasi-steady, aspects. For this reason, we begin with a brief sketch of the Rankine–Hugoniot picture of steady detonations

J. Buckmaster et al. (eds.), Combustion in High-Speed Flows, 493–512.

and deflagrations in the context of a reactive Euler model. Attention is directed towards a significant feature, that is often overlooked in such discussions, which is that steady one-dimensional detonations are not uniquely determined for any given mass-flux.

More specifically, the location of a shock (separating deflagration behaviour from weak detonation) in thermodynamic and chemical phase-space can remain arbitrary unless determined by some unsteady history or by any chemical constraints that eliminate one or other type of behaviour. This indeterminacy plays a key role in some pictures of unsteady detonation initiation, particularly those which consider the asymptotic limit of extreme thermodynamic sensitivity of the chemical reaction rate.

Numerical studies do not necessarily conform to any such limiting asymptotic descriptions and, when they don't, they raise further questions demanding clearer asymptotic answers. Cases of this nature and the questions that they raise are discussed below. Fortunately, some numerical results do manage to conform quite well with asymptotic descriptions, showing that their carefully matching structures were not all derived in vain.

However, one overiding matter does indicate that most of the studies discussed in this article have still only managed to scratch the surface of the underlying physical phenomenon. Even the qualitative nature of their descriptions depends very strongly on the nature of the chemical model. This is a major difference from theoretical descriptions of many slower combustion phenomena which depend crucially on interactions between diffusive processes and chemistry. The interaction of chemistry with a shock wave (and other gasdynamic effects) is totally different, in a manner that leads to a much greater dependence on chemical properties. The reasons for this are clarified, to some extent, by looking at a few examples.

1.1. Model equations

A reasonable model, that is fairly widely accepted (for example Landau and Lifshitz, 1987), for studying the one-dimensional behaviour of a uniform reactive medium undergoing detonation consists of the set of Euler equations for ideal gases with the inclusion of chemical source terms. If there is a single chemically active component that releases energy without change in molecular weight, the equations become

$$V_t - u_\psi = 0 \tag{1}$$

$$u_t + P_\psi = 0 \tag{2}$$

$$VP_t + \gamma P u_\psi = -(\gamma - 1)q_t \tag{3}$$

$$t_c \lambda_t = -\lambda \exp(-T_A/T) \tag{4}$$

$$q = Q\lambda \tag{5}$$

$$PV = RT \tag{6}$$

where V is the specific volume ($V = 1/\rho$), u is the velocity, P is pressure, q is the amount of energy bound up in the chemical component, whose mass-fraction is related to λ, with proportionality coefficient Q. The ratio of principal specific heats is $\gamma > 1$, R is the gas constant of the mixture, t_c^{-1} is a frequency factor and T_A is the activation temperature of the chemical reaction. Time is measured by t and ψ is a Lagrangian mass coordinate defined by

$$\psi = \int_{x_0(t)}^{x} \rho \, dx \quad \text{where} \quad x_0'(t) = u(0, t) \tag{7}$$

so that a particle of constant ψ moves with the local fluid velocity.

Of course, these equations apply only in a piecewise manner between discontinuities such as contact interfaces and shock waves which are admissible structures in such systems. At shock waves, the jump conditions

$$\frac{P_1}{P_0} = \frac{\gamma(2M^2 - 1) + 1}{\gamma + 1} \tag{8}$$

$$\frac{V_1}{V_0} = \frac{\gamma + (2/M^2 - 1)}{\gamma + 1} \tag{9}$$

$$u_1 - u_0 = 2\frac{M^2 - 1}{(\gamma + 1)M} c_0 \tag{10}$$

$$\lambda_1 - \lambda_0 = 0 \tag{11}$$

relate the state behind (subscript 1) to that ahead (subscript 0) of a shock propagating at the Mach number $M > 1$ measured in terms of the chemically frozen (or high frequency) sound-speed

$$c = \sqrt{\gamma PV} = \sqrt{\gamma RT}. \tag{12}$$

2. Steady High-Speed Waves

In order to set the groundwork for later discussions of unsteady processes, a brief review of steady one-dimensional detonations is worthwhile. In fact, the broadest interpretation of the possible steady solutions is highly instructive in illuminating some of the later discussion of unsteady behaviour.

The system of equations (1)–(12) supports *steady-wave solutions* for which the thermodynamic and chemical states (P, V, λ) are related throughout their structure by Rankine–Hugoniot conditions (for example, Fickett and Davis, 1979). These are, the Rayleigh line relation

$$P - P_1 = -m^2(V - V_1) \tag{13}$$

and the Hugoniot relation, for any (partially burnt) chemical state λ

$$\frac{\gamma+1}{\gamma-1}(PV - P_1V_1) + P_1V - V_1P = 2Q(1-\lambda) \tag{14}$$

where the thermodynamic state of the medium (P_1, V_1) is achieved when $\lambda = 1$, and m is the (uniform) mass-flux of material passing through the wave. The Mach number through the wave at any point is given by

$$M = m/\alpha \quad \text{with} \quad \alpha = \sqrt{\gamma P/V} \tag{15}$$

where α is the acoustic impedance, ρc, which is the chemicaly frozen rate of sound propagation through the mass-coordinate ψ.

The Hugoniot relation (14) describes a family of rectangular hyperbolæ which typically intersect the Rayleigh line (13) at either two points or none, depending on the value of m, as sketched in Figure 1.

There are two Chapman–Jouget mass-fluxes $\overline{m}_c > \underline{m}_c > 0$ at which corresponding Rayleigh lines are tangent to the fully burnt Hugoniot curve $(\lambda = 0)$. Consistent steady solutions do not exist for any mass-flux in the range $\overline{m}_c > |m| > \underline{m}_c$.

Those solutions that are found for $|m| \leq \underline{m}_c$ (to the right of the point (P_1, V_1) in Figure 1) describe deflagration waves, in which pressure decreases as λ decreases through the wave; the flow is subsonic, except in the case of a Chapman–Jouget deflagration, $|m| = \underline{m}_c$, for which $M = 1$ only at the very end of the chemical reaction. Because pressure must increase through any shock-wave, shocks are excluded from such deflagration structures. This also ensures that, of any two intersections between the Rayleigh line and a

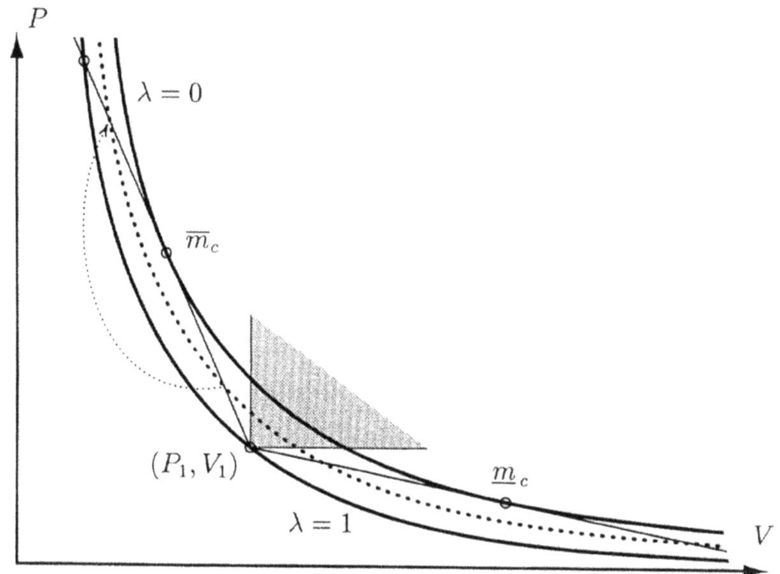

Figure 1: Sketch of a Rankine–Hugoniot diagram illustrating transitions from an unreacted (lower) to a fully reacted (upper) Hugoniot curve. The two Chapman–Jouget Rayleigh lines are also shown (thin lines) as well as a possible shock transition across a partially-reacted Hugoniot curve (dotted). Rayleigh lines are excluded from the grey area.

Hugoniot curve, only the one with the higher pressure is accessible via a deflagration.

2.1. Weak and strong detonations

The solutions that are found for $|m| \geq \overline{m}_c$ are known as detonations. These are supersonic with respect to the sound speed at the unburnt state $(P, V, \lambda) = (P_1, V_1, 1)$, but need not be *locally* supersonic throughout, depending on the existence or not of any shocks in the structure of the wave.

If (for suitable chemical parameters) the reaction at the "initial" or "upstream" state $(P_1, V_1, 1)$ were to be chemically stable (or too slow to be meaningful) then a shock-wave would be absolutely necessary in order to boost the temperature to a level at which chemical energy could be released. The effect of a shock wave at a given value of λ is to raise the pressure and density from one intersection between the Rayleigh line and Hugoniot curve (for that value of λ) to a higher intersection, as illustrated by the thin dotted curve in Figure 1. Once this has happened, the evolution can only then take place via a subsequent decrease in pressure. This corresponds to a deflagration behind the shock wave although the overall effect of the shock and the chemically induced changes in front of it is an increase in pressure. Clearly, after any shock, only the higher-pressure intersection between the Rayleigh line and any Hugoniot curve is accessible. A wave of this sort, involving a shock, is known as a strong detonation.

Without a shock or ahead of any shock, only the lower pressure intersection is accessible and pressures increase as λ decreases. Because of this, a wave involving this kind of evolution throughout, and therefore ending at lower-pressure intersection between a Rayleigh line and the fully-burnt Hugoniot curve is known as a weak detonation.

Provided that the reaction-rate at the "initial" state $(P_1, V_1, 1)$ is non-zero (or at least that this point is chemically unstable), steady detonation waves can exist (at least theoretically) with both weak and strong parts. So long as it is chemically feasible (as in the model (4)), a weak evolution can take place with P increasing as λ decreases to a value λ_s (say) at which a shock causes pressure to jump with a subsequent deflagration evolution from the state created by the shock. Because this picture could hold for any value of λ_s within the range $0 \leq \lambda_s \leq 1$, such a detonation wave is far from uniquely determined.

3. Simple Unsteady Processes

Where there is no "cold" chemical equilibrium state (as in the model (1)–(6)), any given set of initial conditions (with or without a shock) has associated with it an induction time t_i, which measures a delay period before chemical self-ignition sets in. Based on this, there is also an acoustic length $t_i c$. Depending on the kind and length-scales of the initial and boundary conditions, in relation to these characteristic quantities, a number of different sorts of evolution are possible, as identified in several studies (for example, Clarke, 1978, 1979; Clarke and Cant, 1984; Jackson and Kapila, 1985; Majda and Rosales, 1987; Almgren, 1991; Jackson *et al.*, 1989; Dold, 1988,

1989a).

The key illuminating feature of these studies is the use of a large activation temperature approximation, for which T_A is considered to be large compared with any relevant temperature found in any evolution. Asymptotic approximations are then developed using the limit as $T_A/T \to \infty$. This has the physical significance of representing an extreme sensitivity of chemical reaction-rates to temperature changes.

3.1. Induction

When initial disturbances (some representative values of which are characterised below by the subscript 1) are distributed so widely that significant changes in initial induction time $t_i(\psi)$ occur only on the acoustic length-scale $t_{i1}c$, the asymptotic perturbations

$$\frac{T}{T_1} \sim 1 + \frac{T_1}{T_A}\phi(\chi,\tau), \quad \frac{P}{P_1} \sim 1 + \gamma\frac{T_1}{T_A}p(\chi,\tau) \tag{16}$$

$$\lambda \sim 1 - \frac{C_pT_1}{Q}\frac{T_1}{T_A}\ell(\chi,\tau), \quad u \sim u_1 + \sqrt{\gamma P_1 V_1}\frac{T_1}{T_A}v(\chi,\tau) \tag{17}$$

as $T_1/T_A \to \infty$, where

$$\tau = \frac{t}{t_{i1}}, \qquad \chi = \frac{\psi}{t_{i1}\sqrt{\gamma P_1/V_1}} \tag{18}$$

lead to a system of equations which describe exponential chemical nonlinearity in a medium of linearised acoustics, namely

$$\phi_\tau = (\gamma - 1)p_\tau + \ell_\tau \tag{19}$$

$$p_\tau + v_\chi = \ell_\tau \tag{20}$$

$$v_\tau + p_\chi = 0 \tag{21}$$

$$\ell_\tau = e^\phi. \tag{22}$$

Eliminating all variables but ϕ, leads to Clarke's model induction equation (Clarke, 1985)

$$\left(\phi_\tau - \gamma e^\phi\right)_{\tau\tau} = \left(\phi_\tau - e^\phi\right)_{\psi\psi}. \tag{23}$$

This remarkable equation combines several features of reactive gasdynamic fluids subject to small-amplitude disturbances. In situations where different terms predominate, one can identify processes of constant-pressure and the quicker constant-volume blowup; depending on their frequency, one can also identify sound-speeds ranging from the adiabatic (chemically frozen) sound-speed to the isothermal sound-speed.

In the context of detonation-initiation, the principal feature of this equation is its ability to predict blowup; the temperature perturbation ϕ becomes unbounded in a finite time (Clarke and Cant, 1984; Jackson and Kapila, 1985; Dold, 1988; Jackson *et al.*, 1989). In fact, the appearance of a singularity as time proceeds only marks the first point to be encountered on a larger-scale path of singularities in space and time. This path moves supersonically, typically with an initially infinite speed as has been found in analogous reactive–diffusive ignition problems (Dold, 1989b; Henderson and Dold, 1991). Studies establishing this were carried out in (Dold, 1989a; Short, 1992).

This phenomenon arises for very many types of initial and boundary conditions (differing only in detail) and would appear to be the rule rather than the exception whenever initial and boundary conditions are consistent with the asymptotic structure (16)–(18). It can be denoted by the existence of a path of blowup $\tilde{\tau}(\chi)$ which is such that

$$\lim_{\tau \nearrow \tilde{\tau}(\chi)} \phi(\chi, \tau) = \infty \tag{24}$$

with the speed (or more correctly the mass-flux in the dimensionless mass-coordinate χ)

$$\tilde{m}(\chi) = 1 \Big/ \tilde{\tau}'(\chi). \tag{25}$$

The value of $|\tilde{m}(\chi)|$ is necessarily greater than one (the linearised sound-speed at the reference state (P_1, V_1)) and is generally unbounded towards the first point of singularity where $\tilde{\tau}'(\chi) \to 0$.

3.2. Steepening and reaction

Recalling that the acoustic impedance is $\alpha = \sqrt{\gamma P/V}$, equations (2) and (3) can be combined to give

$$P_t \pm \alpha P_\psi \pm \alpha\,(u_t \pm \alpha u_\psi) = -(\gamma - 1)\frac{q_t}{V} \tag{26}$$

in which the three characteristics on which

$$\frac{\partial \psi}{\partial t} = \pm \alpha \quad \text{and} \quad \frac{\partial \psi}{\partial t} = 0 \tag{27}$$

are clearly active. With each characteristic one can identify quantities which, when disturbed, radiate as a leftwards-propagating wave, $P - \alpha u$, radiate as a rightwards-propagating wave, $P + \alpha u$, or are simply convected with the flow, T or λ, representing particle-fixed or "entropy" waves (Majda and Rosales, 1987; Almgren, 1991).

Where short wavelength disturbances are present, being of the order of $\frac{T_1}{T_A} t_{i1} c$ in length but of the same typical amplitude as those leading to Clarke's equation (23), wave-steepening and possible shock formation cannot be neglected. It is most illustrative to consider a single localised disturbance, propagating in the direction, say, of increasing ψ. Characteristics that propagate in the opposite direction and particle-fixed characteristics pass rapidly through this wave so that the information they carry remains practically unchanged. This leads to good algebraic approximations of the kind

$$P - P_1 \sim \alpha(u - u_1) \quad \text{and} \quad P - P_1 \sim -\alpha^2(V - V_1). \tag{28}$$

However, taking suitable perturbations using a dimensionless coordinate z that follows the wave, one also obtains (Clarke, 1978, 1979; Blythe, 1979)

$$\phi_\tau + \phi\phi_z = \ell_z = e^\phi. \tag{29}$$

This reactive-Euler Burgers' equation is relatively easily analysed in terms of a characteristic β that is chosen such that

$$\left.\frac{\partial z}{\partial \tau}\right|_\beta = \phi. \tag{30}$$

The solution then takes the simple form

$$\phi = -\ln\left(\hat{\tau}(\beta) - \tau\right) \tag{31}$$

where the singularity-path $\hat{\tau}(\beta)$, in terms of the characteristic coordinate β, is determined by initial data and the precise choice of the characteristic coordinate. In general, since ϕ increases nonuniformly, there is a tendency towards shock-formation, in which a solution of (30) would involve a coalescence of suitable ranges of the characteristic β.

Essentially, this gives rise to two different ways in which $\phi(z, \tau)$ can grow without bound in finite time, as illustrated in Figure 2. If the first point of blowup in the space of the characteristic β (*i.e.* the minimum of $\hat{\tau}(\beta)$) lies outside any range of β that is absorbed into

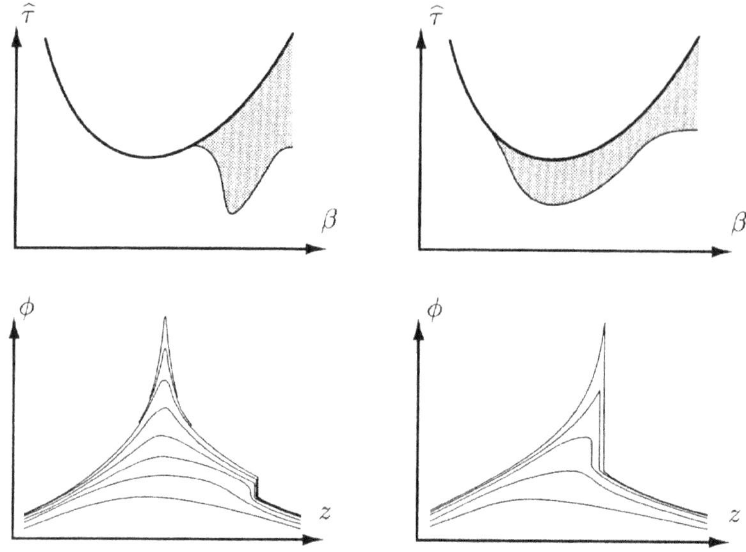

Figure 2: Sketches of paths of blowup $\hat{\tau}$ in the characteristic coordinate β, showing shaded areas of shock annihilation and the corresponding blowup structures seen in physical space in the lower figures. The left-hand figures demonstrate blowup away from a shock while the right-hand figures demonstrate blowup at a shock.

a shock-wave, then it represents a relatively normal type of blowup of the propagating disturbance—similar to those seen in solutions of Clarke's equation (23). On the other hand, if it lies at a value of β that has been annihilated by a shock, then a blowup occurs actually on the shock-wave; the strength of the shock increases without bound, along with the temperature and pressure perturbations behind it.

4. Quasi-Steady Processes

Of course, a growth towards an infinite value of a perturbation variable that satisfies an approximate differential equation, arrived at by neglecting some terms in an asymptotic limit, does not necessarily signify a real physical unboundedness. Rather, it signifies that a physical variable grows out of the range in which its behaviour can realistically be approximated by the perturbation equations.

4.1. Spontaneous flame

This fact was recognised by (Zel'dovich, 1980a, 1980b) in proposing a compelling rule-of-thumb approach that applies to fairly regular initial-value situations. Supposing that $\mathbf{f}_i(\psi)$ represents the vector of initial values $(T_i(\psi), P_i(\psi), u_i(\psi), \lambda_i(\psi))$ then, based solely on these values, one can propose a rough but simple approximation to any path of ignition by calculating a constant-volume self-ignition time at any value of ψ

$$t \approx t_i(\mathbf{f}_i(\psi)) \tag{32}$$

based on a Semenov explosion, assuming that the properties at that point were uniformly distributed in space. The mass-flux passing through this "spontaneous flame" at any point is then given by

$$m = \left.\frac{\partial \psi}{\partial t}\right|_{\text{flame}} \approx 1 \Big/ \frac{\mathrm{d}t_i}{\mathrm{d}\psi}. \tag{33}$$

This approximate relation is extremely good for near uniform mixtures, and corrections can be made for interactions that are a perturbation from locally uniform states. When compared with numerical calculations such corrections are found to provide very accurate estimates (Short and Dold, 1993).

The formula (33) is completely analogous to the formula (25). Indeed, for large enough activation temperatures the path $\tilde{\tau}(\chi)$ can be considered to be a leading-order asymptotic estimate of a more fully developed flame-path that would be described by continuing the perturbation approach used in calculating $\tilde{\tau}(\chi)$ using matched asymptotic expansions (Dold and Kapila, 1993a).

4.2. Transition from weak to strong detonation

The flame that is produced spontaneously by such an induction blowup in fact turns out to be well-described, to leading order, as a quasi-steady detonation as outlined in Section 2. To begin with, when the spontaneous flame travels with unbounded velocity there is no mechanism that could impose a shock on its structure. It must therefore start off its existence as a shockless weak detonation; its movement is brought about, not by any mechanism for self-propagation, but by virtue of the fact that non-uniform induction processes cause some places to self-ignite before others.

Only when this high-speed flame slows down below the Chapman–Jouget mass-flux does a part of the flow through the flame

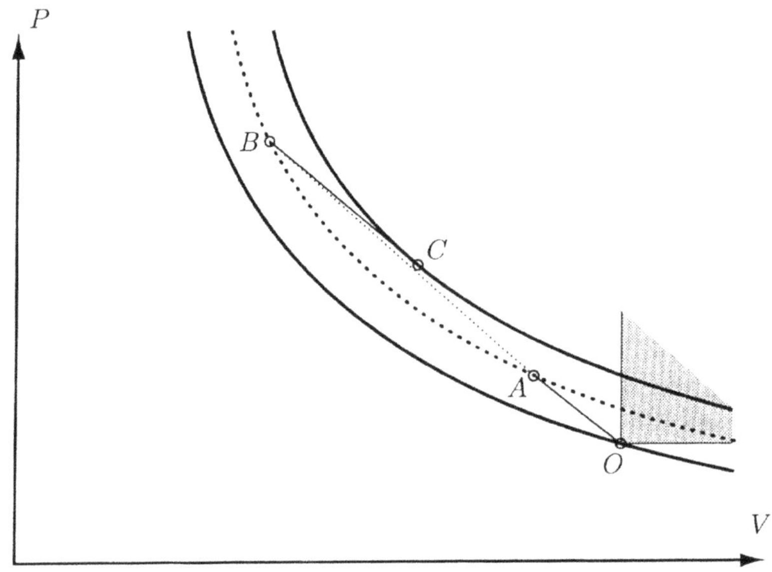

Figure 3: An illustration of quasi-steady waves during an unsteady transition to detonation, as seen in the (P, V) plane. A weak detonation $O \rightarrow A$ is interupted by a faster shock $A \rightarrow B$ which initiates a deflagration $B \rightarrow C$ that ends at a Chapman–Jouget (tangency) point. Each of the Hugoniot curves has to be interpreted as being relevant to the instantaneous starting point of the quasi-steady Rayleigh line that approaches it.

become subsonic. Only then can characteristics begin to cross and, as (Zel'dovich, 1980a, 1980b) identified, can a shock-wave form, leading to a transition to a strong Chapman–Jouget detonation wave. More detailed large activation temperature studies of this situation reveal both when and how such a shock-wave would form, and the manner in which it would advance to create a strong detonation (Dold *et al.*, 1991; Dold and Kapila, 1993b).

In brief, once a shock has formed, it carries a deflagration behind it and so creates a strong detonation very quickly, albeit one in which the chemical conversion has partially proceeded ahead of the shock. Because the earlier slowing down of the weak detonation would have created an expansion behind it, this strong detonation will be constrained to travel at very close to its Chapman–Jouget velocity. If the shock and spontaneous flame were then both to travel at exactly this velocity, one would have created a steady detonation wave containing both weak and strong components as described in Section 2.

More typically, of course, the weak detonation part (the spontaneous flame) would continue to slow down while the strong detonation would maintain a higher Chapman–Jouget speed. The latter would therefore advance through the former, strengthening the shock-wave as it does so and finally producing a predominantly strong detonation. In the pressure–volume state-space, this can be interpreted as a combination of three quasi-steady processes, as indicated by the weak detonation, shock and deflagration Rayleigh lines illustrated in Figure 3.

5. Numerical Simulations

It is well worth comparing theoretical pictures such as this with numerical simulations of the full equations (1)–(12) for finite activation temperatures. In fact, two useful comparisons are available.

A shock-driven initiation problem was calculated by (Singh and Clarke, 1992) and the most salient results in the space of (P, V) are reproduced in Figure 4. These show a wide range of behaviour including ranges in which Rayleigh lines are very closely approximated, indicating that very nearly quasi-steady processes predominate.

It should be noted that Rayleigh lines must have negative slopes while the first few traces in Figure 4, corresponding to particles very close to a piston-face, have positive slope. The evolution of these particles is therefore strictly unsteady. This behaviour does, however, give way to two quasi-steady processes, one a weak detonation evolution (increasing pressure) followed by a deflagration evolution (decreasing pressure) before the reaction is completed and an unsteady expansion ensues. There is also an unsteady transition between the two quasi-steady branches, and it is from this region that a shock is born to complete the generation of a strong detonation.

On the face of it, there would appear to be some conflict with the findings of (Dold *et al.*, 1991) which predict that a shock should be formed very close to the end of the chemical reaction, in the limit of large activation temperature. It is very likely, however, that these results do depend strongly on the actual value of the activation temperature. This is particularly so since the rate of deceleration of the spontaneous flame calculated for *shock-initiation* problems is very

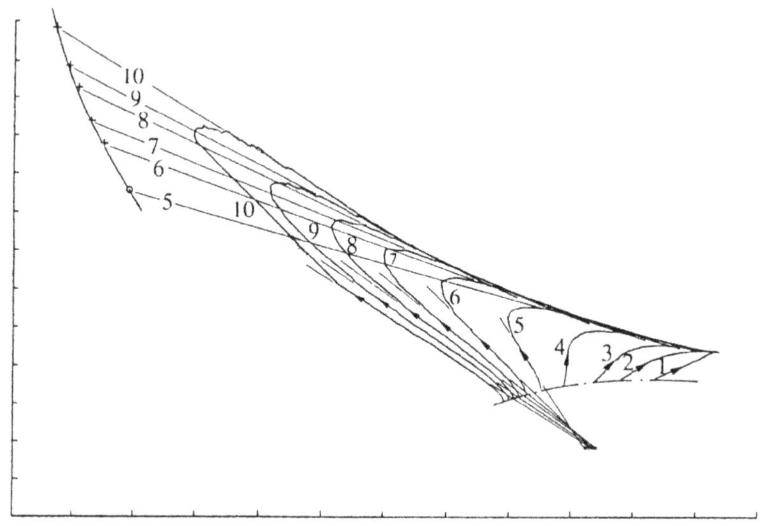

Figure 4: Traces in the (P, V) plane of particle histories close to the piston-face in a shock-initiated evolution to detonation (reproduced from Singh and Clarke, 1992). The straight lines are approximate Rayleigh lines fitting the data over parts of the particle histories, showing both weak detonation and deflagration structures. There are also significant unsteady processes (where no such lines can be fitted). In this evolution, an unsteady process, not a shock, connects the two distinct quasi-steady processes until a shock is born out of the unsteady process around the particle marked 10.

fast indeed. It has been shown (Short, 1992) that most of the slowing down of this path happens in only about $\frac{1}{10}\%$ of the characteristic length $t_i c$. There are thus two large (or small) parameters inherent in the problem and it is likely that a proper theoretical description of the events depicted in Figure 4 would require some kind of combined asymptotic limits. This description remains open, although it is quite likely that a model similar to (29) should describe the shock formation.

This interpretation is reinforced by recent numerical calculations shown in Figure 5. These are obtained for an initial-value problem yielding a much more gradual slowing-down of a spontaneous flame. The calculations carried out at fairly low activation temperatures are clearly close in character to those of Singh and Clarke. However, increasing the activation temperature yields a much more regular

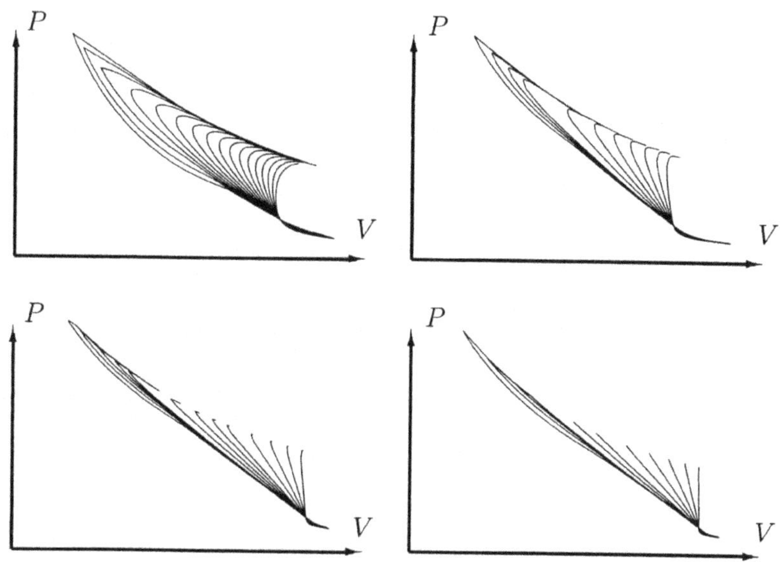

Figure 5: Four different ignition calculations all associated with the same initial-value disturbance but for increasing activation temperature. Traces are plotted in the (P, V) plane of the behaviour around the time of shock formation for values of $T_A/T_1 = 10,\ 12\frac{1}{2},\ 15$ and $17\frac{1}{2}$. The numerical method employed adds some diffusion to shock waves so that these do not appear as straight Rayleigh lines but as curves.

classification into the quasi-steady processes described earlier, with shock formation near to the end of the chemical reaction and evolutions much as described in the previous section.

6. Concluding Remarks

The study of ignition via the reactive Euler equations has numerous steady and unsteady aspects, many of which are yet to be fully understood. It is worth finishing with a description of one of the most peculiar of these.

6.1. A "bouncing-ignition" accumulation of pressure

When one shock catches up with another shock, the resulting Riemann problem produces a stronger shock and a significantly hotter region than had hitherto been found, between the new shock and a contact interface. An expansion cools the gas behind the contact interface, so that ignition must take place first between the shock and the contact surface.

In fact, as is confirmed by calculating the spontaneous flames for this problem using Clarke's equation (Short, 1992), ignition should take place close to the contact interface provided only that the activation temperature remains *relatively* large. This must quickly create a shock and following Chapman–Jouget deflagration that advances into a strong Chapman–Jouget detonation within the gases behind the leading shock-wave.

As for all large activation temperature strong detonations in one dimension, the structure of this detonation consists of a shock-wave, followed by an induction region of little real change, followed by a flame. Thus when the shock-wave of this structure catches up with the leading shock, exactly the same situation is found as at the start, only significantly hotter and more violent. A new Riemann problem produces the same kind of scenario, creating a new ignition of a strong Chapman–Jouget detonation, behind the new leading shock, that can repeat the process over and over.

Each time a new Chapman–Jouget detonation is formed, it is able to increase the post-shock pressure by a finite amount related to the energy of reaction Q. This increases the temperature by a similar finite amount and so reduces the self-ignition time t_i progressively at each step. In this sense it is akin to a bouncing ball, which is able to bounce an infinite number of times in a finite time because the time between bounces decreases geometrically at each bounce. If the chemistry were able to continue getting faster indefinitely, one would produce a spike of infinite pressure and temperature in a finite time.

Of course nature does not behave this way (not entirely), and as temperatures increase (after a few "bounces") to a level comparable with the activation temperature T_A the process must saturate and give way to another form of unsteady behaviour.

One reason for this peculiar ability of the reactive Euler equations to blow up completely is that a shock wave is naturally fed with fresh chemical reactants just as fast as it is able to travel. If that shock is able to raise temperatures to a level at which those reactants are consumed on whatever time-scales are associated with the speed of the shock, it may be possible for the energy liberated to keep pace with the appetite of any growing shock.

Interestingly, a similar large-scale growth of pressure has recently been discovered in considering the fuller developments of the blowup problem on a propagating wave (29) when the blowup takes place actually at the shock (Kapila and Jackson, 1992). This too is a case in which a shock itself is leading the initiation of a chemical reaction and, whatever the speed of the shock, it is always fed with a full supply of fresh reactants at a rate proportional to its speed.

6.2. Closing remarks

It is not entirely clear that such unusual findings as those just outlined are helpful. They may be very important in revealing the nature of brief "spikey" pressure events that (as may be the case with other aspects of real detonations) have not yet been observed in detail. They would thus offer one route by which unstable detonations (as are found for sufficiently large activation temperatures) could behave. They also indicate that detonations experience a much greater degree of sensitivity to the nature of any relevant chemical kinetics than is found in reactive–diffusive, laminar or even turbulent flame problems (Dold and Kapila, 1991); global chemical approximations that appear to provide widely applicable results in these latter forms of combustion (Buckmaster and Ludford, 1982) are probably much less generally applicable in the description of detonation.

There are also many more features of detonations to be found in more than one dimension (not touched on here) so that the field remains both complex and open to many avenues of investigation and improvement.

References

Abousief, G. E. and Toong, T. Y., 1981. "Nonlinear wave–kinetic interactions in irreversibly reacting media," *J. Fluid Mech.* **103**, pp. 1–22.

Almgren, R. F., 1991. "High frequency acoustic waves in a reacting gas," *SIAM J. Appl. Math.* **51**, pp. 351–373.

Blythe, P. A., 1979. "Wave propagation and ignition in a combustible mixture," *17th Symposium (International) on Combustion*, The Combustion Institute, Pittsburgh, pp. 909–916.

Buckmaster, J. D. and Ludford, G. S. S., 1982. "Theory of Laminar Flames," Cambridge University Press.

Clarke, J. F., 1978. "Small amplitude disturbances in an exploding atmosphere," *J. Fluid. Mech.* **89**, pp. 343–356.

Clarke, J. F., 1979. "On the evolution of compression pulses in an exploding atmosphere; initial behaviour," *J. Fluid. Mech.* **94**, pp. 195–208.

Clarke, J. F., 1981. "On the propagation of gasdynamic disturbances in an explosive atmosphere," *Prog. Aeronautics and Astronautics* **76**, pp. 383–402.

Clarke, J. F., 1985. "Finite amplitude waves in combustible gases," in "The Mathematics of Combustion," J.D. Buckmaster, Ed., SIAM Publications, Philadelphia, pp. 183–245.

Clarke, J. F., 1989. "Fast flames, waves and detonation," *Prog. Energy Combust. Sci.* **15**, pp. 241–271.

Clarke, J. F. and Cant, R. S., 1984. "Nonsteady gasdynamic effects in the induction domain behind a strong shock wave," Prog. Astronautics and Aeronautics **95**, pp. 142–163.

Clarke, J. F., Kassoy, D. R. and Riley, N., 1986. "On the direct initiation of a plane detonation," *Proc. Roy. Soc. Lond. A* **408**, pp. 129–148.

Clarke, J. F. and McChesney, M., 1976. "Dynamics of Relaxing Gases," Butterworths, London.

Dold, J. W., 1988. "Dynamic transition of a self-igniting region," in "Mathematical Modelling of Combustion and Related Topics," C-M. Brauner and C. Schmidt-Lainé, Eds., Martinus Nijhoff, Dordrecht, pp. 461–470.

Dold, J. W., 1989a. "Induction period generation of a supersonic flame," *Lecture Notes in Physics* **351**, Springer Verlag, pp. 245–256.

Dold, J. W., 1989b. "Analysis of thermal runaway in the ignition process," *SIAM J. Appl. Math.* **49**, pp. 459–480.

Dold, J. W., 1991. "Emergence of a detonation within a reacting medium, in "Fluid Dynamical Aspects of Combustion Theory," A. Tesei and M. Onofri, Eds., Longman, U.K.

Dold, J. W. and Kapila, A. K., 1991. "Comparison between shock initiations of detonation using thermally-sensitive and chain-branching chemical models," *Combustion and Flame* **85**, pp. 185–194.

Dold, J. W. and Kapila, A. K., 1993a. "Asymptotic analysis of detonation initiation for one-step chemistry: I–Emergence of a weak detonation," submitted.

Dold, J. W. and Kapila, A. K., 1993b. "Asymptotic analysis of detonation initiation for one-step chemistry: II–from a weak structure to ZND," submitted.

Dold, J. W., Kapila, A. K. and Short, M., 1991. "Theoretical description of direct initiation of detonation for one-step chemistry" in "Dynamical Structure of Detonation in Gaseous and Dispersed Media," A. A. Borissov, Ed., Kluwer Academic Publishers, Dordrecht, pp. 109–142.

Fickett, W. and Davis, W. C., 1979. "Detonation," University of California Press, Berkeley.

Henderson, K. L. and Dold, J. W., 1991. "Transition from thermal-runaway to propagating flames," *SIAM J. Appl. Math.* **51**, pp. 1304–1314.

Jackson, T. L. and Kapila, A. K., 1985. "Shock-induced thermal runaway," *SIAM J. Appl. Math.* **45**, pp. 130–137.

Jackson, T. L., Kapila, A. K. and Stewart, D. S., 1989. "Evolution of a reaction center in an explosive material," *SIAM J. Appl. Math.* **49**, pp. 432–458.

Kapila, A. K. and Dold, J. W., 1991a. "Evolution to detonation in a nonuniformly heated reactive medium," *Lecture Notes in Pure And Applied Mathematics* **130**, pp. 161–173.

Kapila, A. K. and Dold, J. W., 1991b. "A theoretical picture of shock-to-detonation transition in a homogeneous explosive" *9th Symposium (International) on Detonation*, pp. 111–118.

Kapila, A. K. and Jackson, T. L., 1992. private communication.

Lee, J. H. S., 1980. "The mechanism of transition from deflagration to detonation in vapour cloud explosions," *Prog. Energy Combust. Sci.* **6**, pp. 359–389.

Landau, L. D. and Lifshitz, E. M., 1987. "Fluid Mechanics," 2nd edition, Pergamon Press, Oxford, p. 491.

Majda, A. J., 1986. "High Mach number combustion," *Lectures in Applied Mathematics* **24**, pp. 109–183.

Majda, A. J. and Rosales, R. R., 1987. "Nonlinear mean field – high frequency wave interactions in the induction zone," *SIAM J. Appl. Math.* **71**, p. 149.

Short, M., 1992. *Ph.D. thesis*, School of Mathematics, University of Bristol.

Short, M. and Dold, J. W., 1993. "Corrections to Zel'dovich's 'spontaneous flame' and the onset of an explosion via non-uniform preheating," *Prog. Astronautics and Aeronautics,* to appear.

Singh, G. and Clarke, J. F., 1992. "Transient phenomena in the initiation of a mechanically driven plane detonation," *Proc. Roy. Soc. Lond. A* **438**, pp. 23–46.

Strehlow, R. A., 1985. "Fundamentals of Combustion," McGraw-Hill.

Zel'dovich, Ya. B. and Kompaneets, A. S., 1960. "Theory of Detonation," Academic Press.

Zel'dovich, Ya. B., Librovich, V. B., Makhviladze, G. M. and Sivashinsky, G. I., 1970. "On the development of a detonation in a non-uniformly preheated gas," *Astronautica Acta.* **15**, pp. 313–321.

Zel'dovich, Ya. B., 1980a. "Regime classification of an exothermic reaction with nonuniform initial conditions," *Combustion and Flame* **39**, pp. 211–214.

Zel'dovich, Ya. B., 1980b. "Flame propagation in a substance reacting at initial temperature," *Combustion and Flame* **39**, pp. 219–224.

WEAKLY NONLINEAR DYNAMICS OF NEAR-CJ DETONATION WAVES

John B. Bdzil[*] *& Rupert Klein*[**]

[*] Group M-7/MS P952, Los Alamos National Laboratory
Los Alamos, New Mexico, 87545

[**] Institut für Technische Mechanik, RWTH
Templergraben 64, 5100 Aachen, Germany

ABSTRACT

The renewed interest in safety issues for both large scale industrial devices and in high speed combustion has driven recent intense efforts to gain a deeper theoretical understanding of detonation wave dynamics. Linear stability analyses, weakly nonlinear bifurcation calculations, as well as full scale multi-dimensional direct numerical simulations have been pursued for a standard model problem based on the reactive Euler equations for an ideal gas with constant specific heat capacities and simplified chemical reaction models. Most of these studies are concerned with overdriven detonations. This is true despite the fact that the majority of all detonations observed in nature are running at speeds close to the Chapman-Jouguet (CJ) limit value. By focusing on overdriven waves one removes an array of difficulties from the analysis associated with the sonic flow conditions in the wake of a CJ-detonation. In particular, the proper formulation of downstream boundary conditions in the CJ-case is a yet unsolved analytical problem.

A proper treatment of perturbations in the back of a Chapman-Jouguet detonation has to account for two distinct weakly nonlinear effects in the forward acoustic wave component. The first is a nonlinear interaction of highly temperature sensitive chemistry with the forward acoustic wave component in a transonic boundary layer near the end of the reaction zone. The second is a cumulative three-wave-resonance in the sense of Majda et al. which is active in the near-sonic burnt gas flow and which is essentially independent of the details of the chemical model. In this work, we consider detonations in mixtures with moderately state sensitive chemical reactions (no large activation energy). Then, the acoustic perturbations do not influence the chemistry at the order considered, and we may concentrate on the second effect: the three-wave resonance.

J. Buckmaster et al. (eds.), Combustion in High-Speed Flows, 513–540.

1. Introduction

The theory of detonation waves has recently received considerable attention in connection with safety issues involving large scale industrial devices, (Shepherd, 1985 and Breitung, 1991), for the storage of condensed phase explosives, as well as in connection with the problem of efficient, controlled high Mach number combustion in hypersonic aircraft engines (Shepherd, 1992). The stability properties of detonation waves play an important role in these practical problems. Drawing on recent advances in detonation wave stability theory, accurate and efficient new numerical tools have been developed for the direct simulation of these problems. In both the theoretical work on detonation stability (Erpenbeck, 1962, 1963, 1964 and Lee and Stewart, 1990) and the direct numerical simulations (Oran and Boris, 1987; Fujiwara and Reddy, 1989; Schoeffel, 1989; Bourlioux, 1991; and Bourlioux and Majda, 1993), the formulation of downstream, i.e., burnt gas side, boundary conditions is crucial for an accurate representation of the phenomena. The standard approach is to suppress the forward acoustic mode that moves toward the reaction zone from the burnt gas region: so as to mimic a combustion wave that is evolving without any perturbations being imposed from outside (Lee and Stewart, 1990). We will refer to this approach hereafter as the "standard radiation condition" and use the expression "forward" to label acoustic perturbations moving in the direction of detonation propagation in the laboratory frame. In the present setting, where the detonation wave runs from right to left, the forward acoustic characteristics travel at speed $(\partial z/\partial t)^{-} = u - c$, with u and c the local flow velocity and speed of sound, respectively.

The physically most interesting class of detonations is the class of near-Chapman-Jouguet(CJ) waves. The simplest theory for these structures has the detonation reaction zone acoustically decoupled from the burnt gas flow. For convenience, in Section 2 we briefly summarize the ZND detonation theory following Fickett and Davis (1979). In Section 3 we show that for the CJ-regime there is a generation of nonzero forward acoustic perturbations in the burnt gases through weakly nonlinear three-wave resonance of the type first discussed in a different context by Majda and Rosales (1984) and Majda et al. (1988). Thus, even when no forward-acoustic perturbations are present in the burnt gas region at some initial time, they are generated spontaneously through interactions of backward traveling perturbations that emerge from the detonation structure as time evolves. This observation casts suspicion on the standard radiation condition, since the radiation condition is equivalent to suppressing the forward acoustic mode in the farfield. Our goal in this paper is to resolve this apparent contradiction using methods of asymptotic analysis.

To see what is involved, one needs to compare two characteristic length scales. One length scale is the distance behind the lead detonation shock at which the boundary conditions for the stability analysis are imposed. The second is the length scale for the three-wave resonance phenomena mentioned above. Let ε be the small perturbation amplitude used as an expansion parameter. (Later on we consider ε to be the growth rate of the most unstable Eigenmode of a marginally stable detonation.) Figure 1 shows a sketch of the spatial distribution of a representative reaction progress variable, $(1 - \lambda)$, in a ZND-detonation. The gas is completely unburnt (i.e., $(1 - \lambda) = 1$) at the lead shock location, $z = 0$. As $z \to \infty$, we approach the burnt gas region and $(1 - \lambda) \to 0$.

Numerical linear stability codes (e.g., Lee and Stewart, 1990) resolve a region $0 \leq z < z_{1/N}$, where $z_{1/N}$ is the location where $(1 - \lambda) = 1/N$ and N is the number of grid points in the discretized linear perturbation equations. Notice that

$$z_{1/N} = O(1) \qquad \text{as} \qquad \varepsilon \to 0\,, \tag{1.1}$$

i.e., in terms of the perturbation amplitude, the downstream boundary condition is imposed at a finite distance. In contrast, the three-wave resonance occurs at distances

$$z = O(1/\varepsilon). \tag{1.2}$$

The key question is: What becomes of the forward acoustic perturbations generated at these large distances from the front as they approach and finally reach the location of the tail of the detonation structure, $z_{1/N}$, where the numerical downstream boundary conditions are applied?

One may answer this question by using the method of matched asymptotic expansions: separate, simplified asymptotic descriptions of the solution are found in four distinct zones and are matched together to get a composite solution. There is a main reaction layer where $(1 - \lambda) = O(1)$, an intermediate transonic layer where

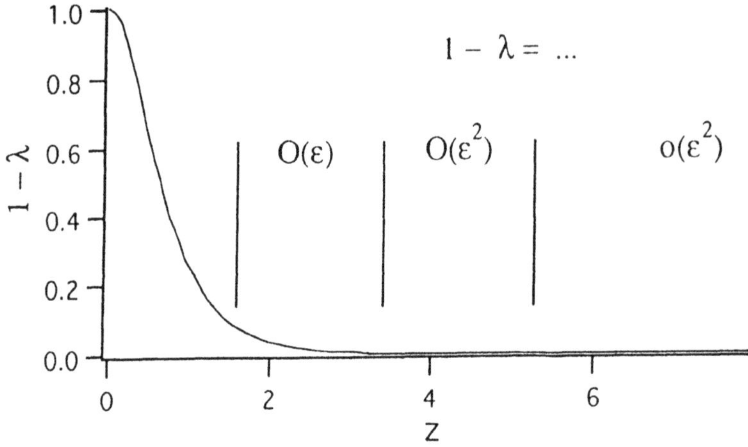

Fig. 1: Spatial distribution of the reaction progress variable in a CJ-ZND detonation and identification of several asymptotic layers.

$(1-\lambda) = O(\varepsilon)$, a burnout layer where $(1-\lambda) = O(\varepsilon^2)$, and the burnt gas region where $(1-\lambda) \ll \varepsilon^2$. The acoustic resonances are dominant in the latter two zones and the present paper focuses on their analysis. The principal facts relevant for the formulation of downstream boundary conditions are readily demonstrated in this paper through a combination of formal asymptotic arguments and numerical solutions of the burnout layer equations. We leave detailed numerical studies of the burnt gas resonance effects, the tedious calculations for the transonic zone and the final matching to the main reaction layer for a later publication (Bdzil and Klein, 1993).

Section 3 of this paper discusses the zone where the chemical activity is negligible; the burnt gas region, $z = O(1/\varepsilon)$. We summarize the equations for weakly nonlinear three-wave resonance and define a model problem suited to exhibit the generation of forward acoustic perturbations. On the time scales considered, however, we find that the perturbations near the end of the reaction zone (e.g., at $z_{1/N}$ from (1.1)) are related directly to the asymptotic solutions only in the burnout layer. The chemical source term appears at order $O(\varepsilon^2)$ in the burnout zone and modifies the acoustic resonance equations. We concentrate on this region in Section 4 and develop formal solutions based on the method of characteristics as well as numerical solutions including shock discontinuities. The main influence of the chemical source term is to continuously accelerate and amplify the forward acoustic perturbations and to establish a reaction tail that matches into the unperturbed CJ-ZND structure. Both the formal characteristic solution and the numerical solutions show the expected behavior – an energy transfer from the backward acoustic mode into forward acoustic perturbations. Furthermore, both approaches predict that the effects of weakly nonlinear acoustic resonance decay as the forward characteristics move upstream into the transonic region.

The tendency of the resonantly generated forward acoustic waves to decay as they move into the reaction-zone tail is demonstrated in detail in Bdzil and Klein (1993), where we construct matched asymptotic solutions for all the layers mentioned above. There we show that the resonance effects are down to order $O(\varepsilon^2)$ at the finite distances where numerical boundary conditions are imposed. Therefore we conclude that:

> The standard radiation condition imposed at some finite distance $z_{1/N}$ behind the lead shock is valid even for Chapman-Jouguet detonations.

Our approach is the first to use rational asymptotic methods to resolve the boundary condition difficulties that arise in the stability problem for plane CJ-detonation. Our success at resolving these difficulties, which have been a part of linear stability analyses of detonation since Erpenbeck (1962, 1963, 1964), suggests this same approach for other problems; notably the confusing "square wave" detonation stability problem and the problem of oblique detonation stability in the CJ-like critical regime.

2. The ZND Detonation Model

We analyze the dynamics of fast combustion waves in the framework of a standard model for gaseous detonations. A plane ZND-detonation (see Fickett and Davis (1979)) consists of a leading inviscid shock wave and a subsequent zone of chemical activity. The lead shock heats up and compresses the gas so that exothermal chemical reactions are turned on. The chemical energy is converted into thermal and kinetic energy, thereby overcoming the dissipation in the lead shock. For a given combustible there is a continuous family of ZND detonations parametrized by the detonation speed, D. The so-called Chapman-Jouguet detonations are those with the smallest possible speed $D = D_{\mathrm{CJ}}$. The main feature of a CJ-detonation needed in our analysis is that forward acoustic perturbations in the burnt gases travel at exactly the same speed as the detonation itself. In other words, the burnt gases move away from the detonation structure at (their own) sonic speed. Figure 2 shows a space-time diagram for a CJ-detonation traveling from right to left with the unburnt gas at rest in the laboratory reference frame. We display the path of the leading shock, indicate the reaction zone and exhibit a family of forward acoustic characteristics.

The ZND detonation structure is described by a traveling wave solution of the reactive Euler equations. Adopting the nondimensional form of the Euler equations used by Lee and Stewart (1990) (see (2.7) for the reference quantities used for nondimensionalization), we have

$$\underline{U}_t + \underline{\underline{A}}(\underline{U})\underline{U}_{\bar{z}} = \underline{C}(\underline{U}), \tag{2.1}$$

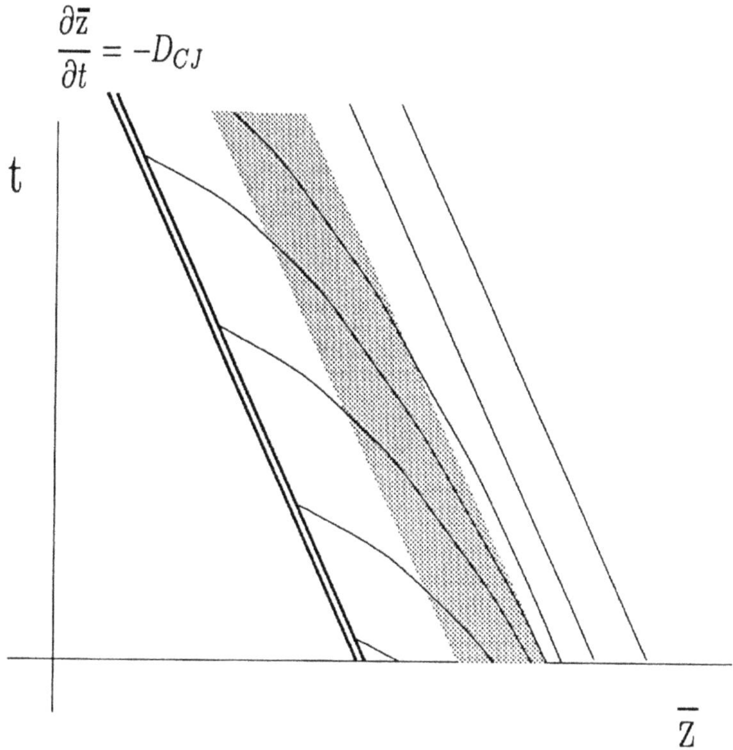

Fig. 2: Wave diagram for forward acoustic perturbations in the wake of a Chapman-Jouguet detonation travelling from right to left in a laboratory frame. The shaded region represents the zone of chemical activity.

where $(\bar{z}, t)$ are laboratory coordinates and $\underline{U}$ is the nondimensional solution vector

$$\underline{U} = (v, \bar{u}, p, \lambda)^t, \tag{2.2}$$

with v the specific volume, $\bar{u}$ the flow velocity in the laboratory frame, p the pressure and λ a reaction progress variable satisfying $\lambda = 0$ in the unburnt and $\lambda = 1$ in the burnt gas. Furthermore

$$\underline{\underline{A}}(\underline{U}) = \begin{pmatrix} \bar{u} & -v & 0 & 0 \\ 0 & \bar{u} & v/\gamma & 0 \\ 0 & \gamma p & \bar{u} & 0 \\ 0 & 0 & 0 & \bar{u} \end{pmatrix} \tag{2.3}$$

and

$$\underline{C}(\underline{U}) = (0, 0, \frac{\gamma - 1}{v} Q r, r)^t, \tag{2.4}$$

where γ is the adiabatic exponent, Q is the specific heat of reaction and

$$r = K(p, v)(1 - \lambda) \tag{2.5}$$

is the reaction rate. For simplicity in the derivations we have assumed an ideal gas with constant specific heats so that $\gamma = \text{const}$. The above equations are supplemented by shock jump conditions at

$$z = \bar{z} + D_{\text{CJ}}\, t = 0 \tag{2.6}$$

(see e.g., Fickett and Davis (1979), Lee and Stewart (1990)). These equations admit a steady, traveling wave solution $\underline{U} = \underline{U}^*(z)$ for $z \geq 0$, that begins with a shock at $z = 0$ (the von Neumann state).

The above equations have been nondimensionalized using the reference quantities

$$\begin{array}{lcr} \text{specific volume} & : & v_{\text{sh}} \\ \text{pressure} & : & P_{\text{sh}} \\ \text{velocity} & : & c_{\text{sh}} = \sqrt{\gamma P_{\text{sh}} v_{\text{sh}}} \\ \text{length} & : & \text{half-reaction length } l_{1/2} \\ \text{time} & : & l_{1/2}/c_{\text{sh}}\,. \end{array} \tag{2.7}$$

The half-reaction length is defined as the distance behind the lead shock where $\lambda = 1/2$ and the subscript "sh" denotes the post-shock (von Neumann) conditions.

The exact solution for the Chapman-Jouguet – ZND detonation structure can be written in terms of the reaction progress variable, $\lambda = \lambda^*(z)$, as, (Lee and Stewart (1990)),

$$\begin{array}{rcl} p^* & = & a^* + (1 - a^*)(1 - \lambda^*)^{1/2}\,, \\ v^* & = & \frac{1-p^*}{\gamma M_{\rm sh}^2} + 1\,, \\ u^* & \equiv & \bar{u}^* + D_{\rm CJ} = v^* M_{\rm sh}\,, \end{array} \tag{2.8}$$

where

$$a^* = \frac{\gamma M_{\rm sh}^2 + 1}{\gamma + 1} \tag{2.9}$$

and $M_{\rm sh}$ is the steady detonation shock Mach number as seen by an observer moving with the mass particles immediately behind the shock (i.e., $M_{\rm sh} = u^*(\lambda^* = 0)$). This quantity is related to the nondimensional heat release parameter, $Q' = (\gamma^2 - 1)Q$, by

$$M_{\rm sh}^2 = 1 + Q' - \sqrt{Q'^2 + 2Q'}. \tag{2.10}$$

Equations (2.8) express the detonation structure solely in terms of the reaction progress variable $\lambda^* = \lambda^*(z)$. The spatial distribution of all quantities follows from solving the fourth equation in (2.1):

$$u^* \lambda_z^* = r(p^*, v^*, \lambda^*)\,, \tag{2.11}$$

with u^*, v^*, p^* and r^* given by (2.8) and (2.5), respectively. In particular, we notice that $u^* - \sqrt{p^* v^*} = 0$ at $(\lambda^* = 1)$, i.e., that the burnt gas flow is sonic.

3. Weakly Nonlinear Resonant Acoustics in the Farfield

3.1. The burnt gas region

Consider a marginally stable CJ-detonation that oscillates with a small slowly varying amplitude in the frame of reference moving with the unperturbed wave as sketched in Fig. 3. Under CJ-conditions, one has $|u_{\rm CJ}| = c_{\rm CJ}$, such that

$$(u - c)_{\rm CJ} = 0\,, \quad u_{\rm CJ} + c_{\rm CJ} = 2u_{\rm CJ} = 2c_{\rm CJ}\,. \tag{3.1}$$

Entropy perturbations generated by the oscillations of the detonation speed and by perturbations of the reaction process travel backwards

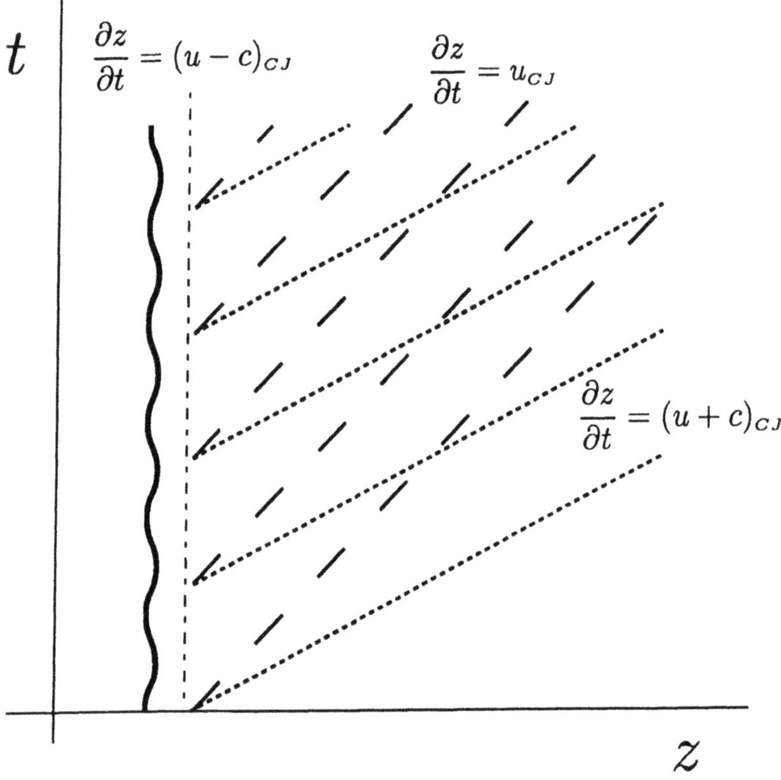

Fig. 3: Characteristics in the burnt gas behind a marginally stable CJ-detonation.

into the burnt gases at speed $(\partial z/\partial t) = u_{CJ}$. Backward facing acoustic perturbations generated in the reaction zone travel into the burnt gases at speed $(\partial z/\partial t) = u_{CJ} + c_{CJ} = 2u_{CJ}$. Since these perturbations are generated in phase within the detonation structure, their associated spatial wave lengths in the burnt gas region differ by a factor of exactly two. From the theory of weakly nonlinear resonant acoustics by Majda and Rosales (1984), one knows that this is the condition for cumulative resonant generation of forward acoustic perturbations. These spontaneously generated forward waves are of the same order of magnitude as the backward traveling perturbations, and cannot be suppressed. To be more specific, we propose the following model problem that allows one to give a quantitative description of the resonance and of its effects near the end of the detonation reaction zone.

A CJ-detonation with a marginally stable structure is perturbed at time $t = 0$. In a suitable frame of reference the wave will typically start to oscillate around its unperturbed position at a frequency given by the imaginary part, α_i, of the Eigenvalue associated with the most unstable Eigenmode. The perturbations of the detonation speed and of the reaction process generate entropy and backward acoustic perturbations. These leave the reaction zone and radiate into the burnt gases. We are interested in the long-time behavior of the radiation field in the back of the wave. We assume the entropy and backward acoustic perturbations to be given in the form of sinusoidal oscillations at the edge of the reaction zone; more precisely at some location $z = z_3$ where $(1 - \lambda^*) = \varepsilon^3$ so that for $z \sim z_3$ according to (2.5) the reaction source terms are negligible up to and including the second order in ε. Then we have to solve the following initial-boundary value problem up to first order in ε for the inert Euler equations,

$$\underline{U}_t + \underline{\underline{A}}(\underline{U})\underline{U}_z = 0 \quad (z > z_3\,, t \geq 0)\,, \tag{3.2}$$

with initial conditions

$$\underline{U}(z,0) = \underline{U}_{CJ} \quad (z > z_3)\,, \tag{3.3}$$

and boundary conditions at $z = z_3$ for the entropy and backward acoustic modes given by

$$\underline{l}^0 \cdot \underline{U}(z_3, t; \varepsilon) = \underline{l}^0 \cdot \underline{U}_{CJ} + \varepsilon S_0(\varepsilon t)\cos(\alpha_i t) \tag{3.4}$$

and

$$\underline{l}^{+} \cdot \underline{U}(z_3, t; \varepsilon) = \underline{l}^{+} \cdot \underline{U}_{\mathrm{CJ}} + \varepsilon W_0(\varepsilon t) \cos(\alpha_i t + \varphi) . \tag{3.5}$$

Here $\underline{U}_{\mathrm{CJ}}$ is the unperturbed burnt gas state, $\underline{l}^0$, $\underline{l}^+$, are the left eigenvectors of the matrix $\underline{\underline{A}}(\underline{U}_{\mathrm{CJ}})$ associated with the eigenvalues u_{CJ} and $(u+c)_{\mathrm{CJ}}$, respectively and $\underline{\underline{A}}$ is the same as in (2.3) except that $\bar{u}$ is replaced by $u = \bar{u} + D_{\mathrm{CJ}}$. (See Appendix A for more details.) Left-multiplication by $\underline{l}^0$, $\underline{l}^+$ filters out the entropy and backward acoustic contributions to the state vectors. Also in (3.4) and (3.5) we allow for slow-time variations of the amplitude functions S_0 and W_0. This ansatz is justified in cases where the linear growth rate, Re(α), of the most unstable Eigenmode is of order $O(\varepsilon)$, i.e.,

$$\alpha = \varepsilon \alpha_r^{(1)} + i\alpha_i , \quad \alpha_r^{(1)} = O(1) \quad (\varepsilon \to 0) . \tag{3.6}$$

This relation defines what we mean by marginal stability. At this stage the precise relations between S_0, W_0 and the phase shift φ in (3.5) remain undetermined. For our model problem we let these quantities be constant and leave the derivation of equations for their long-time evolution for (Bdzil and Klein, 1993).

Closely following the procedure of Majda and Rosales (1984), we introduce a multiple scales perturbation ansatz:

$$\underline{U} = \underline{U}_{\mathrm{CJ}} + \varepsilon[V\underline{r}^{-} + S\underline{r}^{0} + W\underline{r}^{+}] + \varepsilon^2 \underline{U}^{(2)}(x, t; \varepsilon) , \tag{3.7}$$

where the hyperbolic mode amplitudes depend on $(z, t; \varepsilon z, \varepsilon t)$, i.e.,

$$[V, S, W] = [V, S, W](z, t; \varepsilon z, \varepsilon t) . \tag{3.8}$$

The right eigenvectors $\underline{r}^{-}, \underline{r}^{0}$ and $\underline{r}^{+}$ of the matrix $\underline{\underline{A}}$ associated with the eigenvalues $u-c$, u and $u+c$, respectively, are given in Appendix A. At the leading order in ε one finds

$$\begin{aligned} V_t + (u-c)_{\mathrm{CJ}} V_z &\equiv V_t = 0 \\ S_t + u_{\mathrm{CJ}} S_z &= 0 \\ W_t + (u+c)_{\mathrm{CJ}} W_z &= 0 . \end{aligned} \tag{3.9}$$

This gives rise to

$$\begin{aligned} V &= V(x; \varepsilon z, \varepsilon t) \\ S &= S(h; \varepsilon z, \varepsilon t) \\ W &= W(y; \varepsilon z, \varepsilon t) , \end{aligned} \tag{3.10}$$

where the fast scale arguments are defined by

$$\begin{array}{rcl} x & = & z - (u-c)_{CJ}t \equiv z \\ y & = & z - (u+c)_{CJ}t \\ h & = & z - u_{CJ}t \equiv \frac{1}{2}(x+y)\,. \end{array} \tag{3.11}$$

These are the characteristic coordinates for the forward and backward acoustic and the entropy mode, respectively.

At the second order, one first derives equations analogous to (3.9) for the mode amplitudes in $\underline{U}^{(2)}$, but with right hand sides depending on V, S, W and their derivatives with respect to $(\varepsilon z, \varepsilon t)$. Then, by requiring $\underline{U}^{(2)}$ to be bounded even for $z, t = O(1/\varepsilon)$ (i.e., on the space-time scales for the action of weakly nonlinear effects) one obtains secular equations for the evolution of the first order mode amplitudes V, S, W in the slow scale variables

$$\zeta = \varepsilon z \qquad \text{and} \qquad \tau = \varepsilon t\,. \tag{3.12}$$

These equations are

$$\begin{array}{rcl} (V_\tau + (u-c)_{CJ}V_\zeta) - a\left[V - b\langle W\rangle + \langle S\rangle\right]V_x & = & a\langle SW_y\rangle^{(x)}\,, \\ (S_\tau + u_{CJ}S_\zeta) + c_{CJ}\langle W - V\rangle S_h & = & 0\,, \\ (W_\tau + (u+c)_{CJ}W_\zeta) + a\left[W - b\langle V\rangle + \langle S\rangle\right]W_y & = & -a\langle SV_x\rangle^{(y)}\,, \end{array} \tag{3.13}$$

where

$$\begin{array}{rcl} a & = & \frac{\gamma+1}{2}c_{CJ} \\ b & = & \frac{3-\gamma}{\gamma+1} \end{array} \tag{3.14}$$

and $\langle\cdot\rangle$, $\langle\cdot\rangle^{(x)}$ and $\langle\cdot\rangle^{(y)}$ are fast variable averaging symbols. In particular, the superscripts $^{(x)}$ and $^{(y)}$ denote averaging at fixed x and y, respectively. The precise definitions of the averaging operations are given in Appendix B.

Next we introduce new slow-time scale variables on the characteristic curves by

$$\begin{array}{rcl} \eta & = & a\left(\tau - \frac{\zeta}{(u+c)_{CJ}}\right), \\ \xi & = & a\,\frac{\zeta}{2c_{CJ}}\,, \\ \vartheta & = & \frac{\zeta}{u_{CJ}}\,. \end{array} \tag{3.15}$$

Then η measures time along the forward acoustic characteristics $(z, \zeta = const)$, ξ measures time after the passage through the lead shock along backward acoustic characteristics $(z-(u+c)_{CJ}t = const)$, and ϑ measures time along the particle paths ($z - u_{CJ}t = const$). In addition, using the particle path label

$$\chi = \zeta - u_{CJ}\tau = \varepsilon h\,, \tag{3.16}$$

and expressing the slow scale dependencies of V, S and W in terms of ξ, η, χ and ϑ

$$V = V(x;\xi,\eta)\,, \qquad S = S(h;\chi,\vartheta)\,, \qquad W = W(y;\xi,\eta)\,, \tag{3.17}$$

we find that (3.13) becomes

$$\begin{aligned} V_\eta - [V - b\langle W\rangle + \langle S\rangle]\, V_x &= \langle SW_y\rangle^{(x)}\,, \\ S_\vartheta + c_{CJ}\langle W - V\rangle S_h &= 0\,, \\ W_\xi + [W - b\langle V\rangle + \langle S\rangle]\, W_y &= -\langle SV_x\rangle^{(y)}\,. \end{aligned} \tag{3.18}$$

Applying the fast variable averaging to these equations yields

$$\langle V\rangle_\eta = \langle S\rangle_\vartheta = \langle W\rangle_\xi \equiv 0\,, \tag{3.19}$$

so that

$$\langle V\rangle = \langle V\rangle(\xi)\,, \qquad \langle S\rangle = \langle S\rangle(\chi)\,, \qquad \langle W\rangle = \langle W\rangle(\eta)\,. \tag{3.20}$$

These functions are readily determined by the initial-boundary data. Requiring that there be no forward acoustic perturbation in the back of the wave at time $t = \tau = 0$, we have

$$\langle V\rangle \equiv 0\,. \tag{3.21}$$

Then by assuming the exact CJ conditions in the burnt gases for our model problem, we may set

$$\langle S\rangle(\chi) = \bar{S}(\chi) \equiv 0\,; \qquad \langle W\rangle(\eta) = \bar{W}(\eta) \equiv 0\,, \tag{3.22}$$

where we use an overbar to denote the fast variable average. Notice that these conditions are assumed here, not derived. A complete second order analysis, including the detonation reaction zone structure would yield long-time evolution equations for the functions $\bar{S}(\chi)$ and $\bar{W}(\eta)$. It is in this sense that we are proposing a model problem here.

With (3.21) and (3.22), the entropy equation in (3.16) decouples from the acoustic mode equations and we find

$$\begin{aligned} V_\eta - VV_x &= \langle SW_y\rangle^{(x)} \\ W_\xi + WW_y &= -\langle SV_x\rangle^{(y)}\,. \end{aligned} \tag{3.23}$$

The appropriate initial-boundary data for V and W are

$$\begin{aligned} V(x;\xi,0) &= 0 \\ W(y;0,\eta) &= W_0\cos(\alpha_i y)\,, \end{aligned} \tag{3.24}$$

and the entropy mode is explicitly given by

$$S(\frac{1}{2}(x+y);\chi,\vartheta) = S_0\cos(2\alpha_i\frac{x+y}{2}+\varphi)\,, \tag{3.25}$$

where φ is a "constant" phase shift. The system of equations (3.23) through (3.25) models the generation of forward acoustic perturbations in the burnt gases due to resonant energy transfer from the backward acoustic mode. We will describe a numerical study of the solutions to these equations in (Bdzil and Klein, 1993). In the present paper we are mainly interested in the influence that these spontaneously generated forward acoustic waves have on the perturbations at the end of the reaction zone. From the discussion in the introduction we recall that this means determining the forward acoustic perturbations at a finite distance ($z_{1/N} = O(1)$ as ($\varepsilon \to 0$)) behind the lead shock where $(1-\lambda) = 1/N$, with N fixed. In the current section the chemistry was completely negligible. To obtain the desired result on the influence of resonance effects at $z = z_{1/N}$, we have to match this burnt gas region to the main reaction zone, $(1-\lambda) = O(1)$. This matching procedure requires an analysis of two intermediate layers distinguished by different levels of chemical activity:

There is

i) a burnout layer, where $(1-\lambda) = O(\varepsilon^2)$ and

ii) a transonic region, where $(1-\lambda) = O(\varepsilon)$ as $(\varepsilon \to 0)$.

We analyze the burnout layer in detail in Section 4. This analysis will constitute our principal result: that the resonance effects become weaker as one moves upstream towards the main reaction zone.

A complete analysis of the matching through the transonic region is not only tedious, but it does not lead to new insights; thus we leave this discussion for a later more comprehensive report (Bdzil and Klein, 1993).

4. The Burnout Layer

4.1. Asymptotic behavior of the reaction progress equation

In this subsection we are interested in a region behind the detonation structure where the effects of chemical heat release enter at second order and therefore modify the acoustic resonances. To assess the extent of this region, we first consider the reaction progress equation

$$(1-\lambda)_t + u(1-\lambda)_z = -K(p,v)(1-\lambda). \tag{4.1}$$

The analysis will provide the correct scaling for the thickness of this asymptotic layer; it will also yield the appropriate expansion scheme for the reaction progress variable, λ.

Assuming that no chemistry occurs in the lead shock,

$$(1-\lambda) = 1 \qquad \text{at} \qquad z = 0 \tag{4.2}$$

the general solution to (4.1) reads

$$(1-\lambda) = \exp\left(-{}^{(m)}\!\!\int_0^z \frac{K(p,v)}{u}\, dz\right), \tag{4.3}$$

where the integration is along a particle path, $m = \text{const}$, with

$$\left.\frac{dt}{dz}\right|_m = \frac{1}{u(z,t)}. \tag{4.4}$$

With an asymptotic expansion of $K(p,v)$ and u about the ZND profile according to

$$\begin{aligned} K(p,v) &= K(p^*,v^*) + \varepsilon\left((K_p)^* p^{(1)} + (K_v)^* v^{(1)}\right) + \dots \\ u &= u^* + \varepsilon u^{(1)} + \dots, \end{aligned} \tag{4.5}$$

equation (4.3) reads

$$(1-\lambda) = \exp\left(-{}^{(m)}\!\!\int_0^z \frac{K^*}{u^*}\, dz\right) \exp\left(-\varepsilon\, {}^{(m)}\!\!\int_0^z \left[\frac{K}{u}\right]^{(1)} dz\right) \dots, \tag{4.6}$$

where

$$\begin{aligned} \left[\frac{K}{u}\right]^{(1)} = \frac{1}{u^*} \quad & \left[\quad (K_p)^* p^{(1)} + (K_v)^* v^{(1)} + (dK^*/dz)\, z^{(1)}\right] \\ - \quad & \frac{K^*}{(u^*)^2}\left[u^{(1)} + (du^*/dz) z^{(1)}\right], \end{aligned} \tag{4.7}$$

and $z^{(1)}$ is the first order perturbation of the particle path, $m =$ const. Obviously, (4.6) may be written as

$$(1-\lambda) = (1-\lambda)^*(1+O(\varepsilon)) \tag{4.8}$$

and we conclude that in any region where $(1-\lambda) = O(\varepsilon^n)$ the proper expansion is

$$(1-\lambda) = \varepsilon^n \Lambda(z) + \varepsilon^{n+1}\lambda^{(n+1)}(z,t;\varepsilon z,\varepsilon t) + \dots \tag{4.9}$$

Furthermore, as $(1-\lambda)^* \to 0$, $K^* \to K_{\mathrm{CJ}}, u^* \to u_{\mathrm{CJ}}$ so that

$$\Lambda(z) = \exp(-\beta(z-z_n)) \qquad \text{with} \qquad \beta = \frac{K_{\mathrm{CJ}}}{u_{\mathrm{CJ}}} \tag{4.10}$$

and with z_n chosen so that $(1-\lambda^*)(z_n) = \varepsilon^n$. As a consequence, in the burnout layer we have

$$(1-\lambda) = \varepsilon^2 \exp(-\beta(z-z_2))\,(1+O(\varepsilon)) \tag{4.11}$$

and the thickness of the layer is of order $\ln(1/\varepsilon)$. This result determines the appropriate choice of independent variables for the acoustic modes in this region as we will explain in the next subsection.

4.2. Influence of the reaction tail on the acoustic resonance

In the burnout layer considered here, we employ a multiple scales expansion for the hyperbolic modes at first order that is similar to the one used in the burnt gas region. There, we needed to introduce the slow-time variables, (η,ξ,ϑ) on each of the characteristics. In contrast, the residence time of entropy and backward acoustic perturbations in the burnout layer is of order $\ln(1/\varepsilon)$ only, and this is insufficient to allow for accumulation of weakly nonlinear effects. As indicated in Fig. 4, only the forward acoustic characteristics have a long enough residence time in this layer. The graph displays the forward and backward acoustic characteristics and particle paths, and shows the order of magnitude of their passage times through the burnout layer.

The appropriate expansion scheme in the present regime reads

$$\underline{U} = \underline{U}_{\mathrm{CJ}} + \varepsilon[\hat{V}\underline{r}^- + \hat{S}\underline{r}^0 + \hat{W}\underline{r}^+] + \varepsilon^2(\underline{\hat{U}}^{(2)}(z,t;\varepsilon) + \Lambda(z)\underline{r}^\lambda)\,, \tag{4.12}$$

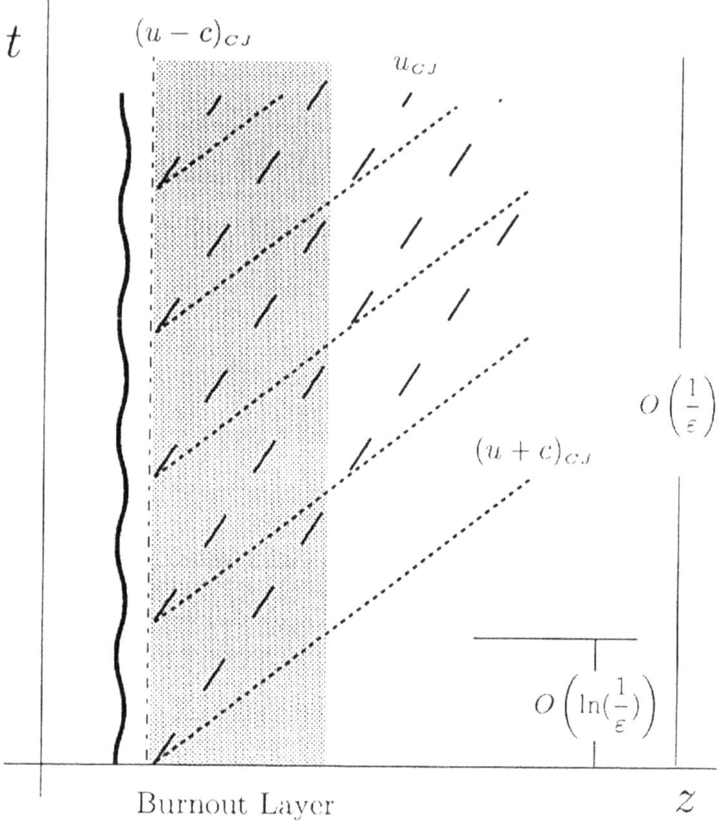

Fig. 4: Characteristics and their residence times in the burnout layer.

where

$$\begin{aligned} \hat{V} &= \hat{V}(\hat{x};\eta) \\ \hat{S} &= \hat{S}(\tfrac{1}{2}(\hat{x}+y);\eta) \\ \hat{W} &= \hat{W}(y;\eta) \end{aligned} \tag{4.13}$$

and

$$\hat{x} = z - z_2\,, \tag{4.14}$$

with z_2 defined in (4.10) and (4.11).

For our model problem we assume that the backward traveling mode amplitudes, W_0 and S_0, the oscillation wave number, α_i, and the phase shift, φ, in (3.4) and (3.5) are independent of the long-time variable, η, and so we set

$$\begin{aligned} \hat{W} &\equiv W_0\cos(\alpha_i y + \psi(\varepsilon)) \\ \hat{S} &\equiv S_0\cos(2\alpha_i\tfrac{(\hat{x}+y)}{2} + \varphi + \psi(\varepsilon))\,. \end{aligned} \tag{4.15}$$

Here $\psi(\varepsilon)$ is an ε-dependent phase shift that makes the boundary data imposed on S and W in the burnt gas region compatible with those for $\hat{S}$ and $\hat{W}$ in the present regime.

The only secular constraint that remains is the long-time evolution equation for $\hat{V}$, namely

$$\begin{aligned} \hat{V}_\eta - \hat{V}\hat{V}_{\hat{x}} &= \mu\Lambda(\hat{x}) + \langle \hat{S}\hat{W}_y\rangle^{(\hat{x})} \\ &= \mu\exp(-\beta\hat{x}) + \tfrac{1}{2}\alpha_i S_0 W_0\sin(\alpha_i\hat{x}+\varphi)\,, \end{aligned} \tag{4.16}$$

where $\mu = (\beta/c_{\mathrm{CJ}}{}^2)\,(Q'/(\gamma+1)^2)$, with Q' defined by (2.10). We use the stationary unperturbed solution, corresponding to the CJ-ZND profile, as the initial condition

$$\hat{V}(\hat{x},0) \equiv \hat{V}^{\mathrm{st}}(\hat{x}) = \sqrt{\frac{2\mu}{\beta}}e^{-\frac{1}{2}\beta\hat{x}}\,. \tag{4.17}$$

Equation (4.16) is the inviscid Burgers' equation driven by an explicit source term resulting from the combined effects of chemical heat release and weakly nonlinear acoustic resonance.

Note that we merely need to solve this burnout layer equation in order to obtain the desired information on the influence of resonant acoustics perturbations on the upstream flow. There is an obvious physical interpretation of this formal result. On the time scales considered, $t = O(1/\varepsilon)$, forward acoustic perturbations of order $O(\varepsilon)$ can travel distances of no more than order $O(1)$ as $\varepsilon \to 0$. The forward acoustics generated in the burnt gas region at distances $O(1/\varepsilon)$

downstream from the detonation structure will thus not arrive in time to have any influence. The only perturbations that make it in time are those generated in the burnout layer. Since the generating backward traveling mode amplitudes, $\hat{W}$ and $\hat{S}$, in this region are given directly by their upstream boundary data, (4.15), one only has to solve the single scalar equation, (4.16), with initial data, (4.17), to determine the evolving forward acoustic mode amplitude.

A formal solution to (4.16) can be found using the method of characteristics. Below we construct this solution. In particular, we derive the behavior of the solution as $\hat{x} \to -\infty$; this is needed for the upstream matching to the transonic region. This solution is valid only as long as characteristic curves do not cross, i.e., for shock-free solutions. To go beyond this, we present additional numerical solutions of (4.16) and (4.17) in Section 4.3. These solutions do reveal the formation of shocks. Nevertheless, they verify the limit behavior of the solution for $\hat{x} \to -\infty$ derived from the formal characteristic analysis.

Let $m = \text{const}$ denote a characteristic of (4.16), so that

$$\left.\frac{d\eta}{d\hat{x}}\right|_m = -\frac{1}{\hat{V}} \qquad \text{with} \qquad \eta(\hat{x}_0(m), m) = 0\,. \tag{4.18}$$

Without loss of generality we may choose $\hat{x}_0(m) = m$. Then (4.16) becomes

$$\left.\frac{1}{2}\frac{d\hat{V}^2}{d\hat{x}}\right|_m = -\mu\exp(-\beta\hat{x}) - \frac{1}{2}\alpha_i S_0 W_0 \sin(\alpha_i\hat{x} + \varphi)\,, \tag{4.19}$$

and the explicit solution in terms of $(\hat{x}, m)$ is

$$\hat{V}(\hat{x}, m) = \left\{2\frac{\mu}{\beta}e^{-\beta\hat{x}} + S_0 W_0\Big[\cos(\alpha_i\hat{x} + \varphi) - \cos(\alpha_i m + \varphi)\Big]\right\}^{1/2}\,. \tag{4.20}$$

To obtain the solution in terms of $(\hat{x}, \eta)$, one needs to construct the family of characteristic curves, $m = \text{const}$, by solving (4.18) numerically, and then using the solution in (4.20). This solution is valid as long as neighboring characteristics do not cross. A crossing signals the formation of a shock, and a solution technique capable of approximating weak solutions to nonlinear hyperbolic equations needs to be employed. We discuss this in Section 4.3.

In the limit $\hat{x} \to -\infty$, (4.20) can be expanded to get

$$\hat{V}(\hat{x},m) = \sqrt{\tfrac{2\mu}{\beta}} \Big\{ e^{-\frac{\beta}{2}\hat{x}} + \\ + e^{\frac{\beta}{2}\hat{x}} \tfrac{\beta}{4\mu} S_0 W_0 \Big[\cos(\alpha_\imath \hat{x} + \varphi) - \cos(\alpha_\imath m + \varphi)\Big]\Big\} + \dots \\ \text{as} \quad \hat{x} \to -\infty. \tag{4.21}$$

The terms in square brackets represent the influence of the initial data and of the acoustic resonance. These are multiplied by an exponential that decays as $\hat{x} \to -\infty$. In contrast, the first term, which corresponds to the background CJ-ZND solution, diverges in this limit. For the matching to the upstream transonic layer we are interested in positions $\hat{x}$ where $(1-\lambda) = O(\varepsilon)$. According to (3.35) and (3.38) this regime corresponds to

$$\hat{x} = \check{x} - \frac{1}{\beta}\ln(\frac{1}{\varepsilon}) \qquad \text{with} \qquad \check{x} = O(1) \qquad \text{as} \qquad (\varepsilon \to 0)\,. \tag{4.22}$$

Substituting (4.22) into (4.21) we find

$$\hat{V}(\hat{x},m) = \sqrt{\tfrac{2\mu}{\beta}} \Big\{ \varepsilon^{1/2} e^{-\frac{\beta}{2}\check{x}} \\ + \varepsilon^{3/2} \quad \tfrac{\beta}{4\mu} e^{\frac{\beta}{2}\check{x}} S_0 W_0 \Big[\cos(\alpha_\imath \check{x} + \varphi + \check{\psi}) - \cos(\alpha_\imath \check{m} + \varphi + \check{\psi})\Big]\Big\} \\ + \quad \dots \\ \text{as} \quad \check{x} = O(1) \tag{4.23}$$

Here $\check{\psi}$ is a phase shift analogous to $\psi(\varepsilon)$ as explained below (4.15), and $\check{m}$ is defined in analogy with $\check{x}$ from (4.22). The immediate conclusions from the formula in (4.23) are that: (i) the deviation of the solution from CJ-conditions is now of order $O(\varepsilon^{1/2})$, and (ii) the resonance effects no longer appear at order $O(\varepsilon)$ in the upstream transonic layer but only at order $O(\varepsilon^{3/2})$.

The above derivations, which are based on the method of characteristics, are valid only as long as the solutions are shock-free; generally this is true only for some finite time. In order to test the validity of the limit represented in (4.23), in the next subsection we present numerical solutions of the burnout layer problem obtained using a modern higher order Godunov-type upwind technique.

A remark on the downstream matching of the burnout layer solution with the burnt gas region is in order. In the burnout layer, the backward traveling mode amplitudes $\hat{S}$ and $\hat{W}$ are frozen with respect to their slow-time scale variables, ϑ and ξ, respectively (see Eq. (4.13)). Therefore, they cannot accommodate the cumulative influence of the acoustic resonance and, as $\hat{x} \to +\infty$, the second order perturbation, $\underline{U}^{(2)}$ diverges. This divergence, however, can be matched to the behavior of the backward traveling mode amplitudes S and W at $\xi = 0$ in the burnt gas region (Bdzil and Klein, 1993).

4.3. Numerical solutions of the burnout layer problem

We use a higher order MUSCL Godunov-type upwind scheme, (see e.g., van Leer (1979) or LeVeque (1990)), to solve the homogeneous inviscid Burgers' equation, $V_\eta - VV_x = 0$ and Strang-type operator splitting to account for the right hand side in (4.16). The sample results shown in Fig. 5 are based on the parameter set

$$\begin{aligned} W_0 = S_0 = 1.0, \qquad & \alpha_t = 2.0, \qquad \varphi = \pi \\ \mu = 0.625, \qquad & \beta = 0.2\,. \end{aligned} \tag{4.24}$$

Figure 5 shows the spatial distributions of the forward acoustic mode amplitude at two different times; $\eta = 0.25$ and $\eta = 1.5$. In both plots the solid line represents the initial data, which in turn coincide with the CJ-ZND background solution. As time evolves, the resonant source term generates oscillations around this background profile (Fig. 5a); later on there is shock formation as seen in Fig. 5b. One common feature of both profiles is that the oscillations become weaker towards large negative $\hat{x}$. The deviations from the CJ-ZND background solution decay in this limit just as predicted by the limit analysis of the characteristic solution in Section 4.2. To quantitatively verify the scalings implied by (4.21), we have computed the solution in the extended region $-20 < \hat{x} < 30$. Figure 6a shows the distribution of $\hat{V}$ in $-20 < \hat{x} < 5$ at the time $\eta = 1.5$. As seen in Fig. 5, the decay of the oscillations to the left continues. Figure 6 shows a scaled deviation from the CJ-ZND solution, $\hat{V}^{\text{st}}$, namely

$$V^* = (\hat{V} - \hat{V}^{\text{st}}) * [\hat{V}^{\text{st}}]^{1/2}\,, \tag{4.25}$$

as a function of $\hat{x}$ at the same time. According to (4.21) this quantity should be of order $O(1)$ as $\hat{x} \to -\infty$ and the plot verifies this

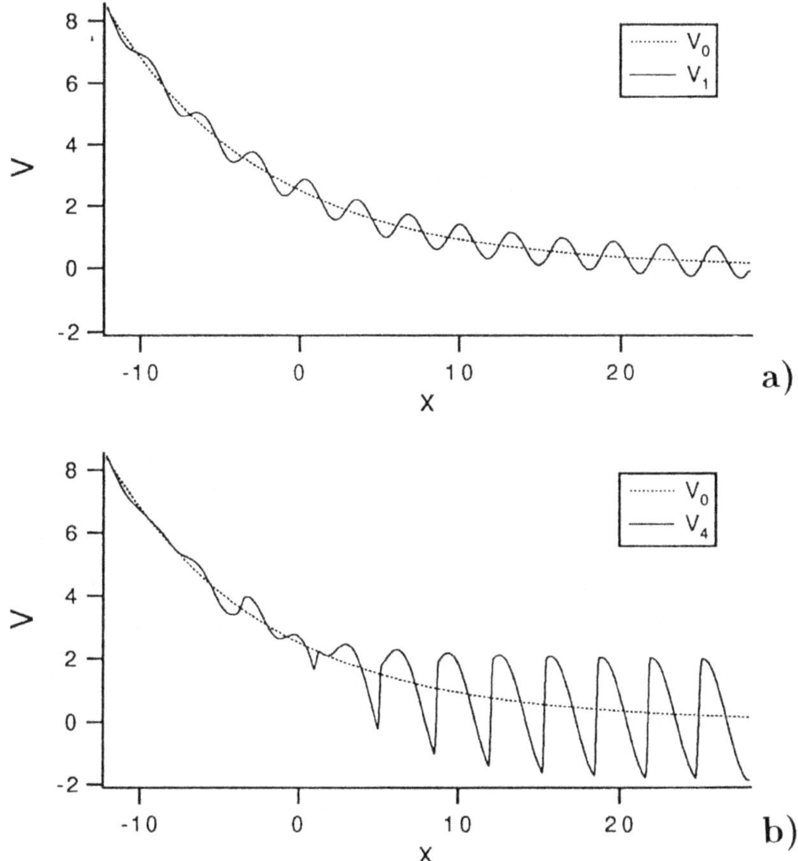

Fig. 5: Spatial profiles of the forward acoustic amplitude in the burnout layer at times: a) $\eta = 0.25$, b) $\eta = 1.5$.

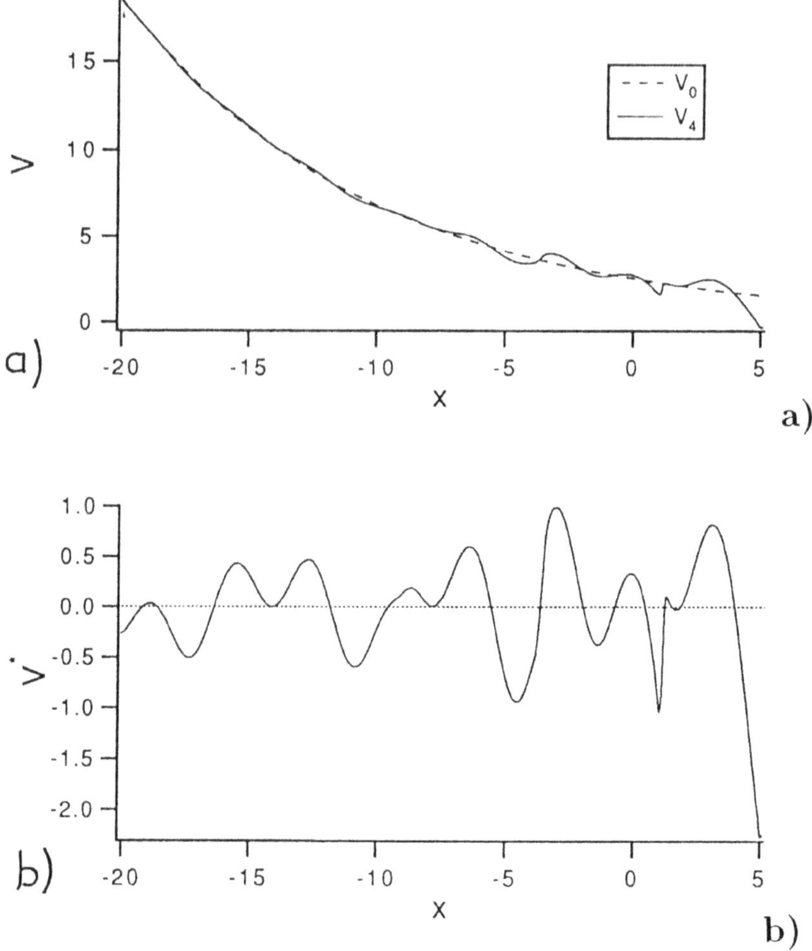

Fig. 6: a) Solution from Fig. 5 at time $\eta = 1.5$ for large negative $\hat{x}$. b) Scaled deviation from the CJ-ZND background profile in the burnout layer according to (4.25), supporting the scaling behavior derived from a characteristic analysis as $\hat{x} \to -\infty$.

tendency. In fact V^* even seems to decay due to shock dissipation. Thus, for shock containing solutions the vanishing of the resonance effects in the upstream direction appears to be at least as pronounced as predicted by the characteristic analysis.

5. Concluding Remarks

In this paper we have pointed out that for marginally stable near–Chapman-Jouguet detonations there is an apparent contradiction between *i)* the standard radiation boundary condition for linear stability analyses and *ii)* the resonant generation of forward acoustic perturbations in the back of the wave. We have identified a simplified asymptotic equation system that describes these weakly nonlinear resonance effects, and defined a model problem that allows quantitative predictions and comparison with results from direct numerical simulations.

It turned out that the proper formulation of the farfield boundary conditions in a linear perturbation analysis of the detonation structure can be derived from solving a simplified system describing the resonance effects only in a so called burnout layer adjacent to the detonation structure. Formal solutions of the burnout layer equations based on the method of characteristics as well as numerical solutions of this burnout layer problem show that the influence of acoustic resonances decays as one leaves the burnout region. A complete analysis, to be presented in Bdzil and Klein (1993), shows that this tendency continues, and that at finite distances behind the lead detonation shock the acoustic resonance effects appear at no more than second order in the linear perturbation amplitude. We conclude that the standard radiation condition is valid even for near-CJ detonations.

Appendix A: Eigenvalue Analysis

The Eigenvalues of the matrix $\underline{\underline{A}}$ in (3.2), where the proper $\underline{\underline{A}}$ is obtained by replacing $\bar{u}$ by u in (2.3), are:

$$a^- = u - c, \qquad a^0 = u, \qquad a^+ = u + c, \qquad a^\lambda = u, \qquad (A.2)$$

with $c = \sqrt{pv}$. Here a^- and a^+ are associated with acoustic wave propagation, a^0 with the advection of entropy perturbations and a^λ with the advection of the chemical reaction progress variable. The associated right Eigenvectors are:

$$\underline{r}^{-} = \begin{pmatrix} -v \\ -c \\ \gamma p \\ 0 \end{pmatrix}, \; \underline{r}^{0} = \begin{pmatrix} (\gamma+1)v \\ 0 \\ 0 \\ 0 \end{pmatrix}, \; \underline{r}^{+} = \begin{pmatrix} -v \\ c \\ \gamma p \\ 0 \end{pmatrix}, \; \underline{r}^{\lambda} = \begin{pmatrix} 0 \\ 0 \\ 0 \\ 1 \end{pmatrix}. \tag{A.3}$$

and the left Eigenvectors are

$$\begin{array}{lcll} l^{-} & = & & [\; 0, \; -1/2c, \; 1/2\gamma p, \; 0, \;] \\ l^{0} & = & \frac{1}{\gamma+1} & [\; 1/v, \; 0, \; 1/\gamma p, \; 0, \;] \\ l^{+} & = & & [\; 0, \; 1/2c, \; 1/2\gamma p, \; 0, \;] \\ l^{\lambda} & = & & [\; 0, \; 0, \; 0, \; 1, \;] \end{array}. \tag{A.4}$$

Appendix B: Fast Variable Averaging Symbols

Let s denote any of the fast variables, x, y, h from (3.11) and let $\sigma = \varepsilon s$ be the related slow variable. Then the fast variable average of a multiple scales function
$f(s, \sigma)$ is defined by

$$\langle f\rangle(\sigma) = \lim_{\Delta\sigma\to 0}\lim_{\varepsilon\to 0} \frac{\varepsilon}{2\Delta\sigma}\int_{\frac{1}{\varepsilon}(\sigma-\Delta\sigma)}^{\frac{1}{\varepsilon}(\sigma+\Delta\sigma)} f(s,\sigma)\, ds\,. \tag{B.1}$$

Notice the specific sequence of limits in this expression. Further let $g(x, y; \xi, \eta)$ with $\xi = \varepsilon x$, $\eta = \varepsilon y$, be some other multiple scales function depending on all of the slow and fast variables. Then

$$\langle g\rangle^{(x)}(x;\xi,\eta) = \lim_{\Delta\eta\to 0}\lim_{\varepsilon\to 0} \frac{\varepsilon}{2\Delta\eta}\int_{\frac{1}{\varepsilon}(\eta-\Delta\eta)}^{\frac{1}{\varepsilon}(\eta+\Delta\eta)} g(x,y;\xi,\eta)\, dy\,. \tag{B.2}$$

and an analogous definition holds for $\langle g\rangle^{(y)}(y;\xi,\eta)$.

References

Bdzil, J. B. and Klein, R., 1993. "Weakly nonlinear dynamics of near-CJ detonation waves," in preparation.

Bourlioux, A. and Majda A. J., 1993. "Theoretical and numerical structure of unstable detonations," *Phil. Trans. Roy. Soc. Sci. Ser. A.*

Breitung, W., 1991. "Conservative estimates for dynamic containment loads from hydrogen combustion," 11th Conference on Structural Mechanics in Reactor Technology, SMIRT, **11**.

Erpenbeck, J. J., 1962. "Stability of steady state equilibrium detonations," *Phys. Fluids* **5**, pp. 604–614.

Erpenbeck, J. J., 1963. "Structure and stability of the square wave detonation," in: 9th Symposium (Intl.) on Combustion, The Combustion Institute, Academic Press, **19**, pp. 442–453.

Erpenbeck, J. J., 1964. "Stability of idealized one-reaction detonations," *Phys. Fluids* **7**, pp. 684–696.

Fickett, W. and Davis, W. C., 1979. *Detonation*, University of California Press, Berkeley.

Fujiwara, T. and Reddy, K. V., 1989. "Propagation mechanisms of detonation – Three dimensional phenomena," in: Proc of 12th ICODERS, Ann Arbor, Michigan.

Lee, H. I. and Stewart, D. S., 1990. "Calculation of linear detonation instability: One-dimensional instability of plane detonation," *JFM* **216**, pp. 103–132.

van Leer, B., 1979. "Towards the ultimate conservative difference scheme V: A second order sequel to Godunov's method," *J. Comp. Phys.* **14**, pp. 361–370.

LeVeque, R. J., 1990. *Numerical Methods for Conservation Laws*, Birkhäuser Verlag.

Majda, A. J. and Rosales, R. R., 1984. "Resonantly interacting weakly nonlinear hyperbolic waves I. A single space variable," *Stud. Appl. Math.* **71**, p. 149.

Majda, A. J., Rosales, R. R., and Schönbeck, M., 1988. "A canonical system of integro-differential equations arising in resonant nonlinear acoustics," *Stud. Appl. Math.* **79**, pp. 205–262.

Oran, E. S. and Boris J. P., 1987. *Numerical Simulation of Reactive Flow*, Elsevier.

Shepherd, J. E., 1985. "Chemical kinetics of hydrogen-air-diluent detonations," in: Dynamics of Shock Waves, Explosions and Detonations, Eds.: J. R. Bowen, N. Manson, A. K. Oppenheim, R. I. Soloukhin, pp. 263–293.

Shepherd, J. E., 1992. "Oblique detonations and propulsion," this volume.

Schoeffel, S. U., 1989. "The mechanism of spinning detonation – Numerical study for rectangular crossection tube," Proc. of 12th ICODERS, Ann Arbor, Michigan.

SOME FUNDAMENTAL PROBLEMS OF DETONATION INSTABILITIES AND ITS RELATION TO ENGINE OPERATION

John H. Lee, Fan Zhang[1], *and Randy S. Chue*

Department of Mechanical Engineering
McGill University
817 Sherbrooke Street West
Montreal, Quebec, CANADA H3A 2K6

ABSTRACT

The correct modeling of stabilization of a detonation wave in a propulsion engine requires detailed considerations of the transient development of the flow field behind the wave front as well as the real three-dimensional structure of the wave itself. Existence of a steady detonation wave depends on the proper matching of the flow field of products to the boundary conditions dictated by the Chapman-Jouguet criterion at the wave front. This may not always be possible since the product flow also has to satisfy boundary conditions as governed by the engine geometry and operating conditions. In this paper, the problem of detonation stability under competing effects of chemical energy release and friction (i.e., boundary layers) as well as area change will be discussed. A number of examples will be given where a steady detonation wave does not exist if mismatch of the product flow field behind the wave front to the C-J sonic condition occurs. Numerical simulations of a pulsating detonation propagating in a moving diffuser will demonstrate the fundamental importance of detonation instabilities in the analysis of detonation wave engine operation.

1. Introduction

The concept of the detonation wave engine (DWE) has stimulated a number of recent numerical simulations of the reactive flow field in the propulsion engine or around a hypersonic projectile (Fujiwara et al., 1989; Oran et al., 1991; Yungster et al., 1991a, 1991b;

[1]The author was supported by an NSERC International Fellowship while on leave from the Shock Wave Laboratory, Technical University of Aachen, 5100 Aachen, Germany.

J. Buckmaster et al. (eds.), Combustion in High-Speed Flows, 541–574.

Nusca, 1991; Soetrisno and Imlay, 1991). These numerical models concentrate on the detailed description of a steady wave configuration (either normal or oblique detonation waves) on the engine or projectile. However, the existence and stability of the stationary detonation wave configuration have not been considered. The correct modeling of a standing detonation wave in the engine requires detailed considerations of the development of the transient flow field as well as the instability of the detonation wave itself.

In gas dynamic flows including shock waves, the shock velocity is determined by the flow condition behind it. In other words, the Rankine Hugoniot relationships across the shock is incomplete and require matching with the flow behind the shock to determine the solution. If a sonic plane exists behind the shock, then a steady shock solution can be matched to a non-steady flow behind the sonic plane since it isolates the transient flow downstream from the steady upstream flow. However, this may not always be possible, and incompatibility of the flow field with the choking condition can result in a longitudinal instability, leading to oscillations in the flow direction. For problems including detonations, one generally assumes the Chapman-Jouguet condition (i.e., $M = 1$) and thus the detonation velocity is uniquely determined for a given fuel-oxidant mixture. If the flow field of the products behind the wave front is not compatible to the steady C-J condition, oscillations in the flow direction can also occur.

An analogous problem is the instability of supersonic diffusers first studied by Oswatisch (1947) and later by Ferri and Nucci (1951), Trimpi (1956a), Trimpi (1956b), and Dailey (1955). At normal operating condition the shock is close to, but downstream of the diffuser throat (where $M = 1$) in the diverging section. The mass flow at the throat corresponds to the free stream stagnation pressure. If the shock is perturbed to a position upstream of the throat, the stagnation pressure loss at the shock results in the inability of the throat section to admit the same mass flow through the diffuser inlet. Thus the shock travels upstream and is expelled from the diffuser inlet where the mass can be spilled. The shock is then swallowed in the diffuser again and cyclic oscillations occur. This instability mechanism can be attributed to the incompatibility of the flow behind the shock with the choking condition. It is of interest to note that to start the supersonic diffuser, one has to accelerate to a higher than normal operating Mach number to compensate for the stagnation pressure loss so that the shock can be swallowed to a position downstream of the throat. Alternatively, one may also increase the throat area temporarily during the starting process to achieve the shock swallowing.

A steady one dimensional detonation wave cannot be always obtained. Beyond a certain value of the activation energy, Fickett and Wood (1966) and later Abouseif and Toong (1982), Moen et al. (1984) and Bourlioux (1991) have demonstrated numerically that one-dimensional detonation wave execute an oscillatory or pulsating behavior as it propagates. Instead of the choking condition arising from the area change, heat addition by chemical reaction drives the flow behind the shock towards the sonic condition at the C-J equilibrium plane in the detonation case. The pulsating detonation is analogous to a normal shock diffuser instability. The formation of a detonation is also analogous to the starting of a supersonic diffuser and requires acceleration of the wave to a higher than normal velocity (i.e., C-J velocity). Again the oscillatory behavior of the wave detonation front can be associated with the incompatibility of the choking condition, i.e., the C-J sonic condition in the product flow, and the boundary condition at the front.

The objective of this paper is to elucidate on the problem of detonation instabilities arising from the matching of the transient flow field of the products with the steady boundary condition as dictated by the C-J criterion at the wave front. The stabilization of a detonation wave in an engine requires the product flow field to satisfy the boundary conditions dictated by the engine geometry. This may not be compatible with the C-J condition at the detonation front and this leads to oscillations. To elucidate this requires an examination of the C-J criterion itself. There exists a number of problems in which a steady C-J wave is incompatible with the transient flow of the product gas. We shall show examples where the incompatibility of the product flow with the C-J sonic choking condition leads to non-existence of a steady detonation wave. These cases include the pulsating detonations, converging and diverging detonation waves, quasi-detonations, detonations in area-changed channels and pulsating detonations in a moving diffuser.

2. The Chapman-Jouguet Criterion

For a given explosive mixture a unique detonation wave speed can be determined from the conservation laws and an additional condition generally referred to as the Chapman-Jouguet criterion. According to Chapman (1899), the desired solution is the minimum velocity solution corresponding to the tangency of the Rayleigh line to the Hugoniot curve. Jouguet (1905) pointed out that this minimum velocity solution also corresponds to sonic conditions at the downstream equilibrium plane. Thus either the minimum velocity or the sonic condition can be used as a criterion to establish a unique detonation solution. Neither Chapman nor Jouguet proved this criterion and it is essentially a postulate. Chapman, however, did argue that since experiments give a unique detonation velocity and that in general, two solutions are obtained for a given wave speed except when the Rayleigh line is tangent to the equilibrium Hugoniot. Hence he justified his choice for the tangency solution. Von Neumann (1942), however, attempted to provide some physical arguments for the Chapman-Jouguet solution. He pointed out that unless supported by a piston, a rarefaction fan always follows the detonation. Thus, if the wave is overdriven (subsonic conditions downstream), the rarefaction waves will penetrate into the reaction zone and attenuate the wave. The weak detonation solution was ruled out from entropy considerations and violation of the second law for a normal Hugoniot. However, there are Hugoniot curves of a form that may not necessarily yield a weak detonation solution that violates the second law. It was G. I. Taylor (1950) who gave a more profound justification of the Chapman-Jouguet solution. Taylor stated that the existence of a steady detonation must depend on the possibility of matching the non-steady flow of the products to the steady boundary condition at the detonation front. This is only possible if the detonation is a Chapman-Jouguet detonation. It may be said that all these arguments are essentially similar, however, it was Taylor who explicitly emphasized that the existence of a steady detonation depends on the product flow field even though the sonic condition of a C-J wave essentially isolates the product flow from the detonation wave front. Strictly speaking, one must also consider the transient development of the detonation to see if the steady Chapman-Jouguet solution can be approached asymptotically. For non-reacting shock waves where there is no C-J criterion to determine the solution, one always consider the downstream flow field to establish the shock conditions.

3. Pulsating Detonations

The one-dimensional ZND model of the detonation wave consists of a normal shock followed by an induction zone and a reaction zone. The termination of the reaction zone is the C-J sonic plane. The ZND model provides a mechanism for the propagation of the detonation wave, i.e., auto-ignition by adiabatic shock compression. However, the ZND model is unstable for the self-sustained C-J detonation. For the one-dimensional case the instability is manifested as oscillation or pulsation in the longitudinal direction of wave motion. This pulsating instability has been demonstrated by a number of numerical simulations (Fickett and Wood, 1966; Abouseif and Toong, 1982; Moen et al., 1984; Bourlioux et al., 1991) and the pulsation occurs even for overdriven waves.

For C-J detonations using a first-order Arrhenius law for the reaction rate, steady ZND detonation waves can be achieved as an asymptotic solution of the transient development providing that the activation energy is below the instability limit (Chue et al., 1992). Increasing the activation energy over the instability limit raises the sensitivity of the mixture chemistry to perturbations in the gas dynamic flow field and this results in an oscillatory detonation front (see Fig. 1). This can be clearly demonstrated by examining closely the flow field behind the shock wave over a oscillating cycle, displayed in Fig. 2 where M denotes the local Mach number with respect to the shock front and β is the reactant mass fraction. A piston rear boundary whose speed equals to the C-J particle velocity is used in the present numerical simulation rather than a closed tube where a rarefaction fan (Taylor wave) follows the detonation front. At time $t = 126.8$ the shock Mach number is close to its minimum, the entire flow field behind the shock front is subsonic and hence the compression waves can catch up with the shock front resulting in the amplification of the shock. As the shock accelerates to its peak value the flow field behind the reaction zone turns into supersonic (i.e., at $t = 129.2$). Then the expansion of the product results in the subsonic matching again at the end of the reaction zone (e.g., at $t = 131.7$) thus causing the shock Mach number to decrease. The average speed of the pulsating wave over a cycle is found to agree with the C-J velocity and the time-averaged sonic condition is satis-

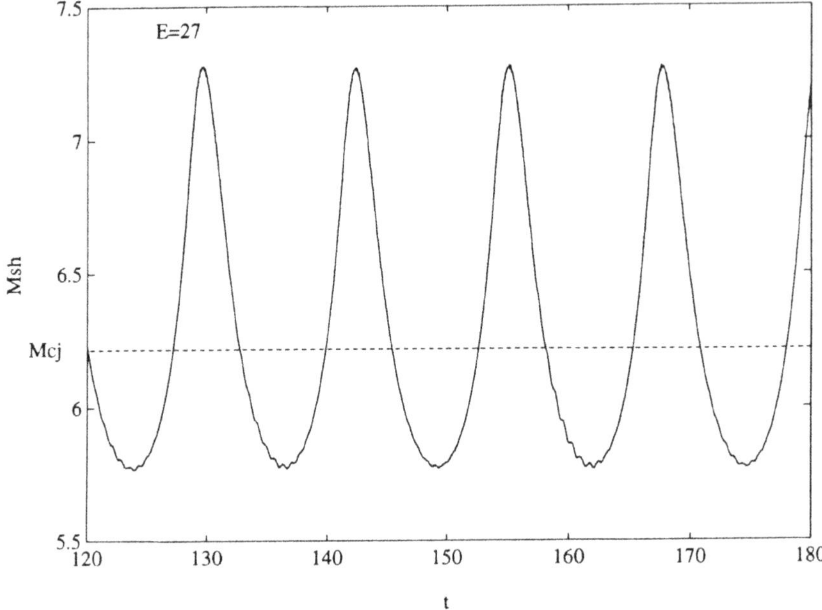

Fig. 1 Time evolution of shock Mach number of a one-dimensional pulsating detonation. Overdrive factor f = 1, specific heat ratio $\gamma = 1.2$, heat of reaction $Q = 50$, activation energy $E = 27$.

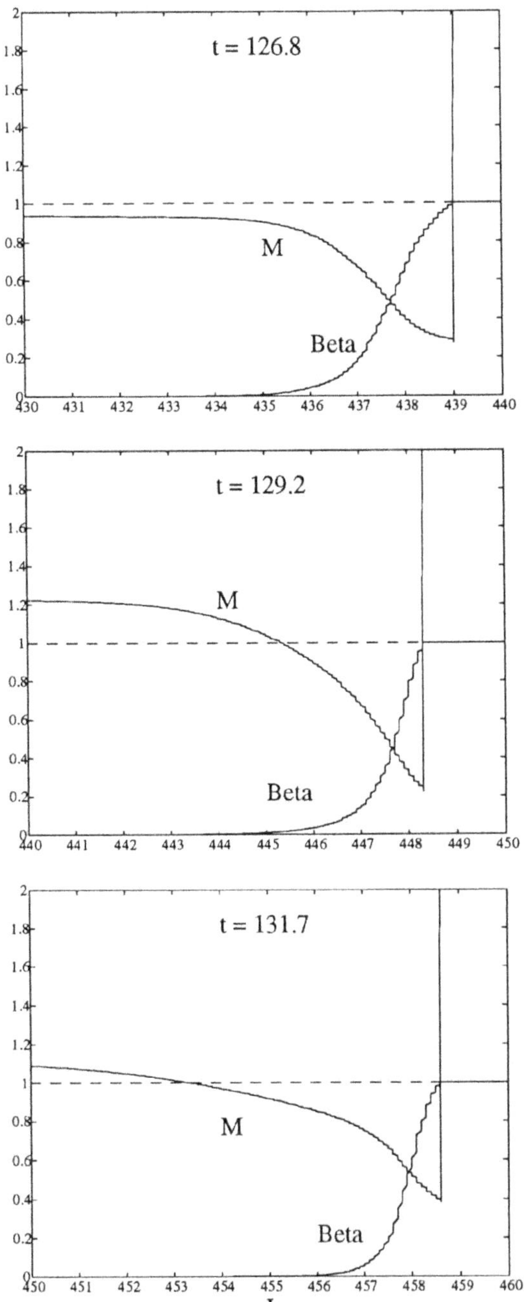

Fig. 2 Spatial profiles of reactant mass fraction β and local flow Mach number M, with respect to shock, of the one-dimensional detonation in Fig. 1.

fied behind the detonation front. The oscillatory cycle is maintained by the mismatch of the flow field behind the wave front to the C-J condition.

4. Converging and Diverging Waves

Examples where imposing the C-J criterion at the front is insufficient to determine the solution and that detailed considerations of the product flow is necessary to establish the existence of steady Chapman-Jouguet detonation waves are best illustrated by the cases of converging and diverging detonations. For converging wave, steady C-J detonations do not exist. A C-J wave boundary condition is incompatible with the transient converging flow of the product gases. A converging detonation must become progressively overdriven as it converges due to the adiabatic compression of the products from the area change. Eventually, the wave is so overdriven that the heat release is itself negligible as compared to the compression work due to the area change and the detonation becomes a strong shock. Thus, near the center of convergence the behavior of the wave is governed by the similarity solution of Guderley (1942) for a strong shock wave which is determined by considering a singularity free flow field behind the front. Further away from the center of convergence the propagation of the converging overdriven detonation can be obtained by perturbing the self-similar solution of Guderley to account for the heat release (Lee, 1967). However, the solution is also determined by using Guderley's criterion of a singularity free flow behind the front. Experiments have confirmed the non-existence of steady converging C-J detonations (Lee and Lee, 1965a).

The existence of steady divergent spherical (or cylindrical) C-J detonation is not a well defined issue. This had been studied by a number of researchers (Taylor, 1950; Courant and Friedrich, 1948; Zeldovich et al., 1980; Manson and Ferrie, 1953; Lee et al., 1965b). The difficulty is centered around the interpretation of the singularity at the detonation front when the transient solution of the product expansion flow is matched to the steady C-J boundary condition at the front. It is perhaps illustrative to review briefly this analysis for diverging detonations.

For a steady C-J detonation, the product flow is isentropic if no further reactions, friction and heat losses are considered. Thus the basic isentropic equations can be written as:

$$\frac{2}{\gamma-1}\frac{\partial c}{\partial t}+c\frac{\partial u}{\partial r}+\frac{2}{\gamma-1}u\frac{\partial c}{\partial r}+\frac{jcu}{r}=0 \tag{1}$$

$$\frac{\partial u}{\partial t}+u\frac{\partial u}{\partial r}+\frac{2}{\gamma-1}c\frac{\partial c}{\partial r}=0 \tag{2}$$

where c (sound speed) and u (particle velocity) are the dependent variables chosen for isentropic flow and j in the conservation of mass equation takes the value of 1 or 2 for cylindrical or spherical symmetry respectively.

Assuming central ignition ($r = 0$) and without any characteristic length scales, the solution is self similar in the variable $\xi = r/Dt$. Non-dimensionalizing the dependent variables u and c by $\phi = u/D$ and $\beta = c/D$ where D is the steady C-J detonation speed, the two conservation equations becomes

$$\frac{2}{\gamma-1}(\phi-\xi)\beta'+\beta\phi'+\frac{j\phi\beta}{\xi}=0 \tag{3}$$

$$\frac{2}{\gamma-1}\beta\beta'+(\phi-\xi)\phi'=0. \tag{4}$$

To make the singularity at the front apparent, we can solve for the derivatives ϕ' and β' and obtained

$$\phi'=\frac{j\phi\beta^2}{\xi}\frac{1}{(\phi-\xi)^2-\beta^2} \tag{5}$$

$$\beta'=-\frac{\gamma-1}{2}\frac{j\phi\beta(\phi-\xi)}{\xi}\frac{1}{(\phi-\xi)^2-\beta^2}. \tag{6}$$

At the detonation front where $r = Dt$ or $\xi = 1$, the Chapman-Jouguet criterion or sonic condition requires that $D = u + c$ or $(\phi_1 - 1)^2 - \beta_1^2 = 0$ at $\xi = 1$. Thus the derivatives ϕ' and β' becomes infinite. For planar waves where $j = 0$ the numerators vanishes and such a singularity does not exist. Carrying out a Taylor expansion around the front, the solution can be written as

$$\phi(\xi)=\phi_1\pm\sqrt{\frac{2j\phi_1\beta_1}{\gamma+1}}(1-\xi)^{\frac{1}{2}}\pm\ldots \tag{7}$$

$$\beta(\xi)=\beta_1\pm\sqrt{\frac{2j\phi_1\beta_1}{\gamma+1}\frac{\gamma-1}{2}}(1-\xi)^{\frac{1}{2}}\pm\ldots \tag{8}$$

where ϕ_1, β_1 are the boundary values for ϕ and β at the detonation front $\xi = 1$. The plus and minus signs denotes the two solutions for a C-J diverging wave where a piston follows the detonation (the positive sign) and free expansion (the negative sign) to satisfy the boundary condition of $u = 0$ at $r = 0$. Because of this singular behavior, the question of the existence of diverging C-J detonations arises. Strictly speaking, such an infinite expansion gradient behind the wave will penetrate into the reaction zone and attenuate the wave. However, Taylor argued that since the detonation front itself is treated as a singularly (discontinuity), the infinite expansion gradient is therefore compatible to the assumption of an infinite compression gradient for the detonation. Here is a case where the matching of the product flow of the boundary condition of a steady C-J front has created some difficulty. Experimentally, the question can not be resolved readily since a finite energy must always be used to initiate the detonation. Lee (1965b, 1965c) considered this initiation energy in his analysis of the problem and concluded that steady C-J diverging detonations do not exist because we have a decaying overdriven detonation throughout. Near the center of symmetry, the initiation energy dominates and we essentially have a strong blast wave. As the strong blast expands, it becomes an overdriven detonation when the heat release is comparable to the initiation energy. Thus Lee viewed the diverging detonation problem as a reacting blast wave problem. Like the inverse of the converging detonation problem, the diverging detonation is also transient starting out as a strong shock and decays asymptotically to a C-J detonation at infinity. Later studies by Levin and Cheryni (1967) and Lee et al. (1969) of the asymptotic decay of diverging detonation had revealed the possibility of the approach to the C-J condition occurring at a finite radius instead of at infinite radius. However, only the solution near the front was considered in these asymptotic decay analyses and the entire product flow field was not considered. Neither was stability analysis of this asymptotic C-J wave carried out. It would be of interest to revisit this question of the existence of the diverging spherical detonation via high resolution numerical simulation with a finite reaction zone. Most likely, one would obtain the pulsating detonation solution as discussed in the previous section which is itself an indication of the non-existence of steady C-J waves resulting from a mismatch of the flow behind the leading shock with the back sonic condition due to heat release.

5. Quasi-Detonation

The existence of a steady C-J detonation in a smooth tube is well established both theoretically and experimentally. In very rough tubes where friction competes with heat release to drive the flow towards the sonic condition, it is not obvious that a steady wave solution exists.

In the past two decades there is a strong interest in the study of the propagation of detonations in very rough tubes (Lee et al., 1984a, 1984b; Knystautas et al., 1986; Peraldi et al., 1986; Teodorczyk et al., 1988, 1991). These studies followed the pioneering work of Shchelkhin (1940) and Lafitte (1923) followed later by Guenoche (1949) and Brochet (1962). Experiments indicated that the "quasi-steady" detonation velocities (i.e., averaged over distances large as compared to the length scale of the roughness) observed can be substantially lower than the theoretical C-J values for the mixtures. Hence these low velocity detonations are referred to as "quasi-detonations" (Lee and Moen, 1980). Theoretical explanations of these quasi-detonations have mostly been based on the Zeldovich model based on heat and momentum losses (Zeldovich and Kompaneets, 1960). Lee (1979) pointed out that heat loss alone cannot account for the very large velocity deficit observed because the dependence of the detonation velocity energy is weak (i.e., $D \sim \sqrt{Q}$) according to C-J theory. Momentum loss consideration shows interesting result. Without momentum loss, one follows the Rayleigh line to the shock Hugoniot and then retraces along the same Rayleigh line until it intersects (or tangent) the equilibrium Hugoniot. With momentum loss, one follows a different integral curve from the shock to the equilibrium Hugoniot. It can be shown that the integral curves has a saddle point behavior and only one solution passes the saddle point without diverging either side of it (Frolov, 1987). This saddle point solution is analogous to the similarity solution for converging shock waves. This saddle point solution being tangent to a partially reacted Hugoniot and thus if accepted as a valid solution, can result in steady waves substantially below the normal Chapman-Jouguet velocity for the mixture (based on complete heat release). Quantitative comparison between theory and experiments is different because of the lack of an adequate friction law for these very rough tube for transient shock flows. However, it appears worthwhile to study this more thoroughly via numerical simulations of the entire flow field rather than limiting the analysis to the conservation laws at the front. Due to the competing effect of friction and heat release on choking behind the leading shock the existence of a steady wave solution is not obvious.

To elucidate on the effect of friction and heat release on the existence of steady detonation waves, a quasi one-dimensional numerical solution is carried out. The dimensionless form of the 1-D governing equations are used and presented below:

$$\frac{\partial \rho}{\partial t} + \frac{\partial(\rho u)}{\partial x} = 0 \tag{9}$$

$$\frac{\partial(\rho u)}{\partial t} + \frac{\partial}{\partial x}(p + \rho u^2) = F_f \tag{10}$$

$$\frac{\partial(\rho e)}{\partial t} + \frac{\partial}{\partial x}(p + \rho e)u = 0 \tag{11}$$

$$\frac{\partial(\rho \beta)}{\partial t} + \frac{\partial(\rho \beta u)}{\partial x} = -w \tag{12}$$

where

$$e = \frac{1}{\gamma - 1}\frac{p}{\rho} + Q\beta + \frac{u^2}{2} \tag{13}$$

$$T = \frac{p}{\rho} \tag{14}$$

$$w = k\rho\beta e^{-E/T} \tag{15}$$

$$F_f = -\frac{2}{D_h} C_f \rho u|u|. \tag{16}$$

Following Zeldovich et al. (1988) the Fanning friction factor C_f is assumed to obey the Blasius formula, i.e.,

$$C_f = \lambda_f = \frac{0.3164}{Re^{0.25}} \tag{17}$$

where

$$Re = \frac{\rho|u|D_h}{\mu} \tag{18}$$

D_h denotes the hydraulic diameter. For the present study the precise friction law is not important since we are interested only in the qualitative result of competing effects of friction and heat release. Equations (9-12) are solved numerically by a higher order extension of Godunov's method with Strang's splitting algorithm.

Combustion is initiated at a closed end $x = 0$ by imposing the post-shocked pressure and temperature of a C-J detonation within a length of two half-reaction thicknesses. The parameter chosen are as follows: activation energy $E = 22$, heat of reaction $Q = 50$ and specific heat ratio $\gamma = 1.2$. This value of E is much below the instability limit of $E = 25$, so that a non-oscillatory ZND detonation can be achieved as a long time solution.

For very high momentum losses ($D_h = 10$), the wave decays after initiation and eventually fails (see Fig. 3). For low friction $D_h = 40$), the shock evolves to a steady asymptotic value below the normal C-J value of $M_{CJ} = 6.2$. This appears to be in accord with experimental observations of small shock velocity deficit in rather smooth tubes due to the wall friction or the boundary layer effect behind the shock front. However, for a range of intermediate values of the friction parameter, steady solution cannot be obtained and an oscillatory detonation occurs (see Fig. 4). Over an oscillating cycle displayed in Fig. 5, the flow at first undergoes subsonic at the end of the reaction zone (i.e., $t = 55$ at which the shock Mach number is close to the minimum) and thus the shock is amplified by the compression waves behind it. Then the shock arrived at its maximum at $t = 69$ while the flow behind the reaction zone is supersonic. Afterwards the expansion of the product gas causes the subsonic matching at the end of the reaction zone, hence decreasing the shock Mach number (e.g., $t = 76$). These results are similar in nature to the pulsating detonation solution discussed previously for the frictionless case when the activation energy is above the stability limit. Only the oscillation frequency and the reaction zone thickness are greater than those in the frictionless case because of the momentum loss. The trivial difference of the near flow field in Fig. 4 and Fig. 2 is caused by the different boundary conditions.[2] Thus, with the presence of momentum loss and the heat release, the incompatibility of the flow field behind the wave front with the C-J choking condition leads to the non-existence of the steady C-J wave and results in an oscillatory motion of the shock front.

---

[2]Since a stationary end wall requires more expansion of the flow field to match it than in the case of a forward moving piston.

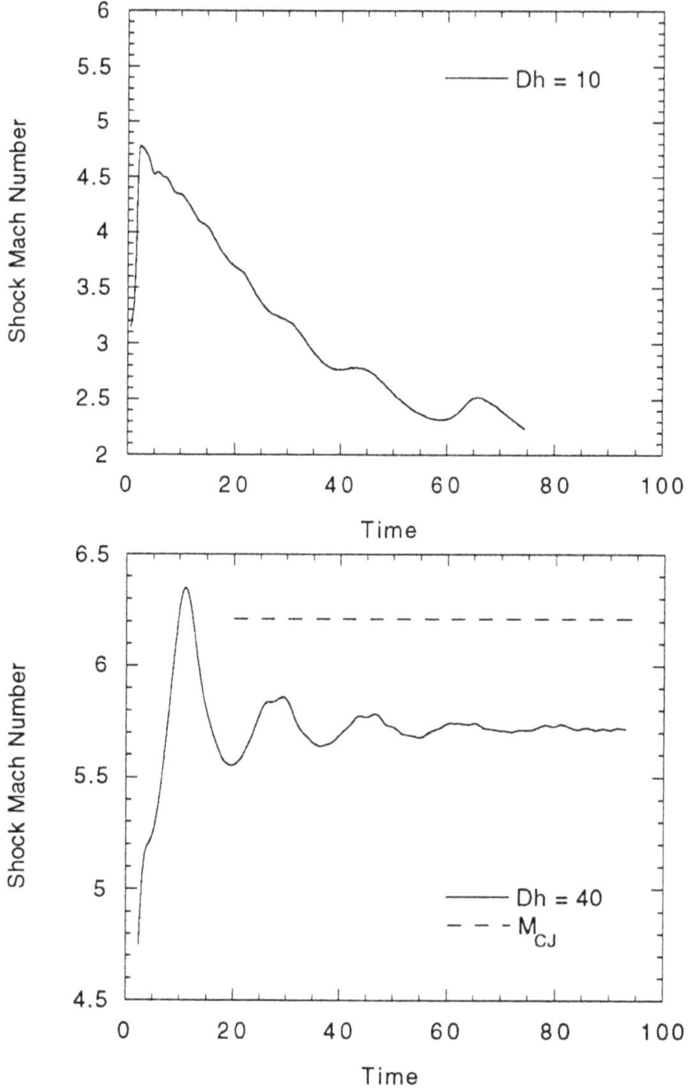

Fig. 3 Time evolution of shock Mach number of a one-dimensional detonation with friction for hydraulic diameter $D_h = 10$ and 40 respectively. Overdrive factor $f = 1$, specific heat ratio $\gamma = 1.2$, heat of reaction $Q = 50$, activation energy $E = 22$.

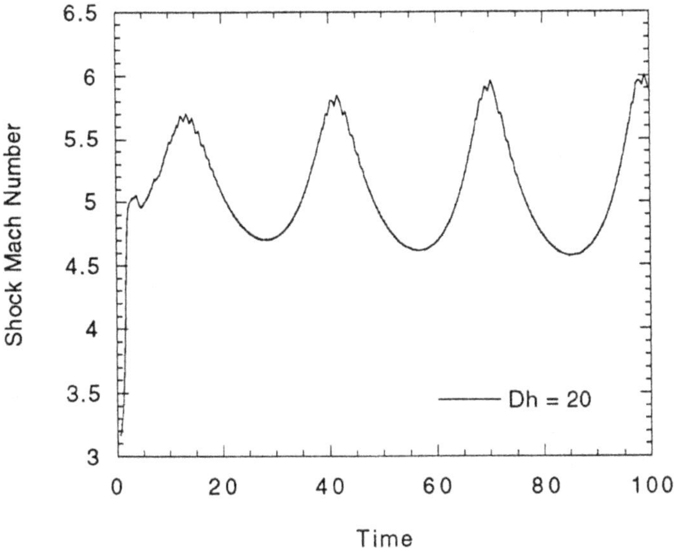

Fig. 4 Time evolution of shock Mach number of a one-dimensional detonation with friction for hydraulic diameter $D_h = 20$. Overdrive factor $f = 1$, specific heat ratio $\gamma = 1.2$, heat of reaction $Q = 50$, activation energy $E = 22$.

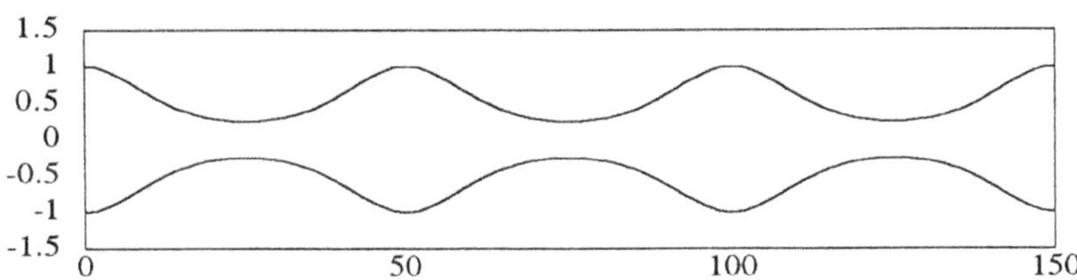

Fig. 6 Periodic area-changed channel described by equation (23).

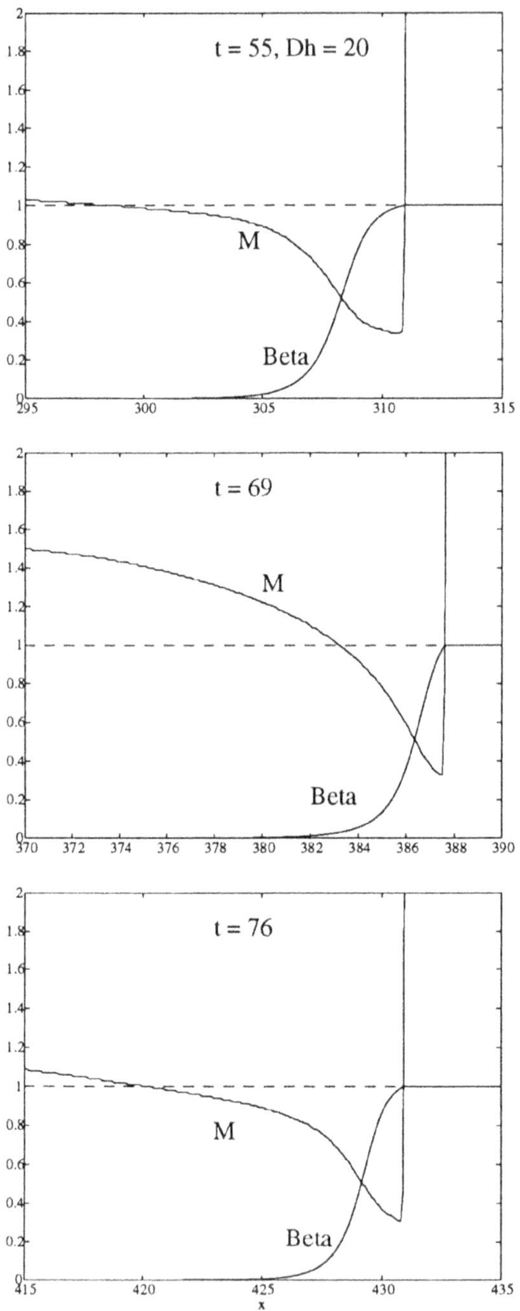

Fig. 5 Spatial profiles of reactant mass fraction β and local flow Mach number M with respect to shock corresponding to Fig. 4.

6. Detonation in Area-Changed Channels

Note that the chemical energy release and the momentum loss are the source terms in the reactive Euler's equations. It is interesting to see competing effects of the other source terms that also tend to drive the flow to the sonic choking condition. The source term of area change is directly associated with the propulsion engine problems in which the diffuser is the key part. The area change in essence represents the boundary condition in the transverse direction, and thus can be regarded as the external force that acts on the flow while the flow has no feed-back influence on it.

To elucidate the competing effect of the area change and the energy release, instead of the friction term the area change terms are put in the governing equations (9-12), i.e.,

$$\frac{\partial \rho}{\partial t} + \frac{\partial(\rho u)}{\partial x} = -\frac{1}{A}\frac{dA}{dx}\rho u \tag{19}$$

$$\frac{\partial(\rho u)}{\partial t} + \frac{\partial}{\partial x}(p + \rho u^2) = -\frac{1}{A}\frac{dA}{dx}(\rho u)^2 \tag{20}$$

$$\frac{\partial(\rho e)}{\partial t} + \frac{\partial}{\partial x}(p + \rho e)u = -\frac{1}{A}\frac{dA}{dx}(p + \rho e)u \tag{21}$$

$$\frac{\partial(\rho\beta)}{\partial t} + \frac{\partial(\rho\beta u)}{\partial x} = -\frac{1}{A}\frac{dA}{dx}\rho\beta u - w \tag{22}$$

in which the area change is simply assumed to be of a periodic sinusoidal form, i.e.,

$$\frac{1}{A}\frac{dA}{dx} = \frac{\pi}{L}\ln\sigma\sin\frac{2\pi x}{L} \tag{23}$$

where L refers to the period and σ is the area ratio of the throat (i.e., the minimum) to the inlet (i.e., the maximum). The variation of area in the channel is shown in Fig. 6. The combustion initiation and the mixture parameters are chosen the same as in the friction case. Hence without the area change the steady C-J wave can be obtained.

For a range of the area ratios as well as the periods of the cyclic area change, steady solution cannot be achieved and an oscillatory detonation occurs (see Fig. 7). The shock Mach number oscillates in the same frequency as that of the area change $\frac{1}{A}\frac{dA}{dx}$ but has a slight time lag. The maximum and the minimum shock Mach number do not appear at the throat as well as the inlet. They occur at the positions approximately symmetric to the throat where the

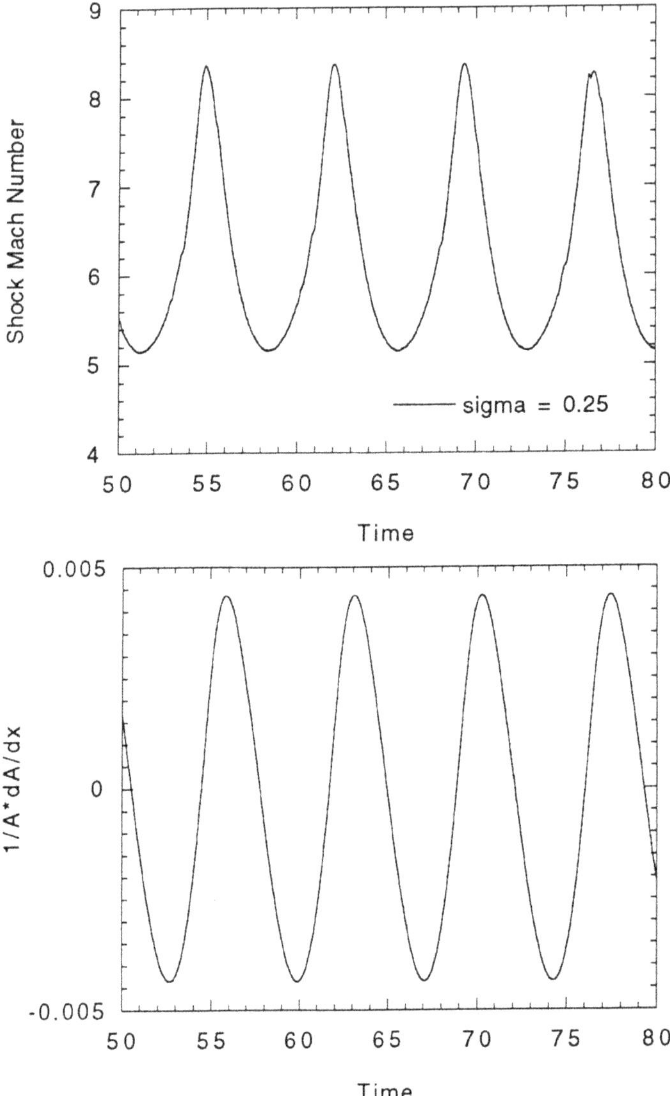

Fig. 7 Time evolution of shock Mach number of a one-dimensional detonation with area change. Area ratio of throat to inlet $\sigma = 0.25$, period of area change L = 50. Overdrive factor $f = 1$, specific heat ratio $\gamma = 1.2$, heat of reaction $Q = 50$, activation energy $E = 22$.

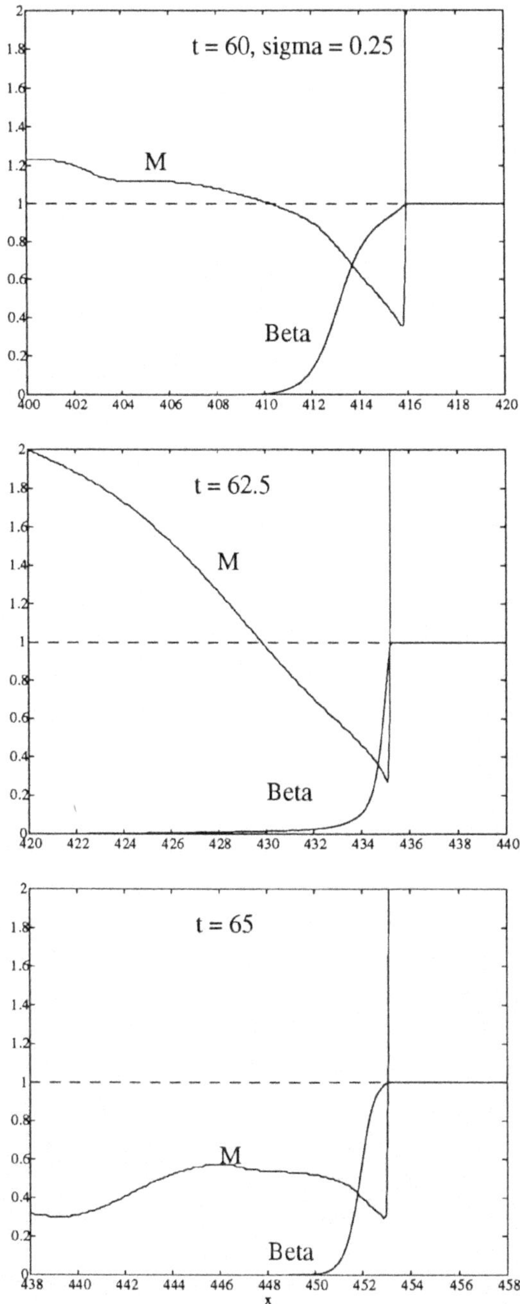

Fig. 8 Spatial profiles of reactant mass fraction β and local flow Mach number M with respect to shock corresponding to Fig. 7.

maximum and the minimum area change appear. It is probably not surprising to obtain an oscillatory solution since the area change acts as a periodic external driving force. However, the frequency of the area change and the area ratio can strongly influence the ignition and the behavior of the detonations because of the inherent unstable frequency of the detonation itself. It is, no doubt, worthwhile for further studies. The main interest now is to elucidate the non-existence of the C-J sonic choking plane when competing effects of heat release and area change take place. Again over a cycle (see Fig. 8), the mismatch of the product flow field behind the wave front to the C-J condition as well as the corresponding response of the shock front are in essence the same as that of the oscillatory detonation solution described previously for either the frictionless or the friction case. To match the near boundary condition, the rear flow field changes more severely than the cases in Fig. 2 and Fig. 4. The diverging area change behind the throat intensifies the flow expansion (e.g., at $t = 62.5$) while the converging area change ahead of a throat chokes the product flow and thus causes the relative flow speed to decrease. Thus the incompatibility of the product flow field to the C-J choking condition is demonstrated and the competing effect of area change and heat release in choking the flow leads to the oscillatory solution of the detonation instead of a steady one.

7. Pulsating Detonation in a Diffuser

Chemical energy release, area change and friction are the basic source terms present in a detonation wave engine. The numerical simulations discussed in the previous sections point out that competing effects between these source terms can give rise to an oscillatory solution. Thus for engine operations, the existence of a stationary detonation wave cannot be postulated a priori without full consideration of the evolution of the detonation in the engine configuration. The diffuser is a key part of a propulsion engine as well as representing the essential features of a ram accelerator. Thus it is of value to consider the transient development of a detonation in a moving diffuser.

To elucidate the competing effects of heat release and area change on the development of the detonation in the diffuser, the quasi-1D dimensionless equations (in the frame of a moving diffuser) that govern the flow inside the diffuser are used. They can be written as

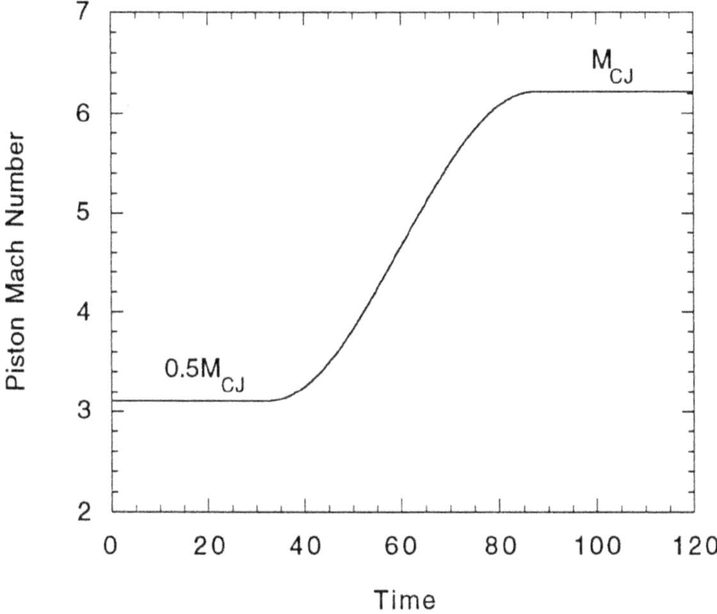

Fig. 9 Prescribed diffuser speed.

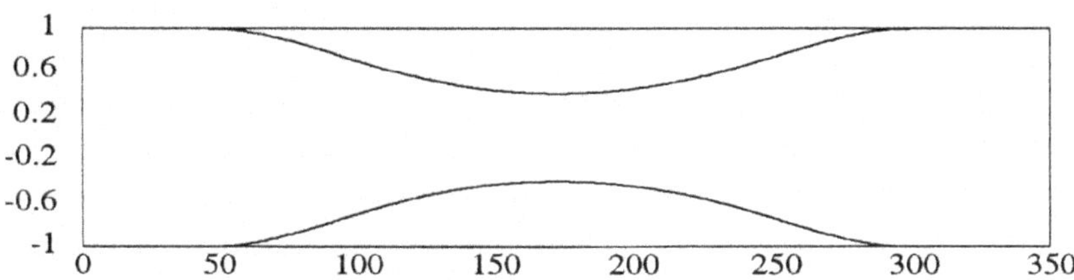

Fig. 10 Sketch of diffuser. Area ratio of throat to inlet $\sigma = 0.4$. Throat position $x_t = 172$ for $E = 22$, and $x_t = 167$ for $E = 27$.

follows:

$$\frac{\partial \rho}{\partial t} + \frac{\partial(\rho u)}{\partial x} = -\frac{1}{A}\frac{dA}{dx}\rho u \tag{24}$$

$$\frac{\partial(\rho u)}{\partial t} + \frac{\partial}{\partial x}(p + \rho u^2) = -\frac{1}{A}\frac{dA}{dx}(\rho u)^2 - \rho a_p \tag{25}$$

$$\frac{\partial(\rho e)}{\partial t} + \frac{\partial}{\partial x}(p + \rho e)u = -\frac{1}{A}\frac{dA}{dx}(p + \rho e)u - \rho u a_p \tag{26}$$

$$\frac{\partial(\rho \beta)}{\partial t} + \frac{\partial(\rho \beta u)}{\partial x} = -\frac{1}{A}\frac{dA}{dx}\rho \beta u - w \tag{27}$$

where a_p is the acceleration of the diffuser and is a prescribed function. In the present study the following form is chosen:

$$a_p = \begin{cases} 0 & 0 \le t \le t_0 \\ \frac{\sqrt{\gamma}}{4} M_{CJ} \frac{\pi}{\tau} cos\left[\pi\left(\frac{t-t_0}{\tau} - \frac{1}{2}\right)\right] & t_0 < t < t_0 + \tau \\ 0 & t \ge t_0 + \tau \end{cases} \tag{28}$$

corresponding to the prescribed function for the acceleration, the velocity of the diffuser u_p is given by:

$$u_p = \begin{cases} \frac{\sqrt{(\gamma)}}{2} M_{CJ} & 0 \le t \le t_0 \\ \frac{\sqrt{\gamma}}{2} M_{CJ} \left\{\frac{3}{2} + \frac{1}{2} sin\left[\pi\left(\frac{t-t_0}{\tau} - \frac{1}{2}\right)\right]\right\} & t_0 < t < t_0 + \tau \\ \sqrt{\gamma} M_{CJ} & t \ge t_0 + \tau. \end{cases} \tag{29}$$

In the present study we prescribe the diffuser motion and investigate the transient development of the detonation due to the start up of the diffuser. The geometry of the diffuser (Fig. 10) is described by Eq. (23) in which the diffuser length and the area ratio of the throat to inlet is chosen to have the same value as the particular ram accelerator model of Herzberg et al. (1991).

In the numerical simulation, a steady flow field is assumed to exist in the diffuser at the time of ignition. Ignition is effected at the diffuser outlet plane by assuming the pressure and temperature there to take on instantaneously the values corresponding to the post-shocked state of a C-J detonation of the mixture for a short duration.

The first case simulated is based on an activation energy of ($E = 22$) which is below the unstable boundary value of $E = 25$. Figure 11 shows the trajectory of the detonation and its velocity (Mach number) subsequent to initiation, where the dashed line is the prescribed Mach number of the diffuser. As can be observed, the detonation propagates from the exit plane of the diffuser towards the throat and stabilizes right at the throat. The detonation after initiation overshoots to an overdriven state ($M = 7.8$) and then decays asymptotically to a steady state value of $M = 6.4$ at the throat. Note that this steady value is slightly higher than the normal C-J ($M = 6.2$) because of the flow convergence of the diffuser.

The pressure profiles behind the detonation as it propagates into the diffuser are shown in Fig. 12. Note that due to the area convergence of the diffuser cross section, the detonation pressure increases progressively as it moves towards the throat because of the increase of the local pressure ahead of the detonation. This is due to the fact that the flow is not choked at the throat but is supersonic throughout.

The profiles of the flow Mach number behind the detonation as it propagates towards the throat ($x = 172$) are shown in Fig. 13 for three different times. Also plotted are the profiles of the reactant mass fraction $\beta(0 < \beta < 1$ and $\beta = 0$ denotes complete reaction). Note that the flow behind the detonation is subsonic as the detonation propagates towards the throat indicating that the detonation is overdriven. When it is finally stabilized at the throat and reaches a steady value (see Fig. 13: $t = 395$) the Chapman-Jouguet condition is realized (i.e., $M = 1, \beta = 0$).

For the case of a higher activation energy of $E = 27$ (which is above the instability limiting value of $E = 25$) the shock trajectory and its Mach number when it travels into the diffuser after initiation are shown in Fig. 14. Note that the behavior is essentially the same as for the stable case of $E = 22$. However, the detonation does not reach an asymptotic steady state velocity and stabilized at the throat section, but oscillates about the throat ($x = 167$).

To demonstrate the effect of the diffuser geometry, a case of more severe area change is shown in Fig. 15. Here the detonation initially propagates into the converging section towards the throat ($x = 164$) after initiation as in the previous cases. However, due to the severe expansion of the flow behind the detonation (because of a more severe area change of the diffuser) the detonation is quenched and failed. Thus it gets swept out of the diffuser by the flow. This illustrates that a stabilized detonation cannot always be met for all diffuser geometries and enhances the point made in this paper that detailed analysis of the transient development of the flow is necessary prior to the assumption of a stationary wave flow configuration.

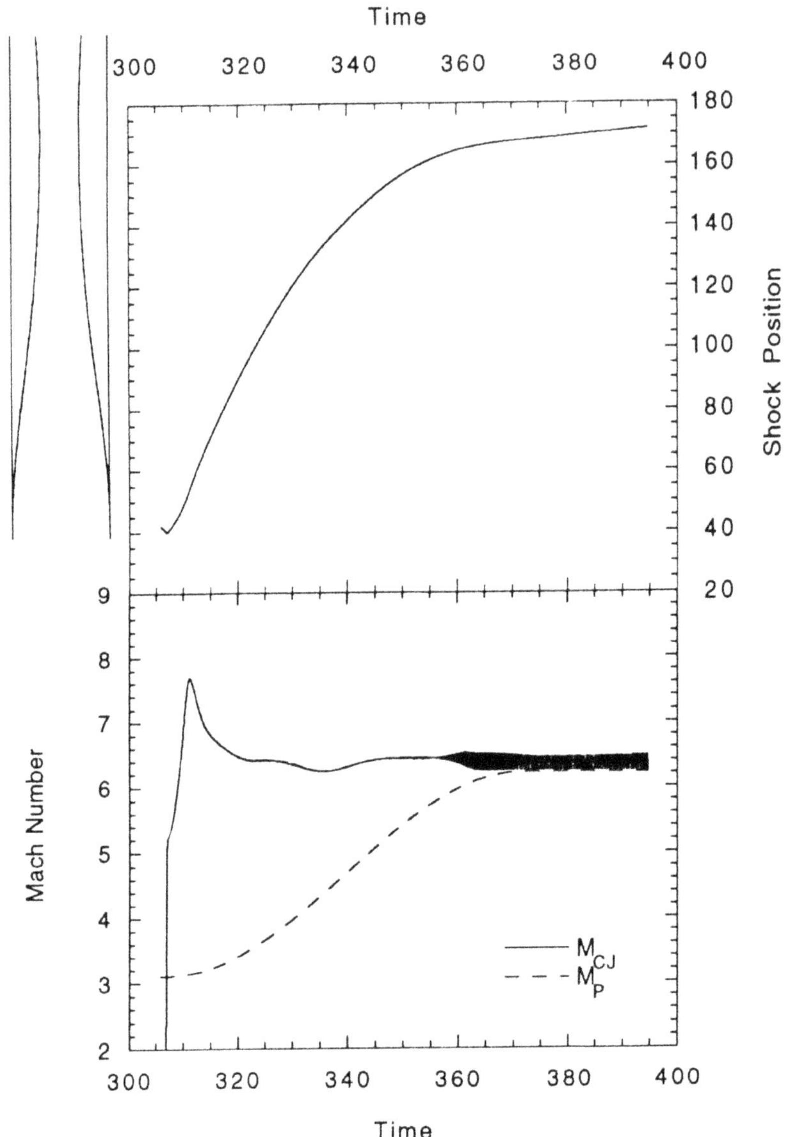

Fig. 11 Time evolution of shock front trajectory and Mach number of a one-dimensional detonation in the moving diffuser and the prescribed diffuser speed. Overdrive factor $f = 1$, specific heat ratio $\gamma = 1.2$, heat of reaction $Q = 50$, activation energy $E = 22$.

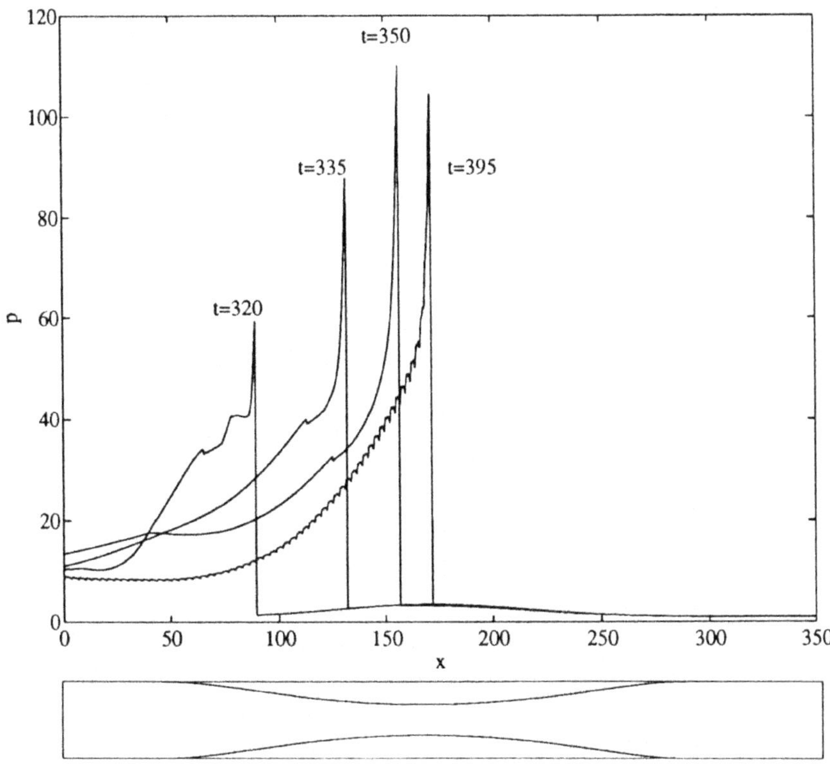

Fig. 12 Spatial profiles of pressure of the one-dimensional detonation in Fig. 11.

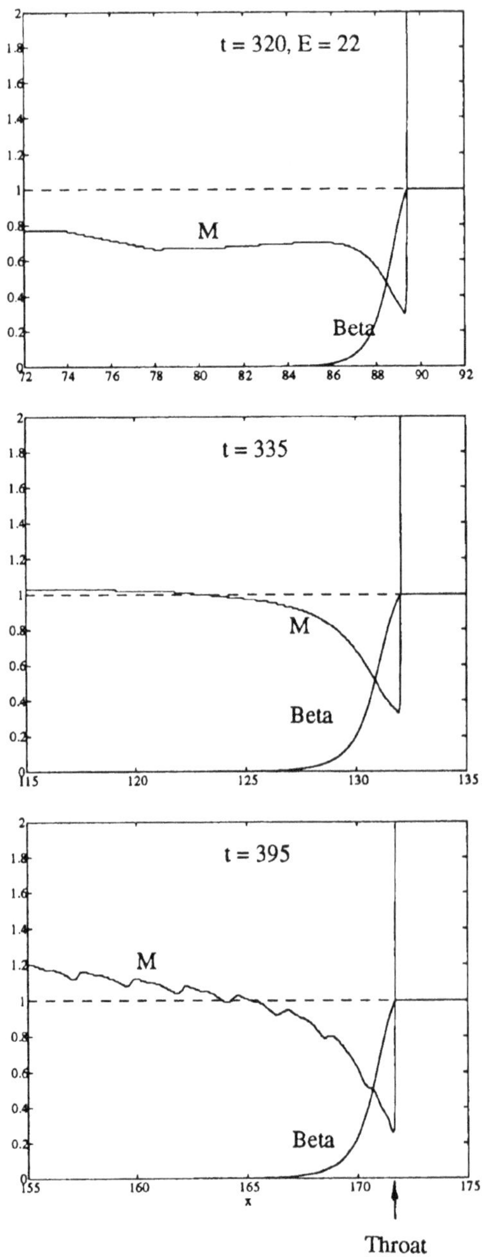

Fig. 13 Spatial profiles of reactant mass fraction β and local flow Mach number M with respect to shock of the one-dimensional detonation in Fig. 11.

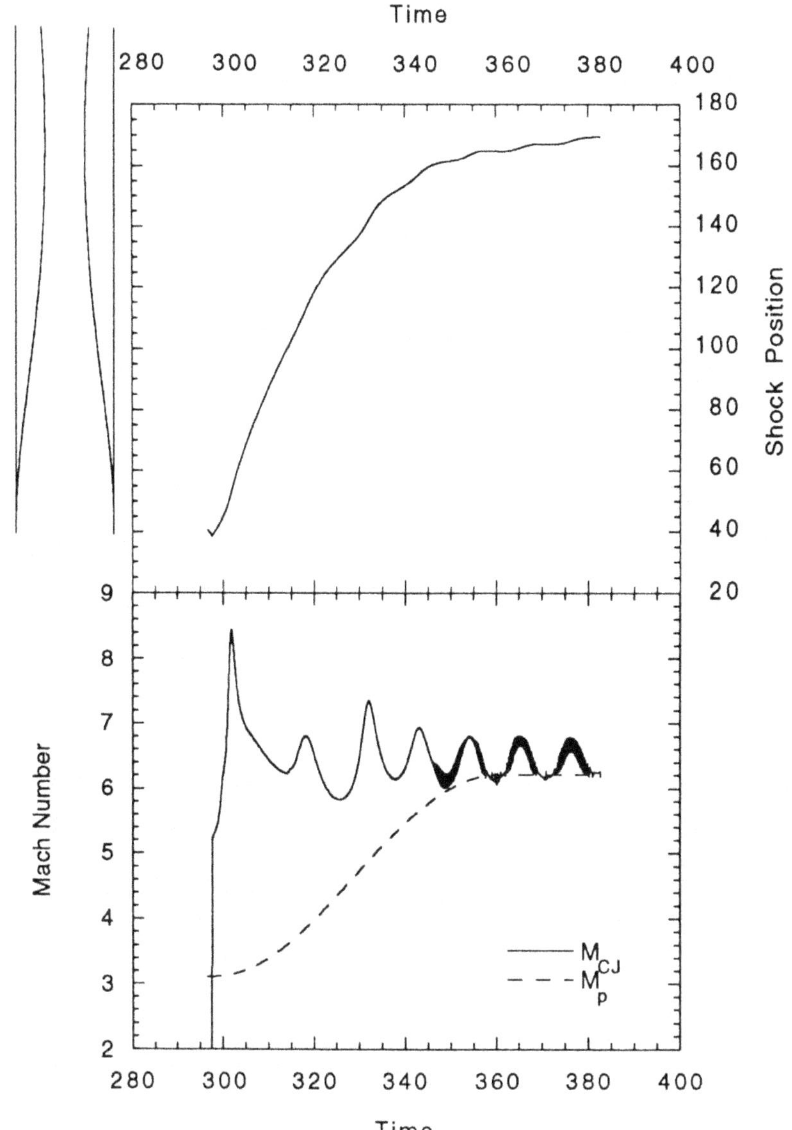

Fig. 14 Time evolution of shock front trajectory and Mach number of a one-dimensional detonation in the moving diffuser and the prescribed diffuser speed. Overdrive factor $f = 1$, specific heat ratio $\gamma = 1.2$, heat of reaction $Q = 50$, activation energy $E = 27$.

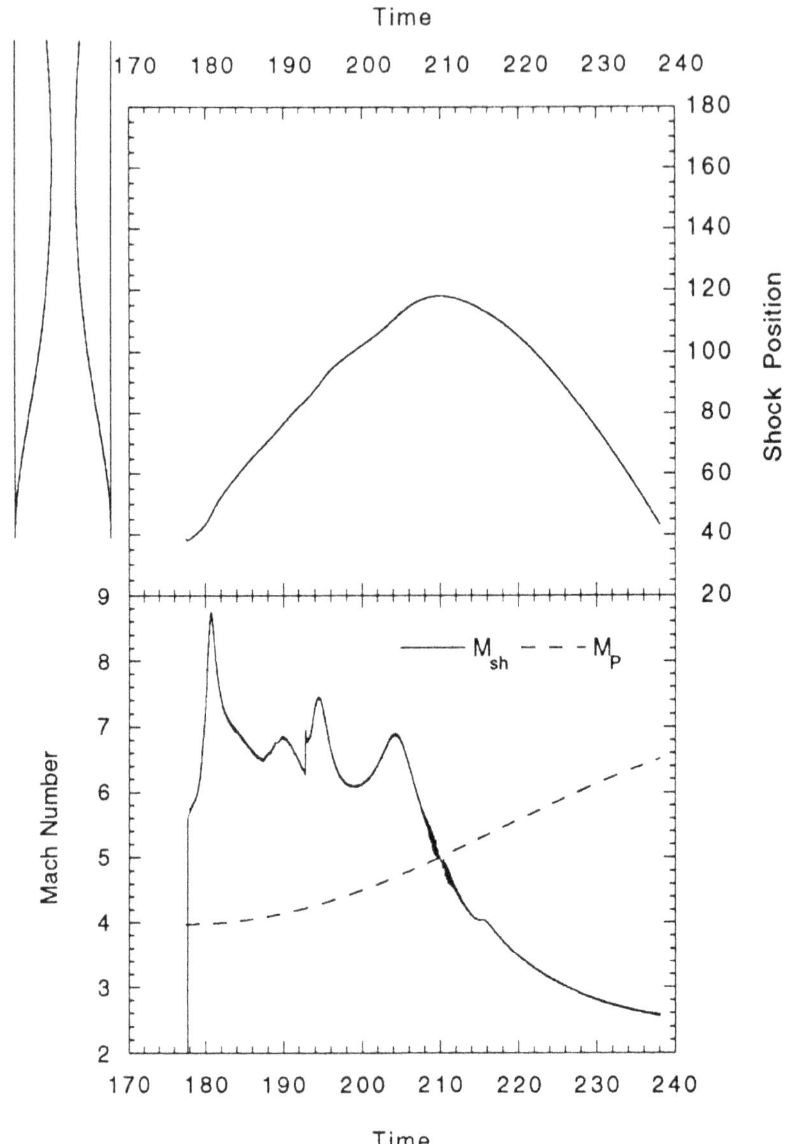

Fig. 15 Time evolution of shock front trajectory and Mach number of a one-dimensional detonation in the moving diffuser with prescribed diffuser speed for area ratio of throat to inlet $\sigma = 0.25$, throat position $x_t = 164$. Overdriven factor $f = 1$, specific heat ratio $\gamma = 1.2$, heat of reaction $Q = 50$, activation energy $E = 30$.

8. Conclusions

The present paper draws attention to the importance of considering the entire flow field as well as the transient development of this flow field when analyzing a stabilized detonation wave engine. The fact that there exists a C-J criterion permits a unique solution for the wave speed to be determined independent of the flow field behind the wave. The paper emphasized that the C-J criterion itself is justified on the basis of matching the transient flow of the product gases to a steady wave solution (i.e., only for a Chapman-Jouguet detonation wave when it is possible to match a transient flow to a steady wave boundary condition). This is well established for planar waves but the existence of steady C-J diverging (spherical or cylindrical) and converging waves is not clear due to the inability to match the transient flow of the product gases to a steady C-J condition at the front. It is also pointed out that area change (as in a diffuser), heat release and friction all tend to drive the flow towards sonic conditions. In a real detonation engine all these three effects are present and competing to render the flow sonic. It is not correct to assume a priori that heat release dominates and thus use the C-J criterion to determine the solution for the detonation wave from the Rankine-Hugoniot equation across the front. For highly exothermic flows, this may well be the case and the flow has to adjust to the C-J condition at the wave. However, if area change and friction are significant, they may play a role in establishing the wave solution (as in the case of detonation in a very rough pipe) and velocities other than the C-J solution is obtained. It is clear that in general that the entire flow must be considered to determine the detonation solution and also the stability of the wave itself.

References

Abouseif, G. E. and Toong, T. Y., 1982. "Theory of unstable one-dimensional detonation," *Combustion and Flame* **45**, pp. 67-94.

Bourlioux, A., Majda, A. J., and Roytburd, V., 1991. "Theoretical and numerical structure for unstable one-dimensional detonations," *SIAM J. Appl. Math.* **51**, pp. 303-343.

Brochet, C. and Manson, N., 1962. "L'effet des parois sur les detonations et les phenomenes vibratoires dans les melanges propane-oxygene-azote," *Compte Rendu a l'Acad. des Scie.* **254**, pp. 3992-3994.

Bruckner, A. P., Burnham, E. A., Knowlen, C., Herzberg, A., and Boydanoff, D. W., 1992. "Initiation of combustion in the thermally choked ram accelerator," in *Shock Waves: Proceedings of the 18th International Symposium of Shock Waves*, ed. Takayama, K., Springer-Verlag, Berlin Heidelberg, pp. 623-630.

Chapman, D. L., 1899. "On the rate of explosion in gases," *Philos. Mag.* **47**, pp. 90-104.

Chue, R. S., Lee, J. H., and Makhviladze, G. M., 1992. "On the structure of a pulsating detonation," Technical report published in the Institute for Problems in Mechanics, Russian Academy of Sciences, Moscow, Russia.

Courant, R. and Friedrich, K. O., 1948. *Supersonic Flow and Shock Waves*, International Publishers, New York.

Dailey, C. L., 1955. "Supersonic diffuser instability," *J. Aeron. Sci.* **22**, pp. 733-749.

Ferri, A. and Nucci, L. M., 1951. "The origin of aerodynamic instability of supersonic inlets at subcritical conditions," NACA RM L50K30.

Fickett, W. and Wood, W. W., 1966. "Flow calculations for pulsating one-dimensional detonations," *Phys. Fluids* **9**, pp. 903-916.

Frolov, S. M., 1987. "Detonation in systems with friction, heat and mass transfer," Ph.D. Thesis, Phys. Techn. Inst., Moskow.

Fujiwara, T. and Matsuo, A., 1989. "Oxyhydrogen oblique detonation supported by two-dimensional wedge," presented at the International Symposium on Computational Fluid Dynamics, Nagoya, Aug. 28.

Guderley, G., 1942. *Luftfahrt-Forsch* **19**, p. 302.

Guenoche, H. and Manson, N., 1949. "Influence des conditions aux limites transversales sur la propagation des ondes de choc et de combustion," *Rev. Inst. Fr. Pet.* **2**, pp. 53-69.

Herzberg, A., Bruckner, A. P., and Knowlen, C., 1991. "Experimental investigation of ram accelerator propulsion modes," *Shock Waves* **1**, pp. 17-25.

Jouguet, E., 1905. "On the propagation of chemical reactions in gases," *J. de Mathematiques Pures et Appliquees* **1**, pp. 347-425.

Knystautas, R., Lee, J. H. S., Peraldi, O., and Chan, C. K., 1986. "Transmission of a flame from a rough to a smooth-walled tube," in *Dynamics of Shock Waves, Explosions and Detonations*, ed. Bowen, J. R., Leyer, J.-C., and Soloukhin, R. I., AIAA, New York, Vol. 106, pp. 37-52.

Lafitte, P., 1923. "Sur la formation de l'onde explosive," *Compte Rendu a l'Acad. des Scie.* **176**, pp. 1392-1394.

Lee, B. H. K., 1967. "Non-uniform propagation of imploding shocks and detonations," *AIAA Journal* **5**, pp. 1997-2003.

Lee, J. H. and Lee, B. H. K., 1965a. "Cylindrical imploding shock waves," *Phys. Fluids* **8**, pp. 2148-2152.

Lee, J. H., Lee, B. H. K., and Shanfield, I., 1965b. "Two-dimensional unconfined gaseous detonation waves," *10th Symposium (International) on Combustion*, The Combustion Institute, Pittsburgh, pp. 805-815.

Lee, J. H., 1965c. "The propagation of shocks and blast waves in a detonating gas," Ph.D. Thesis, McGill University Report No. 65-1.

Lee, J. H., Knystautas, R., and Bach, G. G., 1969. "Theory of explosions," AFOSR Scientific Report, AFOSR 69-3090 TR.

Lee, J. H. S., 1979. "Recent advances in gaseous detonation," AIAA Paper No. 79-0287.

Lee, J. H. S. and Moen, I. O., 1980. "The mechanism of transition from deflagration to detonation in vapor cloud explosions," *Prog. Energy Combust. Sci.* **6**, pp. 359-389.

Lee, J. H., Knystautas, R., and Freiman, A., 1984a. "High speed turbulent deflagrations and transition to detonation in H_2-air mixtures," *Combustion and Flame* **56**, pp. 227-239.

Lee, J. H., Knystautas, R., and Chan, C. K., 1984b. "Turbulent flame propagation in obstacle-filled tubes," *20th Symposium (International) on Combustion*, The Combustion Institute, Pittsburgh, pp. 1663-1672.

Levin, V. A. and Cheryni, G. G., 1967. "Asymptotic laws of behavior of detonation waves," *PMM* **31**, pp. 393-405.

Manson, N. and Ferrie, F., 1953. "Contribution to the study of spherical detonation waves," *4th Symposium (International) on Combustion*, Williams and Wilkins, Baltimore, p. 486.

Moen, I. O., Funk, J. W., Ward, S. A., Rude, G. M., and Thibault, P. A., 1984. "Detonation length scales for fuel-air explosives," in *Dynamics of Shock Waves, Explosions and Detonations*, ed. Bowen, J. R., Manson, N., Oppenheim, A. K., and Soloukhin, R. I., AIAA, New York, Vol. 94, pp. 55-77.

Nusca, J. Michael, 1991. "Numerical simulation of reacting flow in a thermally choked ram accelerator projectile launch system," AIAA Paper No. AIAA-91-2490.

Oran, E. S., Li, C., Kailasanath, K., and Boris, J. P., 1991. "Numerical studies for the ram accelerator," AFOSR-MIPR-91-005.

Oswatisch, K. L., 1947. "Pressure recovery for missiles with reaction propulsion at high supersonic speeds (the efficiency of shock diffusers)," NACA TM 1140.

Peraldi, O., Knystautas, R., and Lee, J. H., 1986. "Criteria for transition to detonation in tubes," *21st Symposium (International) on Combustion*, The Combustion Institute, Pittsburgh, pp. 1629-1637.

Shchelkhin, K. I., 1940. "Influence of tube roughness on the formation and detonation propagation in gas," *Zh. Exp. Teor. Fiz.* **10**, pp. 823-827.

Soetrisno, M. and Imlay, S. T., 1991. "Simulation of the flow field of a ram accelerator," AIAA Paper No. AIAA-91-1915.

Taylor, G. I., 1950. "The dynamics of the combustion products behind plane and spherical detonation fronts in explosives," *Proc. Roy. Soc. Ser. A* **200**, pp. 235-247.

Teodorczyk, A., Lee, J. H. S., and Knystautas, R., 1988. "Propagation mechanism of quasi-detonations," in *22nd Symposium (International) on Combustion*, The Combustion Institute, Pittsburgh, pp. 1723-1731.

Teodorczyk, A., Lee, J. H. S., and Knystautas, R., 1991. "Photographic study of the structure and propagation mechanism of quasi-detonations in rough tubes," in *Dynamics of Detonations and Explosions: Detonation*, ed. Kuhl, A. L., Leyer J.-C., Borisov, A. A., and Sirignano, W. A., Vol. 133 of Progress in Astronautics and Aeronautics, AIAA, Washington, DC, pp. 223-240.

Trimpi, R. L., 1956a. "An analysis of buzzing in supersonic ram-jets by a modified one-dimensional non-stationary wave theory," NACA TN 3695.

Trimpi, R. L., 1956b. "A theory for stability and buzz pulsation amplitude in ramjets and an experimental investigation including scale effects," NACA TR 1265.

von Neumann, J., 1942. "Theory of detonation waves," in *John von Neumann, collected works* **6**, ed. Taub, A. J., Macmillan, New York.

Yungster, S., Eberhardt, S., and Bruckner, A. P., 1991a. "Numerical simulation of hypervelocity projectile," *AIAA Journal* **29**, pp. 187-199, Feb.

Yungster, S. and Bruckner, A. P., 1991b. "Computational studies of a superdetonative ram accelerator mode," *Journal of Propulsion and Power*, in press.

Zeldovich, Ya. B. and Kompaneets, A. S., 1960. *Theory of detonation*, Academic Press, New York.

Zeldovich, Ya. B., Barenblatt, G. I., Librovich, V. B., and Makhviladze, G. M., 1985. *The Mathematical Theory of Combustion and Explosions*, Nanka, Moscow, translated from Russian by McNeill, D. H., Consultants Bureau, New York, 1985.

Zeldovich, Ya. B., Borisov, A. A., Gelfand, B. E., Frolov, S. M., and Mailkov, A. E., 1988. "Nonideal detonation waves in rough tubes," in *Dynamics of Explosion*, ed. Kuhl, A. L., Leyer, J.-C., and Borisov, A. A., AIAA, Washington, DC, Vol. 114, pp. 211-231.

GODUNOV-TYPE SCHEMES APPLIED TO DETONATION FLOWS

James J. Quirk[1]

Institute for Computer Applications in Science and Engineering
Mail Stop 132C, NASA Langley Research Center
Hampton, Virginia 23681

ABSTRACT

Over recent years, a variety of shock-capturing schemes have been developed for the Euler equations of gas dynamics. During this period, it has emerged that one of the more successful strategies is to follow Godunov's lead and utilize a nonlinear building block known as a Riemann problem. Now, although Riemann solver technology is often thought of as being mature, there are in fact several circumstances for which Godunov-type schemes are found wanting. Indeed, one inherent deficiency is so severe that if left unaddressed, it could preclude such schemes from being used to capture detonation fronts in simulations of complex flow phenomena. In this paper, we highlight this particular deficiency along with some other little known weaknesses of Godunov-type schemes, and we outline one strategy that we have used to good effect in order to produce reliable high resolution simulations of both reactive and nonreactive shock wave phenomena. In particular, we present results for simulations of so-called galloping instabilities and detonation cell phenomena.

1. Introduction

Following the rise of computational fluid dynamics, numerical simulations of shock wave phenomena are now commonplace. Such simulations are attractive as replacements for experiments which are either difficult, dangerous, or expensive, and can be done for problems which are not amenable to analytical methods. However, because of the disparate scales involved, simulations are all too often under-resolved and so are of limited use. Indeed, despite the plethora

[1]Research was supported by the National Aeronautics and Space Administration under NASA Contract No. NAS1-19480 while the author was in residence at the Institute for Computer Applications in Science and Engineering (ICASE), NASA Langley Research Center, Hampton, VA 23681.

J. Buckmaster et al. (eds.), Combustion in High-Speed Flows, 575–596.

of numerical schemes that have been developed, only the Godunov-type methods have been shown to produce, genuinely, high fidelity simulations of complex shock wave phenomena (Berger and Colella, 1989; Quirk, 1992 *a*). Consequently, following a survey of the literature, Godunov-type methods are likely to be picked up by people who are not algorithm developers as tools with which to simulate real problems. Unfortunately, despite their undoubted strengths, Godunov-type schemes have certain inherent weaknesses that can, occasionally, result in simulations failing catastrophically. Amongst the cognoscenti, such weaknesses, whilst not always fully understood, can be overcome. However, the nonexpert is severely disadvantaged by the fact that the various failings and their associated fixes go largely unreported. Recently we have attempted to redress this unfortunate state of affairs (Quirk, 1992 *b*), and here we present an abridged version of this work which is targeted directly at people within the combustion community who would like to take advantage of modern shock-capturing schemes for simulating detonation flows.

The rest of this paper is as follows. In Section 2 we present a brief outline of Godunov-type schemes. This is followed by a section containing some specific examples of how different schemes can fail. In Section 4 we describe the strategy that we use to improve the robustness of our flow solvers, following which, we present results for simulations of galloping instabilities and detonation cell phenomena. Finally, in Section 6 we present some conclusions that we have drawn from this work.

2. Outline of Godunov-type Schemes

Many expositions of Godunov's method and its descendants appear in the literature (Holt, 1984; Roe, 1986); here we simply want to present the general gist of such schemes in order to orientate the reader for the material which follows in the remaining sections of this paper.

With reference to Figure 1, a Godunov-type scheme may be viewed as follows. The scheme works with a low-order projection of the flow solution; each mesh cell contains a cell-averaged value for the true solution over the cell. Thus, the numerical representation closely approximates the true solution near discontinuities, and regions of smooth flow are reasonably well approximated by a series of step functions. This discrete system is integrated by first

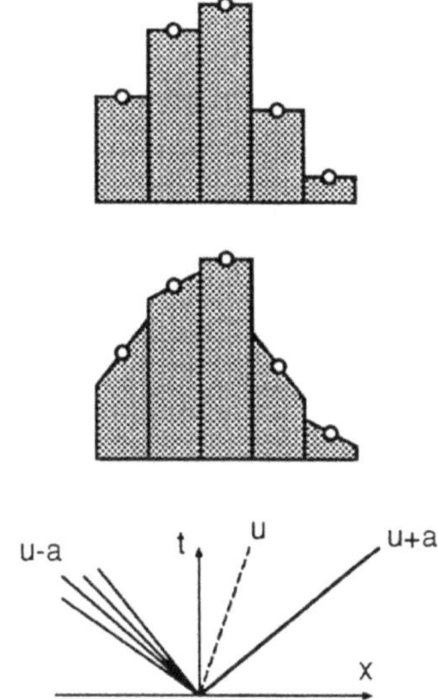

Cell-averaged projection of the flow solution.

Piecewise-linear reconstruction of the flow solution.

Typical Riemann solution showing three centred waves.

Figure 1: A Godunov-type scheme consists of three basic steps: *reconstruction→ evolution →projection.*

reconstructing the flow solution within each cell. This step is effectively an extrapolation process for finding the flow states at the edges of mesh cells given values at the centres of cells. Note that, in general, the reconstructed solution whilst smoother than the projected solution will still be discontinuous at cell interfaces. Next there comes an evolution step. A Riemann problem is solved for each cell interface using the reconstructed states on either side of the interface as input data. Recall that the Riemann problem for any set of conservation laws arises, if initial data are prescribed as two semi-infinite states ($\mathbf{W} = \mathbf{W}_L$ for $x < 0$, $\mathbf{W} = \mathbf{W}_R$ for $x > 0$). The solution to a Riemann problem consists of a number of centred waves. For the one-dimensional Euler equations of gas dynamics there are three waves. The inner wave is a contact discontinuity separating states at different temperatures, and each of the two outer waves may be either a shock wave or an expansion wave. Finally, the different solutions to the separate Riemann problems are averaged so as to find a cell-centred projection of the flow solution at some new time level. If repeated, the sequence *reconstruction→ evolution →projection* results in an accurate and well-behaved scheme for simulating the propagation of shock waves.

Whilst the overall strategy of a Godunov-type scheme is largely clear-cut[2], the same cannot be said of its individual components. For example, more often than not, the Riemann problem for a system of conservation laws does not have an analytic solution and is therefore expensive to compute. Consequently, many workers prefer to compute an approximate solution to the Riemann problem which embodies the spirit of the exact solution but which is cheaper to compute (Roe, 1986; Einfeldt, 1988; Toro, 1991). Indeed, the design of approximate Riemann solvers has become something of an industry in its own right, and there is considerable debate concerning the relative merits of different solvers. Moreover, there are many circumstances for which certain approximate solvers actually prove to be more reliable than the exact solver. Therefore, in general, it is far from obvious which particular Riemann solver is best suited to a given application.

The reconstruction step is similarly open to different interpretation. Firstly, there is a free choice as to which variables are reconstructed, and which quantities are derived from the reconstructed variables. For example, although shock capturing schemes invariably work with the conserved variables, experience shows, at least for the Euler equations, that better results are obtained if the reconstruction is performed using primitive variables. Secondly, there is also a choice as to the order of accuracy of the reconstruction. As is common practice, we employ piecewise-linear slopes (Quirk,1992 *a*); however, Colella and Woodward (1984) use piecewise-parabolas to good effect in their PPM scheme, and more recently Harten *et al.* (1987) have introduced the idea of ENO schemes where the reconstruction step may be carried out to some arbitrary order of accuracy.

It is worth noting, however, that to a large extent the quality of results produced will depend on how well the flow solution is reconstructed near discontinuities and so the notional order of accuracy for some reconstruction process is not necessarily a reliable indicator of its actual performance. Typically, limiter functions (Roe, 1985) are employed so as to ensure that the reconstruction process does not introduce unwanted overshoots near extrema, and it is the properties of the chosen limiter function that largely dictate the quality of

[2]Here we have described the so-called MUSCL approach for producing a high-order Godunov method. An alternative methodology is followed by flux-limited schemes where the reconstruction step is replaced by a procedure which post-processes the Riemann solution.

(a)

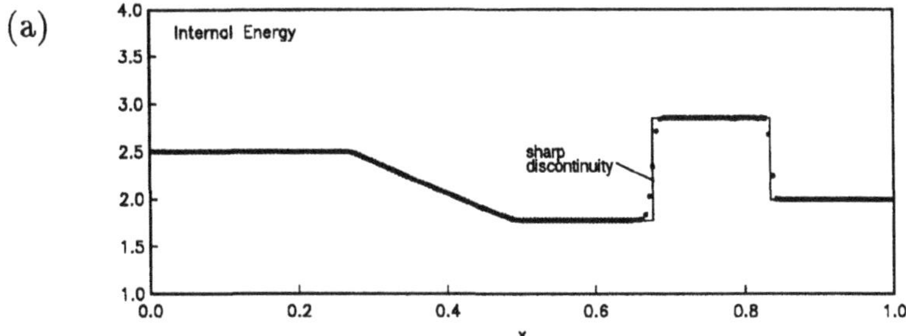

(b)

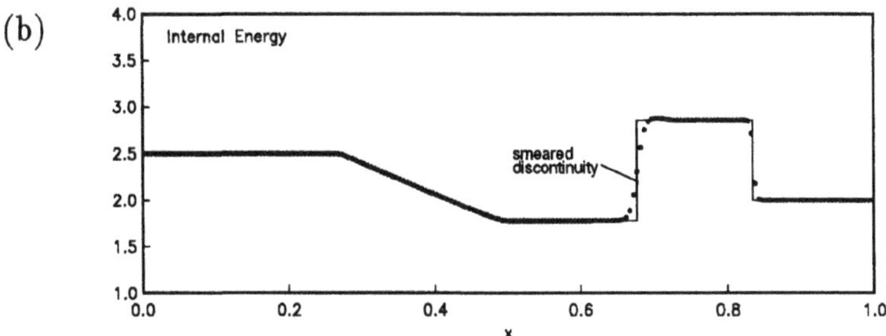

Figure 2: The internal energy profiles computed for Sod's problem using: (a) 2nd order characteristic MUSCL (Quirk, 1992)., (b) 3rd order non-characteristic MUSCL (Anderson *et al.*, 1985).

the flow solution. But a limiter function is only as good as the data upon which it acts. For example, although we employ a piecewise-linear reconstruction which results in a scheme that is nominally only second-order accurate, the slopes are derived from the Riemann solutions which are computed using the cell-centred states as input data. For very high resolution simulations which involve thousands of time steps, this strategy gives markedly better results for weak discontinuities such as contact surfaces than does the third-order reconstruction process proposed by Anderson *et al.*(1985) which does not utilize the same level of characteristic information. While the differences are much less marked for problems that contain just a few time steps, they are still appreciable (see Figure 2).

Everything considered, given the basic framework of a Godunov-type scheme, it is possible to construct many different variations on a theme. Unfortunately, there is no "correct" way of doing things, for the theory which underpins this class of scheme does not always discriminate between the different options that are available. In practice, seemingly innocuous details of the reconstruction process can have a large bearing on the quality of the results produced for the sorts of very detailed simulations that will be necessary in order to unravel the dynamics of detonation phenomena. Thankfully, the large scale flow features are normally insensitive to such changes, and so one's faith in Godunov-type schemes is not unduly undermined by the uncertainties in the precise details of the method.

Before proceeding to the next section which exposes some of the failings of Riemann solvers, so as not to appear to paint an overly pessimistic view of the capabilities of Godunov-type schemes, it is worthwhile showing what can be achieved if due care is taken. Figure 3 shows a snapshot taken from the simulation of a planar shock wave diffracting around a 90° corner (Quirk, 1992 *b*). This picture is similar to a Schlieren image in that the different shades of grey depict the magnitude of the gradient of the density field (the darker the shade, the larger the gradient), and it clearly shows all the salient flow features as identified experimentally by Bazenhova *et al.* (1984).

Briefly, with reference to Figure 4, the diffraction of the incident shock wave (AC) around the corner gives rise to an expansion fan which emanates from (O). The shape of this fan's lead characteristic (AQO) indicates that the flow upstream of the incident shock is supersonic. The expansion fan interacts with the incident shock to form the disturbed shock front (ADMN). The incident shock is sufficiently strong that this disturbed front is kinked; a Mach reflection at the wall gives rise to a triple point at (M). A contact surface (ALO) marks the boundary between fluid that has been induced into motion by the incident shock and fluid which has been processed by the disturbed shock front. Note that the flow is separated from the wall at a point slightly downstream of the apex of the corner, and a slipstream (OS) separates the expanded flow from this region of almost stationary gas. The free end of this slipstream rolls up into a vortex. A secondary shock wave (KTS) matches the pressure of the flow accelerated by the expansion fan to that of the decelerated flow behind the disturbed part of the incident wave. This secondary shock is kinked as a result of its interaction with the slip stream (OS). A secondary contact surface (TL) begins at the point of intersection of the secondary shock wave with the weak shock (OT) which terminates the expansion fan. Lastly, a shock wave (PB) is present so as to decelerate the reversed flow within the separated region as it approaches the point of diffraction.

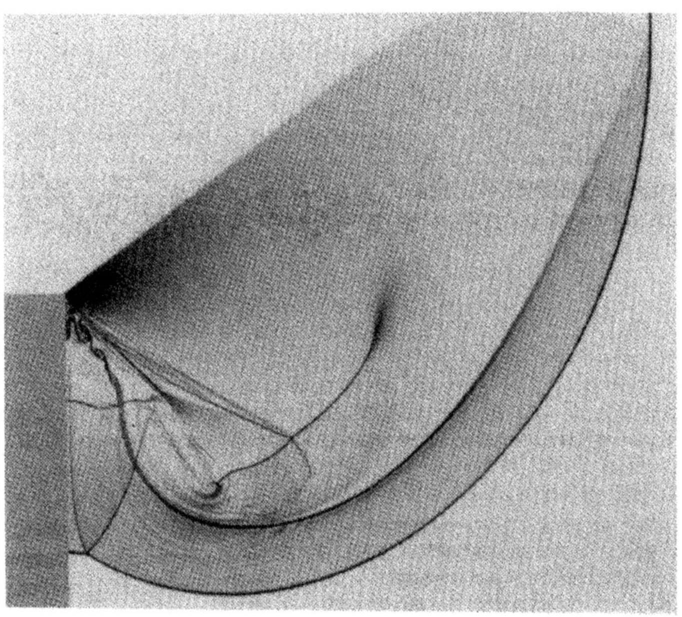

Figure 3: A numerical Schlieren-type image taken from the simulation of a planar shock wave diffracting around a 90° corner.

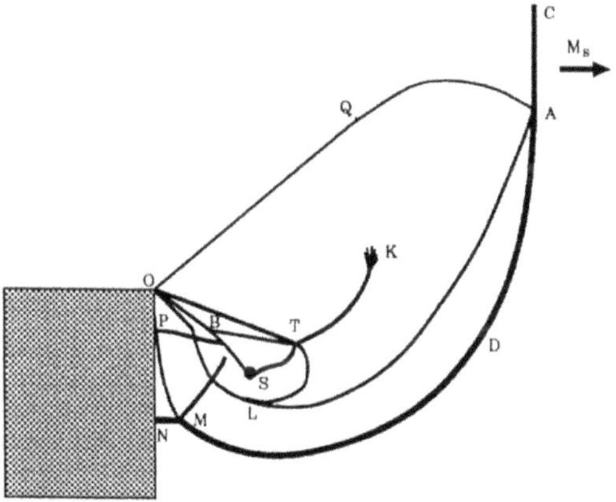

Figure 4: Schematic showing the main flow features of the above Schlieren-type image.

3. Some Failings of Riemann Solvers

We now present some examples where certain Riemann solvers are known to give unreliable results for the Euler equations of gas dynamics. Whilst our catalogue of failings is not exhaustive, it should alert the reader to the types of problems that they might encounter when using Godunov-type methods. It is important to realise that no one failing afflicts all Riemann solvers. Conversely, however, it would appear that no one Riemann solver is completely free of defects.

With reference to Figure 5, Example (a) shows how spurious post-shock oscillations can occur whenever a shock wave moves very slowly during the course of a simulation. Here, a one-dimensional shock wave is moving from left to right (for a Courant number of one, it takes approximately 50 time steps for the shock to traverse a single cell). As the shock moves relative to the mesh, so the smeared numerical shock profile inevitably changes shape. Unfortunately, for many Riemann solvers including the exact solver, the points within the smeared profile do not lie on the Hugoniot curve connecting the pre-shock state to the post-shock state. Thus these two states cannot be connected by a single family of acoustic waves. For such schemes, whenever the shock profile changes, so the supposedly passive wave fields are activated (in this case, the u and $u - a$ wave fields), thus giving rise to low frequency oscillations. A thorough description of this particular failing has been given by Roberts (1990). Note that these oscillations are not the same as the high-frequency, post-shock oscillations that afflict finite-difference shock-capturing schemes.

Example (b) in Figure 5 is taken from a simulation of double Mach reflection which was performed using Roe's scheme; here we show a snapshot of the pressure contours at one instant in time. The Mach stem is inexplicably kinked giving rise to a spurious triple point (S). It should be noted that this kinking is not related to the slight bulging that is often observed experimentally for this type of shock reflection problem. Such bulging arises because the contact discontinuity emanating from the primary triple point is deflected by the reflecting surface resulting in a strong wall jet which effectively pushes out the base of the Mach stem. The mechanism behind this failing is not fully understood, but it would appear to arise from the fact that the Mach stem is closely aligned with the computational grid. Consequently, little or no dissipation is provided by Roe's scheme, in a direction parallel to the stem, to control the kinking.

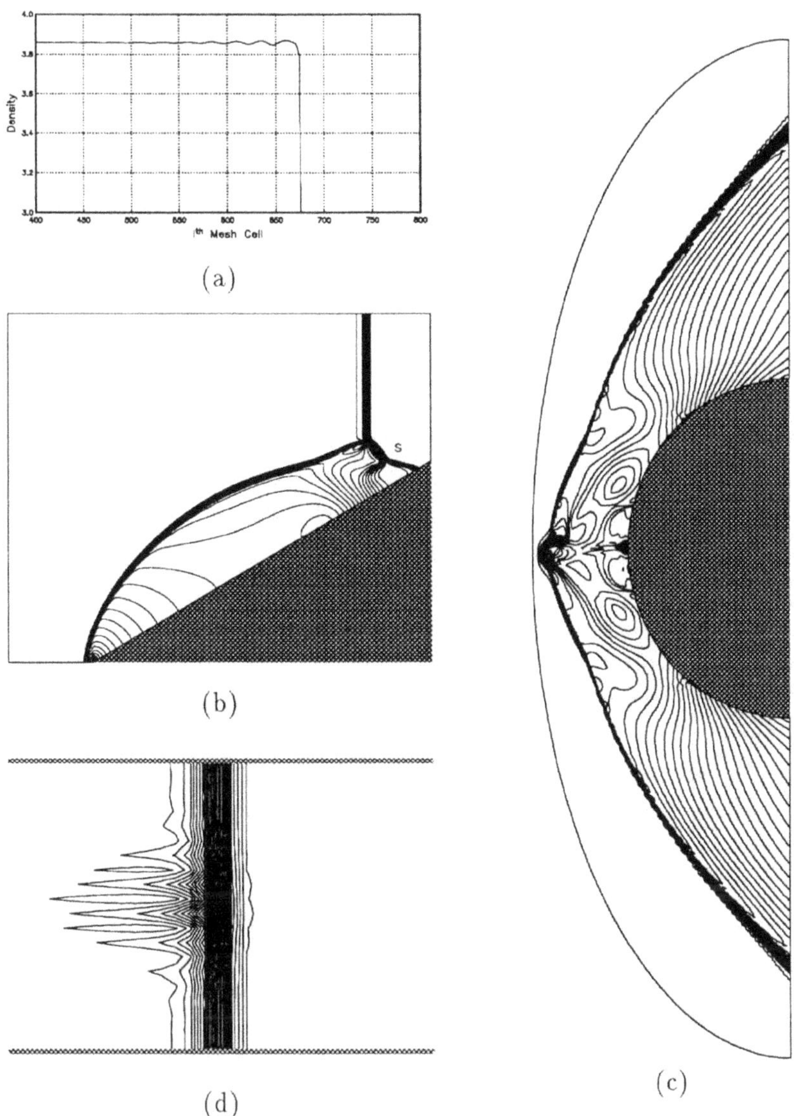

Figure 5: Some Riemann solver failings: (a) Slowly moving shocks., (b) Kinked Mach stems., (c) The Carbuncle phenomena., (d) Odd-Even decoupling.

Generally speaking, the dissipation mechanism provided by many Riemann solvers proves to be inadequate whenever a strong shock wave is aligned with the computational grid. Example (c) shows the so-called carbuncle phenomena (Peery and Imlay, 1988) where some schemes fail to produce a realistic bow shock for a blunt body placed in a high Mach number flow (here we have plotted density contours). Note that along the stagnation line the bow shock is more or less aligned with the body-fitted grid used for the calculation, and so very little dissipation is added normal to the stagnation line in the vicinity of the bow shock; a small amount of auxiliary dissipation applied in this region is generally sufficient to suppress the carbuncle.

Example (d) in Figure 5 shows a particularly insidious failing that can occur when a strong shock is aligned with the grid. Here we have plotted a single snapshot of the density contours taken from the simulation of an initially planar shock wave which is propagating down a duct, from left to right. A nominally uniform Cartesian grid was used for the calculation; the grid centre-line carried a small saw-tooth perturbation. This perturbation acts like a forcing function which causes odd-even decoupling to occur, along the length of the shock, in both the density and pressure fields. Interestingly, within the body of the shock, the decoupling of the density field is out of phase with that of the pressure field. An attempt to explain this failing has been given by Quirk (1992 *b*), but certain details of the mechanism remain unclear. It is clear, however, that this numerical instability only occurs for strong shocks, and that it takes a long time to develop (here, the shock had propagated approximately thirty channel widths before the instability first became apparent). Moreover, as will be shown in Section 5, this particular instability has quite grave consequences for the simulation of two-dimensional detonation fronts; waves produced by the numerical instability can interfere with the genuine transverse waves associated with the propagation of the front.

For many problems, the nature of the flow solution is known *a priori*, and so it is fairly obvious whenever a simulation gives anomalous results. But often no such safety net exists, in which case it becomes very difficult to determine the fidelity of a numerical simulation. One tactic that we routinely employ is to run a simulation two or more times, each time varying the elements of the flow solver. For example, we may change our choice of Riemann solvers, or we may change our choice of variables for the reconstruction process. Admittedly, the fact that two such simulations give similar behaviour is no guarantee that the results are correct, but it is useful for determining which features of the solution are likely to be numerical artifacts. Lastly, it should be noted that many of the failings associated with Godunov-type methods could be circumvented if strong shocks were fitted rather than captured. However, given the complexity of a general purpose shock-fitting scheme, this option is not likely to appeal to the worker who is merely using a Godunov-type scheme as a tool.

4. An Adaptive Riemann Solver

Having exposed some of the weaknesses of Riemann solvers, we now present a simple strategy, that we have found useful, for improving the all-round robustness of Godunov-type schemes. In essence, we select the precise flavour of upwinding to match the local flow data such that a particular Riemann solver is only employed in those situations where it is known to give reliable results. By recognizing the limitations of any one solver it is possible to reap its advantages without suffering its attendant failings.

Our synergetic strategy has a number of attractions, not least of which is that some favoured solver need not be jettisoned simply because it, occasionally, fails. However, it does introduce the difficulty of how to decide when to use one Riemann solver in preference to another. But it has been our experience that this added difficulty is not particularly bothersome, for we tend to combine a single high resolution Riemann solver with just one or two other solvers that prove more reliable under conditions which are fairly well-defined, and so a set of *ad hoc* switching functions suffice. For example, some of the worst failings of Riemann solvers occur in the vicinity of strong shock waves. To overcome such failings we use the HLLE scheme (Einfeldt, 1988). Now it makes little sense to chop and change the choice of Riemann solver used along the length of a shock wave, since to do so would inevitably perturb a planar shock front. Hence, we apply this particular Riemann solver throughout the immediate vicinity of a strong shock. Thus the HLLE switching function need only locate the position of a shock wave, but such functions already exist in the guise of mesh refinement, monitor functions.

A simple test that identifies those cell interfaces which are in the vicinity of a strong shock is to check whether or not

$$\frac{|p_r - p_l|}{min(p_l, p_r)} > \alpha, \tag{1}$$

where α is some threshold parameter which is problem dependent and p_r and p_l refer to the pressures which act on the interface. If this condition is met, the two cells separated by the interface are flagged as lying within a strong shock. Then, when it comes to computing cell-interface fluxes, if the cells either side of an interface have both been flagged as lying within a strong shock, the flux is computed using the HLLE solver. Note that since numerical shocks are invariably smeared over several mesh cells, it is worth locating shocks using a projection of the flow solution on a grid which is coarser than that used for the calculation. On such a grid a shock will appear much less smeared, and so the left-hand side of the above switching function will be a fair indication of its strength. Once a set of cells have been flagged on this coarse mesh, the flags may be prolongated to the actual computational mesh so as to find those cells which lie in the vicinity of a shock wave.

Figure 6 shows how the HLLE solver may be used to correct the tendency of Roe's scheme to produce kinked Mach stems, c.f. Figure 5 (b). For this calculation the HLLE switching function was tuned such that it would only be activated by the incident shock, and the principal Mach stem; the threshold parameter α was simply set to half the strength of the incident shock wave as given by the left-hand side of Equation 1. Note that apart from the region near the Mach stem, these new results are very similar to the old ones. This shows that the HLLE scheme has had no adverse affect on the resolution of Roe's scheme.

Having presented the gist of our strategy, we see little point in trying to sell a particular combination of solvers. Starting with some high resolution Riemann solver, whose choice will inevitably be a matter of personal taste, the correct combination of solvers will depend both on that schemes weaknesses and on the specific application in hand. In turn, the combination of Riemann solvers will dictate the choice of switching functions.

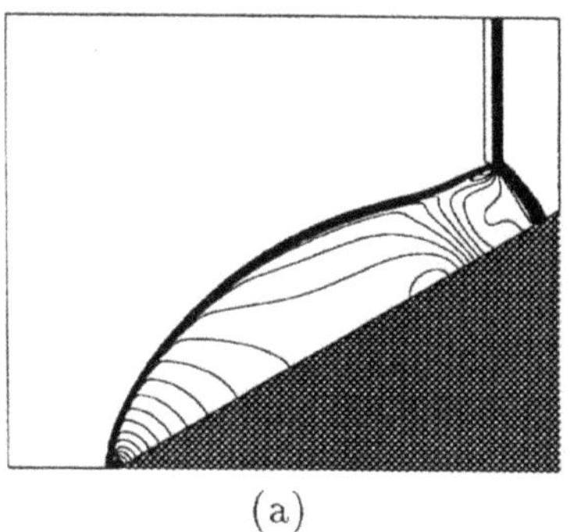

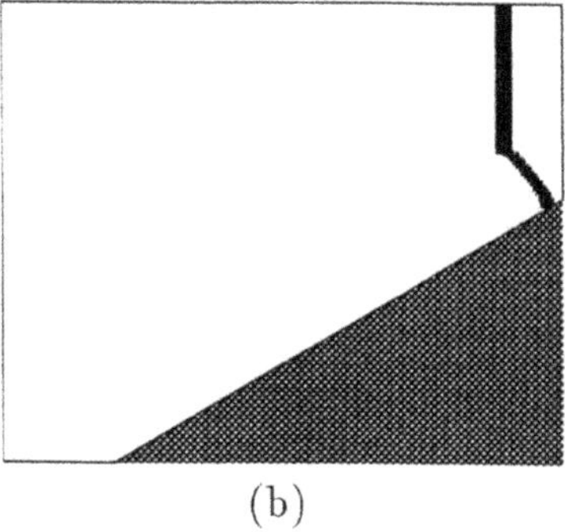

(a) (b)

Figure 6: The HLLE scheme can be used to circumvent the tendency of Roe's method to produce kinked Mach stems: (a) Pressure Contours, (b) HLLE switching function.

5. Galloping Instabilites and Detonation Cells

Almost all of our experience with Godunov-type schemes has been gained from calculations of nonreactive flows. However, as will be shown in this section, the basic lessons that we have learnt remain important when it comes to performing simulations of detonation phenomena. Thus far, for simplicity, we have utilized the so-called Reactive Euler equations for our detonation simulations; a single reactant A is converted to a single product B by a one-step irreversible chemical reaction which is governed by Arrhenius kinetics.

In one space dimension the reactive Euler equations may be written in non-dimensional form as

$$\frac{\partial}{\partial t}\begin{pmatrix}\rho\\ \rho u\\ E\\ \rho Z\end{pmatrix}+\frac{\partial}{\partial x}\begin{pmatrix}\rho u\\ \rho u^2+p\\ (E+p)u\\ \rho u Z\end{pmatrix}=-\begin{pmatrix}0\\ 0\\ 0\\ K\rho Z\exp^{-E^+/T}\end{pmatrix}. \quad (2)$$

Here ρ, u, p, T, E, Z and E^+ are the density, velocity, pressure, temperature, total energy per unit volume, reactant mass fraction, and the activation energy, respectively. Note that K is a free parameter that simply sets the spatial and temporal scales. Typically, K is chosen such that for a ZND wave the half-reaction length (the distance behind the detonation front by which point half of the reactants have been consumed) is scaled to unit length. The following equations are used to close the system (2)

$$\begin{aligned} E &= \rho e + \rho q Z + \tfrac{1}{2}\rho u^2, \\ p &= (\gamma - 1)\rho e, \\ T &= p/\rho. \end{aligned} \tag{3}$$

Here e is the specific internal energy, γ is the ratio of specific heats and q is the heat release parameter for the chemical reaction $A \rightarrow B$.

We integrate the above system of equations, which are of the form $\frac{\partial \mathbf{w}}{\partial t} + \frac{\partial \mathbf{f}}{\partial x} = \mathbf{s}$, using the method of fractional steps

$$\mathbf{w}^{n+2} = \mathcal{L}_s.\mathcal{L}_c.\mathcal{L}_s.\mathbf{w}^n \quad .$$

The source operator, $\mathcal{L}_s$, corresponds to integrating $\frac{\partial \mathbf{w}}{\partial t} = \mathbf{s}$, which in this case reduces to integrating the single ODE, $\frac{dZ}{dt} = -KZ \exp^{E^+/T}$. We assume that the temperature field is frozen for this step, allowing us to use the nominally exact operator

$$Z^{n+1} = Z^n \exp(-K \exp(-E^+/T^n)\Delta t),$$

where Δt is the time step going from time level n to time level $n+1$. For the convective operator, $\mathcal{L}_c$, which corresponds to integrating $\frac{\partial \mathbf{w}}{\partial t} + \frac{\partial \mathbf{f}}{\partial x} = 0$, we employ Hancock's finite-volume scheme (Quirk, 1992a) in conjunction with the adaptive Riemann solver outlined in this paper. Note that the convective operator uses a time step that is twice as large as that used by the source operator. The generalization of this integration strategy to two space dimensions simply consists of replacing the one-dimensional convective operator by its two-dimensional counterpart.

To assess the capabilities of our scheme we have performed simulations of one-dimensional pulsating detonations, or so-called galloping instabilities, for which bench-mark results appear in the literature (Fickett and Wood, 1966; Bourlioux *et al.*, 1991). Here we limit ourselves to presenting the case where the overdrive is 1.6 and the dimensionless parameters that appear in equations (2) and (3) are given by: $\gamma = 1.2$, $E^+ = 50$, $q = 50$ and $K = 230.75$. We have run this problem using several different mesh resolutions in combination with several different approaches for performing the reconstruction step of our Godunov-type scheme. Figure 7 shows the shock pressure history for the case where the computational grid provided 40 mesh cells per half-reaction length, and the reconstruction process operated on a characteristic decomposition of the conserved variables. Qualitatively, this pressure trace is identical in form to that presented by Bourlioux *et al.* (1991) which was found using a scheme that fitted rather than captured the detonation front. Figure 8 shows a convergence study for the variation of the peak shock pressure with mesh spacing for two popular limiter functions, namely Superbee and Minmod (Roe, 1985), in both cases the reconstruction process operated on the conserved variables. Note that a relative mesh spacing of 1 corresponds to having 10 cells per half-reaction length, and so 0.125 corresponds to having 80 cells per half-reaction length. Pleasingly, the grid-converged value for the peak pressure is independent of the reconstruction process. Moreover, it is in close agreement with the value given by Fickett and Wood (1966), and the value found by extrapolating the results of Bourlioux *et al.*(1991).

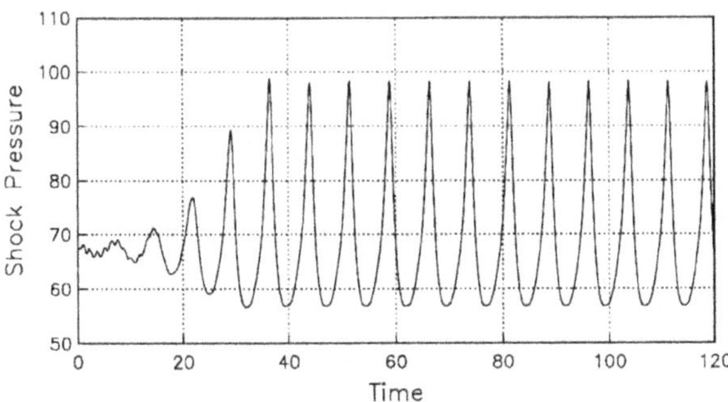

Figure 7: Shock pressure history trace for a calculation with 40 mesh cells per half-reaction length.

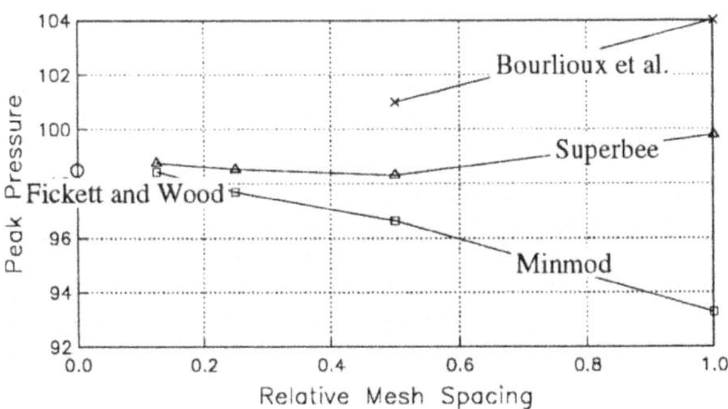

Figure 8: Variation of peak pressure with mesh spacing.

Even for this relatively simple problem, some numerical artifacts can occur if one is not careful. For example, at one stage in the cycle of the pulsating detonation, the profile for the reactant mass fraction is considerably steeper than the profile for the steady ZND detonation wave upon which the mesh spacing is based. Consequently, as shown by Figure 9, a grid which gives 10 mesh points for the half-reaction length of the ZND wave may at times give only half the expected number of cells for the pulsating front. For a compressive limiter function such as Superbee, if there are too few cells covering a steep but continuous profile, the profile will appear as a smeared discontinuity that needs steepening. Hence an under-resolved profile is often artificially steepened by a compressive limiter. Here, such oversteepening causes anomalies to appear in the shock pressure history, see Figure 10. Note that such anomalies are not associated with the lead shock front, and so the question of whether the front is fitted or captured is immaterial.

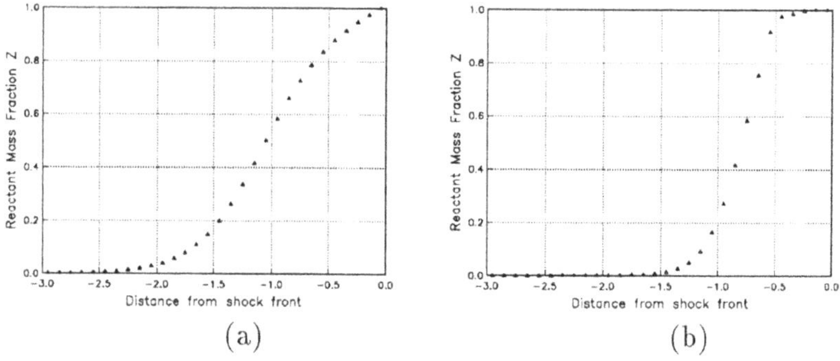

Figure 9: The reaction profile for a pulsating detonation wave may at times be considerably steeper than that for the initial steady ZND wave. (a) ZND wave. (b) Pulsating wave.

Following Bourlioux (1991) we have also conducted a two-dimensional test of our detonation code, namely, the simulation of the transverse waves, or so-called cellular structure, produced by a detonation wave travelling down a narrow channel. For this calculation the overdrive is 1.2, the channel width is 10 half-reaction lengths, and the dimensionless parameters are $\gamma = 1.2$, $q = 50$, $E^{+} = 10$ and $K = 3.124$. The calculation was started with the solution for the ZND wave as initial data. This planar detonation front was perturbed by simply allowing it to ingest a small region of fluid where

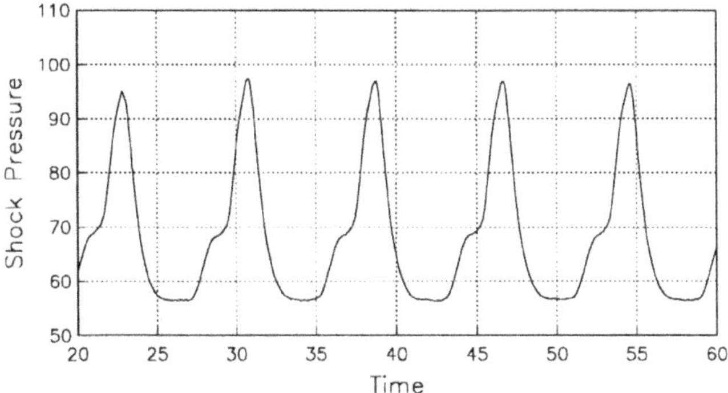

Figure 10: Anomalies often arise when an under-resolved calculation employs a compressive limiter function.

the rate constant K was artificially decreased by 20%. The calculation was then run until the ensuing transverse wave structure was fully developed, at which point a series of Schlieren-type images were taken so that we could compare our results with Bourlioux's. These Schlieren-type images are shown in Figure 11. Note that the calculation applied periodic boundary conditions along the top and the bottom of the channel and that here we have plotted four periods. Qualitatively, at least, these results compare well with Bourlioux's results, and all the salient features of the flow have been resolved (256 mesh cells covered the width of the channel). In particular, the regularity of the repeated vortex patterns are the same for both sets of calculations.

At this juncture, it is worth showing the results that were produced when we employed an exact Riemann solver rather than our adaptive Riemann solver, see Figure 12 (only one period is shown). As might be expected from having seen Figure 5 (d), spurious transverse waves are produced in the vicinity of the lead shock front. These spurious waves prevent the correct transverse wave structure from developing, thus ruining the simulation. It is our contention that most, if not all, of the Riemann solvers that are commonly employed would be similarly unable to *capture* the lead detonation front for this test problem. Whilst some may take this as reason enough why one should always fit the lead shock front, we have shown that if some care is taken, when it comes to detonation simulations, shock-capturing remains a viable alternative to shock-fitting.

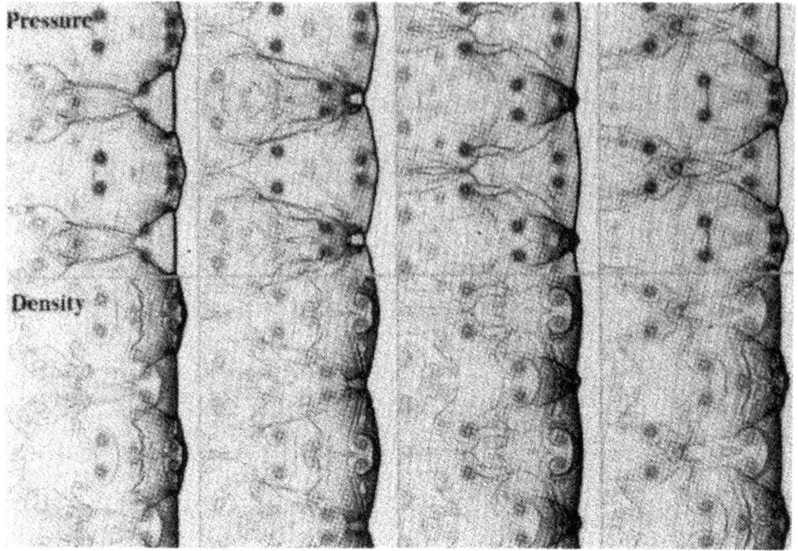

Figure 11: A sequence of four Schlieren-type snapshots which show the transverse wave structure of the detonation front.

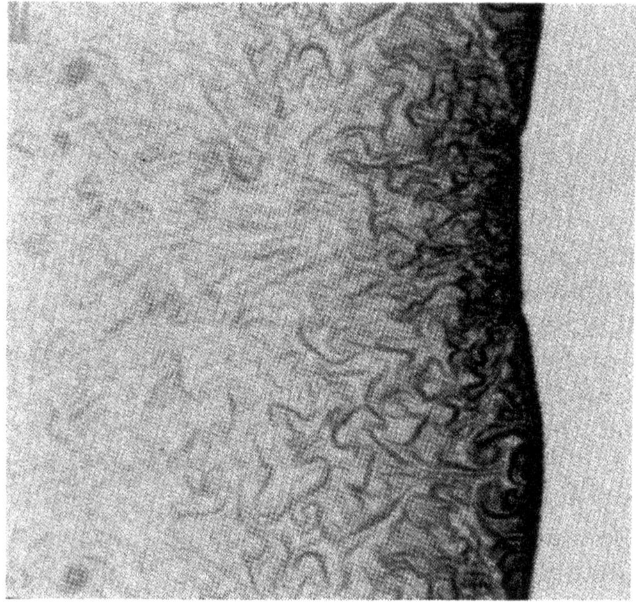

Figure 12: Spurious waves are produced, if the lead shock front is captured using an exact Riemann solver.

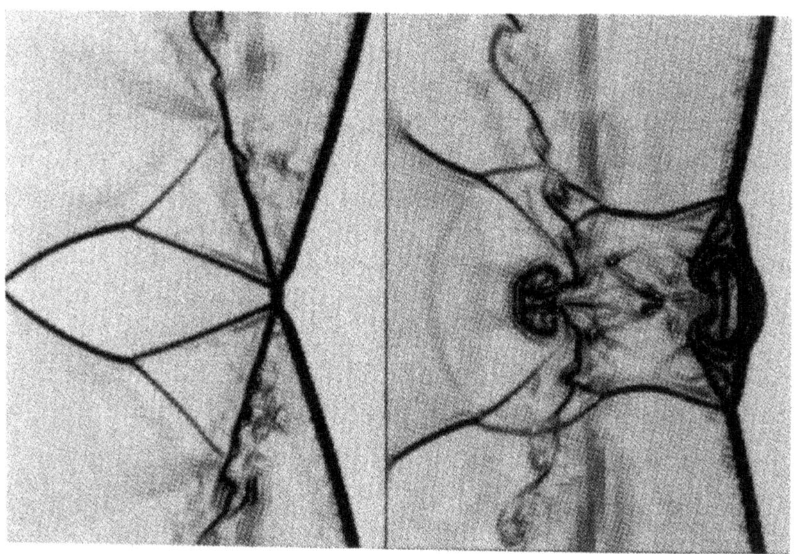

Figure 13: A pair of Schlieren-type images for the temperature field which show the sequence of events when two triple points collide.

In an attempt to unravel some of the dynamics of the detonation wave for this problem, we have rerun this test using a grid that was four times finer than before. In order to achieve such high grid resolutions we employ a relatively sophisticated mesh adaption scheme the details of which are too involved to give here (Quirk, 1991). Figure 13 presents a pair of Schlieren-type images for the temperature field which show the local flow structure just before, and just after, two triple points collide. It would appear that the collision gives rise to the familiar "explosion within an explosion", which in this case, because of the local shock structure, is highly anisotropic. This explosion causes slugs of hot fluid to be shot fore and aft giving rise to vortical structures which are similar to those associated with Rayleigh-Taylor instabilities. Note how the impact of the forward facing jet on the lead shock front causes the front to bulge.

Obviously, given the complexity of the flow field for this test problem there is little chance of validating every detail of the simulation. However, we feel that our computational method has matured to the point where it may be relied upon to provide simulations of sufficient fidelity for fathoming the details of complicated flow mechanisms as is done here.

6. Closing Comments

In this paper, we have shown that Godunov-type schemes do not always live up to their reputation as being models of robustness. Specifically, the upwind dissipation provided by almost all Riemann solvers is inadequate for *capturing* detonation fronts in complex, multi-dimensional flows; a numerical instability can develop along the length of the detonation front which interferes with the genuine transverse waves associated with the propagation of the front. At the very least, we suggest that some artificial dissipation mechanism be used to augment the inherent upwind dissipation so as to suppress this type of numerical instability, albeit at a cost of some loss in resolution. In general, however, to improve the all-round robustness of a Godunov-type scheme without incurring any appreciable loss in resolution, we advocate the use of an adaptive Riemann solver. The shortcomings of any one preferred Riemann solver are circumvented by combining it with one or more complementary solvers, such that an individual Riemann solver is only used in the sorts of situations for which it is known to give reliable results. Admittedly this approach is not as straightforward to implement as would be the addition of an auxiliary dissipation mechanism to one's favourite Riemann solver, but it does prove to be an effective means for producing high fidelity simulations of detonation phenomena. As such, it provides an alternative means of simulating detonation phenomena for workers who might otherwise feel compelled to fit the detonation front solely in order to avoid the numerical difficulties associated with strong shocks.

References

Anderson, W.K., Thomas, J.L. and van Leer, B., 1985. "A comparison of finite-volume flux-vector splittings for the Euler equations,"AIAA paper 85-0122.

Bazhenova, T.V., Gvozdeva, L.G. and Nettleton, M.G., 1984. "Unsteady interactions of shock waves," *Progress in Aerosp. Sci.* Vol. **21**, pp. 249–331.

Berger, M.J. and Colella, P., 1989. "Local adaptive mesh refinement for shock hydrodynamics," *J. Comput. Phys.* **82**, pp. 64–84.

Bourlioux, A., 1991. "Numerical study of unstable detonations," Ph.D. Thesis, Princeton University, Princeton, NJ.

Bourlioux, A., Majda, A.J. and Roytburd V., 1991. "Theoretical and numerical structure for unstable one-dimensional detonations," *SIAM J. Appl. Math.* **51**, No.2, pp. 303–342.

Colella, P. and Woodward, P.R., 1984. "The piecewise parabolic method (PPM) for gas-dynamical simulations," *J. Comput. Phys.* **54**, pp. 174–201.

Einfeldt, B., 1988. "On Godunov-type methods for gas dynamics," *SIAM J. Numer. Anal.* **25**, No.2, pp. 294–318.

Fickett, W. and Wood, W.W., 1966. "Flow calculation for pulsating one-dimensional detonation," *Phys. Fluids* **9**, pp. 903–916.

Harten, A., Enquist, B., Osher, S., and Chakravarthy, S.R., 1987. "Uniformly high order accurate essentially non-oscillatory schemes, III," *J. Comput. Phys.* **71**, pp. 231–303.

Holt, M., 1984. "Numerical methods in fluid dynamics,"Springer, Berlin, 273 pp. 2nd ed.

Peery, K.M. and Imlay, S.T., 1988. "Blunt-body flow simulations," AIAA paper 88-2904.

Quirk, J.J., 1991. "An adaptive grid algorithm for computational shock hydrodynamics,"Ph.D. Thesis, College of Aeronautics, Cranfield Institute of Technology, U.K. .

Quirk, J.J., 1992 *a*. "An alternative to unstructured grids for computing gas dynamic flows around arbitrarily complex two-dimensional bodies,"*ICASE Report No. 92-7.* In press, *Computers & Fluids.*

Quirk, J.J., 1992 *b*. "A contribution to the great Riemann solver debate,"*ICASE Report No. 92-64.* Submitted to *Int. J. Numer. Meth. Fluids.*

Roberts, T.W., 1990. "The behaviour of flux difference splitting schemes near slowly moving shock waves,"*J. Comput. Phys.* **90**, pp. 141–160.

Roe, P.L., 1985. "Some contributions to the modelling of discontinuous flows,"*Lecture Notes in Applied Mathematics.* **22**, Springer-Verlag, pp. 163–193.

Roe, P.L., 1986. "Characteristic-Based schemes for the Euler equations,"*Ann. Rev. Fluid Mech.* **18**, pp. 337–365.

Toro, E.F., 1991. "A linearised Riemann solver for the time-dependent Euler equations of gas dynamics,"*Proc. Roy. Soc. London* A **434**, pp. 683–693.

PANEL DISCUSSION

Panel Discussion Summary

With but minor exceptions, the panel discussion is here reproduced verbatim. Without an indication of the pauses, tone, and body language which, in spoken English, signify erasure, correction, or a new direction, some of the text lacks coherence. Nevertheless, the essential meaning of the speakers is clear.

The text retains chronological integrity and is divided into six sections.

- Introduction
- Numerical Modeling
- Detonations
- Mixing
- Kinetics
- Flame Holding/Concluding Remarks

Some of the flavor of the discussion is identified in the following summary paragraphs.

There appears to be a lack of communication and understanding between the university community and device developers, or, to use different labels, the science people and the practical people. Bushnell argues forcefully that the science community has not been looking at the right problems. Libby points out that the university community looks at what DoD is willing to fund. This situation is unfortunate because, as Hertzberg emphasizes, there is a need for high accuracy if a successful high Mach number engine is going to be developed, and this needs good science. Surely one of the premises for looking at this whole issue again is that our science base and our scientific resources are much better than they were in the 60's. Not that we can't learn a great deal from what was done in the past, and we should be familiar with the earlier work, as Libby notes.

The problem of plane mixing is noted as one of little practical interest which, nonetheless, has received much attention. It is important to examine three-dimensional nonlinear stability issues in turbulent flows. Lee argues that computation together with clean experimental work is needed to provide quantitative understanding

J. Buckmaster et al. (eds.), Combustion in High-Speed Flows, 599–643.

of the various mechanisms. The value of simple experiments designed specifically to aid modeling efforts - the scientific approach - is a recurring theme.

Rogers provides a useful list of some of the problem areas, and identifies the importance of facility simulation, an opinion supported from the floor by Morgan.

The strengths and limitations of turbulence computation are addressed by several participants. Givi emphasizes the importance of DNS in the evaluation of turbulence models. But it is also noted that DNS is usually two-dimensional, that it does not address the laminar flamelet regime, and that there are difficulties in incorporating reduced chemistry schemes.

There are two possible technologies - the Ferri engine, and the oblique detonation wave engine. Bushnell claims that the current Ferri engine does not work, and the "great white hope" is the ODWE. There has recently been a significant improvement in the quality of unsteady, multi-dimensional detonation wave computations, and Kapila suggests that we need to extract macro information from these which can be used by designers. In response to Lee's appeal for calculations that will correctly simulate critical tube diameters or critical expansion ratios, it is suggested that to deal with three-dimensional detonations with many cells we need a subgrid model.

Kinetic modeling is a major issue. Convincing simulation of detonations requires a good, but simple kinetic model. Reduced mechanisms have several flaws, and perhaps a better procedure is to use an adaptable approach, exemplified by Harvey Lam's scheme.

It is noted that in operating an ODWE, it is likely that the mixture entering the wave will consist of solid hydrogen, liquid drops, slush balls, and gases; little is known of this situation.

Introduction

Buckmaster: We have several distinguished panelists whom you see sitting in front of you. We have Dennis Bushnell from NASA, Peyman Givi from SUNY at Buffalo, Abe Hertzberg from the University of Washington, Ash Kapila from RPI, John Lee from McGill, Paul Libby from San Diego, and Clay Rogers from NASA. What we do is really up to the panelists and up to you people. The idea is to try to identify issues that are important for the scientific program that we've been engaged in for the last couple of days. The procedure that we're going to adopt is that I'm going to allow each panelist up to 3 minutes to identify one, two, or three questions that they might like to have debated. At the end of that time, a topic will be chosen and the panelists will discuss it. After they have exhausted it, or after I get bored, whichever comes first, I will stop them and throw it open to discussion from the floor. We will then exhaust that or perhaps decide we're spending too much time, in which case we'll stop and pick a new topic. That's roughly speaking how it will go, but I hope it will have its own dynamics and you will also not be shy when you have an option to talk and will participate. One thing I should say, we do have a fairly large panel, and so not every panelist needs to feel obliged to make a contribution to every issue that's raised. If they don't have anything important to say, better they be quiet than keep on going. Let me start with Dennis Bushnell who is going to raise some issues for our consideration.

Bushnell: It seems to me that the state-of-the-art is similar to where we were in gas dynamics in the 1950's. You're attempting to sort out physics by idealized approximate approaches. Back to Buckmaster's question on reasonable assumptions; the problems, it seems to me anyway, are mainly nonlinear. A suggestion is to attempt to identify, control, and utilize instabilities to enhance various measures of performance. Tony Patera from MIT did that for computer cooling using cavity edgetone instability with marvelous results. It's not enough to throw the research over the fence. I think that you're going to have to learn about applications, and you ignore this advice at your peril in this day of greatly changing resource funding plans and so forth. Suggested research topics, in the presence of "real life" combustor flow characteristics, include mixing/reaction physics and enhancement, turbulence-kinetic interactions, and catalysis via metal oxides for nozzle disassociation mitigation. For the ODWE, forebody solid and liquid hydrogen injection, penetration, atomization, vaporization, and pre-ignition; oblique shock mixture combustion for mixtures of liquid, solid, and gaseous hydrogen and air; performance envelope and estimation for the second stage of a TSTO. Now, what do we have in these combustors? I discussed this a little bit. People asked me what do you mean; let me try again. All of these effects are present simultaneously in scramjet combustors, and they excite numerous additional instabilities. We have 3-D flows putting in longitudinal vortices and additional 3-D modes; longitudi-

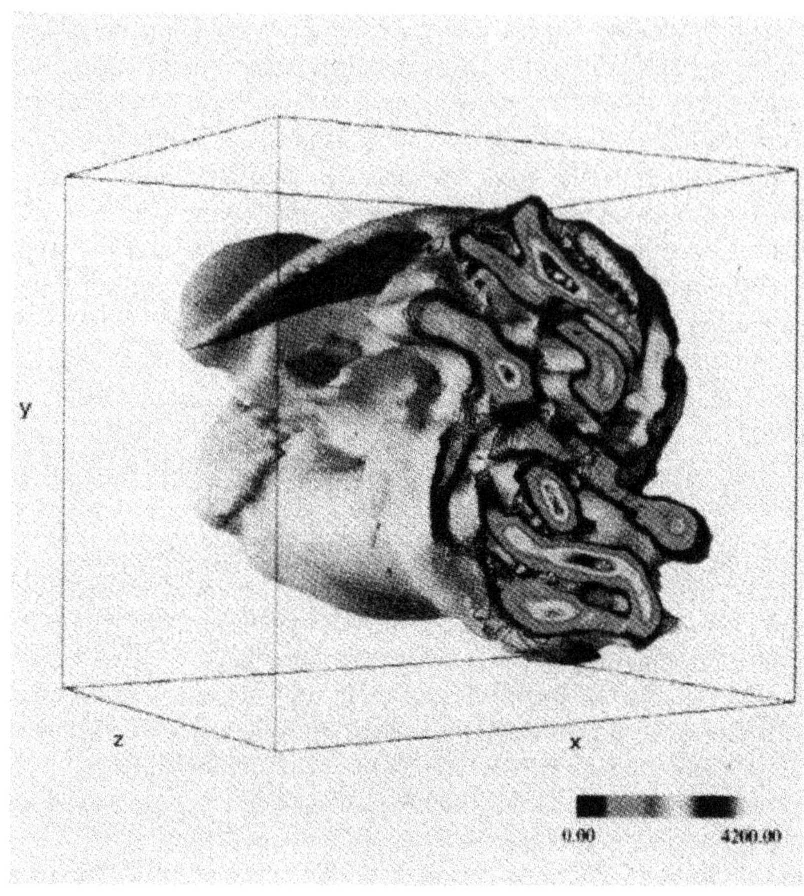

Figure 1. Plot of vorticity magnitude contours in a three dimensional temporally developing mixing layer (Miller *et al.*, 1993b).

nal vortices from the injectors; the incident flow in the injection giving us stirring; Rayleigh instabilities; shock-expansion waves which are unsteady, curved, and producing vorticity. Also, walls giving us wall modes; incident turbulence which gives us stirring; pressure gradients which give us baroclinic effects; streamline curvatures which give us Rayleigh; and nonaxisymmetric injectors which give us azimuthal modes, near-field mixing, and reaction. Thus far, the CFD community that has worked with us on NASP and other things has been able to use conventional turbulence models, no compressibility corrections, and are getting results that are far too good. Good to the point where they're unbelievable. Now, one asks why aren't we worried about this horrible problem that the research community is telling us is in the 2-D simple shear layers. The reason is that we haven't yet run up against it, and possibly because there's all these other bits in real combustors that are terribly interesting and very rich in terms of physics. In other words, let's do mixing enhancement and mixing and reacting physics, for those "real" conditions.

Buckmaster: Dennis is the only man I know who can put 20 slides of information on only 2 slides. Who would be next on the alphabetical list? Thank you Dennis. Peyman Givi.

Givi: I'd like to suggest that we try to discuss numerical simulations of turbulence. I'd like to start with this picture (Fig. 1), simply because it seems that at least in the past 2 or 3 years, despite the advances that have been made in DNS, people seem to be somewhat shy of showing color contours! I feel that in the areas of direct numerical simulations or large-eddy simulations, we have made worthy progress. We listened to a few talks on the first day on the importance of such simulations, but for the past 2 days we have gotten a lot of beating as far as numerical simulations are concerned. My suggestion is that when you see color contours, don't panic. Simulation results are very difficult to obtain, but the important thing is the kind of physical information you would get from such simulations. For example, Sharath Girimaji gave a talk on PDF modeling. If you have a single-point PDF equation, which is extremely important and useful for compressible turbulent reacting flow modeling, he indicated that the modeling of the conditional dissipation or the conditional diffusion is extremely important. Now, if you come up with the closure that Bob Kraichnan came up with, which has been

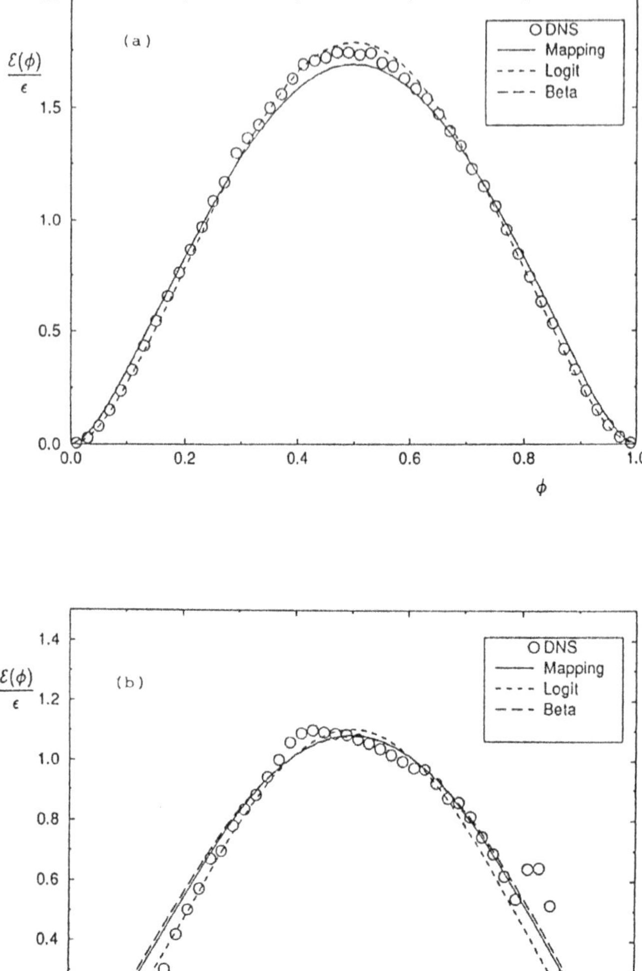

Figure 2. The normalized conditional expected dissipation of a conserved scalar in isotropic turbulence. Comparison of several models with DNS data. (a) The initial stage of mixing, (b) The final stage of mixing (Miller *et al.*, 1993a).

known only for the past 2 or 3 years, the approach seems to be satisfactory. Now, if you use direct numerical simulations, it points to the problems that the closure may have. For example, if you plot the conditional dissipation, the curve, the solid curve being the model and the dots being DNS results, you see the problem, especially at the final stages of mixing, as shown in the bottom figure (Fig. 2). The model cannot capture the physics, as indicated by DNS. Moreover, there are some other closures in which there are no other ways of testing them. For example, the conditional moment method proposed by Bilger is a cure for the problem that Bob Pitz was talking about in regard to the laminar diffusion flamelet model. For the conditional moment method, we try to solve a transport equation. Now, the DNS data are very useful in trying to understand some of the limitations associated with both models. For example, one of the key assumptions is that if you use the conditional moment method, the average of the reaction rate conditioned on the scalar value is very close to the average of these quantities for some of the models. This is shown in Figure 3. One of them is the model, and the other one is the exact type, ...

Buckmaster: Peyman, I hate to interrupt, but you've taken 3 minutes, and I wonder whether you could identify ...

Givi: O.K. ... whereas, the conventional models show a large departure. So, to come back to this figure again, I think that DNS and turbulent combustion have come a long way, and I think we should have some discussions in regard to numerical simulation, in general, and also on the role of turbulence modeling in high-speed reacting flows.

Buckmaster: Thank you very much. Abe Hertzberg.

Hertzberg: I'd like to show one figure that illustrates my point. You know, there is the presumption of an adversarial role between the experimentalist and the theoretician. God forbid, we take ourselves seriously at that. But, there are times when questions have to be asked. Dennis didn't bring it up, but we are now working on trying to use hypersonics in an area where the unknowns are increasing, the advantages of doing this flight regime are smaller, and the need for accuracy is beyond our capabilities. And that is really the bottom line. Nobody can say in this room, not even an experimentalist,

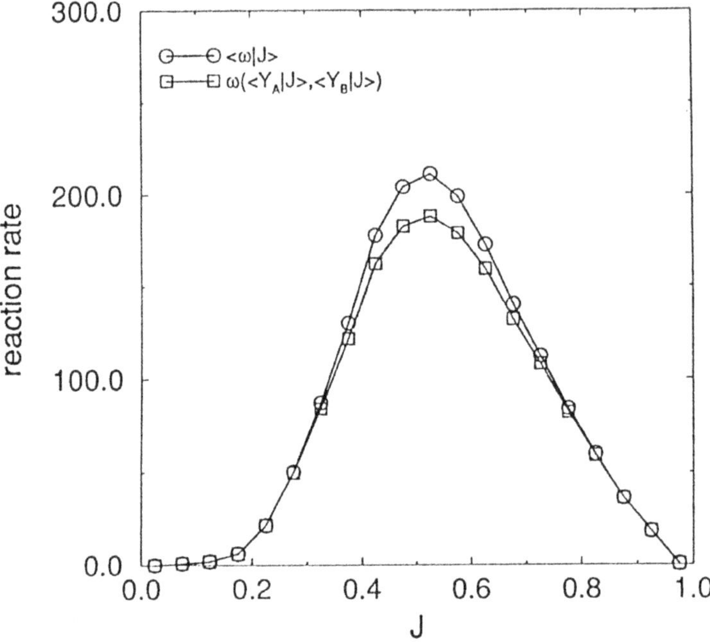

Figure 3. Comparison of the average value of the reaction rate with the magnitude of reaction rate based on the average values of the scalar variables. (Averaging by conditioning on the magnitude of the mixture fraction J. (Miller *et al.*, 1993b)).

that the enormous power of the computing machines, the modern computing machines, as given into use by CFDers, is really amazing. And I think we will see great growths in the future, but you do have to take some problems relating to the physics into account. Now, these are some results obtained by AMTEC. This is a result obtained for 14% over C-J velocity. This shows some burning on the nose, and what's shown here are the temperature contours and the H_2O mass fractions. We then apply this to a point on this curve here, which is the thrust history, acceleration history of the vehicle. There are the AMTEC results. There's that curve. Now this particular run fits this curve amazingly. However, I've got a problem. It's too good. I mean, I know what's happening to the projectile there, and here it goes, to let's say, 1.14 or slightly more. This is one of the points up there. Now in the last 2 weeks, I've been arguing for an experiment; I finally got the students to do it, and we're now out here to 1.3 $\times$ C-J velocity. The thrust was still increasing as the vehicle left the acceleration tube. Which is right, did the vehicle not understand the code, or does the code pertain to the vehicle, or has there been some curve fitting? It's a question. It's just, going back to the old reentry days, you could always tell me what happened when I looked at a reentry signature, but you could never predict it.

Buckmaster: Abe, that's the end of it.

Hertzberg: I didn't want the old reentry days to take us. So, I take it that you would like to discuss this issue of the predictions of these projectiles with things like John Lee has said and the experimental results.

Buckmaster: Thank you very much Abe. Ash Kapila.

Kapila: Let me just start out with a general remark. When we were trying to put this conference together last December, we felt that it would be a good idea if, at the end of the meeting, we identified, at least for the modelers and the fine mathematicians, certain relevant problems or bits of problems that they could take back with them and do some work on. John Buckmaster has identified, very ably, what the difficulties are in getting to that stage, but I'm hoping that with the discussion today we can get around that and make some progress. I have several remarks about three different areas. Numerical approaches – I think it's fairly clear that as far as the practical people are concerned, people who design and build these machines and make sure that they work, that a lot of numerical work is going to be needed before a final design is made. The kind of questions that they would be interested in are: given a certain type of configuration, given a certain parameter range, and I'm going to confine my comments, incidentally, to the detonation aspect of the problem; so, given a configuration and a parameter, is the detonation established or not, is the detonation stabilized where it should be, what is the thrust that is given to the vehicle, and then one would like to do parameter studies for optimization and design. Now, as far as the numerical work that we have heard about at this meeting is concerned, these are detailed calculations done in exquisite detail of model problems, and even though there are a lot of numerical issues to be resolved, the dynamics seem to have been reasonably well treated, but they asked questions about what kind of complications come in when one deals with complex chemistry. I think one of the needs is to extract from these detailed model calculations some global and some macro information that can then be used by the designer to put into numerical codes when the time comes to do practical computations. That is one area that would be worth concentrating on. As far as the analytical approaches are concerned, we have heard at this meeting of work done by using asymptotic ideas on model problems and using relatively simple or reduced numerics, and there is still work to be done on problems having to do with initiation of detonation under a variety of conditions, especially conditions where the starting mixtures are nonuniform, because it is very unlikely in these detonation engines that you're going to have perfect mixing before the detonation starts. So then there's the question of initiation, the question of evolution, or the lack thereof, under a given set of situations, and identification of critical parameters, under which you will or will not get a detonation to evolve. The related issues of stability both are in microscale level as well as the structural or catastrophic instabilities that might, in fact, be triggered by these small-scale disturbances. Then there is a question of reduced kinetics. What are the appropriate kinetics to use for the systems of interest, and how can we, in a sensible way, reduce the kinetic so that it is useful for practical computations? Another way in which the analytical approach on small-scale, well-defined model problems is useful, is in producing test results that the numerical people can then use. As an aid in not only understanding the physics of the situation, these results could also be used to interpret numerical results, as Bill Dold showed this morning on how the analysis could be used by John Clark's numerical results, for example, to see exactly what was going on in a given situation. The final point, about which I know very little, has to do with experimental approaches which are extremely useful for a variety of things, and among those, exposing weaknesses of modeling. As John mentioned yesterday, modeling necessarily requires making sometimes quite drastic assumptions. I think what would be very useful is to suggest and devise "simple" experiments, relatively clean experiments whose results one could use in association with numerical or analytical work, and as examples of these experiments, the context, I mention the work of Strehlow, having to do with the SWACER (**S**hock **W**ave **A**mplification due to **C**oherent **E**nergy **R**elease) mechanism and also experiments by Lee and Toppings. The very nice experiments of Oppenheim, often done in tubes, very clearly demonstrated how the DDT process occurs in gases, and experiments that are currently being done in places like Los Alamos, where they're trying to understand safety characteristics of high-energy explosives. There again, the experiments are designed very carefully, to be as one-dimensional as possible, and to put in as good diagnostics machinery as possible so that one can look at the detailed results from those experiments and then compare them and also help the modelers in producing not only analytical studies, but also numerical codes.

Buckmaster: Thank you very much Ash. John Lee.

Lee: We've heard in the last 3 days some formidable practical problems from Dennis Bushnell and Griff Anderson that have to be resolved. We've listened to a lot of recent developments on computing capabilities, and we've heard a lot of modeling capabilities. What I'd like to do is to bring up the question of how to combine these things, how to really solve the practical problems. So let us consider spherical engines. That's essentially what it is, that we have fuel and we have air, and essentially we have products coming out. So that pretty well sums it up. Now, in designing an engine, what you really want to know is how big, how long the residence time, what turbulence it requires for mixing, and so on. So, if you put some propellers in here, and then you start churning up, then there's this good old mixing problem. In my opinion, the key problem to resolve is really to describe the mixing problem. It turns out that it's not so simple. If you have a low-speed incompressible thing, then you essentially just generate turbulence and vorticity and you just simply have the kind of things that I've seen recently. Pick a box and calculate the mixing rate. This vorticity is essentially dominated by shear. But when you start increasing the flow velocity of these things up to supersonic and hypersonic speeds, then instead of vorticity being generated mainly by shear, now you essentially have waves that are generated so you can consider a number of oscillators that generate waves, together with the turbulence. Now the waves themselves produce vorticity, and there's a whole list of mechanisms that one can write down. Dennis kind of put that down but he whipped that out before everybody got the chance to really see all of these things, like shock/shock interactions produce vorticity and shock oscillations produce big vorticity which you see in James Quirk's thing; the shear-layer developed Kelvin-Helmholtz instability. Then starts vortex pairing and breaks down to the smaller scales; then the shock waves, themselves, when they interact with a density gradient field which produces vorticity from the baroclinic terms. Then when you have density interface, which you're bound to have when you have combustion, then you have the Rayleigh-Taylor instability when it traverses that shock front. Now, none of these things are quantitatively understood. So if you break down this whole problem into the basic ingredients that contribute eventually to the big mixing problem, you can write down a whole list of all these very important fundamental physics. Now, I would personally like to see the computing power with the resolutions and experiments, clean experiments, to individually address these things and numerically try to describe these things on a quantitative level and then slowly put these things together to eventually attack that thing. But it would be complete nonsense to simply pick an engine and start numerically simulating it, because even if you get some results, you can't interpret these. Whether they are good or bad, you don't know. There's no guidelines, because you don't understand the basic physics.

Buckmaster: Thank you. Paul Libby is next.

Libby: John, when they asked me to do this job, they didn't make clear that I was supposed to set forth a research project. But I thought that since I'm certainly one of the senior people here, I don't know, I thought it might be appropriate to point out, I guess Dennis Bushnell inferred this quickly on Monday morning, that supersonic combustion has been around a long time, you know. I lifted the front piece of three papers that were published in the 60's and that's one of them. This appeared at the Stockholm meeting, the third ICAS meeting. I think the sequence of those meetings died some years ago, but the first one was in Madrid, the third was, I don't know where the second one was, but the third was in Stockholm. When young people, everyone seems young these days to me, but when young people come in and tell me they're working in supersonic combustion, I ask them if they happen to know this paper, and universally they, of course, don't know. But I encourage some of you who might be interested in the background of this problem, not the ODWE that we heard so much about at this meeting, but the so-called Ferri engine. I'm delighted to know that that name has sort of stuck on that, the whole scheme of a non-premixed supersonic combustion. I refer you to this paper that there are actually experiments burning invitiated air at Mach 2 or so. There was never any discussion about how the flow isn't going to mix at all, and apparently now this question of mixing is a very fashionable one. That paper was done in 1962. Here is Ferri's Lancaster Memorial Lecture which was in 1964. You'll find some very interesting things about flight paths and how supersonic combustion must be brought to bear or else the engine doesn't look right at all. At any rate, then finally, Ferri wrote another statement which appeared in the Journal of Aircraft in 1967, I think it is, 1968. So the point I want to make is that there is a history of this subject, at least as far as the so-called Ferri engine, and the results in the experimental and theoretical work in those papers are worth following up. Now I want to comment on Dennis Bushnell's diatride, and I used that word advisedly on Monday morning. There is one point that I have to agree with him on, perhaps more than one, but at least one, and that is the matter of the two-dimensional mixing layer being very fashionable these days. He blames the academic community for that interest, but I have to point out to him that, by large, the academic community responds to funding agencies such as AFOSR and ONR. Those are the people that said that they were interested in that problem, and naturally, we'll do anything for money, so we went ahead and investigated. But I agree with him, that in terms of an engine, Ferri and his group were burning, I repeat burning, hydrogen at Mach 2 in an axisymmetric configuration without any concern whether the fuel and oxidizer mixed. But Dennis also blamed us for not contributing to NASP. Now, I made some small contributions in this area 30 years ago and nobody ever asked me to come to a workshop to find out about the NASP problems at all. The academic community was simply excluded from this whole project as far as I can tell, as far as I know and, therefore, I don't think the blame that Dennis leveled at us is justified. Now, a couple of other comments.

Buckmaster: You are a little short on time Paul. If you could get your comments quickly, I would appreciate it.

Libby: I'll make them supersonically. One comment that I would like to complain about is the CFD people when they show us results in these multicolored figures; in fact, Peyman just showed us a figure. We don't know what the colors mean, we don't know what the variable being displayed is. It would really be very useful if you know before they tell us what wonderful things are in those figures that they would tell us what variable is being used and what the colors mean. That would help very much. Anyway, I don't have time, apparently.

Buckmaster: Perhaps we'll have time later on during the rest of the discussion. If I could make a remark at this point. I mean, one thing that has struck me during these last couple of days is the really large gulf in this particular business between the theoreticians and the experimentalists, and there's been some finger pointing which, perhaps in the end, is not particularly productive. Perhaps we should be focusing on ways that gulf could be narrowed rather than identify those who may or may not be responsible. I hope during this discussion this afternoon, we might be able to deal with that issue, with how we can communicate because this is not something which is not common in all endeavors of this nature. I mentioned yesterday that in the microgravity program there aren't too many theoreticians in that program, but those who are have very strong and close links with the experimental program. I think that it works extremely well. I haven't got the feeling that in this business it's working, and maybe its the nature of the beast. It may be interesting to hear whether people have opinions about that. But, let's go on next to Clay Rogers.

Rogers: Okay. I feel a little bit in awe after the people that have spoken before, but let me try to put this back on the track that I felt this workshop was to take when I was involved in trying to get it initiated. That was that we would have focus on research and the conduction and simulation and analysis of combustion processes and hypervelocity regime, with an objective to develop, or provide a genesis of models that would capture and/or reflect the basic understanding of a physical and chemical phenomena in scramjets and ODWEs. This would be such things as the ignition and structure like in flame holding; unsteady behavior like the stability of a flame in a reacting environment; detonations that may include shock-induced combustion phenomenon; reacting shear layers with regard to how they may be used to enhance the mixing process and how you can tend to eliminate or minimize flow losses that are necessary in achieving any mixing; facilities simulations, in a lot of the hypervelocity regimes, we use pulse facilities and short test times. There is an issue of flow quality, both with the level of oxygen dissociation and how well the simulating flight, and in acquiring measurements in those regimes; and turbulence which might include the issue of mass momentum, the turbulence Schmidt-number issue, and chemistry, with an outcome being that there would be some stimulation between interaction of modelers and experimentalists that could provide some means of augmenting our understanding of combustion phenomenon. Now, with regard to some issues for scramjet combustion, things like chemical kinetics that Dennis has eluded to; facility operations/simulation, which would involve both how to model the hardware involved and how well it simulates flight conditions; fuel mixing determination, perhaps the best figure of merit that we can achieve and a critical figure of merit because it directly impacts any realized heat release, and some performance deficit that must be incurred to achieve it; and measurement technology. What do you measure, how do you measure it, and what do you get from a measurement, and a way of comparing what I would classify as engineering data results which may be integral properties like direct measurement of thrust or overall level of reaction that would occur versus some more scientific measurements where you would be looking for an instantaneous flow disturbance in an effort to formulate your turbulence models. With regard to test facility simulation, there are certain issues and concerns. One is the flow starting problem and the operation and characteristics of a pulse facility that was eluded to during this meeting, the quality and effects on the observed performance that this might have in testing a model. The proximity of the facility to flight in testing scramjets, the operation performance of the model, the modeling of the accomplished mixing I think is a paramount issue that needs to come out of any measurements or out of any modeling that is done. Combustion; fluid dynamics; hypervelocity shear layers, both inert and reacting, need study experiments and analyses at high convective Mach numbers. John Erdos showed some results taken in a high-pulse facility that did not agree with the apparent expected theory as put forth. One last chart, since I'm running out of time on this, some modeling and computational activities. In experiments, the observations are solving the correct equations. Whatever it is that is there, that's what you get. You don't have to worry about whether you have the right turbulence model, the chemistry model, because you get what you got. The challenge then is to relate the observed data to some figure of merit in an experiment. And by relating, I mean to model the flow process from some well-defined boundary condition. Too often, data is acquired without measuring enough to know what the flow is. You don't know what's going on; all you get is an image or some data in the flow field. I would include in this defining the equations of motion you're going to use, including any turbulence models, any chemistry models, any mixing models, and the measurement strategy that is used to acquire the data. Current subjects for numerical studies, this is not meant to be an exhaustive list by any means, but the operation for pulse facility, as I have indicated, both in regard to comparisons of the facilities, reflected shock tunnels, and shock expansion tubes; the flow starting process in that and what it does to the quality of the flow; and how well it affects it in simulating flight conditions is of interest. Pressure length scaling experiments are going on as we speak at the University of Queensland, and have been, and will perhaps answer some questions and provide some modeling data; nozzle recombination issues with regard to chemistry; non-intrusive data interpretation. I've called it WIG WIZ, as opposed to WIZZY WIG, because what do I get for what I see in an image is what I want to know. If I'm trying to analyze a numerical PLIF image or a MIE-scattering image, what is that image telling me about that flow and what assumptions do I have to make about the flow in order to extract a meaningful quantity from it? Numerical generation of such images may be a good way of satisfying or leading us along the right path. I'll have to continue later.

Numerical Modeling

Buckmaster: Now we need to perhaps focus on some of these topics that have been identified. And I wonder, Peyman, since you are such a major player, computationally, whether you would like to respond to Clayton's remarks about numerical modeling, where it stands, where you think it can go in helping him in some of the issues that he identified. Perhaps others on the panel might have some remarks, and then perhaps we could ask some of the computational people out there in the audience for their input.

Givi: There is nothing magical about numerics and computations. You're just dealing with a system of nonlinear equations. If you could come up with analytical solutions for the nonlinear problem, go ahead with it. In trying to use a numerical scheme to try to solve an engineering problem or a basic research problem, it is important to understand the limitations associated with that scheme. There was some discussion about the Courant number, and there is nothing magical about this number either. When you're using an explicit method, you have to satisfy the Courant number condition. As far as the whole idea of numerical computation is concerned, where to go with it depends on what type of physics you are interested in looking at, and what type of issues are of your concern. I think that numerical computations have played a major role in the analysis of turbulence and turbulent combustion, and I think that current efforts on direct numerical simulations and large-eddy simulations should continue aggressively.

However, I'd like to point out that an important part of this type of simulation is to couple it with turbulence models. For those of you who are following turbulence literature, at least for the past 20 years, you notice that in the early 80's or before that, the primary work in this area was on turbulence modeling. I mean all of us were doing modeling of turbulence in one way or the other. But with an interest in DNS around that time, and with generation of color pictures, a lot of you were told by many sources, including some of the funding agencies, that there is no point in modeling anymore, saying that they are not going to support anything anymore in turbulence modeling. If there's any suggestions or proposals on turbulence modeling, forget it. Let's just do direct simulations which can capture the whole thing. But, the trend is changing a bit now. In fact, in one such meeting last year, I was delighted to hear that they are coming back to turbulence modeling. Some of the assumptions made in modeling are extremely difficult to evaluate by experimental data. Some of the modeling that, in fact people at ICASE are involved in, in compressible turbulence, or some of the things that we have been doing in modeling of reacting turbulence, are extremely difficult to assess in an experiment. If you want isotropic turbulence, an experiment is very difficult. In this regard, the results generated by numerical simulations have proven very useful. So, I think in the areas of turbulent combustion and compressible turbulence, DNS can be extremely useful in trying to validate closures. And I'd like to see if anyone else agrees with me on this comment.

Buckmaster: Dennis, do you have any response from listening to Clayton about the value of numerical work in combustion?

Bushnell: No, I simply would like to make a clarification about your remarks John, and that is you set the problem as being experimentalists vs. theoreticians. I would state it as being more science versus application. Okay. I think that's more than a play on words. This meeting is an attempt, among many other attempts, to have what really are two very different worlds come together. This is just my perception of what the problem is.

Buckmaster: Well, one question that I would like to perhaps have people discuss, and I'd like Dennis to tell us what, perhaps to tell us whether, in fact, he thinks that modelists who don't do computation are of any value to your kind of problem. But since we are presently talking about the numerical stuff, perhaps some of the numerical people in the audience have some contributions to make about what they see as the role of computation in this business. And I think Scott has something.

Stewart: Peyman mentioned the necessary role of modeling. I want to emphasize some more, but from a different view. It looks to me by doing just real simple practical things on present computers and sort of even looking at what technological advancement we're going to get in the computers that are coming on down the line in the next, say 5 to 10 years, things like a connection machine and so on that we sort of got a bottleneck. The punch line is that you can simulate only very, very, very little tiny pieces of physical space in a chemically reacting flow of combusting compressible reacting flows. There is not much you can do, so this is entitled the economics of multiscale calculations, and the premise of this is arithmetic. I mean the mathematics of this is arithmetic, and basically there is a calculation here that goes TCPU, the CPU time, the number of cells in your computational domain times the CPU per cell cycle; that's a machine hardware-dependent number typically, not always. It can be algorithmic dependent times the number of times you simulate. That depends on the physical simulation that you're interested in. So, you can image how this goes. You sort of figure out how many cells you need and how long you gotta go and what kind of resolution you need. I'm not going to take you through all of that. But the thing that is interesting is that if you say you have a one-macro scale like a combustor size, say, call that 'L', then you have the first subscale, call that 'l' here. You imagine that these are disparate, so there's epsilon smaller than one. Maybe you have even another scale. We'll call it δ, and δ is either order one or its also smaller. So you could have two small scales. We've seen all kinds of examples of that. In fact, Peyman's slide just showed that if you looked carefully just a moment ago. You can estimate how long it will take to do this thing, like on a Cray YMP. For example, using maybe a uniform mesh. Well, it's sort of scary, and I might point out if you do the estimates for like a compressible reacting flow like for the Euler equations, you come down to about a millimeter cube. That's what you might be able to do with about a thousand cells in a size; and a CM-5 is going to give us 5/12, so nobody's going to do anything more than that. Now, what's the scale of the Enterprise? How many people are actually going to get to do this kind of stuff? Not very many. We're talking a national laboratory accelerator type facility with nobody able to use them except a team that has a community code to do one. So you get my drift. In other words, the point is that you can't proceed without analysis because there is (a) no way to get the size you want and (b) you can't interpret the data practically in any meaningful way. Now here you go. This is too much stuff, as John pointed out. But notice numbers down here in a $1000 \times 1000 \times 1000$; that's not crazy for a CM-5 right now. A number like 10^6 hours, and I'll point out that 10^6 hours is a hundred years for one physical experiment. Now the premise of this is the following: is that you have one or two embedded scales and you have one event that sweeps from bottom to top. So alpha is the number of sweeps, the number of interactions you have to have for some sort of sweep-over the domain. That's how you sort of calculate the final point. And you get numbers like, depending on whether you have coarse resolution or fine resolution, coarse resolution is like 100 and 100. Say you have a one-scale calculation, ten interactions – 100×100 is about an hour. That's about right. That turns out that the things that I think Robert or James was showing this morning, they're 4 hours on a Cray. That's not atypical; that's reasonable. So the point is that you can't live without modeling. The tyranny of the amount of data that you get is unbelievable. I mean you have to do something with all this data, so direct numerical simulation is not, in and of itself, going to solve the problems that we want to solve, and also, there are very few, if any, resolved gas kinetics calculations that exist in the literature. They really don't exist by any really secure scientific demonstration. They just aren't there. We've seen demonstrations in this meeting to illustrate that. I point that out. Now, the issue, of course, is what is the proper role of DNS. I think Peyman indicated that it's mechanisms, looking at specific mechanisms, looking at prototyping sort of certain interactions using scientific calculations. I offer another solution to this dilemma that we have here, which is the possibility of having intelligent hierarchies of calculations. Now, AMR is an example of that in a modern algorithmic technique that's actually a multigridding technique that will reduce this by one order of magnitude. But it doesn't really go away, because any simulation that takes more than 8 hours is not something that a laboratory or research group can do. You can't have a parametric study with any run costing more than a days work, basically. So this dilemma doesn't go away very quickly, and this kind of simulation has to be used with that type of an intelligence. There really needs to be a lot of attention paid to how this is done. Thanks.

Buckmaster: Is there anyone else out there? Dick? I think somebody at the end had their hand up first. Do you have some remark that you'd like to make?

Sarkar: Just a quick thought on what Peyman Givi said. That model is only as good as what you put into that. Essentially what's happened is that DNS gives you an amount of data and can indicate new physical mechanisms at all times, but one should not fail to address Dennis Bushnell's critique which is that real-life experimental stuff is primarily three-dimensional and lower complexities. Most of the DNS simulations that you can do right now are very simplex 2-D kinds of configurations. So, it is possible, and I think it should be, that we move on to a little bit more complicated inhomogeneous direct simulations having three-dimensional effects. And as we understand the physical mechanisms peculiar to those particular flow configurations, the turbulence models used in real-life applied computations will be more successful, I think.

Buckmaster: Yes, Dennis.

Bushnell: I'd like to make a comment. You asked me what was the role of theory. What's of absolute paramount importance are the concepts, the physics, and the ideas. The experiments, theory, modeling, and computation are simply tools, all of which you use wherever you can use them; however, they mix and match in order to work the problem, which is to explore and optimize the concepts, physics, and ideas. So, the answer is yes.

Buckmaster: One issue might be what is the definition of modeling. I know Clayton said to me the other day, "Don't we all do modeling?," so I think that when Scott talks about modeling and when I talk about modeling and Clayton talks about modeling, they're talking about sometimes different things. Paul, you have a comment?

Libby: Well, I just wanted to make a comment about this business of DNS. You know that if you try to apply DNS to turbulent combustion, we're still locked into resolving the Kolmogorov length scale. What people do is to slow down the chemistry essentially so the chemical length scales are essentially the same as the Kolmogorov length scale. I don't wish to criticize that methodology because it does provide very useful information. But we've heard a lot at this meeting about laminar flamelets, and laminar flamelets, in fact, provide one of the more convincing methods of attacking turbulent reacting flows, both nonpremixed and premixed. But you know the whole basis of the flamelet theory is that the thickness of the flame is small compared to the Kolmogorov length. At least that, shall we say, relatively popular representation of the chemical behavior, and one which incidentally is believed to prevail in many applied systems. In fact, your automobile sitting out there runs pretty much on the basis of a laminar flamelet, and is not handled very well by slowing down the chemistry. And what I have been (I'm not a CFD guy), but what I keep mumbling at cocktail parties to my CFD friends is that why not do a surface tracking calculation of a flamelet, with a flamelet advancing into the reactants at a specified rate or rate which depends on the local strain field. I think that would be an alternative representation of turbulent reacting flows which would be extremely useful in eliminating this.

Detonations

Buckmaster: I was wondering whether we should perhaps shift our direction a little, and talk about detonations, perhaps since we have a lot of talks about detonation. Dennis made a remark yesterday, I think, that he did not believe the stabilization of oblique detonation waves, for example, was an issue. I'd like to know whether that is a shared opinion. There is a theoretical detonation community, some of which is represented here. We heard talks; I'd be interested to know, I think, whether the practical people believe that there are any of those ideas or approaches that might have some impact on their understanding of detonation waves and the ODWE. So, who would like to start on the panel with those kinds of issues?

Bushnell: Let me try to clarify what you just said. It was in my opening talk, and the comment was made that a few calculations, which Ajay Kumar and company did, indicate that if you overdrive the wave enough that stability didn't seem to be a major problem for our particular application. Now, that doesn't mean that for other types of engines and so forth it isn't a problem.

Buckmaster: Alright. Did you have something you wanted to say?

Hertzberg: Well, not on this issue. I agree with him that every experiment I've ever made, every analysis of those, saying that the only thing that works is if it's well overdriven. It's not a detonation wave; it's merely the heat addition behind the shock wave. That's a very simple problem. If the scale is large enough, you don't have to worry about interaction. I just have one question. How do we know that you're right? You know, we have a very harsh task master in our experiments. Particularly, when we're trying to do anything new. Now, are we to believe you or not? And I have never resolved that question in my own mind.

Buckmaster: Who is the "we" here?

Hertzberg: The "we" is the honorable machine with the superbly brilliant technicians surrounding it and feeding it like a lady Queen Anne or something. What comes out, it may be honey, or it may be poison. We have made so many mistakes on the basis of large computer programs. For example, take our economy.

Buckmaster: I think it's safe to say we've gotten off the subject here.

Hertzberg: How do we know? Am I to believe you?

Buckmaster: What it seems to me that this is the importance of, for example, work like Anne Bourlioux's. There's been a great deal of work that's been done in the past years on detonation simulations, but no clear tests on whether those numbers are correct or not. It's an extremely important thing when what she did was to look at the predictions of linear stability theory, then look at the predictions of weakly nonstability theory, and very carefully compare the predictions of those theories with the numerical simulations and got that wonderful agreement. That's the kind of scientific approach to computing that's needed to satisfy people like you when you are asked "are we getting accurate answers." When you say that wrong answers have been achieved, that's with people who are simply trying to throw everything in, bar the kitchen sink, and sometimes that as well, and then crossing their fingers that everything is okay.

Hassan: I'd like to talk about computations. Dennis mentioned earlier today that he is amazed at how CFD codes are predicting things which they are not supposed to predict. Did I quote you right?

Bushnell: We have been obscenely accurate in the NASP calculations compared to the best experimental data on often very complex flows.

Hassan: I don't think that you should be essentially disturbed by that fact, Dennis, because if we accept the Navier-Stokes equations as the equations that govern all the defined processes that were put on the screen for the last 3 days, then we have the physics at hand.

The only thing that we don't know is, of course, the turbulence. And the only way we're going to know that is by essentially doing some experiments like what Bob Pitz shows over here. I think these are the kinds of experiments that you can take a code and essentially develop it to the stage where you can have confidence in. How can you essentially take a code which you have no confidence in and design an experiment? There is no way; there is no possible way. But I think the way we understand this business is essentially to keep improving our CFD codes and by essentially doing the kind of experiments that will help us to validate these codes. And I think we'll be in great shape. We can design an experiment that needs to be designed.

Riggins: In order to have an accurate numerical simulation of an existing experiment, you need a lot of information from the experimentalists that is often not forthcoming with any degree of accuracy at all. I can't tell you how many times I've seen inflow problems not being able to be specified by the experimentalist. We're expected to run that inflow.

Hassan: I understand that, but that doesn't mean that ...

Hertzberg: He's talking to me?

Riggins: Yes.

Hertzberg: Okay, I'm trying to talk to you. Now that's really the problem. Essentially, there is a dichotomy, and it's adversarial, and I claimed that we're slowly drifting apart rather than getting together.

Hassan: That's not true here. I think that some people are doing good experiments and they would like to essentially work with other people who are doing computations, and there is a lot of that going on, but we don't have enough good experiments. I'd like to see us do more of these things, because I think that modeling of CFD computations are in good state. I'm not talking about DNS, I'm talking about CFD computations based on Navier-Stokes equations with modeled turbulence in it.

Lee: I'd like to return to the detonation problem. You mentioned the fact that in the detonation field there is a very solid analytical modeling work that's been done and the stability analysis presented here. Essentially, all of these things assume the existence of a second wave to start out and look at all the instability modes coming up from that. In the numerical simulations, we've seen some very, very detailed structure on what's going on behind the wave. And you mentioned, John, that Anne has done the stability analysis and has verified it numerically. The bottom line is to still verify with experiments. But, you can't verify the kinds of things that James Quirk or Anne showed on the board. You can't do an experiment to measure the kind of detailed flow structure to verify the numerical results, but experimentally you do measure a few things, and I'll mention again that for the past 15 years since simulations of detonations came along, the simplest thing that the experimentalist needs to know in detonation is not being predicted by the simulation. For example, how big a tube can be to pass a detonation wave through. Now, the reason why they are not producing this result is that they always have periodic boundary conditions which are reflecting stuff off of unreal things. So let's do a problem where they reflect things off real walls and then reduce the tube size. Then if you come up with a result, say, if the tube is less than half a cell, you can't detonate, that's great. I've done the experiment, I can say "hey, this is right." Or I can pick up a tube and suddenly let it expand. So how big should the tube be so that the detonations still continue to propagate rather than quench? That's a very simple experiment in which we measure the tube diameter very accurately. Now we should be able to simulate this kind of stuff, so that we have a multicell detonation coming out of the tube and you can predict 13 cells, that's great. Now then, these are the things that you can verify. So the bottom line is this: experiments produce only limited information; numerical simulations produce too much information. So what we need is an algorithm to somehow go in there and reduce this data into some basic information amenable to experimental measurements.

Buckmaster: Well, Anne, Scott, why can't you do or why haven't you done these calculations that John was just pointing out?

Bourlioux: Well, we identified that yesterday. Certainly, we can do that calculation in detail; we need the chemistry.

Hassan: But you can always assume chemistry. We've got all kinds of ways of limiting that.

Stewart: First, there is a theory for loss of solution in tubes; I mean there is a critical diameter theory. It comes from theory.

Lee: But it's wrong.

Stewart: No, no it isn't wrong. It was published too. The point I wanted to make was this. You know CFD does work for lots of applications. I just gave a sort of bleak picture of how many cells we needed and how it was impossible to resolve everything. The point is that you don't need all the information to equate. So the question is defining how you want to use the CFD calculation and what are its limits to answer your particular physical question. I'll give you a great example. Paul Thibault showed this wonderful simulation, the code verification, where they had the detonation going into the box with all the obstacles and showed that their code did a very good job. Now I would venture the reason that worked well is because all he had to do was a pretty good job on the lead shock. So, in some sense, it's an easy problem. You get most of the physics by the fact that the algorithm captures the lead wave. And, if that's basically all you need, then you're basically in good shape. But sometimes that's not all you need. You were talking about detailed kinetics, then you've got a different ... So you have to very carefully define the issue, the scientific or the engineering or the design point. I think that's what Gill was saying before going off, and then you ask what you can and cannot do. I should also finally say that CFD calculations, and when they are good and when they are bad, is poorly defined. Nobody really knows yet how well they really do. I mean James Quirk showed that this morning.

Buckmaster: Did you have a comment to make?

Thibault: First of all, addressing John Lee's issue. There have been attempts to model the critical tube diameter. You have to understand that there's 13 cells, and this is a three-dimensional problem. It's a humongous problem. From the minimum history I have on modeling turbulence and from what I know on detonation modeling, I would just mention that turbulence modeling people started with turbulence models, and now they're saying "let's go with direct numerical simulations." Well historically, in detonations we've always been doing direct numerical simulations, and we're actually in prehistoric times because we have not discovered a statistical model yet, and this is what we need. We need a subgrid model. We haven't even conceived yet of the possibility of using one, so I think for designing engines, this is a big problem because we can't, and it effects the stability, because, as it was said, if the length scale is small compared to the engine, it probably will be stable. Then the CFD code won't be able to model it, because it's going to be too small. So we need a subgrid model. It's our problem.

Buckmaster: Clayton, I think you had your hand up.

Rogers: Yeah. I think that some of you have missed another subtle point that Professor Lee may have made regarding measurement of things that make some sense and being able to compute similar parameters that you can compare with what you can measure. We went through an exercise. I say "we" in an editorial sense. Actually, it was Dave Riggins and Bob Bitner who did a lot of it at NASA in applying a CFD code to the problems in a supersonic combustion ramjet engine flying at Mach 17 and trying to look, using CFD, to try to assess what is it that we can measure that will allow us to determine the performance of this beast. What things that we can observe in the flow are sensitive to things like radical variations in the turbulent mixing? And they went through and varied the turbulence Schmidt number from 0.2 to 1.0 to see what affect it would have on various parameters that were measurable, directly observable, and came out with the fact that things that are routinely advocated as being the panacea of all the supersonic reacting flow, things like OH PLIF measurements, instantaneous OH PLIF measurements are of no value because they do not show any change over this range, but things like fuel mixing parameters do. Fuel number density does have an effect and can be observed, the amount of water that is being produced does have an effect and can be observed. The idea was to lead us onto a path, which I think is an admirable use of CFD, to model the flow process we want to study to direct us into things we need to look for because they are the things that we are going to be able to observe that are going to make a difference in the answer we get and lead us to getting a scramjet that performs well enough to keep the pointed end going forward.

Mixing

Buckmaster: Perhaps we could talk about mixing. Several of the panel members, in their introductory remarks, use the word "mixing." Again, we always appeal to Dennis for the inspiration for much of our discussion. Yesterday he said that mixing was not an issue, but then I think later on in the morning Phil Drummond said that he thought that it was. Paul Libby, who I think, from a fairly good perspective, is quite right when he is saying that much of the funding seems to have been focused toward mixing. I get the impression that every university in this country has somebody who's doing a plane mixing experiment, so I think it's quite natural then for mathematicians and numerical people in universities seeing their colleagues doing this stuff to assume that plane mixing is an interesting problem. You told us that we shouldn't be doing plane mixing. Perhaps other people disagree with you, and we would like to hear their opinion and what should we be doing and perhaps we could say the real problem, isolate specific ingredients of the real problem that those here could look at in the future. So, Dennis.

Bushnell: Yeah. Let me try to clarify. If you give me my chart back. The issue on the mixing is that it's not as bad as you think it is, but it's not as good as we'd like it to be, okay. We are getting mixing because of all these bits and pieces in here, and the reason why I'm surprised about the CFD calculations being so good is that none of the physics is in the CFD calculations. If we're getting good numbers, it's only because we're making compensating errors.

Grosch: Or, maybe they're not very important Dennis.

Bushnell: Wait a minute. I have been thinking into this a little bit, and I have on my table piles of reports which address individually these various pieces, and it turns out that longitudinal vortices, for instance, increase mixing by a factor of two every day of the week.

Kumar: Why do you think that that's not in the code?

Bushnell: It's not in the code with the Baldwin-Lomax turbulence model.

Drummond: It's in the code to a certain scale. We're really back to the scale issue that we were talking about a few minutes ago.

Bushnell: Baldwin-Lomax doesn't have the curvature stuff in it that's actually causing the mixing. The point I'm trying to make, what was asked for, is what should be worked for the free-mixing problem. I made the point 2 days ago that we have looked too long at simple zero pressure gradient 2-D free shear waves. We understand these, and we understood them the third or fourth time they were studied.

Libby: Dennis, that's wrong.

Buckmaster: I think we should let Dennis talk.

Libby: Then we never get a chance to rebuttal. It's not right.

Bushnell: The issue on what needs to be done has to do with trying to sort through each of these individual parameters. Do we really understand how to parameterize these, what their effects on the physics are, then start combining them, two, three, four different combinations. See if you get any synergisms. Because, yes we do get mixing, but we also have far too many losses. We need to reduce the distance to "mix and burn." The only way that we're going to do that in the conventional diffusive burning engine is, in fact, to enhance the mixing. But we want to enhance the mixing without killing the thrust. You don't want to make a lot of turbulence by simple drag. You would rather make turbulence by promoting those instabilities which turbulent flows are sensitive to, as opposed to just the laminar flows. Some instabilities are present in both laminar and turbulent flows; some are present only in turbulent flows because you're talking about 3-D, nonlinear, finite-amplitude instabilities. So, it's in that way of trying to be creative in terms of trying to get mixing enhancement as opposed to looking at 2-D mean shear layers and trying to find out why they don't mix very fast, because if they don't mix very fast, we're not really interested in them. We're interested in trying to get faster mixing in real engine flows.

Buckmaster: Let's leave it with the panel for a while, and after they've exhausted themselves, we'll be happy to pick it up from the audience.

Libby: Can I make a couple of comments? In the first place, I agree, and I repeat my statement from before, that the two-dimensional mixing layer is probably not of great practical interest, but it is of fundamental interest. The statement we just heard that we know pretty much about all we need to know, I just don't agree with. For example, the Stanford people have shown the following extremely interesting results. You know that from the Brown-Roshko pictures we have this view of the two-dimensional mixing layer showing pairs of vortices in which essentially you have a vortex on one side coming over to the other side so it looks like a series of rollers. What the Stanford people have shown is that if the Mach number is high enough and there's enough heat release, then the mixing on the two sides of the mixing layer are decoupled, and so, one may have a certain spacing of the vortices on one stream and half that spacing on the other stream. This is extremely interesting and of great fundamental interest, certainly something that has not been completely understood. Okay, so I think the two-dimensional mixing layer is not very well understood, but it's not of great practical importance; it's a fundamental interest. Now, two other points. First, we've heard from Dennis that he's been surprised by how well the CFD codes have done under some circumstances. I don't know what those circumstances are, but I can tell you that if you have large strong pressure gradients, the turbulence model is completely irrelevant. The entire field is determined by essentially the mean of the Euler equations, and there's no Reynolds stresses in that.

Bushnell: But you don't get combustion unless you micro-mix.

Libby: The second point I'd like to make is that we've heard that the turbulence modeling for reacting flows is in good shape. I don't think that's true whatsoever. For example, at a meeting 2 or 3 years ago at this place, I pointed out that if one takes the $k - \epsilon$ model, which is a rather popular model, and tries to apply that to variable density turbulence, as we did, we did it for the case of a turbulent jet impinging on a wall, let's say, with premixed reactants when the chemistry was very active, so the mean density at the wall was very low compared to that of the reacting stream. We predicted that one of the components of the turbulent kinetic energy was negative. Now you know that that's absolutely wrong. We got that by simply converting the usual expression between the exchange coefficient and the turbulent kinetic energy divided by ϵ by simply replacing ρ with $\overline{\rho}$, which makes the exchange coefficient proportionate to the mean density. To get away from that pathology, we had to make the exchange coefficient inversely proportionate to the mean density. That pathology went away. We don't even know what the effect of mean density variation is on the turbulent exchange coefficient. It will take some very good, detailed experiments combined with moment methods to clarify that issue, so I don't think it's useful to go around saying we know everything we need to know about turbulence modeling; we just don't.

Buckmaster: Does anyone else on the panel have any remarks they'd like to make about this mixing issue? Phil, did you want to say something about this?

Drummond: Paul has covered one of my points, so I won't go back again with that, but I think that really the important point to remember is we don't have a real good understanding of what this engine flow looks like. We need a flight experiment in order to understand whether many of the features that Dennis has pointed out here are really important. I think there's one prevailing line of thought, and that's the feeling perhaps that the engine flow field on the small scale will be relatively turbulent; it probably will provide a fair amount of small-scale mixing. The real issue may well be to accomplish the large-scale mixing, get the large blobs of fuel and air relatively close together, relative to the small scale, and then allow it to mix out on the small-scale turbulence within the engine. That, I think, points out the importance of looking at mixing layer flows, which, by the way, aren't just plain old mixing layers. I mean jets have mixing layers on edges of their airstreams as well. So that's also in the categorization. We have to make those flows mix together on a larger scale and then mix them on a smaller scale as a result of the small-scale turbulence. Now this may also explain, too, as we look at the way codes behave as why they also indicate very good mixing. Some of these things are, on those smaller scales, pretty dissipative anyway. We might, perhaps, be accomplishing that small-scale mixing numerically. The larger scales may be physically more important. In that case, you can do a pretty good job by understanding and properly predicting the large-scale mixing and modeling the small-scale mixing. The real question that I think has to be answered, and it's going to be very difficult to answer, is what the real state of the flow is in the device that you're interested in and what the state of the flow is in the small scale. That's not clear at all yet.

Rogers: Can I respond to Phil? Phil, first, I don't think we need to do a flight experiment to understand, to make the scramjet work or for you to model how a scramjet works. Surely, that's desirable to flight tests. Well, I can argue that from a practical engineering viewpoint the optimum scramjet would be one that has mixing controlled combustion. Hopefully whatever flies would be mixing controlled combustion, mixing limited, because I want the thing to light first time every time today, tomorrow, and the next day without having to worry about flame stability and ignition processes and everything else. If I'm going to figure you have a pilot in there, he's going to want it to work, right? Then, from an end product that's going to be a mixing controlled process, and from that end, I don't have to understand details about how the fuel gets micromixed with the air in order for me to apply it and make the thing work. All I have to know is how to control the overall process. I do not have to understand the turbulence, I do not have to understand the details of the kinetics, I don't have to understand any of that; all I need to know is how to control it. If I put the fuel in at this location at this part of the airstream, I know by the time I get down to the nozzle that a certain fraction of it has released its energy and has energized the flow, and when it expands through the nozzle I'm going to get kinetic energy back and it's going to keep the pointy end going forward like it's supposed to.

Auslender: When you go and design these engines, I think two requirements need to be put in perspective. I don't think anyone expects that CFD is going to go down and get the microscales, just from a practical short-term requirement. And the requirement to do subgrid modeling is, therefore, essential. Now, the question is "where are you going to hang your hat so that you can make the design step that you started on." Now, we're way off in Mach number and enthalpy. Given that, we have a long attack to get to the problem, and how clever one is, is kind of user-dependent. But it's pretty clear though that all the perturbations around the fields would be nice if we were closer. So my comment would be that the community would work up in the high-end regime around the problem of interest in a spirit that we can always come away from the problem easier. We have really shied away as a community of getting the experimental data. All be it minimal, it could solely make an estimate to answer Dennis' question of how far away are we. I mean do we really know, as a community, that we're not going to get mixing? For example, almost all experiments of the class that we're running at Brooklyn Polytechnic, they burn. Many of the modern-day ones burn in all facilities that we have available. So, it's very hard to make an estimate of where we are until we assess where the new database will be. So we just ask where the new database would be. I think part of the push in the community should be on emphasizing a higher speed database, to then develop all these very intricate models that are spiraling around. I mean it's a massive question.

Buckmaster: Scott, yes.

Stewart: I just wanted to sort of ask a question here. Isn't it true that when they design a new airplane, there is a master designer that sort of has the concept idea, typically? In other words, there's sometimes a design process that's external to all the details, like you said. In other words, they just kind of put the macroscopic chunks and sort of try to fit that into its mission.

Rogers: You're thinking of Mr. Kelly back at Lockheed.

Stewart: Well, yeah. But the point is that's how it works. There's a prototype and then there's a working group of concepts, engine or device, and then it's fine tuned. For example, people in explosive technology, the basic guts of it, works very well. Using it to cut things very precisely works not hardly at all. So there's a tremendous difference between, you know, sort of a coarse aspect of explosive engineering and precision application of that explosive. Those things are like totally different. What I would offer is that what everybody says is sort of right and that the organizing principle is really a clear identification of, in this case, for NASP, of an engine that's going to work. You know, you're not asking the theoretical or computational community to come up with a geewhiz idea that's going to provide the design concepts. I don't think that's around. I think it is much more around in terms of the people that are in the business of putting all the hardware together. It's sort of like using what they have and what they can afford and what they can have contractors do.

Kinetics

Buckmaster: I guess this is inevitable; we tend to deal more with the generalities rather than the specifics. I really had hoped we'd try to focus a little harder on the specific scientific issues that people should be looking at. Another question that came up that was mentioned by at least one of the panelists was on chemical kinetics. Of course we have the talk by Trevino on reduced chemistry. Are there kinetic issues that people should be looking at? Is reduced chemistry something which the computational people think is a useful tool despite the algebraic difficulties that Paul Libby discussed. I mean, in the low-Mach number business, there are a lot of people working very hard on reduced chemistry, and it's, in some sense, a controversial subject, because some people think that you don't gain a great deal, not from simplifying the chemistry, but then using that afterwards. I mean reducing your 50 equations to 5 equations might have some value, but then people who are trying to do analysis with that are making questionable assumptions and numerical people are having difficulty using it because of the algebraic relationship they have to deal with, and when mass fractions vanish, they get singularities, etc. So I wonder whether we could perhaps discuss kinetics for a little while if anybody has any thoughts about that. Do you Clayton?

Rogers: I'll kick this off. It goes with regard to what Aaron has said about hanging a hat on a flight regime, certainly with regard to the NASP program, where we're talking about hypervelocity conditions, say above Mach 12 or 14, maybe up to 18. In that regime, there's very little known about how the mixing processes occur; it's very difficult to simulate the conditions on the ground, and even in the simulations, there's so many other unknowns that assuming you have complete reaction or equilibrium chemistry is a small additional uncertainty to everything else. However, there are some chemical kinetics issues that surface because of trying to do ground testing at these conditions, and one of them I eluded to in my opening remarks has to do with facilities simulation capabilities. To produce these kinds of conditions on the ground, we use reflected shock tunnels which stagnate the flow at tremendous temperatures and pressures, causing massive dissociations, and you end up with a test gas that you're trying to do combustion work in that may be as much as 60 or

75 percent dissociated oxygen. That is only about a third of the O_2; the rest of it is O or NO. Trying to take measurements of the combusting flow raises issues about how do I handle the kinetics. One of the problems that we run into, and if I may put this up briefly, this is the list of simulation parameters that Griff showed, and I will point out to you Damkohler's second number, which has to do with the reaction energy relative to the flow enthalpy. If you're talking about subsonic flows, you're talking about a value on the order of one; if you're talking about flow at Mach 18, you're talking about something that's on the order of a tenth or less, a very small fraction of the total energy that you're adding due to the combustion. If you take into consideration that the temperature is high and static pressure is low in pulse facilities, you run into difficulties where the fuel reacts but you don't get any heat release out of it because all of the heat manifests itself in dissociated products and you don't get any pressure rise, so you get no thrust and the pointy end starts going backwards. So, from that standpoint, kinetics is an issue in simulating the flow in a ground test facility and understanding how to interpret the results that you get in that facility.

Buckmaster: Anyone? Yes, Dennis.

Bushnell: The kinetics are important for facilities. They are important in the diffusive burning engine at high Mach number because it's essentially running out of performance at the high Mach number and we've got to be terribly accurate. The gentleman from the University of Washington was exactly right when he said that. More to the point, if there is a great white hope at the high Mach number right now, it's the detonation wave engine. And there you're talking about kinetics associated with probably burning a mixture of solid hydrogen, liquid hydrogen drops, gaseous hydrogen, air, and I don't know anything about that. So I don't know what to put into my systems studies to even start to estimate the performance to find out whether it's better than a diffusive burning scramjet. So that in terms of a go, no-go situation as opposed to a yeah, we need to get better by a factor of 5 or 10 percent; that's where the kinetic issue is, as far as I'm concerned.

Buckmaster: Does anyone else on the panel want to address this issue?

Hertzberg: Facilities will always be a problem. You know that; I know that. I do believe, though, that the shock tunnel facility can be improved an order of magnitude. I don't know that that will solve the problem. That really reflects on our understanding.

Rogers: We could improve the facility two orders of magnitude by putting a free-piston driver on the HYPULSE expansion tube.

Hertzberg: If we are going to propose to build us a vehicle we better have a way of testing it. I do believe that the requirements for performance at this upper end are so close to the existing known sources of drag that we've got a problem. Is it going to have an ISP of 2,000 or 200? I don't know, but I know I wouldn't build it if it had 200. I admit that I've been taking a deliberately adversarial role here just to bring these things out. We have so much fun talking to each other about the elegance, the beauty, the marvelous charts that come out of the codes.

Buckmaster: Dennis identified a problem that deals with detonation with solid and liquid hydrogen. Is this something which people are looking at or modeling, and is this something they should be doing?

Rogers: I think that the issue is that if you're going to fly an ODWE on an airplane, you have to get the fuel-air premixed. How do you get it to premix on a vehicle that's 300 feet long before it enters the combustion chamber? How do you do it? So you inject slush balls of hydrogen way upstream and they vaporize, so it may come through the shock wave when it's still solid.

Audience: That's a good question. If you inject slush balls of hydrogen, what are you going to use to cool the engine?

Bushnell: You don't have a combustor to cool, compared to what you had before. That's the whole reason for going to the detonation wave engine.

Rogers: In scramjets, the combustor takes a large part of the heat load.

Bushnell: But in the detonation wave engine, you have to fuel the whole shock layer. In one approach, the outer parts of it you fuel with solid hydrogen, the intermediate parts with liquid hydrogen, and the inner parts you do with gaseous hydrogen. But you have to fuel the whole entering air stream, the whole shock layer.

Audience: Are you going to do that without incurring drag that's going to kill you?

Bushnell: I don't know. All I know is that the present engine doesn't have enough performance to do the job. And if we're orders of magnitude more accurate about what we know, maybe we can pull it off, and that's what we're trying to do. This other thing, the detonation wave engine, we're so ignorant about that we still have hope. It at least hasn't been worked to death. We haven't spent $2.5 billion looking at it. We have hardly spent 2 manyears looking at it. Let's find out what it can do. He (Buckmaster) asked about kinetic problems, and there is a kinetic problem associated with that great white hope.

Buckmaster: Any other remarks about kinetics problems? Yes.

Anderson: Well, I think that hydrocarbon fuels were also mentioned as an interest that we talked about during reactions. I remember kinetics models that were 130 or 150 reactions or more. I guess you asked the question I'd like to hear the answer to, "don't simplified kinetic schemes or even analytic correlations of the results of kinetic calculations have a place?" Ferri published correlations of ignition delay and reaction time for hydrogen-air based on Pergament's detailed kinetic calculations. We've implemented those things in one-dimensional codes to know whether ignition delay was significant or if reaction would keep up with mixing. That seems a lot simpler approach than including very complex kinetic schemes with lots of additional equations in the CFD codes. I talked about the computational study of combustors where we used local equilibrium because we couldn't afford to use the code with kinetics in it.

Buckmaster: Peyman? Are you interested in using these chemical models?

Givi: As a matter of fact, one of the important issues in DNS is that associated with the kinetic mechanism particular to nonequilibrium flames. I personally do not know under nonequilibrium conditions how to come up with reduced mechanism, and we are interested to know what would be the good mechanism to use in these simulations.

Buckmaster: Oh, but there are people who are very good at coming up with reduced mechanisms and already have for hydrogen, for a whole family for hydrocarbons, and are actively working on them.

Givi: A major problem that we had with reduced mechanisms for hydrocarbon fuels was the ratio of the different Damkholer numbers that were used in those kinetics. If we have one- or two-step chemistry models, depending on the type of problem at hand, we cannot artificially reduce the magnitude of the Damkholer numbers involved in reactions, as Paul Libby indicated, and to look at some of the problems of selectivity. If you have two reactions competing with each other, how do they compete for the amount of product being generated? In chemical engineering, this is an extremely important problem. With DNS we have been able to come up with very nice results. Unfortunately, in combustion, we have not been able to make much progress yet, simply because we do not exactly know how to reduce the magnitude of the Damkholer numbers, even when we have only a few steps.

Thibault: Just on the point of detonations, of course for hydrocarbons, one problem that we have is that a lot of reduced reactions don't model the basic Hugoniot curve. They don't have the correct gammas, they don't go endothermic when they're overdriven. It's a big problem when you're going to oblique detonation engines. So they're missing a key ingredient not in the rates, but even in the basic thermodynamics, and I think that's one issue that should be addressed. It's not just the rates near the CJ point that are important, it's the whole Hugoniot curve. I think that has to be addressed.

Buckmaster: Well, my understanding of people who do reduced chemistry is they have to do the reduction which is valid for specific configurations that they're looking at, so they'll have one set of reduced chemistry, say for premixed flame and low Mach number combustion where most of this work is being done, and another set for diffusion flames. You're saying there are other special cases.

Thibault: Detonations should cover a wide range of Mach numbers when you're talking about these applications here. They're going way above CJ, and I think that is a concern for us to model.

Kapila: Dennis mentioned the lack of information about kinetics, particularly with solid and liquid hydrogen under the conditions of operation of the engine. Does he or anyone else have suggestions about how that information can be obtained?

Thibault: Swithenbank had told me that he looked at it a little bit and that it was doable. Those are the only people that I know of that have even touched that problem. That's a facility problem. Right now, we can only work enthalpies in the Mach 12 range with full pressures. In order to get up in to the Mach 15 range, we need 120,000 to 140,000 psi; to get in the Mach 20 range, we need 1.2 million psi. The only way we know to get that is with one of those pretested drag and expansion tube gadgets. But the plea was to the modelers, to the theoreticians, to the computationalists to attack that. The experimentalist needs to do some facility modifications before they can do that.

Buckmaster: Bill, yes.

Dold: Back to the area of the reduced mechanisms and reduced mechanisms that are flexible that you can use in a variety of different of applications. There are halfway houses that you do not need to use the entire scheme, a way in which the chemistry can be treated in a manner that identifies the appropriate reduced mechanism for the problem that you're looking at. The work of Harvey Lam which gives you the computational method for reduced mechanisms is just such a way. In terms of numerical modeling, too, the need to take some species out of your problem and put it into an algebraic subproblem, and is exactly the case when the species becomes a very stiff component in solving the system of equations. It's not a tremendously difficult thing to do, provided, of course, you have a good mechanism. With a complete mechanism, the reduction can be done in a way that adapts itself to the problem and can save enormously on the computation.

Buckmaster: Can that be applied to spatially dependent problems?

Dold: Yes.

Buckmaster: Because I've only seen the applications that will do spatially homogeneous situations.

Dold: The problem with applying to spatially dependent problems is that you then change your set of ODE's across some boundary where you change your reduced mechanism, and this means that there has to be some method of tying in or applying boundary conditions in some sense that would come in.

Flame Holding/Concluding Remarks

Buckmaster: Let me ask a different question. Is flame holding an issue to the experimentalists? This is something that was, perhaps, only lightly approached by Linan's talk when he talked about these leading-edge flames down stream of splitter plates. Is flame holding an issue, do these flames go out, are there things that modelers should be looking at in that connection?

Bushnell: Burt Northam probably knows that as much as anybody. Burt, you want to comment on that? Is flame holding an issue in these engines?

Lezberg: I think that depends on the static temperature coming in there. If the delays are short enough, then you don't need flame holding.

Northam: You need some recirculation to get residence times even in the low-speed flows.

Lezberg: I thought we were just talking about the very high-speed flows where static temperatures are well above 1000°K.

Buckmaster: What I know will embarrass a modeler, but let me ask you a blunt question. Linan gave a talk which I found very interesting. I thought that whole issue of flames near, you know, the initial point of the flame was very interesting. Is there any supersonic application of that, or are you people simply not interested in that kind of thing? You simply see that as one of the idealized university-style exercises which are irrelevant?

Northam: The lift-off in the coaxial jet, is that what you're referring to?

Buckmaster: Well, were you at Linan's talk? I mean that was one of the things he drew. This whole general issue of what the leading-edge structure of the flame looks like with its pair of premixed branches and the trailing branch and the possible locations that it has. Those are the kinds of problems which naturally tend to attract people like Linan, myself, etc. Are you interested in those things?

Northam: We are trying to use that to understand the effects of vitiation and possibly the effects of recombination of nozzle problems by catalyzing the flame. By having catalyzed it, you have modified the strain rate, too.

Rogers: I would think that from a standpoint of trying to do the process of trying to design an engine, that the role of kinetics in regard to flameholding would be to identify the set of conditions or the envelope where it was a problem so I could design my combustor in the region where it wasn't a problem. You know, experiments that identify or tell me that a problem at these conditions under these circumstances would be beneficial or theoretical/computational studies using chemistry models that would identify a region where it was a problem would be telling where not to build it or where not to fly it. But it's sort of the inverse, trying to avoid it rather than trying to implement it.

Linan: The main lesson obtained from this high activation energy asymptotics way of looking at the chemical reactions and the large activation energies are there always, even in the very complex kinetic scheme, is that the thing goes diffusion controlled or is frozen. The main thing is to identify conditions where combustion occurs in the diffusion controlled environments, because otherwise the engine would not work. That is the main thing, and it does not involve a lot of complex calculations. What is really the problem is mixing, and I do not understand it how, by hand waiving, one can solve the turbulence problem, especially the turbulence problem with very high speeds. That mixing is the real problem.

Buckmaster: Yes?

Dold: I have a question that relates to what Linan just said as well. Are there good experimental results for the compressible turbulent situation? You have the hot wire anometer that used to give us the results about low-speed turbulence. Are there good experimental results about this kind of turbulence?

Givi: Yes, there are. In fact, DNS, to a large extent, has been promoted because of some of these experiments. Some of the experiments have tried to understand the global features in reacting flows. What they're trying to do is perform it in a very simple manner for which the DNS can also be conducted. Phil Drummond gave a detailed presentation on the effects of heat release. For example, the experiments at Caltech have been performed with very simple reactions, like almost one step, which provide heat release without Arrhenius kinetics or anything that can make numerical simulation complicated. So there are a lot of experiments that are being performed in order to see some of the nice characteristics of the turbulence – the effect of large scales, the effects of chemical reactions – which not only have been useful for DNS, as we have been performing, but have also been extremely useful for a lot of studies such as those by Chet Grosch and Tom Jackson on linear stability analysis.

Dold: But these are numerical experiments.

Givi: No, no, these are conventional laboratory experiments. Some of the work at Caltech, for example, on turbulent planar mixing layers with chemical reactions.

Dold: With compressibility?

Givi: Yes.

Buckmaster: Dennis?

Bushnell: Yeah, let me give another version. Hot wires, once you get above Mach 2.5 or 3, only respond to mass flow and total temperature fluctuations; you can't get velocity fluctuations out of them. The particulates in the L-V no longer follow the flow, particularly through shocks, so you can't use an L-V. The lasers don't fire fast enough to do PIV, even if you could find small enough particles. The only thing which really works in the megahertz range, which is what you need to do real compressible turbulence at high Mach number, is the electron beam. That's been looked at Calspan; the problem is it only works in nitrogen and it's density limited because of beam broadening. There is a tremendous need to take some of the smoke and mirror stuff that people are developing where they get a few points and then over minutes to hours they get a lot of points and they make RMS plots. But you know, the question is "where is your spectrum." The answer is that it's not there, and it's not there for a very good reason. We need, desperately, technology to measure real turbulence in high-speed compressible flows. We really don't have it.

Buckmaster: John, yes.

Lee: Let me answer your question in a simple way. If you want to know mixing, high-speed mixing, with compressibility effects and shock waves and so on, you've seen that the detonation wave is a beautiful example that within the structure, you got all the ingredients of compressible turbulence mixing. Yesterday I pointed out that if you simply measure the relaxation length, in other words, starting with an inlet flow of Mach 6 or Mach 7, just how long does it take for all these things to mix and burn and to come out at the other end. Now you have seen the numerical simulation and they produce all these fine vorticities and so on. Well, you've got to put in a dissipation mechanism, so either the comments that Paul Thibault made on saying that we need a subgrid model in there to do that, or, somehow, you have to put in the diffusive terms into that solution and calculate this length. But if I do this experimentally, it is very simple to measure this length. If I just measure one length scale, never mind about all the fancy diagnostics, just measure one length scale, I get a global length scale of how fast it takes to mix things at Mach 6 with all the ingredients.

Dold: Would you trust the model that came out in just the detonation simulations for mixing, Dennis?

Bushnell: Oh no. No, there isn't any mean shear in that kind of flow. The essence of the turbulence, low speed or high speed, is that, in fact, it still contains mean shear. You can modify that shear-induced turbulence with all these other mechanisms which we displayed, but the fundamental issue is inhomogeneous turbulence.

Lee: Well, there's really no problem to add mean shear. This experiment has been done by just putting an obstacle in the flow and let the detonation wave go through it. You generate tremendous mean shear.

Buckmaster: Well, we're down to the last few minutes. Does anyone have a key question which they have been itching to have answered and addressed by this panel? Yes?

Morgan: Most of the discussion this afternoon has been philosophical and long term. In Australia, we've got more short-term requirements. We don't believe in velocity anyway. The current capability of CFD has been, perhaps, underestimated by the discussion of all the weaknesses its got. I can give a couple of examples where we used the CFD usefully in parallel with our experiments. And as for the help in continuing this process in some new experiments that we've got, if you look at the parallel experiments that we did in the GASL expansion tube, looking at lack of oxygen, the planning was done using what falls into the category of applied mathematics approach. The results afterward were analyzed by a parabolic code with 8-reaction chemistry. You get very good agreement with the data. We also repeated some selected examples with the full kinetic mechanism for hydrogen and got approximately the same result as for the axisymmetric situation where we had a code that could handle it. But we've now got data for the same model with discrete hole injectors causing three-dimensional flow which, I think, is a very suitable data set for the application of 3-D computer codes, which I think are around, but we don't have them. Also, the pressure length scaling problem which I discussed in the talk is a two-dimensional situation. A lot of data is coming out of it, and I think that also is amenable to some CFD work. I'd be interested to work with anyone who'd like to have the data analyzed and would like to use it with their programs. It is also an area where you could do a lot of analytical work for comparison. The problem Clay Rogers mentioned about the verification of the test conditions of facilities. We've done some simple boy scout arithmetic concepts. We do know approximately what we get, but I think there are now programs that can do a much better job. I'd like to support Clay Rogers' suggestion of more existing codes that are applied to flows coming out of shock tubes, expansion tubes, and nozzles.

Buckmaster: Thank you. Well, perhaps, I've got the feeling that we've sort of wound down to an end. So, perhaps, we'll close this session. I'd like to say one thing that is noncontroversial, believe it or not. I'd like to thank the organizers and the arrangements; they were all superb. Thank you very much.

References

Miller, R. S., Frankel, S. H., Madnia, C. K., and Givi, P. "Johnson-Edgeworth translation for probability modeling of binary scald mixing in turbulent flows," *Comb. Sci. Tech.*, in press (1993a).

Miller, R. S., Madnia, C. K., and Givi, P. "Structure of turbulent reacting mixing layer," submitted for publication (1993b).

www.ingramcontent.com/pod-product-compliance
Ingram Content Group UK Ltd.
Pitfield, Milton Keynes, MK11 3LW, UK
UKHW021902190726
13853UKWH00003B/1385

* 9 7 8 9 4 0 1 1 1 0 5 1 8 *